RADARS
VOLUME 6

FREQUENCY AGILITY AND DIVERSITY

Artech

DAVID K. BARTON

THE ARTECH RADAR LIBRARY
David K. Barton, Editor

Electronic Scanning Radar Systems (ESRS) Design Handbook
Peter J. Kahrilas
Consulting Engineer, Raytheon Company

Phased Array Antennas
Arthur A. Oliner
Head of the Department of Electronics and Electrophysics,
Polytechnic Institute of New York
George H. Knittel
Staff Member, MIT Lincoln Laboratory

Radar Detection and Tracking Systems
Shahen A. Hovanessian
Senior Scientist, Hughes Aircraft Company

Radar Signal Simulation
Richard L. Mitchell
Vice-President, MARK Resources

Radar System Analysis
David K. Barton
Consulting Scientist, Raytheon Company

Radar Technology
Eli Brookner
Consulting Scientist, Raytheon Company

RADARS – in six volumes
I. Monopulse Radar
II. The Radar Equation
III. Pulse Compression
IV. Radar Resolution and Multipath Effects
V. Radar Clutter
VI. Frequency Agility and Diversity
David K. Barton,
Consulting Scientist, Raytheon Company

Significant Phased Array Papers
Robert C. Hansen
President, R.C. Hansen, Inc.

Synthetic Aperture Radar
John J. Kovaly
Consulting Engineer, Raytheon Company

Affiliated Title:
RF Radiometer Handbook
C.W. McLeish and G. Evans
Staff Members, National Research Council of Canada

RADARS
VOLUME 6

FREQUENCY AGILITY AND DIVERSITY

DAVID K. BARTON
Consulting Scientist
Raytheon Company
Bedford, Mass.

Artech

Printed and bound in the United States of America.

Library of Congress Card Catalog Number: 74-82597
International Standard Book Number: 0-89006-067-3

International Standard Book Number (Volume One): 0-89006-030-4
International Standard Book Number (Volume Two): 0-89006-031-2
International Standard Book Number (Volume Three): 0-89006-032-0
International Standard Book Number (Volume Four): 0-89006-033-9
International Standard Book Number (Volume Five): 0-89006-034-7
International Standard Book Number (Volumes One through Six): 0-89006-060-6

TABLE OF CONTENTS

Introduction

Frequency diversity communications systems long have been used to minimize the fading in a propagation path that generally is caused by multipath reflections from an underlying surface. In radar, frequency diversity has similar objectives - to avoid fading caused by the propagation path or the target reflection characteristics, and to evade jamming which may be concentrated at one frequency or another. By avoiding fading, higher probabilities of detection are achieved, and measurement errors caused by target glint, amplitude modulation, and multipath are reduced.

Radar systems can obtain diversity advantages either by using two or more parallel transmitting and receiving channels, which generally share a common antenna, or by sequential use of many different frequencies (a technique often known as frequency agility). Parallel diversity systems offer improved reliability, because of their redundancy, and higher performance, in that the average powers of the transmitters effectively are added together in the final process of signal integration. Also, the information rate is increased by parallel operation of the two or more receivers. A frequency agile system is generally simpler, using a single transmitter and receiver combination retuned to a different channel on each successive pulse or group of pulses. Doppler processing for MTI requires coherent generation of pulses in a group at constant frequency, followed by rapid tuning to a new channel.

In this volume, both types of system are discussed from theoretical and practical viewpoints. The editor's summaries of applicable theories are supported by the most important papers published by other engineers in the field; each is reprinted in its entirety. The first section of the book covers general theory and problems of multifrequency radar, surveying the benefits and penalties of operation on two or more channels. Detailed theory of detection using diversity channels is discussed in the second and third sections, including data on decorrelation of target signals. Decorrelation of sea clutter is discussed in the fourth section. Contributions to the accuracy of tracking and measurement are covered by the fifth section, which primarily discusses target glint and includes effects of propagation and amplitude fading. The final section deals with practical system examples along with circuits and components for frequency diversity and agility systems. Annotated bibliographies in each section provide a guide to further data in the literature of radar and propagation.

Section I General Papers

Frequency diversity and agility radar systems received little attention in US open literature until the late 1960s, but they had been discussed by several foreign authors. Thus, the earliest paper listed in the bibliography appeared in France, and the first and most thorough analysis of diversity-agility detection performance (see Section 2) was published in the Soviet Union. While American authors produced early results in the area of glint reduction for tracking radars (see Section 5), many English-speaking engineers received their introductions to the general advantages of agility through the papers of Gustafson and Ås, both of the Swedish Philips organization, and of Ray in the United States. These two papers are presented at the beginning of section, both for their historical importance and because they have served since as the basic references in the field.

The Gustafson and Ås paper introduces all the important advantages of frequency-agile radar, with quantitative comparison of theoretical and experimental data on detection performance, samples of experimental data on glint reduction, and qualitative remarks both on vulnerability to interference and ECM and on detection improvement in a clutter environment. The Ray paper contains more complete theoretical data, which has been used extensively in subsequent work. A discrepancy between Ray's Eqs. (3)-(7) and later analyses can be explained by the fact that the bandwidth, W, in Goldman's book (used as a reference) represents the one-sided spectral width, while $2D'/c$ represents the total delay spread of the target, analogous to 2W. Thus, the frequency change for 0.5 correlation should be $\Delta f = 90/D'$ (with Δf in megahertz and D' in meters), which is consistent with a decorrelation frequency interval, f_c, of $c/2D' = 150/D'$. The remaining portions of the analysis are not affected by this factor, and they correctly assess the advantages of agile systems in detection, tracking, and mapping.

The final paper in this section, by Whitlock, Shepherd, and Quigley, has not been widely available but deserves attention as a basic source of target data - a relatively scarce commodity. The decorrelation frequency of 13 MHz measured on the Sea Hawk aircraft (length $\approx$ wing span = 12 m) is in excellent agreement with theory. Lack of consistency in measurements of the Comet jet is attributed to turbine modulation, which introduces temporal decorrelation. The value of frequency agility in avoiding distortion of the beam pattern (important in search radar beam splitting for azimuth measurement) is shown in the final figures of this paper.

A general problem in evaluating diversity or agility systems is that the fluctuations of signal amplitude, phase, and apparent angle of arrival normally are observed in fixed-frequency systems as functions of time or aspect angle. Thus, the Swerling Case II or Case IV target model has inherent "time-diversity" on a pulse-to-pulse basis, making frequency diversity redundant. Communications links use space diversity to obtain decorrelation of fading; radars obtain equivalent diversity by averaging over paths at different angles as the target moves over the surface or changes its heading angle.

The relationship between time and space (angle diversity) is established by the path of the target (or of the radar, if it is on a moving platform). A corresponding interrelationship between aspect angle and radar frequency is illustrated in Fig. 1-1, taken from an unpublished Air Force report. The cross-section lobes and nulls are seen to be ridges in the frequency-angle plane, and the effect of frequency shift is to scan the reflectivity pattern across the radar line of sight, much in a way a frequency-scanned antenna moves its beam. It is this systematic relationship between lobe position and frequency which introduces large measurement errors in frequency-scanning height finders (Barton and Ward, *Handbook of Radar Measurement*, p. 227).

Figure 1-1

The photograph illustrates the improvement that is gained by using frequency stepping. Depicted is a portion of the actual measured RCS of the upper section of the Vela Hotel satellite for various separate frequencies and the cross section obtained by averaging over eight frequencies. The first curve shows the effects of averaging over several of the separate curves in the rear of the photograph. (Photograph courtesy US Air Force.)

When comparisons are made between fixed-frequency diversity systems, the potential diversity in time or space, available at the fixed frequency, always enters as a confusing factor. In the literature there are repeated references to the fact that frequency diversity may be unnecessary or even harmful in cases in which an adequate number of independent samples already have been provided by the relative motion of the radar-target path or factors such as turbine or propeller modulation.

In addition to the reprinted papers in the general area of diversity and agility radar, there are several sources listed in the bibliography which deserve comment. The paper by Ruttenberg and Chanzit, already reprinted in Volume 3 of the Artech *Radars* set, describes a system which uses the transmitter-receiver techniques of pulse-to-pulse frequency agility to achieve high-resolution pulse compression. While not as straightforward as for the wideband FM or continuously radiated step-frequency waveform, the basic resolution properties are similar for targets which remain static during the period of the pulse sequence. This technique can ease greatly some of the problems otherwise introduced by the requirement for wide instantaneous bandwidth in the transmitter, antenna, and receiver. The applications are quite different from those of other diversity and agility systems, however, which is why the paper was included in the *Pulse Compression* volume.

The listed paper by Barnett is one of a series of studies in the use of frequency diversity to minimize fading on point-to-point microwave links. The predominant causes of this fading are reflections from the ground and from atmospheric layers of differing refractive indices. An early experimental study (DeLange, 1953, reprinted in Volume 4 of *Radars*) of the frequency response of such a path indicated multipath delays up to 10 ns, but most of the reflected signal was within 1 ns of the direct signal. Other data on delay-Doppler distributions of multipath are discussed or listed as references in that volume. Application of these data to radar detection of aircraft and other moving targets is difficult because fluctuation rates on the moving path are much higher than for fixed paths, providing inherent time and space diversity which may mask the effects of frequency diversity.

Finally, before proceeding to the details of detection theory and glint, it is interesting to note that the Soviet open literature contains not only major sections of books describing and analyzing frequency diversity and agility radar, but an entire book (Vishin, 1973, listed below) devoted largely to this subject. That book covers both parallel (diversity) channels and sequential (agile) tuning of a single channel, the latter with options for change either within a pulse, from pulse to pulse, or from pulse group to pulse group. While not all of the detailed presentation in the Vishin book is consistent with the analyses of Bakut and others, the attention to the subject and the evident appreciation of the advantages of diversity and agility in search radar are in marked contrast to the treatment in American literature.

ANNOTATED BIBLIOGRAPHY

Herscovici, S.; and Detape, A.: "Performance of a Diversity Radar," *Ann. de Radioelectricite, Vol. 12, No. 50*, October 1957, pp. 285-298 (in French). (See Comments in Section 3 of this volume.)

Gustafsson, B. G.; and Ås, B.-O.: "System Properties of Jumping-Frequency Radars," *Philips Telecommunications Review, Vol. 25, No. 1*, July 1964, pp. 70-76. Discusses the several advantageous features of pulse-to-pulse agility in radar. Reprint 1-1.

Ray, H.: "Improving Radar Range and Angle Detection with Frequency Agility," *Proceedings of the Eleventh IRE East Coast Conference on Aerospace and Navigational Electronics*, October 1964, pp. 1.3.6-1 to 1.3.6-7. Derives the correlation frequency of a target, and discusses advantages of agility for detection and measurement. Reprinted in *Microwave Journal*, May 1966, and in this volume as Reprint 1-2.

Ruttenberg, K.; and Chanzit, L.: "High Range Resolution by Means of Pulse-to-Pulse Frequency Shifting," *IEEE 1968 Eascon Record*, pp. 47-51. Describes method of obtaining resolution in a system by using an agile magnetron rather than intrapulse FM. Reprinted in *Radars* (Volume 3 - *Pulse Compression*), Artech House, 1975.

Carpentier, M. H.: *Radars: New Concepts*, Gordon and Breach, 1968 (translated from the French second edition). Contains brief discussion of dual-diversity and frequency-agile radars and their detection performance.

Whitlock, W. S.; Shepherd, A. M.; and Quigley, A. L. C.: "Some Measurements of the Effects of Frequency Agility on Aircraft Radar Returns," *AGARD Conference Proceedings No. 66 - Advanced Radar Systems*, May 1970. C-band data on Sea Hawk and Comet aircraft are presented, showing decorrelation for 20 MHz shift in radar frequency. Reprint 1-3.

Barnett, W. T.: "Microwave Line-of-Sight Propagation With and Without Frequency Diversity," *Bell System Technical Journal, Vol. 49, No. 8*, October 1970, pp. 1827-1872. Data from two 50-km relay paths at 4 and 6 GHz were analyzed to determine probability distributions of fade depths from 20 to 40 dB. Dual diversity with frequency separations of 30 to 400MHz reduced the probability of deep fades by factors of 10 to 500.

Grigorin-Ryabov, V. V. (ed.): *Radar Systems: Theory and Design Principles*, Soviet Radio, 1970 (in Russian). Chapter 16 deals with multichannel radar systems, including multiple-frequency radar. (No translation available).

Vishin, G. M.: *Multifrequency Radar*, Voyenizdat, 1973 (in Russian). Discusses various aspects of frequency agile and diversity radars as well as multiple-frequency MTI. Data on detection improvement are based on previous results of Bakut (1963) and Grigorin-Ryabov (1970). Several radar systems using multiple frequency transmissions for different purposes are described, and advantages in ECM and clutter are summarized.

System Properties of Jumping-Frequency Radars

B.G. Gustafson and B.O. Ås

Philips Telecommunication Review, Vol. 25, No. 1 (July 1964)
Pages 70-76

621.396.969.1

Summary *Different types of centimetric radar sets have been developed in which the carrier frequency of every pulse is varied in a random manner. This gives remarkable improvements in system performance as compared with radar sets with fixed or slowly varying frequency. The following system properties have been investigated and the results are given: a) range increase due to frequency-diversity properties, b) tracking-error reduction, and c) increased resistance to jamming. Other differences from radar sets on fixed frequencies are anticipated and discussed, such as new aspects of radio frequency interference and sea-clutter smoothing.*

Introduction

In pulsed radar systems, when the echo signal received is of the same order of magnitude as the noise, the statistical character of noise makes it possible for a peak in the noise to stand out so prominently as to suggest that it is in fact a target, with the consequent possibility that a false alarm will be given. Another phenomenon which is liable to occur is an echo signal which vanishes as a result of fading due to attitude dependence; the result of such fading is a decrease in range. A third phenomenon of importance in tracking-radar is glint. The observed echo signal of a moving target is subject to small angle variations because of changes in the position of the centre of the scattering cross-section. This has an unfavourable effect on the accuracy of the tracking-radar's servo system and therefore also increases the effect of noise in the fire control.

As in other fields of telecommunication, the use of more than one carrier frequency in centimetric radar offers considerable advantages with regard to the phenomena referred to above.

A family of radar sets has been developed in which the carrier frequency of every pulse is varied independently of the carrier frequency of preceding pulses. The pulse repetition rate is kept as high as is normal for similar radar sets of current types. As this pulse repetition frequency is high compared with the fluctuation spectrum of the above-mentioned phenomena, which produce negative effects in normal radar operation, a radar of this kind behaves in many respects as if it operated simultaneously on many frequencies. In this article we will call it 'jumping-frequency radar' as opposed to 'fixed-frequency radar', in which the frequency cannot be changed at all or at least not fast in comparison with the fluctuation effects mentioned above.

The article describes some experiments and deductions made during the development and testing of the jumping-frequency radar. For convenience we will use the abbreviations FFO and JFO in the text to designate fixed-frequency operation and jumping-frequency operation.

Range increase due to diversity

With the aid of the theories outlined by Marcum and Swerling [1, 2], a comparison with regard to the probability of detection can be made between targets possessing different properties. The following models are considered:

a non-fluctuating target (used for comparison).

b scan-to-scan fluctuation. This is the model generally used for air targets and fixed-frequency radars; the target area is assumed constant during one scan but uncorrelated from scan to scan.

c pulse-to-pulse fluctuation. In this case each return is independent of the preceding one as is the case with jumping-frequency operation.

On the assumption that the back-scattering cross-section of an aircraft has an exponential distribution (echo returns Rayleigh-distributed) the curves in Fig. 1 have been calculated for the above three target models. They show the probability of detection P_D versus average signal-to-noise ratio for $N = 20$ (hits per scan per 3 dB beam-width of the antenna) and for the false-alarm probability $P_f = 2 \times 10^{-5}$. Examination of Fig. 1 shows that for 90% probability of detection there is a 'fluctuation loss' [3] of some 8 dB for a fixed-frequency radar detecting a fluctuating target, such as an aircraft, compared with the detection of a non-fluctuating

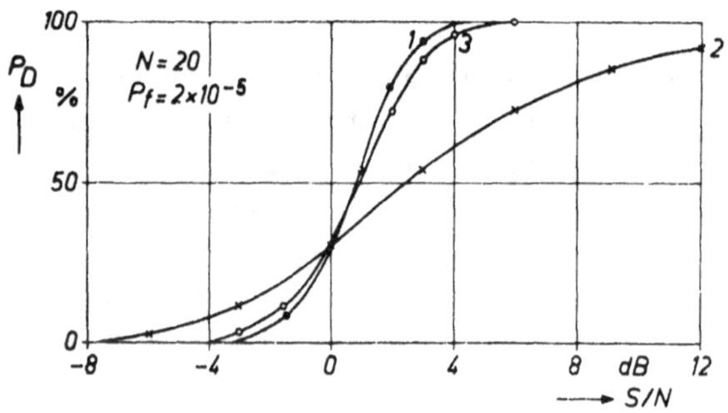

Fig. 1. Probability of detection P_D as a function of signal-to-noise ratio S/N with hits/scan N and false-alarm probability P_f as parameters. Curve labelled 1 is for non-fluctuating targets, 2 for scan-to-scan fluctuations and 3 for pulse-to-pulse fluctuations.

target, whereas the same loss for a jumping-frequency radar is only 0.5dB. This means that if a fixed-frequency and a jumping-frequency radar have the same pulse length, noise figure, etc., the output power of the fixed-frequency radar has to be 7.5 dB (about 6 times) greater than that of the jumping-frequency radar to obtain the same range performance. Stated in different terms, this means that if the output power is the same for both radars, the jumping-frequency radar gives more than 50% increase in range.

For other values of N ($N > 4$) and P_f, the benefit is approximately the same [4].

Experimental results in searching mode

In order to verify the expected gain, field measurements were made using real targets. A radar transmitter-receiver capable of both FFO and JFO with the same power output, etc., was used and connected to a continuously scanning search antenna. As a measure of P_D the blip/scan ratio was recorded using a P.P.I. and an observer to determine whether an echo existed or not. The target aircraft flew on a straight course towards and from the radar site and recordings were made both as they approached and as they receded.

Fig.2 shows the result of these series of measurements. Three different types of aircraft were available, one jet plane (a small fighter) and two propeller planes (a small single-engine and a medium-size twin-engine). For comparison, the theoretical curves 2 and 3 from Fig.1 are also shown, S/N having been converted

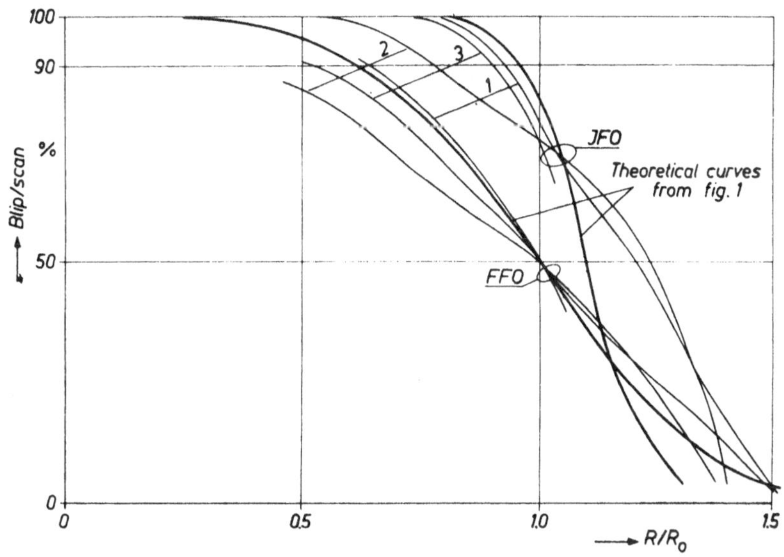

Fig.2. Results of experimental determination of blip-to-scan ratio versus range for air targets, and comparison with theoretical results from Fig.1. R = radar range; R_0 = radar range giving 50% blip-to-scan ratio on fixed frequency. Curves 1 give the result of the measurements with a jet plane (a small fighter), curves 2 with a small one-engined propeller aircraft and curves 3 with a medium-sized twin-engined propeller aircraft.

to range. The abscissa represents the normalized range, the normalization factor for each set of curves being the range which gives 50% blip-to-scan ratio on fixed-frequency radar.

The agreement between the theoretically expected and experimentally obtained values is reasonably good. The differences present can be traced to two main causes. First, since the target was flying a straight course at constant speed, the observer at the display had *a priori* knowledge of where to expect the blip once the target had been detected. Second, the whole process is statistical and as such would require a very large number of observations. In these trials, each curve is the result of some hundreds of observations, which is a reasonably large number but still not sufficient to determine the curves with great accuracy.

Experimental results in tracking mode

Trials have been made to determine the improvement in performance of a tracking radar with a jumping-frequency transmitter. Some thirty measurements, half in FFO and half in JFO, on aircraft of different types on receding courses yielded approximately 20% longer range to first drop-out of tracking when JFO was used. The increase in range to be expected in tracking depends, among other factors, on the time constants in the tracking loop.

Tracking error reduction

In tracking radars a number of sources give rise to tracking errors. Among these sources are target glint, refraction errors in radomes and disturbances in the propagation medium. These sources generally produce slow fluctuations within the passband of the tracking circuits, thus giving rise to noisy output signals.

If jumping-frequency operation with rapidly varying frequencies is used instead of one single frequency, the errors are transformed to random and rapid variations, which to a considerable extent are filtered out by the tracking circuits. This has been verified by trials in the case of glint, as will be explained below. The results of these tests are that glint is greatly reduced, resulting in better tracking accuracy for a given filtering time-constant and the possibility of using shorter time-constants resulting in earlier detection and evaluation of target manoeuvres. It will also greatly improve the stability of – or in some cases even make possible – certain types of missile guidance systems.

Glint reduction measured

With a transmitter-receiver capable of both jumping-frequency and fixed-frequency operation connected to an anti-aircraft fire-control director fitted with conical scan and automatic tracking circuits, a set of measurements of tracking accuracy have been made against air targets flying a linear track with a shortest range to the director of approximately 1 km.

Fig.3 shows one set of recordings of error signals obtained during these trials. The target in this case was a small jet plane. Recordings were made starting at a range of 6 km on the approach and ending at about 4 km on the receding track. Side-by-side in the figure are shown the error voltages, as obtained from the demodulator circuits of the tracking loops in the director, for range, bearing and elevation in FFO and JFO. The first curve (*a*) shows the unfiltered error voltages. Below it are range markers indicating range to target in kilometres (*b*) The remaining eight curves (c_1 to c_8) are the result of octave-passband frequency analysis of the error voltages; the first passband, however, being from a very low frequency up to 1/4 c/s, is of less interest.

An examination of Fig.3 reveals that the error signals in the frequency bands up to 2-4 c/s, i.e. the frequency bands of importance for the tracking mechanism, are much smaller in JFO. This applies to both range and angles, being naturally most evident in angles at short range.

Measurements of the same kind have also been performed on sea targets, see Fig.4. In this case, the evaluation was made using a standard 16 mm camera fixed to a fire-control tracking antenna. The film recording was later studied frame by frame and the position of the target on the frame measured in milliradians. Fig.4 shows the distribution of measurements at intervals of 0.5 milliradians. The recording time for each run was approximately one minute and the target distance approximately 10 km. For comparison, the targets are shown in their 'observed' size, thus in milliradians, below the distribution curves.

Increased resistance to jamming

Changing the carrier frequency of the radar transmission from pulse to pulse decreases the effectiveness of jamming by several orders of magnitude. If we compare a fixed, or slowly tunable, frequency radar subject to narrow-band jamming, with a jumping-frequency radar subject to wide-band jamming, we find that for the same amount of available jamming power, the effectiveness of the jamming decreases as the ratio of the receiver bandwidth to the bandwidth used by the transmitter, which for a typical radar set of this type means some 20 dB. Comparing the case of a FFR and a JFR set both subject to wide-band jamming with the same number of watts per Mc/s, we find the jumping-frequency set gives the same improvement in performance as under jamming-free conditions (see 'Range increase due to diversity' above).

Narrow-band jamming of a JFR set is almost completely unsuccessful, regardless of the jamming power used. Those pulses of the radar that happen to fall within the jammed band, are, of course, strongly jammed. All the jamming can achieve, however, is to saturate the receiver on those frequencies and, since the percentage of pulses falling within the jamming band is small, practically no degradation in performance of the JFR takes place.

Random selection of frequency from pulse to pulse also makes efficient deceptive jamming, e.g. generation of false echoes, difficult. The only kind of repeater jamming it is possible to use is the wide-band type, in which case it is very difficult to produce appreciable delays in retransmission of a received pulse.

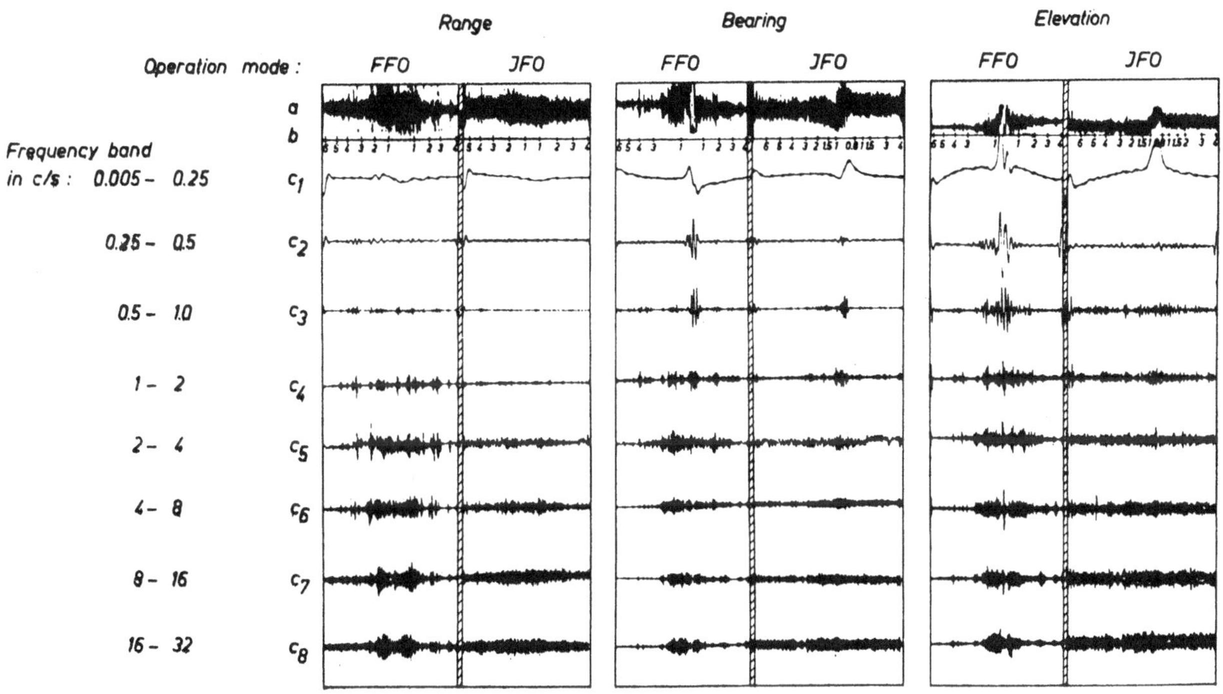

Fig. 3. Reduction of glint when using jumping-frequency radar. The error signals in range, bearing and elevation were registered during automatic tracking of a jet plane with a conical-scan tracking radar. The target was flying a straight course with a shortest range of approximately 1 km to the radar. a = unfiltered error signals; b = range scale in km; c_1 to c_8 = frequency analysis of error voltages a.

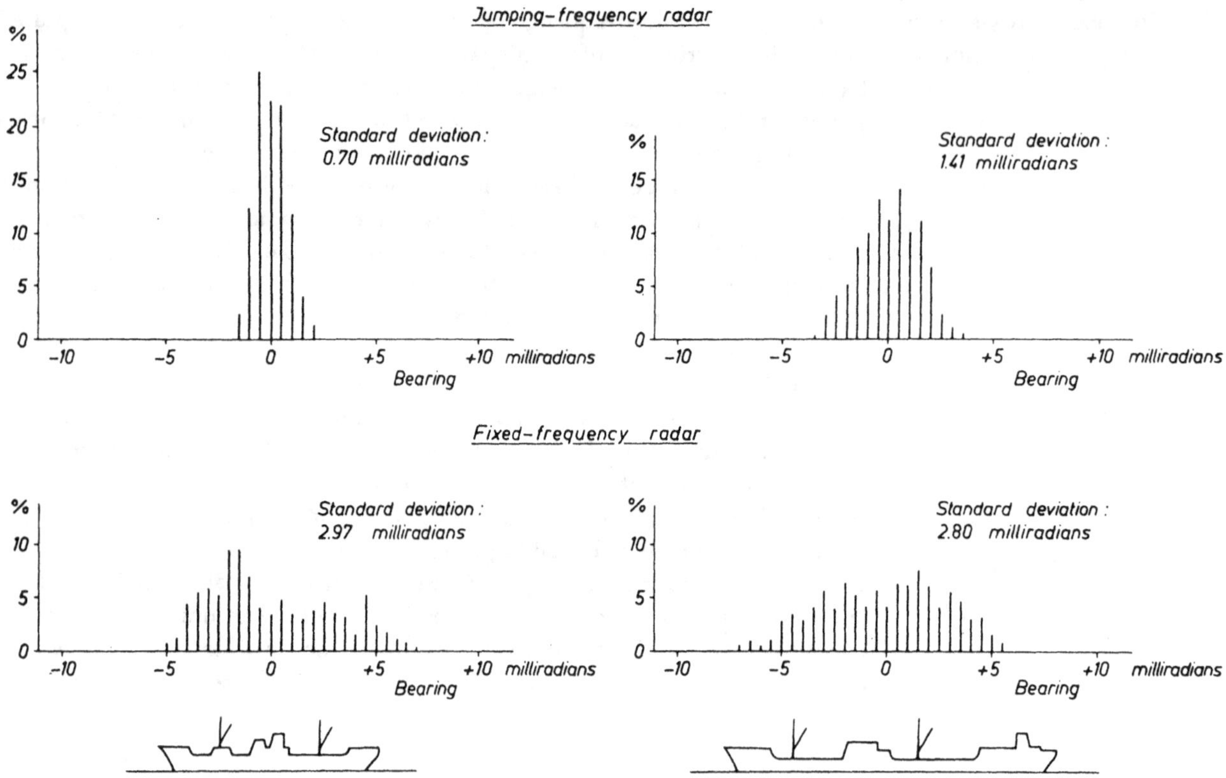

Fig.4. Reduction of glint using jumping-frequency radar to track sea targets, as compared with fixed-frequency radar. Relative distribution of target positions, recorded with a 16 mm film camera and measured at intervals of 0.5 milliradians, is shown together with the silhouettes of the targets, which are reproduced in their apparent actual size in milliradians.

Anticipated additional features

The three sections above all deal with established facts in favour of a jumping-frequency radar system. Below are discussed a few points currently under investigation.

New views on RFI

Widespread use of JFR transmitters is expected to bring about a reduction in RFI (radio frequency interference) between adjacent radar sets. This belief is based on a paper by Costas [5], which deals with advantages to be gained in radio communication by using wide-band techniques instead of the now common narrow-band transmitters and receivers. In the summary of his paper, Costas draws the conclusion that the most efficient use of a particular portion of the radio spectrum is obtained when transmitters and receivers assigned to operate in that band all make use of the whole available spectrum, instead of, as is now the case, different narrow-band channels being allocated to potential users of the band. It is also shown that the jamming resistance of a system is directly proportional to system bandwidth for a given data rate. His final conclusion is that for military communication applications wide-band techniques are often desirable and sometimes mandatory.

The degree of mutual interference between radar sets depends, of course, very closely on the number of radars in a given area, the geographical locations of the sets within the area, the properties of the aerials, etc. We believe, however, that many of the benefits of wide-band techniques as reported for communication will also apply in the case of jumping-frequency radar.

Sea-clutter smoothing Sea targets are, in general, subject to the same fading mechanism as air targets. The same improvement in range against such targets is thus also expected. To this is added another possible advantage.

In one scan with fixed frequency operation the target may give a strong return amidst sea clutter made up of a large number of small returns forming a 'checker-board' background. In the next scan the target may have faded, giving only a weak return, while, in the meantime, the background sea-clutter may also have changed to a different 'checker-board', making detection of the target difficult. With the diversity effect experienced with JFR, the total return from the target will be constant from scan to scan. It is believed that the 'checker-board' pattern from the sea clutter will be smoothed out to a certain extent with jumping-frequency radar, resulting in the presentation of a steady target against a relatively steady background.

Applications

From the above it can be concluded that a radar of this type has many advantages from the system point of view. It may be added that the transmitter-receiver units developed show a slight increase in hardware complexity compared with more conventional types. We consequently believe that this type of radar technique will be used extensively together with broadband antennas in many search and tracking systems in the future.

There exists, however, another group of applications where the ease of frequency choice can be utilized in another way, namely 'Frescan' (frequency scanning) In such applications the transmitter-receiver is used with antennas radiating in a direction which depends on the frequency. The jumping-frequency principle then opens up the possibility of effecting sector scanning by frequency variation and therefore without moving parts in the antenna. This permits, *inter alia*, such rapid scanning that continuous flicker-free presentation is obtained. It should, however, be noted that with such a system some of the advantages pointed out above are lost since a radar of this kind is in effect a fixed-frequency radar for any particular direction.

Acknowledgement

We wish to express our gratitude to all those outside and within the Philips concern who have made it possible for us to obtain the results reported above.

References

1 J.I.Marcum: A statistical theory of target detection by pulsed radar, *IRE Transactions on Information Theory*, Vol. IT-6, No.2, April 1960, Special Monograph Issue.

2 P.Swerling: Probability of detection for fluctuating targets, *IRE Transactions on Information Theory*, Vol. IT-6, No.2, April 1960, Special Monograph Issue.

3 W.M.Hall: Prediction of pulse radar performance, *Proceedings of the IRE*, 44, 2, p.224; 1956.

4 M.Carpentier: Radars, théories modernes. Dunod, Paris, 1963, pp. 104–114.

5 J.P.Costas: Poisson, Shannon, and the radio amateur, *Proceedings of the IRE*, 47, 12, p.2058; 1959.

Improving Radar Range and Angle Detection with Frequency Agility

H. Ray

Microwave Journal, Vol. 9, No. 5 (May 1966)
Pages 63-68

INTRODUCTION

The term "frequency agility" implies a transmitting device with the capability to operate at different frequencies (over a band of frequencies) on a pulse-to-pulse basis. This definition places no limitations on the tuning process. Tuning may be done in a predictable fashion or may be completely random within the band. Similarly, it is not essential for the transmitter to be tunable. The transmitting device may be a broadband amplifier with the output frequency determined by an internal source. The important prerequisite is that each transmitted signal must be at a different frequency from the preceding one.

The particular technique is limited to pulsed radar systems, although agility could be a property of continuous wave systems in which the pulse period is defined as a specific time interval. In a frequency agility system, the shift in transmitter frequency must be accomplished during the interpulse period. This requirement implies the need for some method of tuning.

Most suitable transmitters fall into one of two categories. The first includes devices which are mechanically tuned such as magnetrons and cavity tuned klystrons. The second includes such devices as traveling-wave amplifiers and oscillators (both forward and backward), voltage-tuned magnetrons, and crossed field devices, all of which are or may be electronically tuned. These categories are not mutually exclusive. Klystrons, for example, may be used in both ways under certain conditions.

Further restrictions will be placed on tuning capability when the benefits of frequency agile sources are discussed. There is a tendency to place a lower limit on the pulse-to-pulse frequency change necessary to achieve any real benefit from tunable transmitter characteristics.

In a discussion of frequency-agility radar, there are at least three questions which should be considered:

- First, what constitutes a radar target, and what effect, if any, has frequency agility on target characteristics?
- Second, is there any relationship between target characteristics and the amount of frequency change?
- Third, is radar system performance enhanced by frequency agility and, if so, when? Are there cases when performance degradation can occur?

In a limited discussion of this sort, these questions cannot be explored in depth; however, some insight into the basic problems can be provided.

RADAR TARGETS

The question of exactly what constitutes a radar target is a basic one. For radar design purposes, targets are equated to spherical bodies, and each is assigned some equivalent area (usually in square meters). In most cases, the typical target is somewhat larger than the wavelength of the frequency used. Thus, the effective area is some constant multiplied by the actual synthetic target size.

For evaluation of fixed frequency systems of equivalent types, this target basis does not introduce any real problem. It is obvious, however, that targets cannot be assumed to be that uncomplicated. A real target is a complex arrangement of various types of scatterers. The equivalent area becomes a statistical property which depends not only on wavelength but also on relative aspect. For two conducting elements spaced "1" units apart, the amplitude of signal return will depend on the phase recombination of the reradiated signal from these elements.

Neither can it be assumed that a target is a random spacing of simple elements. Real targets are made up of a discrete spacing of various scattering elements. The number of radiating elements activated will depend, in part, upon the wavelength of the illuminating signal. The result is an amplitude pattern for real targets which is variable with aspect. A change in illumination frequency, by effecting the scattering elements of the target, produces similar amplitude fluctuations at any given aspect. To achieve optimum measure of the actual target cross section, regardless of aspect, it would be necessary to take an average of a number of independent samples of the fluctuating return.

If frequency agility is to be used for this purpose, it would be instructive to determine the parameters $\triangle f$, transmission frequency change, and target dimensions, relative to the radar resolution cell, which relate to decorrelation (independence) of the fluctuating amplitude returns.

Generally, the amount of frequency shift necessary for decorrelation of

HOWARD K. RAY received a BSEE degree in 1957, cum laude, from the University of Illinois. Following graduation, he joined the staff of Sandia Corp. in Albuquerque, New Mexico. Since joining General Electric in 1961, Mr. Ray has been investigating the theory and application of frequency agility radar for airborne weapons systems. He is a member of Tau Beta Pi, Eta Kappa Nu and IEEE.

* *This paper was originally presented at the 11th Annual East Coast Conference on Aerospace and Navigational Electronics, October 21-23, 1964 and published in the conference proceedings.*

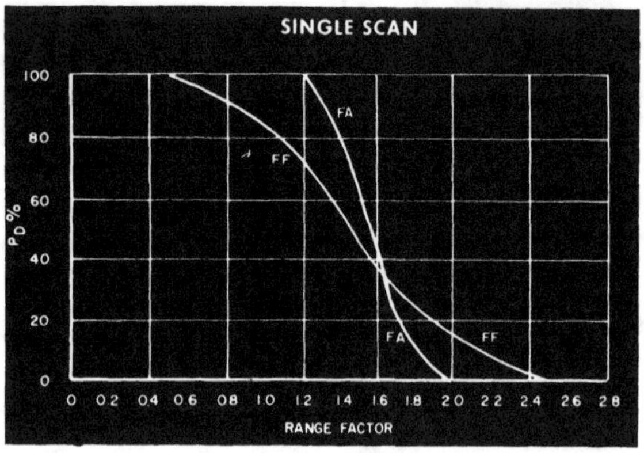

Figure 1 — Probability of detection of a target in one scan.

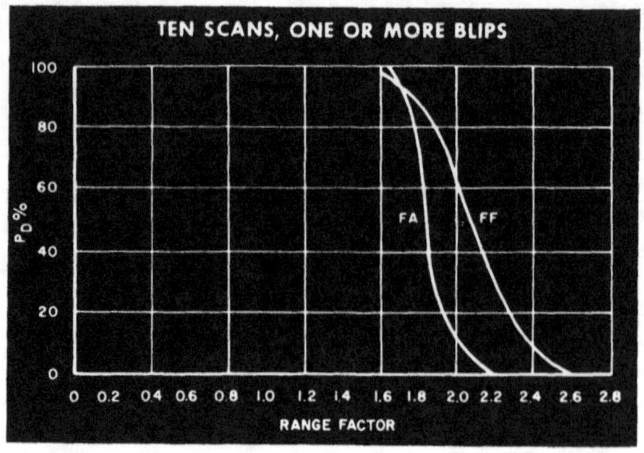

Figure 2 — Probability of detection of a target in ten scans.

returns from a target is a function of the characteristic range dimensions of the target complex. These dimensions can be determined by either of two factors: (1) the physical size of the target itself; or (2) the range resolution of the radar. Initially, it is assumed that this characteristic radial dimension is D′, that x is the radial distance from the nearest part of the target complex to an element of the complex, that f is the transmitted frequency, and that $\triangle f$ is a small change in f. For CW transmission, then, the phase θ of the echo from any element, referred to the phase of the echo of the first element is

$$\theta = \frac{4\pi xf}{C} \qquad (1)$$

and the derivative of θ with respect to f is

$$\frac{d\theta}{df} = \frac{4\pi x}{C} \qquad (2)$$

This expression forms a spectrum for rate of change of phase with respect to transmitted frequency. Since rate of change of phase may be considered to be frequency, and since the target complex for generality is assumed to be randomly located scatterers within the range of x=O to x=D′, the result is a power spectrum having limits of O and $\frac{4\pi D'}{C}$ radians/cycle.

For noise with a bandwidth of W, the autocorrelation function as given by Goldman[8] is

$$\theta(\tau) = \frac{\sin 2\pi W\tau}{2\pi W\tau} \qquad (3)$$

Using this expression, the autocorrelation coefficient for the resultant signal as a function of frequency change is

$$\theta(\triangle f) = \frac{\sin 4\pi D' \triangle f/C}{4\pi D' \triangle f/C} \qquad (4)$$

For $\theta(\triangle f)$ to be equal to 0.5, which seems to be a reasonable amount of decorrelation, it is then necessary that

$$4\pi D' \triangle f/C = \frac{1 \cdot 2\pi}{2} \qquad (5)$$

and

$$\triangle f = \frac{1 \cdot 2C}{8D'} = \frac{45}{D'} \text{MHz} \qquad (6)$$

Thus, for a target with a small range dimension in comparison to the range resolution of the radar, the frequency change necessary to reduce the correlation coefficient to 0.5 is 45 MHz divided by the target size in meters. For the target model, it would be necessary to have an infinite frequency change to achieve complete decorrelation. Physically, this may be justified by noting that complete decorrelation implies complete randomness of the phases of the returns from the various scatterers which comprise the target complex. Because the target model assumes random spacing of the scatterers, there is a possibility of scatterers with a spacing approaching zero. Therefore, a frequency change approaching infinity would be necessary to give the required phase shift. For this reason, it is realistic to accept some finite amount of correlation as representative of the decorrelated case.

Where the target complex has a size approaching, or larger than, the range resolution of the radar, i.e., extended targets like the ground in a mapping radar, the characteristic dimension D′ is generally determined by and equal to the range resolution of the system. The exact relationships of the pulse length, IF bandwidth, and video bandwidth can result in some difference between the range resolution and D′, but this does not seem significant in any practical case. Using the normal relationship that $B = \frac{1.2}{\tau}$ and a range resolution equal to $\frac{\tau C}{2}$, it is found that where the target dimension is established by the pulse length, the $\triangle f$ necessary is

$$\triangle f = \frac{B}{4} = \frac{1}{3.3\tau} \qquad (7)$$

Although the elements of the target complex may be frequency sensitive, the required change in frequency to achieve decorrelation seems to be independent of the frequency used. D′ could be expressed in terms of the wavelength, with no loss of generality.

DETECTION RANGE INCREASE

The performance of search radars is normally presented as a detection probability on a given target at a given range. In general, for such a specification, the figure for range is the only one with a clear meaning.

A detection probability of 80 per cent may sound good, but what about the other 20 per cent? Is detection of these targets accomplished at slightly less range, at considerably less range, or are they never detected? Because this question is crucial in many practical applications, it will be considered in some detail here. The relative importance of the various parts of the curves which follow must be evaluated separately for each situation. Generally, the portion between 80 and 98 per cent is the most important.

Discussion of Calculations

References 2 and 4 by P. Swerling discuss the theory of fixed frequency and frequency agility radar, respectively. The data from these two articles was used as the basic value comparing range with per cent of detection probability (P_D) in Figures 1, 2 and 3. The result was that the frequency agility system shows itself superior in detection range at higher probabilities.

The crossover point on the curves is profoundly affected by the number of pulses integrated. (Compare Figures 1 and 2.) This phenomenon can be explained by considering first, that at the crossover point either fixed frequency (FF) or frequency agility (FA) will provide sufficient return energy to operate beyond a detection threshold with equal likelihood.

Second, above the crossover point, frequency agility is more likely to detect the target than is fixed frequency radar. Frequency agility is assumed to provide independent samples of cross section pulse-to-pulse; over N pulses, we are sampling, therefore, x number of probable values of cross section. This sampling is more likely to exceed a given threshold, when integrated, than the dependent samples of fixed frequency detection. When fixed frequency presents a favorable return, all subsequent returns in that pulse group are apt to be favorable. Any excess energy over that required to exceed a detection threshold is useless. However, when unfavorable returns are presented, all returns in the pulse group are unfavorable, and the integration of the pulse group is less likely to exceed a given threshold.

The third point — fixed frequency's advantage below the crossover — can be thought of in the same manner. In this case, the "average" effect of frequency agility is less likely to be sufficient for exceeding a given threshold, while the occasional favorable group of returns produced by fixed frequency has a higher likelihood of exceeding a given threshold.

It can be concluded that the effect of frequency agility is not to modify or change the character of a group of given pulse returns. The peaks and valleys are the same as with fixed frequency, but the correlation interval of the pulse group is modified.

The only questionable assumption in terms of Swerling's analyses is: Does frequency agility provide independent sampling? If the character of the target and the amount of pulse-to-pulse frequency change is sufficient, we believe it does. Theoretical proof can only be based on heuristic methods. Complex, real life targets preclude any precise analysis of correlation because of specific frequency changes pulse-to-pulse. The experiment would not offer "positive proof" because the results might also be explainable by another theory. Such a measurement program would probably only indicate a "good and sufficient" frequency change per pulse period against a given class of targets. It should be noted that practical testing of this nature does not prove the effects of frequency agility. Testing would only establish a parameter of a method, whatever the cause, whereby essentially independent amplitude samples of a complex target area can be obtained. Even here, a comparison with Swerling's theory would depend greatly on the detection criteria used and the effects of integration.

CLOSING VELOCITY 0.1 PER FRAME TIME — 2 OR MORE BLIPS CUMULATIVE PROBABILITY

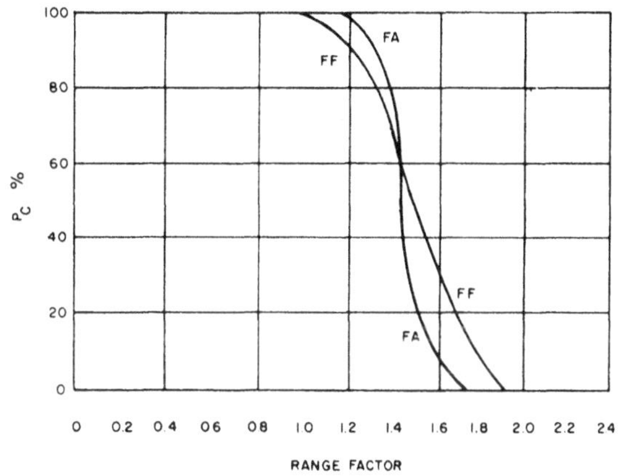

Figure 3 — Probability of detection of a target closing on the radar.

Single Scan

The probability of detection of a target in one scan is shown as a function of normalized range in Figure 1 for both fixed frequency and frequency agility radars.*

Figure 1 shows clearly that with only one scan and a need for high probability of detection, the frequency agile system gives much better range performance (38 per cent more at $P_D = 0.85$). In simplest form, this figure shows the improvement which frequency agility can provide. To accomplish this same improvement by increasing transmitter power would require more than tripling the size of the transmitter.

Multiple Scans

If it is assumed that the antenna will scan the target a number of times and that detection will occur if the target is seen one or more times in all the scans, the situation in Figure 1 can be changed considerably to that depicted in Figure 2. However, it would be well to take a close look at how realistic this definition of detection is.

It is accurate to assert that detection can occur if a target is seen one or more times out of three scans. But if a human operator is used, it is doubtful that seeing something once out of thirty scans could be considered recognition and detection. As a practical matter, the operator has other functions which prohibit his giving full attention to the radar scope. This may be desirable. Staring at a radar display is not the sort of thing which promotes alertness. One sighting out of thirty or more scans does not constitute detection, and even one out of ten is not really dependable.

Cumulative Detection

In the cases cited here, a stationary target has been assumed. For most practical cases, the target is closing on the radar, and there is a need to know the range at which it will be detected. This is obviously a function of the target closing rate, and we would expect the range capability of the radar to be reduced for successively higher closing rates. In Figure 3 no particular range is specified; however, a range factor of one was arbitrarily assigned to the 85 per cent probability for a single scan with a fixed frequency radar. Accordingly, closing rates are based on the distance closed between scans, i.e., if a target is closing at a speed of 10 miles per minute on a radar with a 100 mile range and a 60-second frame time, the closing rate would be 0.1. For the purposes of ascribing a detection range advantage to frequency agility, particular attention should be paid to the curve in Figure 3 showing a cumulative probability of having two or more "detections" as a function of range. It is felt that this approaches the most practical concept of what really constitutes detection in this context.

ANGLE ACCURACY IMPROVEMENT

The major problem in a forward-looking aircraft radar is that the return from any complex target fluctuates

* *Figures 1, 2 and 3 are calculated using a false alarm rate of 10^{-8} and 10 hits per scan.*

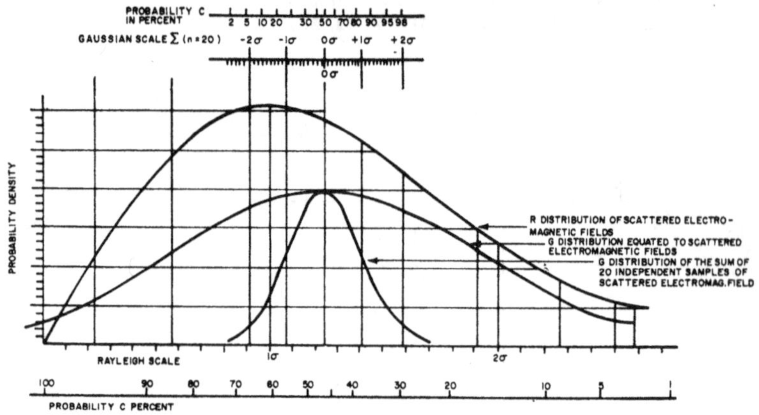

Figure 4 — Probability density functions.

from pulse to pulse as a function of position, look angles, beamwidths, pulse width and frequency.[9] The returns from most targets have a high autocorrelation function for the "dwell" time of the beam in one scan and nearly zero autocorrelation from scan-to-scan when the radar operates at a fixed frequency. This means that the integration of the hits on the indicator, on any element of the target during one scan with fixed frequency, might be considerably different from the average for that element. With mapping radars, we recognize buildings and valley sides because they are brighter than the average terrain, and we recognize water because it is darker than the average terrain. It is a degradation in legibility and resolution to have pieces of terrain on the indicator appearing randomly to be brighter or darker than average. The more accurately the mean signal amplitude can be resolved from display intensities the greater the resolution of the radar.

The first step quantitatively in relating resolution performance is to describe target fluctuation; the second step is to describe amplitude resolution, including fluctuations, in a form suitable for transformations; and the third step is to apply the amplitude resolution description to amplitude modulation of angle and range resolution functions. To simplify the process, it is assumed that there will be no saturation and that an essentially linear system will exist.*

Target Fluctuation Description

Electromagnetic back scatter obtained from radar ground mapping of complex targets usually follows a Rayleigh distribution. The Rayleigh distribution plot is shown in Figure 4 which has a probability density function given by

$$p(x) = \frac{x}{\sigma^2} e^{-1/2 \frac{x^2}{\sigma^2}} \qquad (8)$$

In a single frequency radar, the returned signals in one scan will fall in a very narrow band of amplitude based upon such variables as glint angle and frequency. In the next scan, the aircraft will have moved, the glint angle will have changed, and the signals received will fall in some other very narrow band of amplitude. Using the Rayleigh probability distribution as a description of this fluctuation, the probability of the received amplitude's falling in a certain band can be described from the mean value.

In a frequency agility radar, the amplitude of each returned signal in one scan will fall randomly on the Rayleigh probability distribution so that the signals received in one scan will be a random collection of fluctuations. This collection of random fluctuation samples is averaged on the display, resulting in the amplitude (intensity) variance being reduced by the reciprocal of the square root of the scanning density.

A radar ground map display will integrate the amplitude at each cathode ray tube surface element and will display an intensity equal to the integration of the scanning density number of pulses. In a single frequency radar, the scanning density number of pulses is of the same amplitude and so the measurement of the significant mean value is related to a confidence factor of probability. In a frequency agility radar, each pulse in the scanning density number of pulses falls randomly in amplitude and so a measurement of the significant mean value is related to a confidence factor of probability which has a variance reduced by one over the square root of the scanning density.

The new distribution, which is illustrated in Figure 4, is for an effective scanning density of 20 hits. The mathematical transformation of the Rayleigh distribution to the sum of 20 independent samples of the Rayleigh distribution can be treated as a Gaussian distribution.

The first step in transforming the Rayleigh distribution to an equivalent Gaussian distribution is to start with the Gaussian distribution conditions where

$$\sigma_G^2 = rms^2 - avg^2 \qquad (9)$$

and the Gaussian probability density is given as

$$p(x) = \frac{1}{\sigma(2\pi)^{1/2}} e^{-1/2 \frac{x^2}{\sigma^2}} \qquad (10)$$

From the Rayleigh σ scale we can determine the mean radial error (mre) and rms values as

$$mre = \left(\frac{\pi}{2}\right)^{1/2} \sigma \qquad (11)$$

and

$$rms = 2\text{-}1/2\ \sigma \qquad (12)$$

Substituting these values in Equation (9), we can derive a Gaussian distribution equivalent to the Rayleigh. Thus

$$\sigma_G^2 = (2^{1/2}\sigma_R)^2 - \left[\sigma_R\left(\frac{\pi}{2}\right)^{1/2}\right]^2 \qquad (13)$$

or

$$\sigma_G = 0.657\,\sigma_R \qquad (14)$$

and the desired transformation is completed.

The second step is to translate the Gaussian probability-density mode equivalent of the Rayleigh probability density mode to the probability model of the sum of 20 independent samples. This new probability will have the same average, but the standard error or σ of the new Gaussian probability model will have a new σ which can be expressed as

$$\sigma_g \Sigma\,(n=20) = \frac{\sigma_R \Sigma\,(n=1)}{\sqrt{n=20}}$$

Figure 4 illustrates this transformation. The bottom scale is the Rayleigh probability in per cent, and the next higher scale is the Rayleigh amplitude scale in σ_R. The top curve is the Rayleigh probability density. This is the Rayleigh probability function which describes the single, random radar electromagnetic backscatter.

* *A radar with a log IF and a CRT with ten shades of grey or better are applicable to the example.*

The middle curve is the Gaussian equivalent of the Rayleigh probability density. The probability of this function is not shown because this is only an intermediate step.

The second scale from the top is the transformed Gaussian scale reduced by $\frac{1}{\sqrt{20}}$. The very narrow bottom curve is the same as the Gaussian equivalent to the Rayleigh distribution, but reduced in scale about the mean by $\frac{1}{\sqrt{20}}$. This reduced variance scale is the desired Gaussian probability-density function which describes the sum of 20 independent samples of the Rayleigh probability density.

The second scale from the bottom of Figure 4 still describes the amplitude of the radar return signal for a single pulse. The bottom scale describes the probability of a single pulse. The top scale describes the amplitude probability density of the sum of 20 independent samples.

AMPLITUDE RESOLUTION

The question of what the amplitude resolution is when described as a probability density function is essentially subjective. If the ground map is the usual confusing target complex, we must rely on an arbitrary assumption of probability resolution. For a complex target area, a resolution probability of 80 per cent would be necessary for good interpretation. Extending the 90 per cent and 10 per cent points of the Rayleigh probability scale to the Rayleigh σ scale shows an 80 per cent probability of resolving the radar amplitude of between $0.46\,\sigma$ and $2.2\,\sigma$. This results in an amplitude resolution of $\frac{2.2\,\sigma}{0.46\,\sigma} = 4.78$. In decibels, this becomes 20 log 4.78=13.6 dB.

Therefore, for a single frequency radar, the amplitude resolution at an 80 per cent probability of confidence is only 4.78 or 13.6 dB.

With frequency agility radar, the same resolution criteria of at least 80 per cent probability for interpretation will be used. The amplitude scale that includes 80 per cent probability on the top scale for $\Sigma(n=20)$, and the input Rayleigh σ scale (next to the bottom) shows an 80 per cent probability of resolving an amplitude of between $0.98\,\sigma$ and $1.47\,\sigma$. This results in an amplitude resolution of $\frac{1.47\,\sigma}{0.98\,\sigma} = 1.5$. In decibels, this becomes 20 log 1.5=3.5 dB. Thus, an amplitude resolution improvement of $\frac{4.78}{1.5} = 3.18$, or in decibels, 20 log 3.18=10.1 dB is obtained.

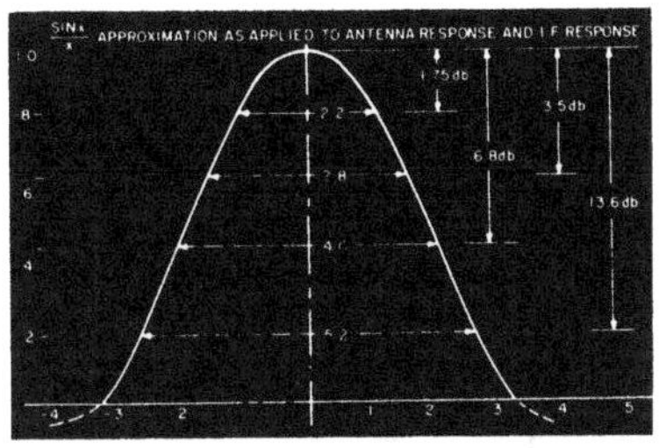

Figure 5 — Plot of $\frac{\sin X}{X}$.

Before completing the subject of amplitude resolution, we should consider the validity of the derived resolution. Most important is the validity of the Rayleigh distribution describing the signal fluctuation. Most likely the fluctuations will be greater or less, depending upon the area being ground mapped by the radar.

Another consideration is the response of the human observer. Practically all high amplitude flashes or fluctuations tend to lower the observer's response even though these effects are compressed by the logarithmic receiver. The result is a tendency to accentuate the detrimental large amplitude fluctuations of the Rayleigh transformation to the Gaussian probability. Because of the different shapes of the probability density functions, the derived resolution improvement changes according to the assumed criteria of resolution. The assumed 80 per cent probability confidence for complex target resolution gave an amplitude resolution improvement of 10.1 dB. Other assumed probabilities would conclude resolution improvement lower and higher than the fundamental reciprocal of the square root of the scanning density, which, in this case, is

$$\frac{1}{\sqrt{20}} = \frac{1}{4.5} \text{ or 13 dB improvement.}$$

This latter number is perhaps the best fundamental relation to work from.

RANGE AND ANGLE RESOLUTION

To describe the resolution improvement quantitatively, amplitude resolution must be related to the antenna pattern of amplitude vs. angle and the receiver response of amplitude vs. time. Both of these responses can be approximated by a $\frac{\sin X}{X}$ characteristic for comparison, although more sophisticated functions are required to derive absolute quantities.

Because the angular resolution must pass through the antenna response twice, the effect is squared. The resolution in dB can also be compared at half the amplitude resolution. This is shown in Figure 5, which is a plot of $\frac{\sin X}{X}$. The amplitude resolution improvement of 6.8 dB to 1.75 dB results in an angular resolution improvement of $\frac{4.0}{2.2} = 1.82$.

The radar receiver response could be interpreted as one way with relative response as shown by the dashed response-width time at 3.5 dB and 13.6 dB. The time or range resolution improvement of $\frac{5.2}{2.8} = 1.86$ results from the amplitude resolution improvement.

Care must be used in the application of these improvement figures for two reasons. First, the definition used here for the word "resolution" is not in strict accord with the classical definition, which is based on targets with equal amplitude. The word as it has been used in this paper connotes the elimination of resolution loss because, although the targets may have the same return on the average, the returns in any one scan will vary from that average. The variation will be more for the fixed frequency system. It is more accurate to say that frequency agility does not really improve resolution; rather it reduces the loss in resolution of a fixed frequency system.

The second reason for advising caution in the use of these improvement figures is the usual assumption that targets can be resolved when they are separated by a distance so that the power received from the point between them is 3 dB less than that received for either target.[7] Since radar indicators are often unable to show a 3 dB difference as a distinguishable shade of grey, no improvements should be claimed when it is not possible to display it on the indicator. In brief, while it is possible to achieve about a 2 to 1 improvement in range and angle resolution through the use of frequency agility, it is also possible that this benefit can be lost through improper design or use of the indicator.

ANGULAR ERRORS FROM SCINTILLATION

In certain cases, severe angular errors

can result from target scintillation. There will be no error if a target does not scintillate or change amplitude during the dwell time of the antenna (ignoring internal, systematic errors). The effects will also average out, and there will be little error if the target scintillates many times (the returns are decorrelated) during the scan.

If, however, the target scintillates only slightly so that the returns are partly correlated, angular errors will result. For example, if reflection is weak while the antenna is just to the target's left and strong while the antenna is to the target's right, the displayed return on the indicator will appear to be shifted to the right.

This effect is described in the following case:

1. Assume that an antenna pattern of beamwidth 1.5 to 2 degrees can be represented by the relationship

$$G = \frac{\cos \pi\, \theta/25}{1-\theta^2/156.25}$$

where θ is in milliradians from beam center.

2. A target consisting of two scatterers separated by a dimension D approximately normal to the line of sight will have a return consisting of two vectors with a phase difference of

$\frac{2\pi 2D}{\lambda}\phi$ where ϕ is the azimuth angle.

The total return will be a voltage having an amplitude (normalized) of

$$\cos\left[\frac{1}{2}\left(\frac{2\pi 2D\phi}{\lambda}\right)\right] \text{or} \cos\left[\frac{2\pi D\phi}{\lambda}\right]$$

If a factor is included for an initial separation in range of the points by a fraction of a wavelength, the amplitude will be

$$\cos\left[\frac{2\pi D\phi + K}{\lambda}\right]$$

where K can have any value from 0 to 2π.

3. The angle ϕ is a function of tangential velocity of the aircraft, range, and time. For this case, ϕ can be described as

$$\phi = \frac{TV \sin \phi_0}{R}$$

or the target return can be described as having an amplitude of

$$\cos\left[\frac{2\pi\, DVT \sin \phi_0}{\lambda R} + K\right]$$

4. It is obvious that the antenna beam position is a function of time and may be expressed as $\theta = \dot{\theta}T$. From this, $T = \theta/\dot{\theta}$. Substituting in the expression above, the target reflectivity can be expressed as

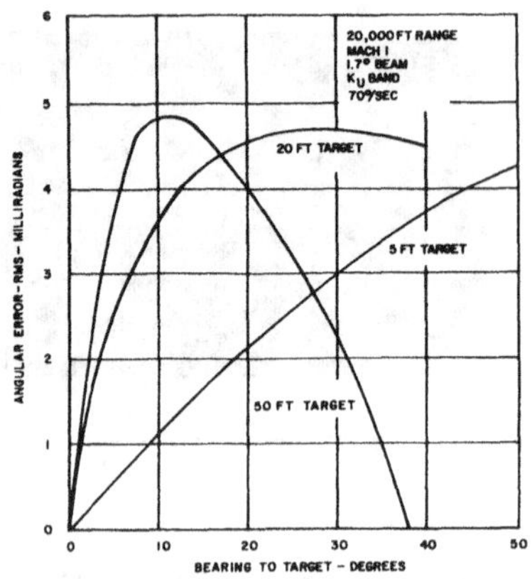

Figure 6 — Bearing angle errors without frequency agility.

$$A_t = \cos\left[\frac{2\pi\, DV\, \theta \sin \phi_0}{\lambda\, R\, \dot{\theta}} + K\right]$$

5. The received voltage amplitude is then G^2A_t and the power is $G^4(A_t)^2$ or

$$P(\theta) = \frac{\cos^4(\pi\theta/25)\cos^2\left(\dfrac{2\pi\, DV\, \theta \sin\phi_0}{\lambda\, R\, \dot{\theta}} + K\right)}{(1-\theta/156.25)^4}$$

6. The apparent center of the target is then at an angle from the true position of

$$\delta\theta = \int \frac{\theta P(\theta)d\theta}{P(\theta)d\theta}$$

for any specific value of K.

7. By substituting specific values and calculating for various values of K, the standard deviation may be calculated from

$$\sigma = \sqrt{\frac{\Sigma\, \delta^2}{N_K}}$$

A typical set of radar characteristics has been postulated in this manner, and the rms values of the resulting angular error calculated on a computer. An example of the calculation is plotted in Figure 6. For those particular situations involving high accuracy mapping or position fixing, frequency agility may offer a substantial advantage.

CONCLUSION

The frequency agility technique for radar energy transmission has been analyzed here to indicate what benefits and advantages it may provide. The analysis applies to the three major types of radar — search, tracking and mapping — to demonstrate this technique's ability to improve their performance.

It is apparent that certain types of radars can and should be equipped with frequency agility. These include:

(1) tracking radars for seekers where "wander" at short range is important;

(2) conical scan radars in which target scintillation has high frequency components;

(3) search radars used to detect fast targets;

(4) mapping radars where target recognition is important;

(5) high altitude radars with high PRF's.

Generalized statements for most applications are subject to error, and it is therefore necessary to analyze the particular application. In all applications it is usually necessary to appraise the quantitative effects of the use of frequency agility. Although this analysis is not exhaustive, it does provide some insight into the workings of the frequency agility technique.

ACKNOWLEDGMENT

The contributions of General Electric's D. Deno and J. J. O'Leary to the technique of frequency agility radar and to this paper are acknowledged with gratitude by the author.

REFERENCES

1. Marcum, J. I., "A Statistical Theory of Target Detection by Pulsed Radar," *IRE Trans.*, Vol. IT-6, April 1960, pp. 59-154.
2. Swerling, P., "Probability of Detection for Fluctuating Targets," *IRE Trans.*, Vol. IT-6, April 1960, pp. 269-308.
3. Schwartz, M., "Effects of Signal Fluctuation on the Detection of Pulsed Signals in Noise," *IRE Trans.*, Vol. IT-2, June 1956.
4. Swerling, P., "Detection of Fluctuating Pulsed Signals in the Presence of Noise," *IRE Trans.*, Vol. IT-3, September 1957, pp. 175-178.
5. Skolnik, M. I., *Introduction to Radar Systems*, McGraw-Hill Book Co., Inc., 1962, pp. 50-56.
6. Jonsen, G. L., "On the Improvement of Detection Range Using Frequency Agile Techniques," *Boeing Document No. DG-6835*, Boeing Airplane Co., Renton, Washington, March 31, 1964.
7. Hammerle, K. J. and J. L. Unger, "Target Resolution Improvement from Frequency Agility," *Boeing Document No. D2-90248*, Boeing Airplane Co., Renton, Washington, January 2, 1963.
8. Goldman, S., *Information Theory*, Prentice-Hall, 1954.
9. Dunn, J. H., D. D. Howard and A. M. King, "Phenomena of Scintillation Noise in Radar-Tracking Systems," *IRE Proc.*, Vol. 47, No. 5, May 1959, pp. 855-863.

Some Measurements of the Effects of Frequency Agility on Aircraft Radar Returns

W.S. Whitlock, A.M. Sheperd, and A.L.C. Quigley

AGARD Conference Proceedings, No. 66 on Advanced Radar Systems
May 25-29, 1970
Reprinted by permission

SUMMARY

A C-band radar with pulse-to-pulse frequency agility is used to investigate the effects of frequency change on aircraft echo amplitude and scanned echo patterns. Decorrelation effects of pulse-to-pulse frequency changes from zero to 198 MHz have been measured using Sea Hawk and Comet aircraft. The aircraft's radar cross-section changed by as much as 20 dB for a frequency change of 20 MHz. The frequency cross-correlation coefficient is zero for a frequency change of 20 MHz and is negative for larger changes. No significant difference between the frequency correlation coefficient was found for the two aircraft studied. Some effects of frequency agility on scanned echo patterns and echo fading are presented.

1. INTRODUCTION

The radar cross-section of reflecting objects is, in general, dependant on the frequency of the incident radiation, hence a change in radar transmitted frequency may result in a change of echo amplitude. This dependance of radar cross-section on frequency may be exploited in a number of ways; for example it may be advantageous to induce changes in the aircraft echo amplitude by appropriate changes in radar frequency in order to reduce echo fading. As experimental results relating aircraft radar cross-section with frequency change are of particular value to the radar designer, some appropriate measurements have been carried out using a radar with a pulse-to-pulse frequency agile facility. The results of these measurements are detailed in this paper.

2. THE RADAR AND DATA ACQUISITION SYSTEM

The salient parameters of the radar system are as follows:-

Transmitted Frequency	-	5.35 to 5.85 GHz
Transmitted Peak Power	-	1 MW
Pulse Length	-	2 microseconds
P.R.F.	-	196 p/s
Antenna Gain	-	38 dB
Antenna Beam Width	-	$1.5^\circ \times 1.5^\circ$ (3 dB, two-way)
Polarisation	-	Horizontal
Receiver	-	Logarithmic Response
I.F.	-	30 MHz
Noise Figure	-	6 dB

A block diagram of the system is shown in Fig. 6.1.

The radar system is fully coherent and the transmitted frequencies are produced by multiplication from a frequency synthesiser operating in the range 84 to 92 MHz. This facility is backed by a special purpose control computer giving a choice of pulse-to-pulse frequency increments, at C-band, between zero and 198 MHz. Random or pattern frequency jumping is possible since all frequencies in the band are defined by any one of 128 different 7-digit binary numbers.

Two modes of pulse-to-pulse frequency agility were used, viz:-

(a) 'Alternate' mode in which only two frequencies were transmitted in sequence f_1, f_2; f_1, f_2; etc., see Figure 6.2. The frequency change, $\Delta f = f_1 - f_2$, could be varied between zero and 198 MHz.

(b) 'Staircase' mode in which successive pulses were transmitted at decreasing frequencies in equal steps for 8, 16 or 32 contiguous pulses before reversing back to the original frequency. As before the frequency change per step could be varied between zero and an upper value which was limited by the number of steps and the 500 MHz system bandwidth.

When carrying out frequency agility experiments it is essential to ensure that the radar performance is not frequency sensitive. For example, it was known that the main t.w.t. output power could vary by $\pm$ 4 dB over the frequency band. In order to reduce this power variation an electronically controlled p-i-n diode attenuator was introduced in the t.w.t. transmitter chain so that the transmitted power levels could be equalised at 80 points in the frequency band. This equalisation is automatic with frequency selection.

It was recognised that a further source of error could arise from the gain/frequency characteristics of the complete receiving channel. This characteristic was found to be cyclic across the 500 MHz band with a gain variation of $\pm$ 2 dB and the technique used to compensate for this gain variation is described below.

The radar data was acquired on-line and transferred to paper tape for subsequent analysis and storage. The video output from a logarithmic receiver was sampled at the range of interest before conversion into a 7-digit binary signal. Up to 10,000 binary video samples could be stored in a computer before transfer to paper tape. In order to compensate for changes in receiver gain the actual signal/noise ratio was measured for each video signal. Firstly, the signal amplitude was measured, then an automatic measurement was made of the mean noise level over a period of 3 ms following the receipt of the signal but before the next frequency change was initiated. The computer was programmed to store the difference between the two logarithmic values.

3. DESIGN OF EXPERIMENTS

3.1 Basic Effects of Frequency Agility

The average value of an aircraft's radar cross-section generally varies little with frequency over the band 1 GHz to 10 GHz, Reference 1. Nevertheless it can be demonstrated that a small change in transmitted frequency can produce a significant change in aircraft cross-section. This apparent anomaly arises because the reflectivity of an aircraft shows fine structure variations with frequency which are normally smoothed in gross measurements. Consideration of a simple model illustrates some of the salient features of this frequency/cross-section relationship.

Let the aircraft be represented by two identical spherical scatterers spaced a distance 'd' apart. Other aircraft models, Reference 2, have been proposed but are not necessarily more realistic. The amplitudes of the two reflected components received by an illuminating radar may be expressed in the form:

$$a_1 = K \text{ Sin } (\omega t + \phi + \frac{2\pi d \text{ Cos } \theta}{\lambda})$$

$$a_2 = K \text{ Sin } (\omega t + \phi - \frac{2\pi d \text{ Cos } \theta}{\lambda})$$

Where K is a constant derived from the radar equation, ϕ an arbitrary phase angle and θ the angle between the line joining the reflectors and the radar line of sight, i.e. the aspect angle. The resultant received power $P_{(r)}$ is proportional to:

$$\left[\text{Cos } (\frac{2\pi d \text{ Cos } \theta}{\lambda})\right]^2$$

or expressed explicitly in terms of frequency:

$$P_{(r)} \propto \left[\text{Cos } (\frac{2\pi df \text{ Cos } \theta}{c})\right]^2 \qquad (1)$$

It will be noted that $P_{(r)}$ varies periodically with θ. This produces the well-known scintillation characteristics of aircraft echoes when changes in θ of 0.1° may change the aircraft cross-section by 20 dB or more. Further, the received power varies periodically with frequency; in fact, for a given aspect angle the aircraft cross-section varies as the product of the separation of the scatterers and the frequency. Using this simple model it can be shown that for a small change in frequency the maximum cross-section change occurs when $\Delta f \approx 10$ MHz.

This simple model has been used to highlight two important effects:

(a) The magnitude of the change in aircraft cross-section due to a change in frequency is aspect dependent.

(b) The maximum change in cross-section should occur when the change in frequency is about 10 MHz.

Initial experimental results did indicate, in fact, that for a fixed change of frequency the cross-section change of a particular aircraft was not reproducible unless the aspect remained virtually constant. In the design of the radar experiments it was necessary to decide on the parameters which would express the quantitative effects of frequency agility in a form which was meaningful and had value to system designers. The choice of such parameters is discussed in the next section.

3.2 Pulse-to-Pulse Amplitude Changes

Given time, aircraft echoes will become decorrelated even at fixed frequency due to random aspect changes. Alternatively the aircraft echoes may be decorrelated by rapid frequency changes. In practice time and frequency decorrelation may both operate on the received signal. These effects may be expressed quantitatively through the use of correlation coefficients, but without computational assistance the derivation of these coefficients is laborious. In the early stages of the experimental studies a computer facility was not available so some alternative parameter was sought which could be derived simply and which expressed quantitatively the decorrelating effects of frequency agility.

Consider a transmitter operating in the alternate frequency agile mode, then two interlaced signal pulse trains will be received, each train corresponding to one of the two transmitted frequencies. Let these pulse trains be identified as p and q, where p = q = 1 to n, where n is the total number of pulses in each train. Let Δa be the change in amplitude between successive pulses, then the mean square of Δa is given by:

$$\overline{\Delta a^2} = \overline{(a_p - a_q)^2}$$

where a_p and a_q are the amplitude deviations from the mean. Hence

$$\overline{\Delta a^2} = \overline{a_p^2} + \overline{a_q^2} - 2\overline{a_p a_q} = 2\left[\overline{a^2} - \overline{a_p a_q}\right] \text{ if } \overline{a^2} = \overline{a_p^2} = \overline{a_q^2}$$

$$= 2\left[\overline{a^2} - \rho . \overline{a^2}\right]$$

where ρ is the cross-correlation coefficient. Therefore,

$$\frac{\overline{\Delta a^2}}{\overline{a^2}} = 2\left[1 - \rho\right] \qquad (2)$$

so that the mean square pulse-to-pulse amplitude difference, which is simple to compute, is related to the cross-correlation coefficient through Equation (2) so long as the basic assumptions are valid. The correlation coefficient may be, of course, both time and frequency dependent.

It is of interest to note that when pulse samples are taken from a source of Gaussian noise and uncorrelated then the r.m.s. pulse-to-pulse difference after processing through a narrow-band logarithmic receiver, can be theoretically derived, Appendix I, and is given by:

$$\left(\overline{\Delta a^2}\right)^{\frac{1}{2}} = \frac{10\pi}{\sqrt{3}\,\log_e 10} = 7.88 \text{ dB}$$

The standard deviation of pulses which have a log-Rayleigh distribution of amplitude has also been derived, Reference 3, vis:

$$\left(\overline{a^2}\right)^{\frac{1}{2}} = \frac{10\,\pi}{\sqrt{6}\,\log_e 10} = 5.57 \text{ dB}$$

Therefore for pulses originating from Gaussian noise and uncorrelated then $(\overline{\Delta a^2})/(\overline{a^2}) = 2$, which agrees with Equation 2 as $\rho = 0$.

The object of this analysis is to show that $\overline{\Delta a^2}$ is a simply computed parameter which may be used as a measure of decorrelation of a pulse train, even when only two discrete frequencies are transmitted. Where the pulse train is noise-like and decorrelated from pulse-to-pulse and the train is processed by an ideal logarithmic receiver than $(\overline{\Delta a^2})^{\frac{1}{2}} = 7.88$ dB. It is important to note that few logarithmic receivers have ideal characteristics so far as their response to noise-like signals is concerned, Reference 4.

3.3 Methods of Aircraft Echo Measurement

When frequency agility data from aircraft was required it was usual to operate the radar in a 'search-light' mode with manual range gating and tracking. The antenna was elevated to 1.5° and the aircraft were directed to fly up or down the stationary beam from a maximum range of 80 miles to a minimum range of 30 miles. Continuous data runs of over five minutes at a nominally head-on aspect were possible although in practice each run was sub-divided into 10 second (2048 pulses) periods, each period corresponding to a particular value of Δf.

When scanned echo patterns were required then both manual range gating and azimuth gating were applied simultaneously with the antenna rotating at 6 rev/min. When the gates were following correctly a computer was switched to acquire data and the scanned echo pattern was passed into the store for subsequent print-out by a line printer.

4. RESULTS

4.1 Initial Experiments

Early work using frequency agility was directed at land clutter and Figure 6.3 shows values of $\sqrt{\overline{\Delta a^2}}$ obtained when the resolution cell contained agricultural land. Figure 6.3 shows that $\sqrt{\overline{\Delta a^2}}$ reaches a maximum when $\Delta f \simeq 1$/pulse length. Using the same data, it was found that the auto-correlation coefficient for one pulse lag reached a minimum when $\Delta f = 1/\tau$.

Initial aircraft experiments using unidentified aircraft targets of opportunity yielded results similar to those also shown in Figure 6.3. For all aircraft trials the alternate frequency jumping mode was used. It was found that the curve of $\sqrt{\overline{\Delta a^2}}$ against Δf was very variable from run to run even for the same aircraft, presumably due to random changes in aspect. It became apparent that, for consistent data, it would be necessary to make observations against controlled aircraft whose aspects would be better defined.

4.2 Sea Hawk and Comet Measurements

Using aircraft controlled from A.S.W.E., two experimental runs were carried out; one using a Sea Hawk GA6 and the other a Comet 4. The object of choosing these very different aircraft types was to try to resolve any dependence of frequency agility response on aircraft size. The Sea Hawk wing-span is 12m whilst that of the Comet is 35m. Figure 6.4 shows the values of $\sqrt{\overline{\Delta a^2}}$ plotted against the frequency separation, Δf, between transmitted pulses for the Sea Hawk in both opening and closing flight. Each point on the graph corresponds to 2048 pulses, i.e. about 10s of record. Maximum decorrelation appears to occur at about 13 MHz. Amplitude probability density functions were computed about a zero mean and Figure 6.5 shows a case of a bi-model distribution due to a shift in mean cross-section of 6 dB as the frequency is changed by 128 MHz.

The Comet experiments produced results very different from those obtained using the Sea Hawk for the Comet returns were found to be largely decorrelated even at fixed frequency. Rain clutter (15 dB above noise) corrupted part of the recordings although the Comet return was generally 30 dB above noise. However, it is believed that the rapid modulation on the Comet return, corresponding to a head-on aspect, could be due to engine turbine-blade motion. Figures 6.6 and 6.7 illustrate the effect of this intrinsic decorrelation process which tends to hide the effects due to frequency agility.

A penalty of using a relatively low P.R.F. (196 p/s) for frequency agile experiments is that there is generally a significant degree of time decorrelation superimposed on any decorrelation due to frequency change. This problem is clearly demonstrated in Figure 6.8 which shows values of the auto-correlation coefficient derived from the Comet data using an alternate frequency jumping mode. The value of coefficient falls rapidly for one pulse lag due to the time-decorrelating process intrinsic to the aircraft, then the coefficient alternates between two values due to changes in frequency. The upper dotted line in Figure 6.8 corresponds to the auto-correlation curve for the fixed frequency case (every other pulse is at the same frequency). This upper curve, therefore, reflects only time decorrelation, $\rho(\tau)$. However, the lower curve traces values of the composite correlation coefficient, $\rho(\tau, f)$. In an attempt to isolate $\rho(f)$ it is assumed that:

$$\rho(\tau, f) = \rho(\tau) \cdot \rho(f) \qquad (3)$$

Having made this assumption, then using all the aircraft data tapes, the auto-correlation coefficients corresponding to all pulse lags between 1 and 10 were computed for the various values of Δf. Using this reduced data the mean values of $\rho(\tau, f)$ and $\rho(\tau)$ were calculated and hence the values of $\rho(f)$ using Equation 3. The results of this approach are plotted in Figure 6.9 and show no appreciable difference between the Comet and Sea Hawk so far as their frequency dependence is concerned despite their different fuselage lengths. This analysis also shows that $\rho(f) = 0$ for $\Delta f \approx 20$ MHz and that the correlation coefficient in the frequency domain is negative for larger frequency changes. This is in accord with the concept that an aircraft consists of a few dominant scatterers.

4.3 Scanned Aircraft Echo Amplitude Patterns

Further experiments were performed in order to simulate aircraft echo amplitude patterns as would be obtained when using a frequency scanned antenna. In fact a mechanically azimuth-scanned antenna was used. The data acquisition technique is described in Section 3.3 for targets which were unidentified aircraft. Using fixed frequency, the ideal signal pattern should only reflect the two-way antenna radiation pattern, as shown in Figure 6.10(a) which is a facsimile of a computer print-out. However, Figure 6.10(a) depicts a rare occasion for it is more probable that a fade will coincide with the time the beam is incident on the aircraft, as is shown in Figure 6.10(b). When alternate frequency jumping is used the echo patterns show considerable modification. For the no-fading case, Figure 6.11(a), the two patterns corresponding to the two frequencies can be resolved clearly. When fading is concurrent with scanning the effect of the two frequencies is clearly shown in Figure 6.11(b) where the long fades have been eliminated. From pulse-to-pulse it can be seen that a change in frequency of 64 MHz can change the amplitude of the return by as much as 25 dB. Other records show numerous examples of large changes in amplitude for relatively small changes in frequency. Where a staircase form of frequency jumping was used it was found that the shape of the echo pattern was distorted by the incremental frequency changes so that if beam shape is used as a detection criteria the effect of frequency agility may degrade detection performance.

5. CONCLUSIONS

Experimental investigations have confirmed that the magnitude of an aircraft's radar cross-section is a sensitive function of transmitted frequency. Cross-section changes greater than 20 dB have been found to occur for a pulse-to-pulse frequency change as small as 20 MHz whereas the average change in cross-section was found to be about 6 dB for a 20 MHz change in frequency. For aircraft, a complex periodic cross-section/frequency relationship exists which is analogous to the cross-section/aspect relationship; therefore a change in frequency may result in the decorrelation of the aircraft return, as would a change in aspect. Frequency correlation coefficients (subtracted mean) have been calculated from data corresponding to nominally head-on aspects of a Sea Hawk and Comet aircraft. It is found that the coefficient is zero for a pulse-to-pulse frequency change of about 20 MHz but for frequency changes between 30-200 MHz the coefficients are negative. Some theories suggest that the frequency change necessary for maximum decorrelation should be inversely proportional to aircraft size, or scatterer separation, but no significant difference was found between the frequency change necessary to decorrelate the returns from a Sea Hawk (wing span 12m) and a Comet 4 (wing span 35m) aircraft.

The main advantage of frequency agility as an aid to aircraft detection derives from the considerable reduction in the occurrence of long fades of the target echo. This advantage may be obtained using only two frequencies separated by 20 MHz; then, on average, successive returns will be decorrelated and fade duration will be confined to one pulse interval. However, in many cases, for example, where the natural fading rate is high and the signal/noise ratio large, the use of frequency agility will not necessarily improve the probability of detection.

6. REFERENCES

(1) Olin, I.D., and F.D. Queen; "Dynamic Measurement of Radar Cross Sections", Proc. I.E.E.E. 53(8), 954-961, August 1965.

(2) Ray, H.; "Improving Radar Range and Angle Detection with Frequency Agility", Microwave Journal, 9, No. 5, May 1966 p.63.

(3) Smith, M.H.A.; "The Rayleigh and Log-Rayleigh Probability Density Functions", A.S.W.E. Laboratory Note XRE-67-8, August 1967.

(4) Reiss, A., and G.W. Carver; "Response of Logarithmic Amplifiers to Noise Signals", Electronic Letters, 4, No. 13, June 1968.

(5) Bromwich, T.J.I'A; "An Introduction to the Theory of Infinite Series", Macmillan, London, 1947, 2nd Edition, p. 498.

(6) Dwight, H.B.; "Tables of Integrals and other Mathematical Data", Macmillan, N.Y. 1961, 4th Edition, p. 13.

APPENDIX I

PULSE-TO-PULSE DIFFERENCES OF SAMPLED NOISE AT THE OUTPUT OF A LOGARITHMIC RECEIVER

M.H.A. Smith, A.S.W.E.

1. DERIVATION OF THE PULSE-TO-PULSE DIFFERENCE PROBABILITY DENSITY FUNCTION (P.D.F.)

Let the amplitude of the logarithmic receiver output sample be 'y' dB (with respect to a datum which will be defined later) and let the probability density for this amplitude be $f(y)$. By definition, therefore, the probability that the output of the receiver will lie within the range 'y' to '$y + \delta y$' is $f(y).\delta y$. Similarly, the probability that the output will lie within the range '$y + D$' to '$y + D + \delta y$' is $f(y + D).\delta y$.

It is assumed that successive samples of the receiver output are uncorrelated. This assumption is true for a noise waveform if the sampling rate is much less than the receiver bandwidth. The probability after a pulse has occurred in the amplitude range 'y' to '$y + \delta y$' that the next pulse will occur in the range '$y + D$' to '$y + D + \delta y$' is now simply the product of the probabilities of the two individual occurrences, i.e. $f(y).\delta y.f(y + D).\delta y$.

If the amplitude difference between two successive pulses is D dB, then the first pulse can take on any value whatsoever in the permissible range of 'y', provided only that the second pulse, which is specified to be the larger in amplitude by D dB, also lies within the permissible range of 'y'. Let the p.d.f. of 'D' be denoted by $p(D)$, then the probability that the difference between successive pulses lies between 'D' and '$D + \delta y$' is

$$p(D).\delta y = \sum_{y} f(y).\delta y.f(y + D).\delta y$$

where the summation is to be performed over all permissible values of 'y'. Since δy can be considered as a factor this equation can be rewritten.

$$p(D) = \sum_{y} f(y).f(y + D).\delta y$$

or, in the limit,

$$p(D) = \int_{y} f(y).f(y + D).dy \qquad \ldots\ldots (1)$$

Now Reference 3 shows that the p.d.f. of the noise output from a logarithmic receiver can be written

$$f(y) = K.z.\exp(-z), \qquad \ldots\ldots (2)$$

where $z = 10^{(y/10)}$,

and $K = \ln(10)/10$,

giving $y = 10 \log z = \ln(z)/K$.

The datum from which 'y' is measured is 2.50 dB above the mean logarithmic noise level and is in fact the modal value of the logarithmic noise output p.d.f.

For an ideal logarithmic receiver 'y' can take on any value from $-\infty$ to $+\infty$.

Thus

$$p(D) = \int_{-\infty}^{\infty} f(y).f(y + D).dy \qquad \ldots\ldots (1, \text{bis})$$

$$= \int_{0}^{\infty} (K\ z\ \exp(-\ z))(K\ z'\ \exp(-\ z'))(1/Kz).dz$$

where $z' = 10^{(y + D)/10} = 10^{(y/10)}.10^{(D/10)}$

If we now denote the linear ratio of the two successive pulse amplitudes as 'r', then

$$r = 10^{(D/10)}$$

$$z' = r\ z$$

and $p(D) = K\ r \int_{0}^{\infty} z.\exp(-\ z(1 + r)).dz$

therefore $p(D) = K\ r/(1 + r)^2, \qquad \ldots\ldots (3)$

where $r = 10^{(D/10)}$

and $K = \ln(10)/10$

The mode of this p.d.f. occurs for $r = 1$ or $D = 0$ dB, that is, the most probable pulse-to-pulse difference is zero, and the modal value is seen to be $K/4$ (i.e. 0.05756). If the most probable pulse-to-pulse difference were not zero then there would be an upward or downward trend in the mean noise level, and it is tacitly assumed that this is not so.

The total area under the p.d.f. is

$$\int_{-\infty}^{\infty} p(D).dD = \int_{0}^{\infty} (r/K).p(r).dr$$

$$= \int_{0}^{\infty} 1/(1 + r)^2 \, dr$$

$$= 1$$

This value is to be expected as it includes all possible values of 'D'.

2. STANDARD DEVIATION OF THE PULSE-TO-PULSE DIFFERENCE P.D.F.

The p.d.f. of the pulse-to-pulse difference, D, is given by

$$p(D) = Kr/(1 + r)^2 \qquad \text{..... (3, bis)}$$

where

$$r = 10^{(D/10)}$$

If the standard deviation of the p.d.f. is denoted by σ then

$$\sigma^2 = \int_{-\infty}^{\infty} p(D).D^2.dD$$

$$= \int_{-\infty}^{0} p(D).D^2.dD, \text{ from considerations of symmetry}$$

Now

$$D = \ln(r)/K$$

and

$$dD = (1/Kr)dr$$

so that

$$\sigma^2 = (2/K^2) \int_{0}^{1} (\ln r)^2/(1 + r)^2 \, dr$$

$$= (2/K^2)I \qquad \text{..... (4)}$$

where I denotes the definite integral.

This integral is not easily evaluated since the logarithmic term becomes infinite at the lower limit and the power series expansion for the denominator is not valid at the upper limit. However, in Reference 5 if we substitute $(\ln r)^2$ for ln x and set the parameter 'p' to zero we obtain

$$I = 2 \sum_{n=0}^{\infty} (-1)^n/(n + 1)^2$$

Therefore

$$I/2 = 1/1^2 - 1/2^2 + 1/3^2 - - - - -$$

$$= \pi^2/12 \text{ by Reference 6}$$

Substituting back into Equation (5)

$$\sigma^2 = \pi^2/(3\,K^2)$$

i.e.

$$\sigma = \pi/(\sqrt{3}\,K)$$

$$= 10\pi/(\sqrt{3}.\ln 10) \qquad \text{..... (5)}$$

$$\simeq \underline{7.877 \text{ dB}}$$

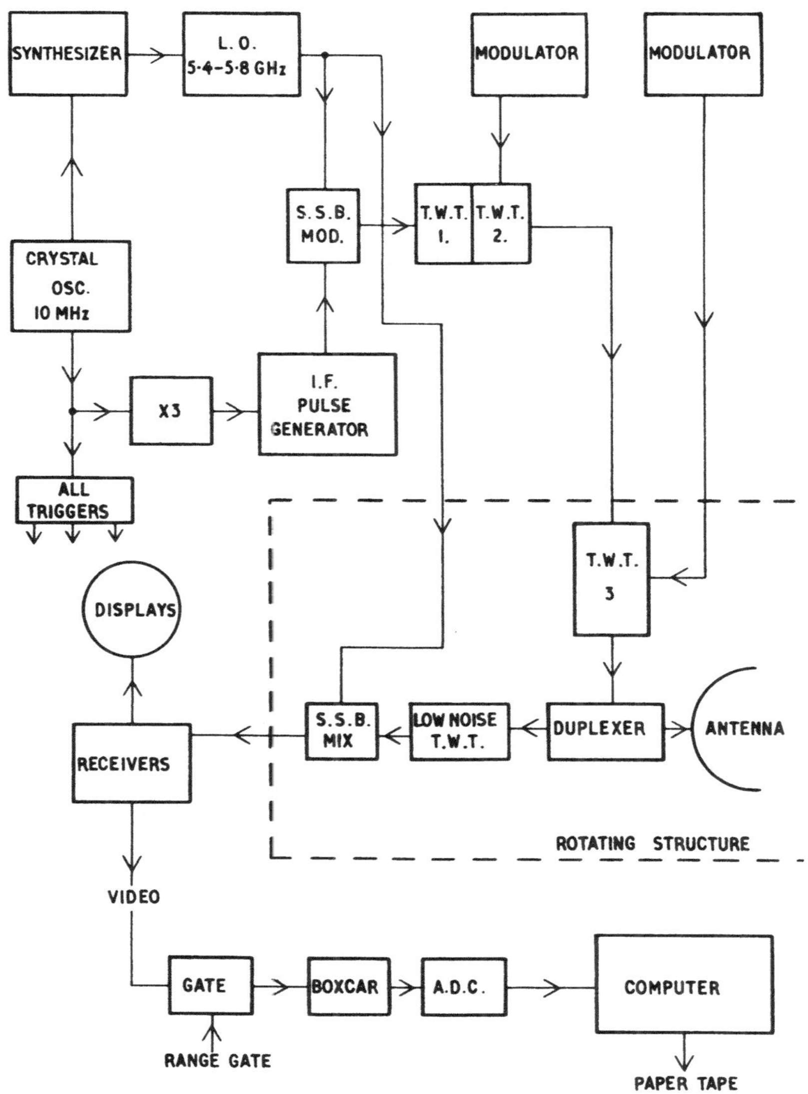

Figure 1 Radar System Block Diagram

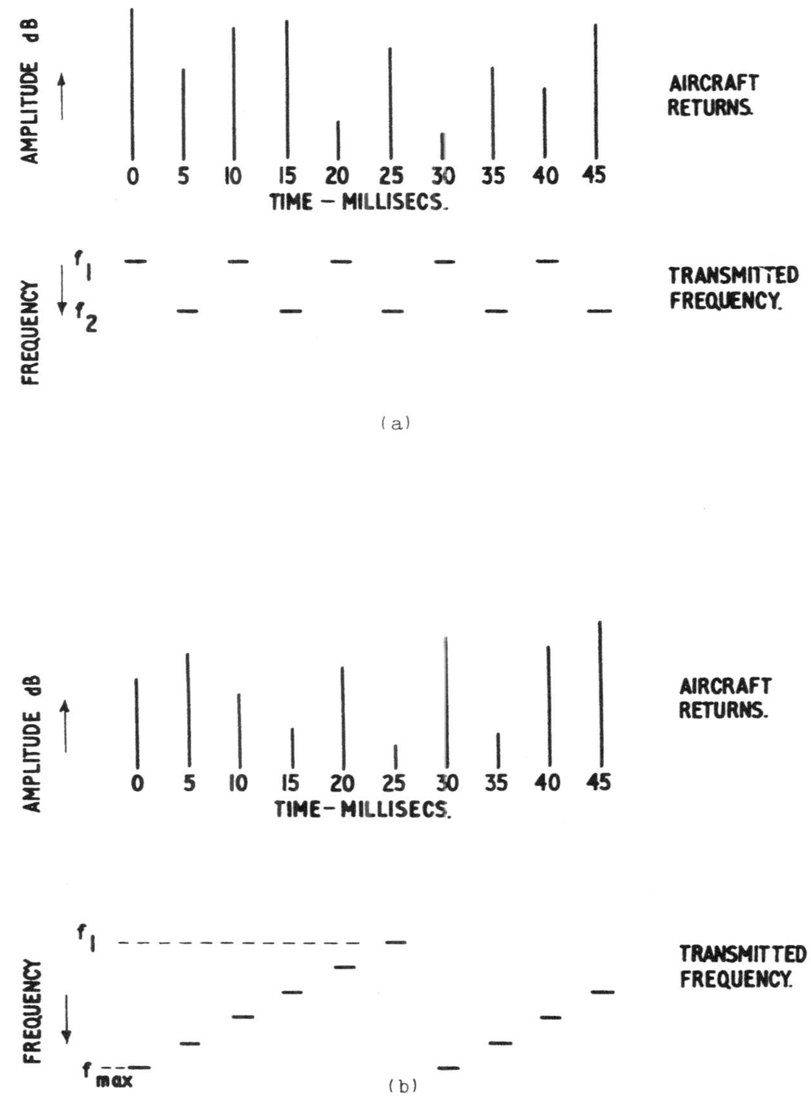

Figure 2 Frequency Agile Modes (a) Alternate Mode; (b) Staircase Mode

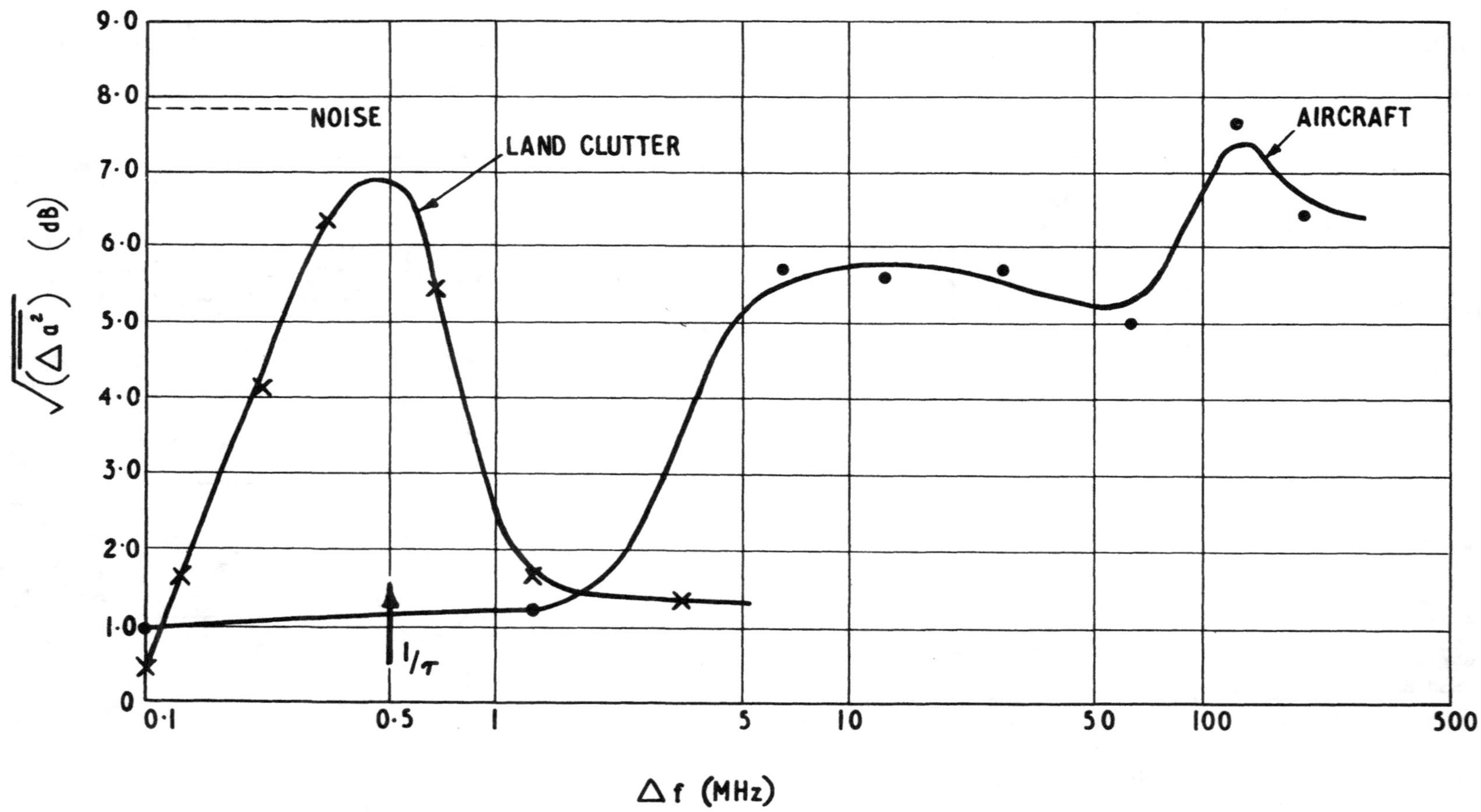

Figure 3 R.M.S. Pulse-to-Pulse Amplitude Change versus Frequency Change for Land Clutter and Aircraft

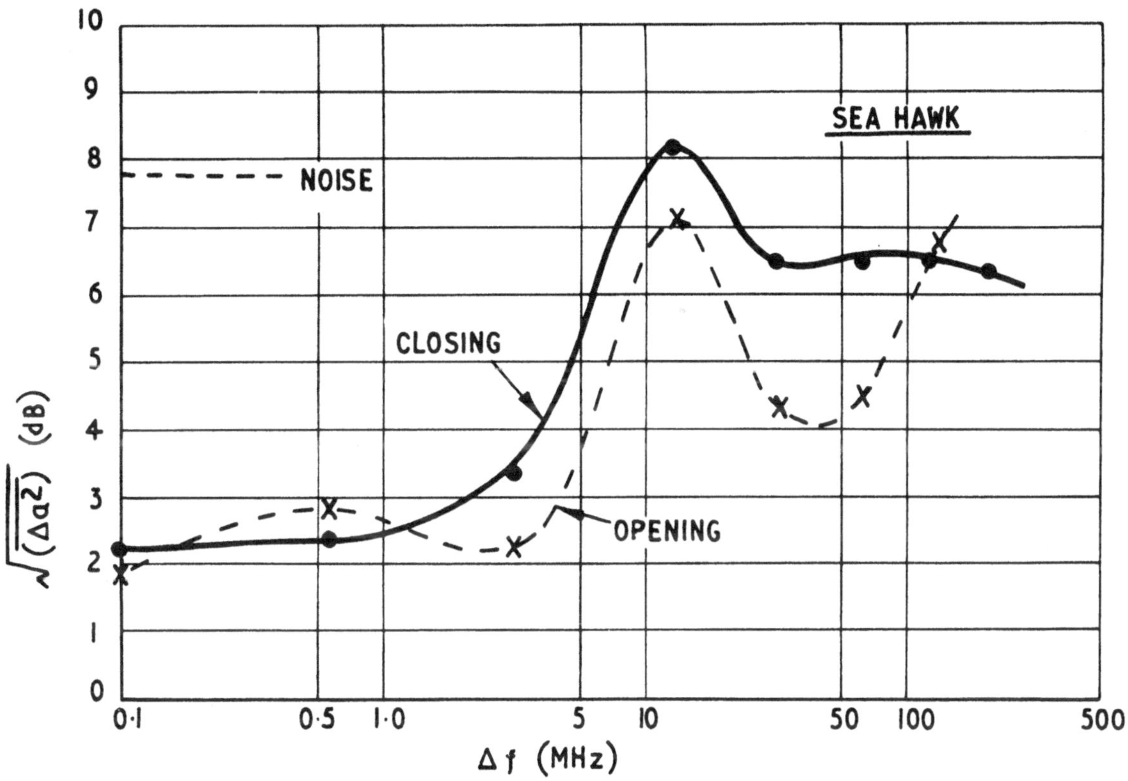

Figure 4   R.M.S. Pulse-to-Pulse Amplitude Change for a Sea Hawk Aircraft

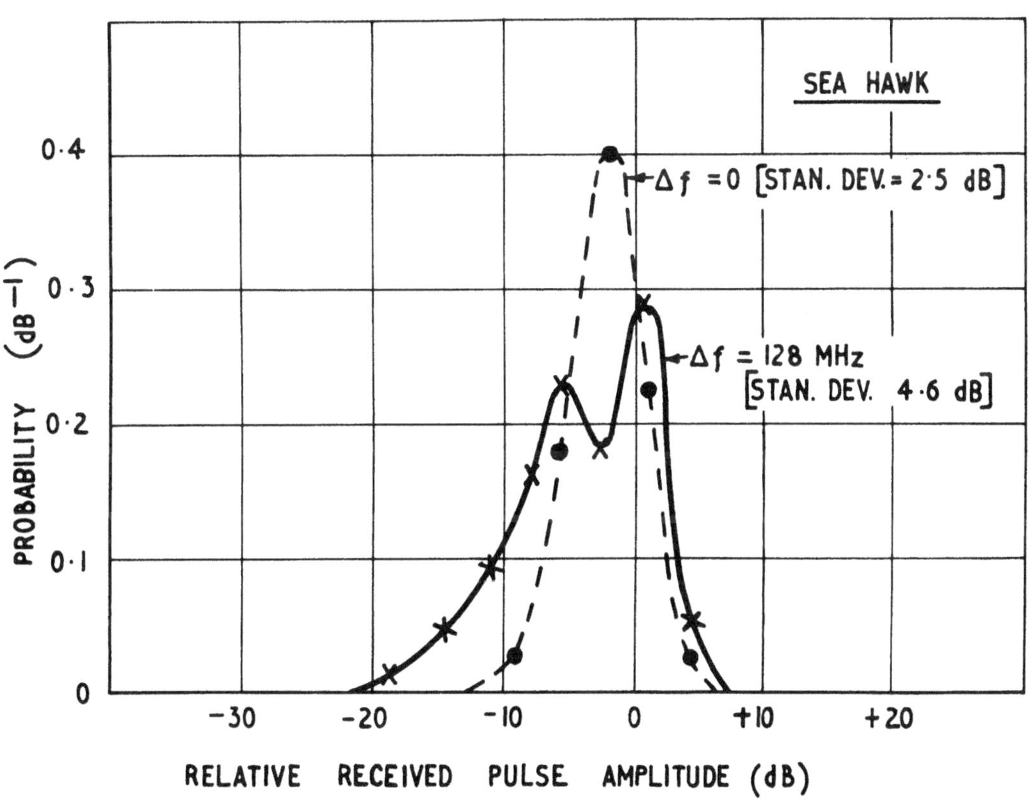

Figure 5   Echo Amplitude Probability Density Function for a Sea Hawk Aircraft

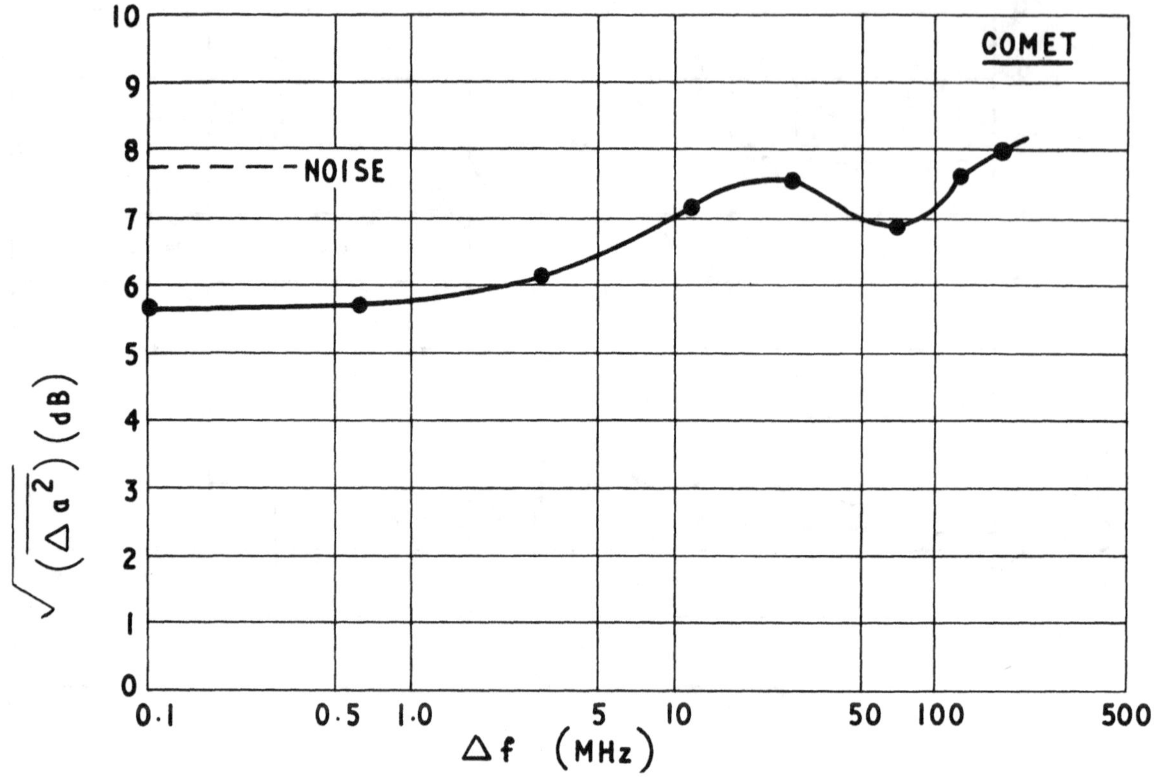

Figure 6 R.M.S. Pulse-to-Pulse Amplitude Change for a Comet Aircraft

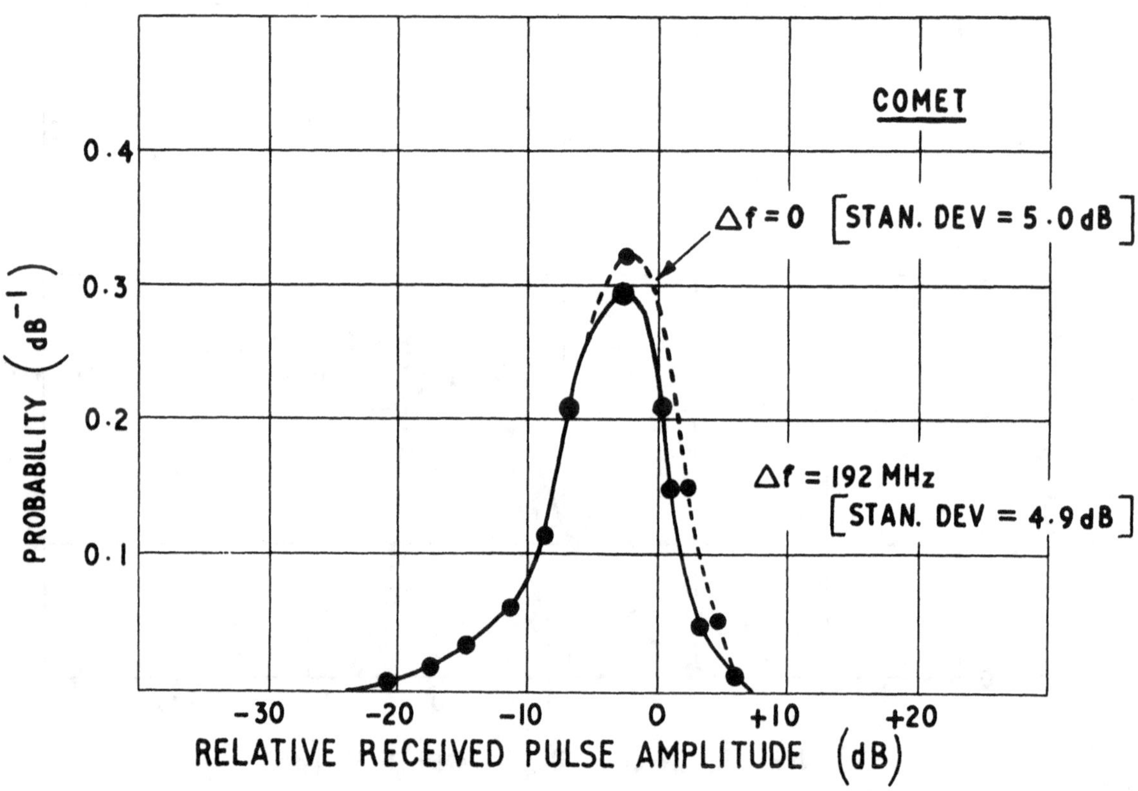

Figure 7 Echo Amplitude Probability Density Function for a Comet Aircraft

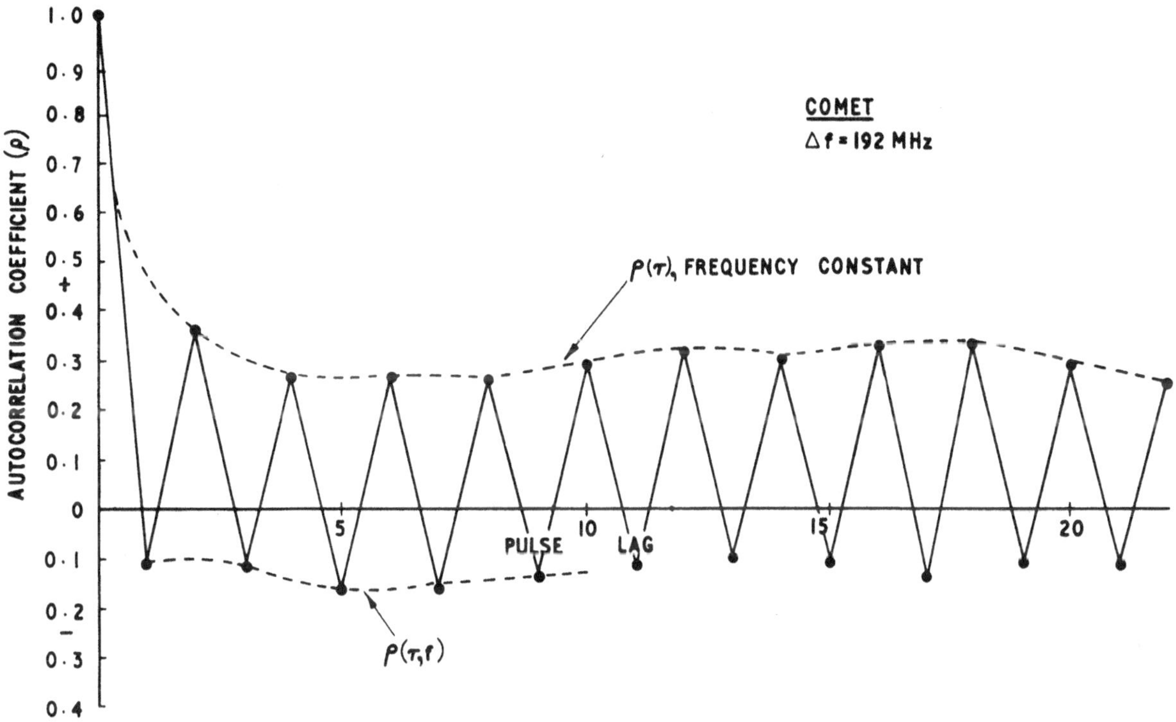

Figure 8 An Auto-Correlation Function of the Comet Aircraft Echoes with Frequency Agility

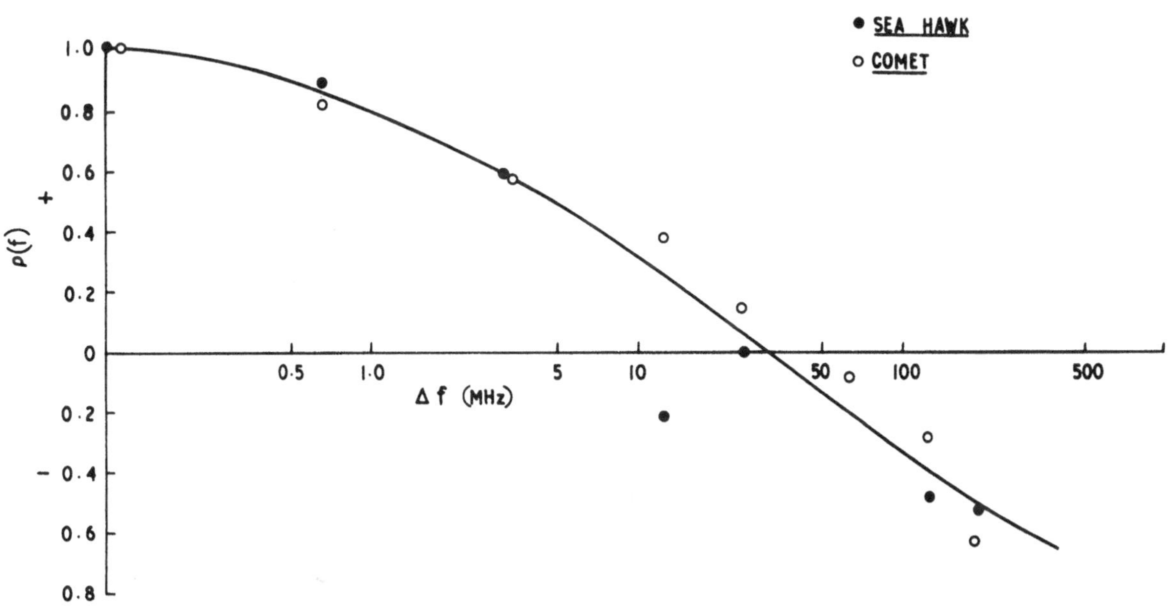

Figure 9 Derived Frequency Correlation Coefficients

31

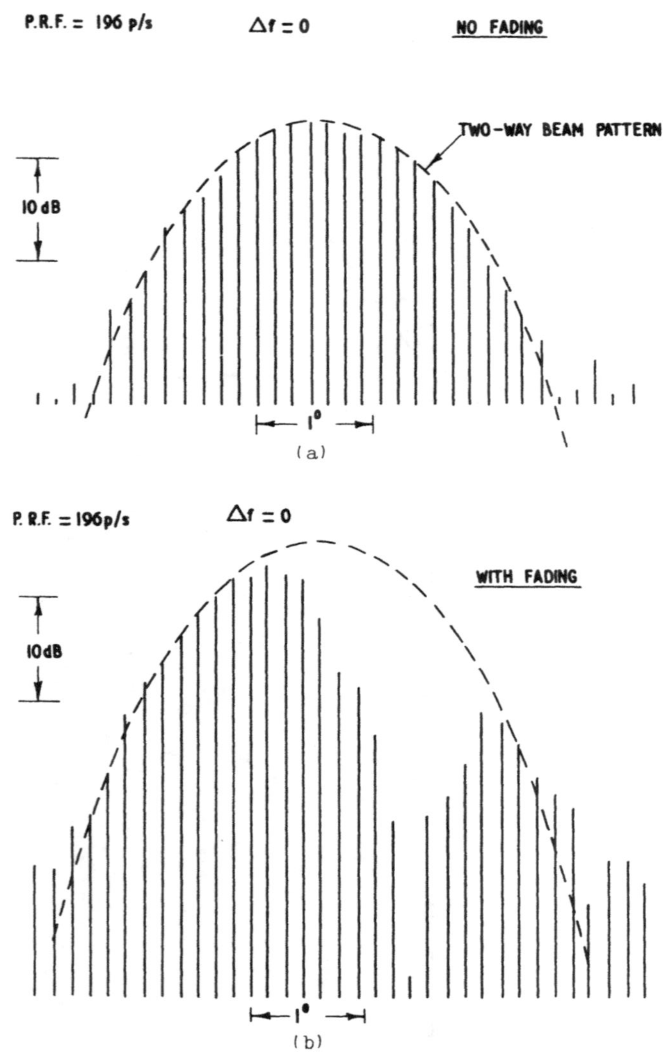

Figure 10 Scanned Echo Amplitude Patterns - Fixed Frequency

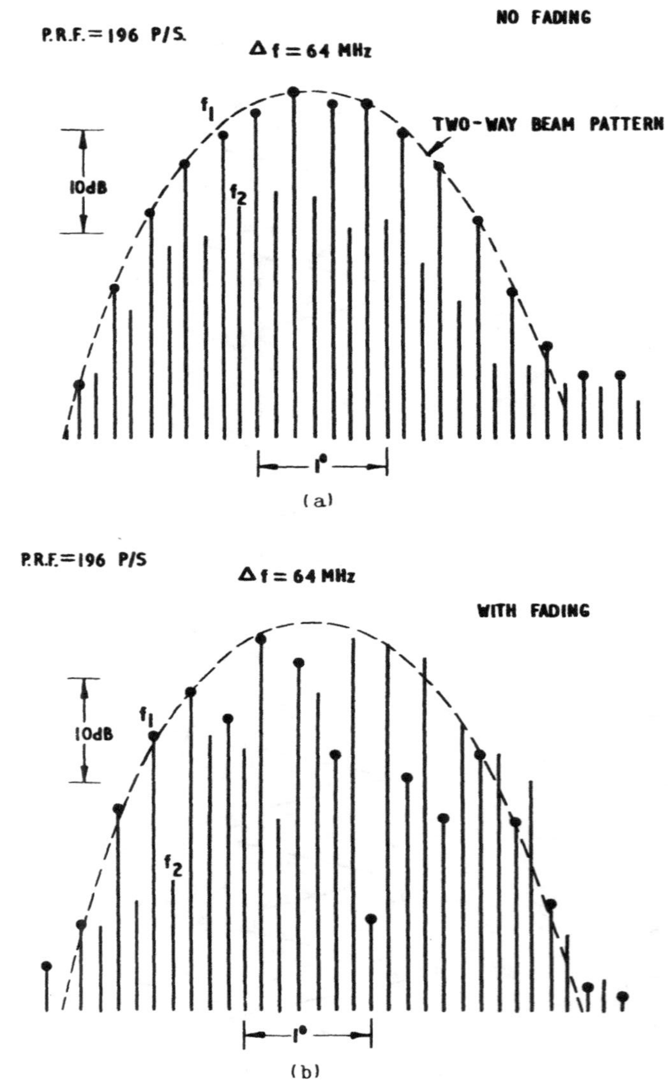

Figure 11 Scanned Echo Amplitude Patterns - Frequency Change 64 MHz

Section II Diversity Detection Theory

BACKGROUND

The basic theory of detection for fluctuating targets was developed by Swerling at RAND Corporation in the early 1950s and reported in the classic 1953 monograph (reprinted in 1960 *IEEE Transactions on Information Theory* and subsequently in Haykin's *Detection and Estimation,* Halsted Press, 1976). In his early paper on prediction of pulse radar performance (see Volume 2 of *Radars*), Hall compared the signal-to-noise ratio required for the Rayleigh fluctuating target (Swerling Case I) with that for the steady target, giving a curve for "fluctuation loss" which depends upon probability of detection. Barton (*Radar System Analysis,* 1964) gave fluctuation loss curves for Case I targets, and subsequently refined them to show a slight dependence on false-alarm probability (1969 paper; see bibliography). Fig. 2-1 gives these curves, recently plotted from exact computer results on steady and Swerling target models. The theory of diversity detection of targets in Gaussian noise is concerned with modifying the value of this fluctuation loss, which is defined (*IEEE Dictionary,* 1977) as:

> "The apparent loss in radar detectability or measurement accuracy for a target of given average echo return power due to target fluctuation. It may be measured as the increase in required average echo return power of a fluctuating target to achieve the same radar detection performance as for a target of constant echo return."

The rapidly fluctuating targets (Swerling Cases II and IV) are characterized by pulse-to-pulse decorrelation (time diversity), with fluctuation loss decreasing rapidly as the number of integrated pulses increases. Barton (1969) gave an empirical relationship for the loss with n_e independent pulses, in terms of the single-sample loss of Fig. 2-1:

$$10 \log_{10} L_f(n_e) \approx \frac{10 \log_{10} L_f(1)}{n_e}$$

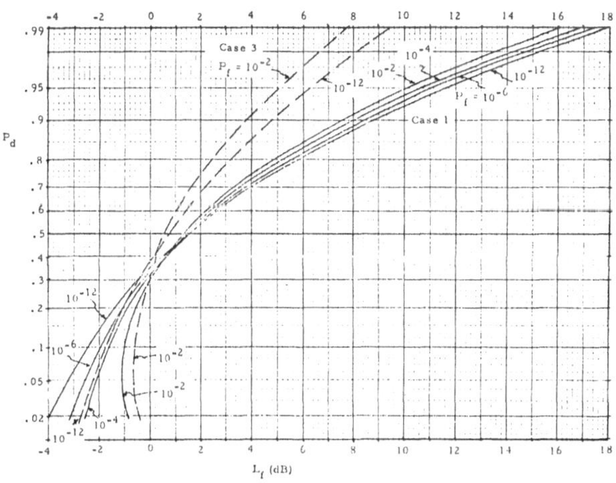

Figure 2-1

Fluctuation loss vs. detection probability for Swerling targets at different false-alarm probabilities.

By separating the effects of fluctuation from those of n-pulse video integration and from the basic detectability factor, it became possible to calculate the detectability factor (required snr) for cases in which n pulses containing $n_e < n$ independent samples were integrated. Thus, $n_e = 1$ represents the well-known Swerling Case I or Case III; $n_e = n$, Case II or Case IV; and $n_e = 2$, the dual-diversity system with independent channels. Other intermediate n_e values give results for general diversity and agility operation, due to either time decorrelation, frequency shift of the radar, or a combination of both. In the concluding section of Volume 2, the procedure is applied to the general case of the chi-squared target distribution with 2K degrees of freedom:

$$10 \log_{10} L_f(n_e, K) = \frac{10}{K n_e} \log_{10} L_f(1)$$

It is possible to find the net advantage of one or another waveform or processor by comparing final detectability factors under stated conditions:

$$D_x(n, n_e) = [D_o(1) \, L_i(n) \, L_f(n_e) \, L \, . \, . \, . \, .]/n$$

where $D_o(1)$ is the basic (single-pulse) detectability factor for a steady target at the required P_d and P_f, $L_i(n)$ is the integration loss, $L_f(n_e)$ is the fluctuation loss, and L . . . . represents other losses in the process. As will be shown, different comparisons result from differing assumptions as to how a system is implemented and used, thereby explaining the divergent opinions as to the advantages of frequency diversity.

REVIEW OF SOVIET WORKS

The theory of diversity detection had been expounded fully in Soviet texts by 1963. The well-known Wainstein and Zubakov text (1960, published in English translation by Prentice-Hall in 1962) covers the detection of Rayleigh fluctuating targets, and Fal'kovich (1961; see bibliography) notes that this theory applies when the independent samples (i.e., Swerling Case II) result from frequency separation. Bakut (1963; see bibliography) presents the complete theory with respect to both noise and clutter environments, including partially correlated target samples. Because his work is not familiar to most engineers whose basic language is English, a brief summary of his analysis is given here.

In Soviet detection studies, all targets are assumed to be fluctuating Gaussian signals (Rayleigh amplitude distribution), with a spectral width, Δf_c. Fast fluctuation relative to the available integration time, T, is characterized by $\Delta f_c T \gg 1$, while slow fluctuation gives $\Delta f_c T \ll 1$. For n-pulse noncoherent integration with interpulse period T_r of T/n, the case of $\Delta f_c T > n$ corresponds to Swerling Case II, while $\Delta f_c T \ll 1$ corresponds to Case I. Swerling Cases III and IV are not considered, except as the equivalent statistics arise in discussing frequency diversity. The case $1 < \Delta f_c T < n$ is covered at least approximately in Bakut, thereby giving the analysis the strength to express results for partially correlated targets and diversity systems.

The basic procedure used in Bakut involves synthesis of the optimum (in the maximum-likelihood sense) receiver structure for a given signal, and calculation of the distributions of noise and signal-plus-noise at the receiver output. The required total energy ratio, q_o, then is expressed in terms of integrals of the probability functions. The ratio q_o is equivalent to E/N_o in the American literature, and is related to the n-pulse detectability factor by

$$q_o = E/N_o = nE_1/N_o = nD_x(n,n_e)$$

where E_1/N_o is the single-pulse energy ratio and $D_x(n,n_e)$ is the final detectability factor after allowing for processing and fluctuation losses.

It can be shown that all the formulas for q_o in Bakut represent special cases, with varying degrees of approximation, of one expression which never appears in the book:

$$q_o = \frac{K_{2n_o}^{-1}(F) - K_{2n_o}^{-1}(D)}{\frac{1}{m_o}K_{2m_o}^{-1}(D)} \qquad \begin{cases} D = P_d \\ F = P_f \end{cases} \qquad \text{(A)}$$

Here q_o is the required energy ratio for a signal centered in the resolution cell under observation, n_o is the number of independent noise samples integrated, and m_o is the number of independent signal samples integrated. The function $K_\nu^{-1}(p) = \chi^2$ is the inverse of the probability integral of the chi-squared distribution with ν degrees of freedom

$$K_\nu(\chi^2) = [2^{\nu/2}\Gamma(\nu/2)]^{-1} \int_{\chi^2}^{\infty} t^{(\nu/2)+1} e^{-t/2} dt = p \qquad (4.4.14)$$

where Γ denotes the gamma function. (Equation numbers are those in the Bakut volume). For $\nu = 1$, this integral is identical to a suitably scaled integral of the normal distribution

$$\Phi(x) = \frac{1}{\sqrt{2\pi}} \int_{-\infty}^{x} e^{-y^2/2}\, dy = \frac{1}{2} + \frac{1}{2}\,\mathrm{erf}\left(\frac{x}{\sqrt{2}}\right) = p$$

where erf ( ) is the error function. Thus for $\nu = 1$,

$$K_1(\chi^2) = 2\Phi(-\chi) = p$$

$$K_1^{-1}(p) = [\Phi^{-1}(p/2)]^2$$

where $\Phi^{-1}(p) = x$ is the inverse function as used by DiFranco and Rubin in *Radar Detection* (Prentice-Hall, 1968). For $\nu = 2$, corresponding to the Rayleigh distribution, we have

$$K_2(\chi^2) = \exp(-\chi^2/2) = p$$

$$K_2^{-1}(p) = -2 \ln p = 2 \ln(1/p)$$

while for larger ν the values are tabulated (see, for instance, Abramowitz and Stegun, *Handbook of Mathematical Functions*, US Government Printing Office, 1964, pp. 978-985). An approximation for large ν is

$$K_{2n}^{-1}(p) \approx 2\sqrt{n}\,[\Phi^{-1}(1-p)+\sqrt{n}\,]$$

Keeping these relationships in mind, the basic Eq. (A) can be applied to find q_o for almost any type of signal and integration process, including slow, medium, and fast target fluctuation, coherent and incoherent integration, matched and mismatched filters, diversity and agility systems, and scan-to-scan accumulation of detection probability.

A. Slow fluctuation, $\Delta f_c T \ll 1$ (Swerling Case 1)

Coherent integration, $m_o = n_o = 1$

$$q_o = \frac{K_2^{-1}(F) - K_2^{-1}(D)}{K_2^{-1}(D)} = \frac{\ln(1/F)}{\ln(1/D)} - 1 \qquad (4.4.22)$$

Incoherent integration, $m_o = 1$, $n_o = n$

$$q_o = \frac{K_{2n}^{-1}(F) - 2(n-1)}{2\ln(1/D)} - 1 \qquad (5.3.6)$$

$$q_o \approx \frac{\sqrt{n}\,\Phi^{-1}(1-F)+1}{\ln(1/D)}$$

B. Fast fluctuation, $\Delta f_c T \gg 1$, $\Delta f_c T > n$ (Swerling Case II)

Coherent integration, $m_o = n_o = \Delta f_c T$

$$q_o = \frac{K_{2\Delta f_c T}^{-1}(F) - K_{2\Delta f_c T}^{-1}(D)}{K_{2\Delta f_c T}^{-1}(D)}\,\Delta f_c T \qquad (4.4.23)$$

$$q_o \approx \frac{\Phi^{-1}(1-F)+\Phi^{-1}(D)}{\sqrt{\Delta f_c T} - \Phi^{-1}(D)}\,\Delta f_c T \qquad (4.4.16)$$

Incoherent integration, $m_o = n_o = n$

$$q_o = \frac{K_{2n}^{-1}(F) - K_{2n}^{-1}(D)}{K_{2n}^{-1}(D)}\,n \qquad \text{(see Equation (4.6.3))}$$

$$\approx \frac{\Phi^{-1}(1-F)+\Phi^{-1}(D)}{\sqrt{n} - \Phi^{-1}(D)}\,n \qquad (5.3.16)$$

$$\approx [\Phi^{-1}(1-F)+\Phi^{-1}(D)]\sqrt{n}$$

C. Intermediate fluctuation, $1 < \Delta f_c T < n$

Coherent integration — use Equation (4.4.23) above.

Incoherent integration, $n_o = n \gg 1$, $m_o = \Delta f_c T/a \gg 1$

$$q_o \approx \frac{\Phi^{-1}(1-F)+\Phi^{-1}(D)}{\sqrt{\Delta f_c T/a} - \Phi^{-1}(D)}\sqrt{n\Delta f_c T/a} \qquad (5.3.18)$$

(where $0.5 < a < 1$ is a coefficient depending on the shape of the fluctuation spectrum)

Table 2-1

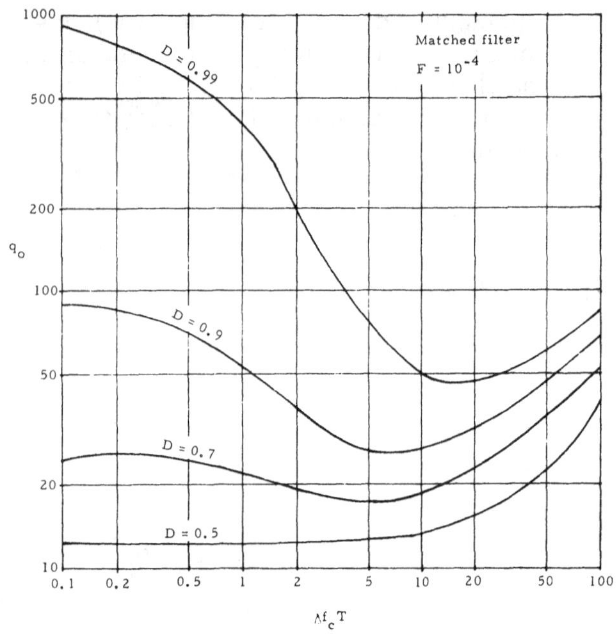

Figure 4.7

Dependence of threshold signal-to-noise ratio on width of spectrum of signal fluctuations for coherent system ($\Delta f_c T$ corresponds to $n_e - 1$ in Barton).

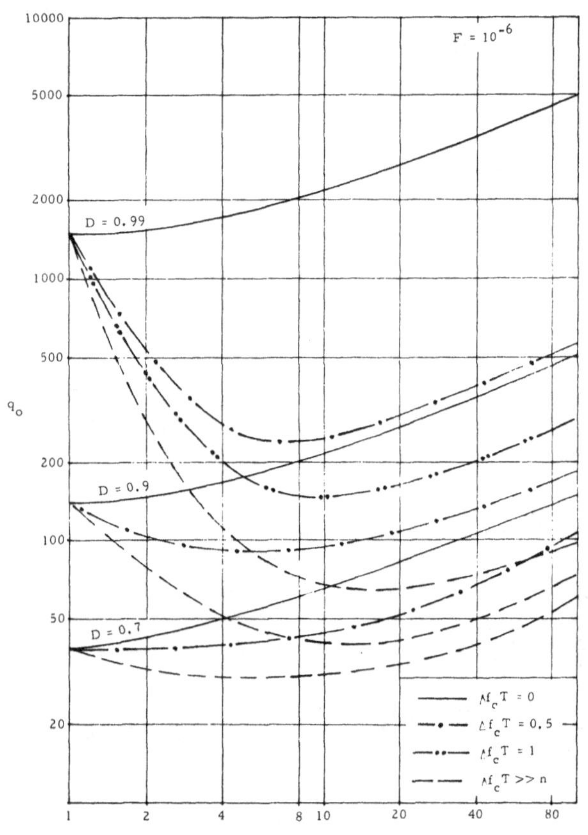

Figure 5.1

Dependence of threshold signal-to-noise ratio on number of pulses integrated noncoherently.

As a reference for consideration of diversity and agility systems, Table 2-1 summarizes the equations for coherent and incoherent operation at fixed frequency over an available signal time, T, equal to nT_r, where all the signals are integrated for one threshold decision having false-alarm probability F and detection probability D. The approximations for Swerling Cases I and II are in agreement with forms used in DiFranco and Rubin and elsewhere. Figures 4.7 and 5.1, reproduced here, indicate the reduction in required energy when the target fluctuation is such as to provide more than one signal sample during the integration time.

Subdivision of the integration time into several shorter intervals, spaced far enough apart to obtain independent signal samples and too far apart to permit convenient integration, represents the "time diversity" approach to overcoming fluctuation loss. Table 2-2 gives the equations; Fig. 4.16 plots the total energy requirements as a function of the number of intervals (scans) used.

Finally, Table 2-3 gives the equations for m-fold frequency diversity, and Fig. 4.13 shows the energy requirement for the slowly fluctuating target at different values of detection probability. The same expressions can be used for multiple-channel diversity as for sequential agility, but the total number of pulses or samples available in a given time will be increased. Thus, in a pulsed system with interpulse period T_r, $n = T/T_r$ for an agile system and $n = mT/T_r$ for an m-channel diversity system. A comparison of four types of diversity is shown in Fig. 5.2, for an incoherent radar observing a slowly fluctuating target with $T/T_r = 100$. In this case, the agile system achieves the best performance, because it increases the number of independent target samples, m, without increasing n and the integration loss. Time diversity, obtained by subdividing the interval into m separate integrations and detection decisions, is next best, while both parallel diversity with integration and parallel detection decisions are less efficient in energy. For $T/T_r < 10$, the results favor parallel diversity because larger values of m are available with smaller integration loss.

When the dominant source of background interference is clutter or uncancelled clutter residue from an MTI system, the use of parallel diversity channels is even more advantageous, for three reasons:

A. Slow fluctuation, $\Delta f_c T/m \ll 1$ (Swerling Case I)

Cumulative over m scans, matched filter

$$q_o = m \left[\frac{\ln(m/F)}{\ln\left(\frac{1}{1 - \sqrt[m]{1-D}}\right)} \right] - 1 \qquad (4.7.2)$$

Noncoherent, m sequential tests, each with n/m pulses
$n_o = n/m$
$m_o = 1$

$$q_o \approx \left[\frac{K^{-1}_{2n/m}(F) - 2(n/m - 1)}{2\ln\left(\frac{1}{1 - \sqrt[m]{1-D}}\right)} \right] m \qquad (5.4.5)$$

B. Fast fluctuation, $\Delta f_c T/m \gg 1$ or $\Delta f_c T > n$

Use Equation (4.4.23), (4.4.16), or (5.3.16) from Table 1 with m/F replacing F and $1 - \sqrt[m]{1-D}$ replacing D.

C. Intermediate fluctuation, $1 < \Delta f_c T/m < n$

Use Equation (A) with m/F replacing F, $1 - \sqrt[m]{1-D}$ replacing D, and q_o/m replacing q_o.

Table 2-2

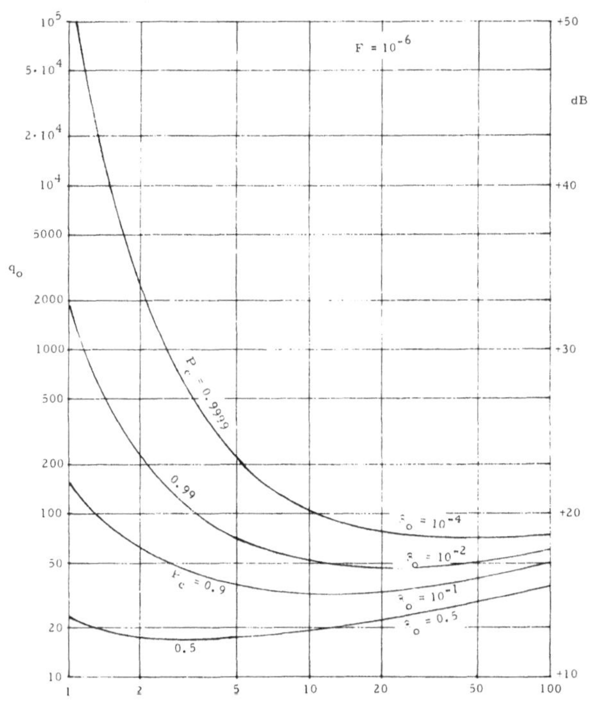

Figure 4.16

Dependence of threshold signal-to-noise ratio on the number of independent threshold tests ($P_c = 1 - \beta_o$ is cumulative probability of detection).

(1) Even though transmitted power is divided among channels, the single-channel S/C ratio remains the same, rather than being reduced by m as in the case of the S/N ratio. Thus, the integrated S/C ratio at the output is improved by the factor m relative to the integrated S/N ratio, for equal input ratios.

(2) The larger number of pulses available in parallel diversity permits better MTI performance or Doppler resolution.

(3) The use of separated frequencies tends to fill in MTI blind speeds, an advantage not provided by time diversity (subdividing the search time into several scans).

These advantages are pointed out in Section 4.11 of Bakut, which also discusses such issues as loss due to use of in-phase detectors only (as opposed to I and Q channels), effectiveness of noncoherent MTI, and the avoidance of blind speeds by varying repetition period.

In summary, the 1963 text by Bakut et al. represents a comprehensive treatment of radar detection theory which has not yet been matched in English-language texts, although most of its results have now appeared in one or another form. The Bakut material is mathematically difficult and the available translation is not in optimum form, thereby discouraging use by readers of English, but the work deserves attention and recognition beyond what it has received.

A. Slow fluctuation, $\Delta f_c T \ll 1$		
Coherent integration, $n_o = m_o = m$	$q_o = \dfrac{K_{2m}^{-1}(F) - K_{2m}^{-1}(D)}{K_{2m}^{-1}(D)}$	(4.6.3)
Incoherent integration, $n_o = n \gg 1$, $m_o = m$	$q_o \approx \dfrac{\Phi^{-1}(1 - F) + \Phi^{-1}(D)}{K_{2m}^{-1}(D)} 2m\sqrt{n}$	
For $m \gg 1$	$q_o = \dfrac{\Phi^{-1}(1 - F) + \Phi^{-1}(D)}{\sqrt{m} - \Phi^{-1}(D)} \sqrt{mn}$	
B. Fast fluctuation, $\Delta f_c T \gg 1$ *or* $\Delta f_c T > n$		
Coherent integration, $n \geqslant n_o = m_o = m\Delta f_c T \gg 1$	$q_o \approx \dfrac{\Phi^{-1}(1 - F) + \Phi^{-1}(D)}{\sqrt{m\Delta f_c T} - \Phi^{-1}(D)} m\Delta f_c T$	
Incoherent integration, $n_o = m_o = n$	$q_o = \dfrac{K_{2n}^{-1}(F) - K_{2n}^{-1}(D)}{K_{2n}^{-1}(D)} n$	(see Equations (4.6.3) and (5.3.16))
$m = n \gg 1$	$q_o \approx \dfrac{\Phi^{-1}(1 - F) + \Phi^{-1}(D)}{\sqrt{n} - \Phi^{-1}(D)} n$	
C. Intermediate, $1 < \Delta f_c T < n$		
Coherent integration, $n \geqslant n_o = m_o = m\Delta f_c T$	$q_o = \dfrac{K_{2m\Delta f_c T}^{-1}(F) - K_{2m\Delta f_c T}^{-1}(D)}{\dfrac{1}{m\Delta f_c T} K_{2m\Delta f_c T}^{-1}(D)}$	(4.6.4)
Noncoherent integration, $n_o = n \gg 1$, $m_o = m\Delta f_c T/a \gg 1$	$q_o = \dfrac{\Phi^{-1}(1 - F) + \Phi^{-1}(D)}{\sqrt{m\Delta f_c T/a} - \Phi^{-1}(D)} \sqrt{nm\Delta f_c T/a}$	

Table 2-3

ENGLISH-LANGUAGE LITERATURE

Although Swerling's early results for slow and fast target fluctuations were generalized in a 1957 paper to cover partially correlated signals, no numerical results applicable to frequency diversity became available until the late 1960s. The first references in English to advantages of agility and diversity in target detection (as opposed to glint reduction) appear to have been the papers by Gustafson and Ås (Reprint 1-1) and by Ray (Reprint 1-2), and the classified symposium paper by Deutch (recently declassified and presented here as Reprint 2-1). In all these works, quantitative data were based on completely uncorrelated target samples (Swerling Case II statistics). The Deutch material has extra significance in pointing out the added advantages of frequency diversity subpulses in reducing clutter power and in providing integration gain in clutter. However, if a train of such pulses, repeating the same subpulse frequencies, is integrated, the detection performance must be analyzed using methods from Bakut or the later literature discussed below. The same prob-

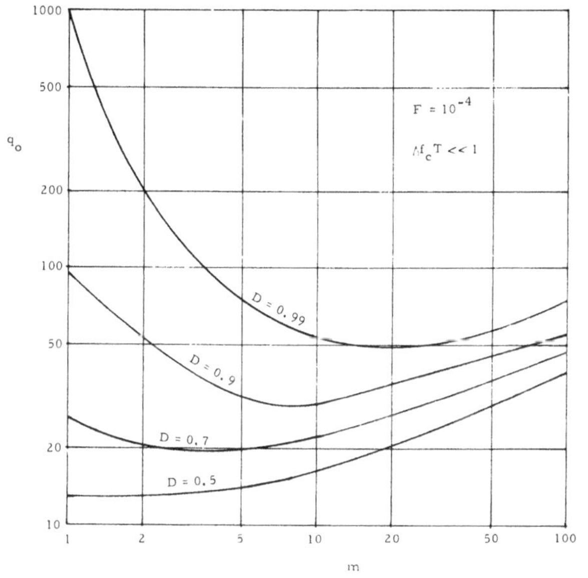

Figure 4.13

Dependence of threshold signal-to-noise ratio on number of carrier frequencies.

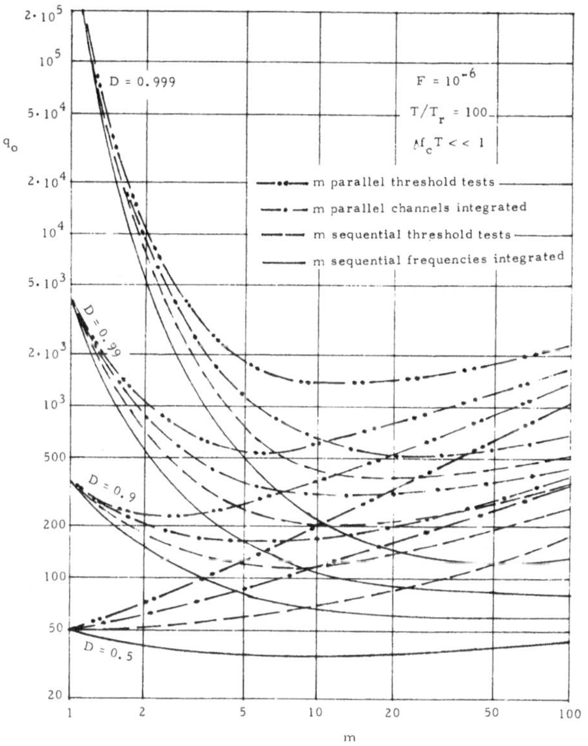

Figure 5.2

Dependence of threshold signal-to-noise ratio on the number of independently fluctuating samples with different types of diversity and processing.

lem pertains to the evaluation of the performance of dual-diversity radars through the use of multiple-pulse integration.

An early, unpublished British report by James (1963; see bibliography) provided analysis and data on detection of partially correlated signals; this work served as the justification for the later empirical procedure described earlier in this section (Barton, 1969). The general case of n pulses collected at m frequencies appears to have appeared for the first time in English in the Grasso and Guarguaglini paper (Reprint 2-2), based on the same authors' 1967 paper in Italian. Data presented on dual-diversity systems show the interesting result that optimum performance is obtained with approximately a 2:1 ratio between numbers of pulses in the two channels, although the performance remains essentially the same for high detection probabilities and equal numbers of pulses. A subsequent IEEE correspondence item by these authors (1968; see bibliography) introduced the material to engineers in the United States and gave an example of results for full diversity ($n_e = n$ in our American terminology).

An aspect of frequency-agile radar operation not discussed elsewhere is covered by Hansen (Reprint 2-3). Noting that most real targets would fluctuate slowly in the absence of frequency agility, he shows that a continuing random frequency shift is not optimum for target detection. By comparing individual pulses to a fixed threshold and stopping the frequency shift when this threshold is exceeded, about a 2 dB improvement in detection can be obtained.

The brief item by Horn and Wallace (1969; see bibliography) uses the frequency correlation function of a target with uniform scatterer distribution to evaluate correlation coefficients over groups of 10 and 100 pulses, and compares detection ranges for agile systems with different products of bandwidth and target length. Their results cover ratios of bandwidth to correlation frequency ($\Delta f/f_c$) from 0.67 to infinity.

The next major Western work on diversity detection, other than that by Barton (1969) discussed at the beginning of this section, is represented by the studies of Poelman and Bennée, reprinted in their entirety in Section 3. Drawing on the early work of Swerling, of Herscovici and Detape, and of Grasso and Guarguaglini, the SHAPE group derived the general expressions for detection probability with m channels, each receiving n_h hits, for a total of m n_h pulses integrated. Extensive results are presented for $m \leqslant 4$ at the two values of P_d used in the standard SHAPE range calculations – $P_d = 0.5$ and 0.8. Results also are given for logical combining procedures rather than linear video integration, and consideration is given to practical losses in channel diplexing and recombination. Comparisons of their results and conclusions with other work are presented in Section 3.

The only major American work on diversity detection is by Vannicola (Reprint 2-4), who obtained exact expressions for detection probability when using M channels, each with N pulses; the result is expressed as summations of terms involving the incomplete gamma functions. The analysis is similar to Poelman's work (Reprint 3-1), but considers only linear integration without diplexing losses. For the first time, this paper presented extensive plots of detectability factor vs. P_d for different M and N. A subsequent correspondence item by Poelman (1974; see bibliography) points out the relationship of the previous European literature and the SHAPE studies reprinted here in Section 3.

Finally in this section, a paper by Lind (Reprint 2-5) gives a refined procedure for calculating the number of independent target samples in random-frequency agile systems, including combined effects of time decorrelation and frequency decorrelation. He applies these results to detection in noise and clutter backgrounds.

ANNOTATED BIBLIOGRAPHY

Dax, P. R.: "The Statistics of Radar Video after Linear and Nonlinear Mixing," *Proceedings of the NEC, Vol. 15,* 1959, pp. 850-859. Considers the performance of dual diversity radar with equal signals in both channels, but different noise levels because of jamming.

Fal'kovich, S. A.: *Reception of Radio Signals in a Background of Fluctuating Interference,* Soviet Radio, 1961 (in Russian; translation TT-63-11517). In the context of fluctuating target detection (p. 215), the author refers to the fact that independent signal samples can be obtained by use of frequency separation, thereby greatly reducing the required radar energy for high detection probability.

James, P. W.: "Detection of Fading Targets in Noise: The Case in Which the Fading is Neither Fast nor Slow," *Technical Note RAD 828,* Royal Aircraft Establishment, United Kingdom, January 1963, AD406158.

Bakut, P. A.; et al.: *Problems in the Statistical Theory of Radar* (Volume 1), Soviet Radio, 1963 (in Russian; translation AD608462). As part of a comprehensive analysis of detection performance (Chapter 4), one entire section is devoted to frequency diversity systems with m parallel channels each receiving one or more samples from a fluctuating target (Section 4.6); advantages of diversity in MTI for detecting targets in clutter are presented as being even greater (Section 4.11.6). Use of diversity in noncoherent radar systems also is discussed (Section 5.4). See introductory material to Part 2 of this reprint volume.

Deutch, H. M.: "Intrapulse Frequency Diversity for Detection Improvement with Fluctuating Targets or Targets in Clutter," *Proceedings of the Twelfth Annual Tri-Service Radar Symposium,* 1966, pp. 283-312. Use of subpulses of different frequencies, with summing after envelope detection, is shown to provide reduction in target fluctation loss, reduction in size of the competing clutter cell, and integration gain with respect to clutter. Reprint 2-1.

Grasso, G.; and Guarguaglini, P. F.: "General Detection Procedure for a Multifrequency Radar," *Alta Frequenza, Vol. 37, No. 2,* February 1968, pp. 132-138. Analysis of dual-diversity detection of fluctuating targets with arbitrary numbers of pulses on each frequency. Reprint 2-2.

Guarguaglini, P. F.: "A Unified Analysis of Diversity Radar Systems," *IEEE Transaction on Aerospace and Electronic Systems, Vol. AES-4, No. 2,* March 1968, pp. 318-320c. Derives detectability factor and fluctuation loss for general case of fluctuating target having chi-squared distribution with 2 m degrees of freedom, when N total pulses are received in M diversity channels.

Hansen, V. G.: "A Sequential Logic for Improving Signal Detectability in Frequency-Agile Search Radars," *IEEE Transaction on Aerospace and Electronic Systems, Vol. AES-4, No. 5,* September 1968, pp. 763-773. Considers sequential detection based on holding fixed frequency of an agile radar, once a threshold has been crossed, to avoid target fading and to improve angular accuracy. Reprint 2-3.

Horn, K. L.; and Wallace, N. D.: "Detection of Slowly Fading Targets with Frequency Agility," *Proceedings of the IEEE, Vol. 57, No. 5,* May 1969, pp. 817-818c. Presents detection performance curves for Rayleigh fluctuating targets for varying values of the product of agile bandwidth and target radial depth (corresponding to partially correlated echoes).

Barton, D. K.: "Simple Procedures for Radar Detection Calculations," *IEEE Transaction on Aerospace and Electronic Systems, Vol. AES-5, No. 5,* September 1969, pp. 837-846. Gives general procedure for finding detectability factor, including fluctuation loss and diversity gain effects for Swerling Case I and Case III targets. Presents empirical relationship for fluctuation loss with n_e diversity samples:

$$L_f(n_e) = [L_f(1)]^{1/n_e}$$

Reprinted in *Radars (Volume 2 - The Radar Equation),* Artech House, 1974.

Poelman, A. J.: "Performance Evaluation of Frequency Diversity Radars," SHAPE Technical Centre, The Hague, Netherlands, April 1970. Derives the detectability factor for Swerling Case I and Case III targets when two to four frequency channels are combined by non-coherent integration, and compares results with single-frequency and rapidly fluctuating cases. Reprint 3-1.

Poelman, A. J.; and Bennée, E. J.: "Further Notes on Performance Evaluation of Frequency Diversity Radars," SHAPE Technical Centre, The Hague, Netherlands, May 1973. Extends results of Poelman (1970) to triple-channel systems using linear addition and logical combination of threshold crossings in ECM environment. Reprint 3-2.

Poelman, A. J.; and Bennée, E. J.: "The Performance of Triple-Channel Frequency Diversity Radar Systems," SHAPE Technical Centre, The Hague, Netherlands, November 1973. Derives and compares performance of five methods for combining outputs in triple-diversity radar systems. Reprint 3-3.

Vannicola, V. C.: "Detection of Slow Fluctuating Targets with Frequency Diversity Channels," *IEEE Transactions on Aerospace and Electronic Systems, Vol. AES-10, No. 1,* January 1974, pp. 43-52. Derives exact expressions for probability of detection of a Rayleigh target when M-channel frequency diversity is used with N pulses per channel and all MN pulses are integrated after square-law detection. Reprint 2-4.

Poelman, A. J.: "Comments on 'Detectability of Slow Fluctuating Targets with Frequency Diversity'," *IEEE Transactions on Aerospace and Electronic Systems, Vol. AES-10, No. 4,* July 1974, pp. 544-545c. Calls attention to European literature which preceded Vannicola (1974), including reports reprinted in Section 3 of this volume.

Lind, G.: "Frequency Agility Radar Range Calculation Using Number of Independent Pulses," *IEEE Transactions on Aerospace and Electronic Systems, Vol. AES-12, No. 6,* November 1976, pp. 811-815c. Derives number of independent target samples resulting from combined temporal and frequency decorrelation, and applies results to detection in noise and clutter backgrounds. Reprint 2-5.

Intrapulse Frequency Diversity for Detection Improvement with Fluctuating Targets or Targets in Clutter

H.M. Deutch

Proceedings of the 12th Annual Tri-Service Radar Symposium (1966)
Pages 283-312

ABSTRACT

Intrapulse frequency diversity is the subdivision of a pulse into equal, contiguous subpulses, each at a different frequency. The minimum frequency difference between any two subpulses is made greater than the reciprocal of the subpulse time extent or width. The transmitted bandspan is therefore greater than the product of the number of subpulses and the reciprocal of the subpulse width.

The received subpulses are summed after envelope detection. The subpulse width, not the transmitted bandspan, determines the effective received signal bandwidth and range resolution. A pulse compression ratio equal to the number of subpulses is obtained noncoherently.

The use of intrapulse frequency diversity in the absence of clutter results in:

1. A loss for a nonfluctuating target equal to the noncoherent integration loss for the number of subpulses used.
2. A gain for a fluctuating target. An optimum number of subpulses exist, which minimizes the signal-to-noise ratio required for detection.

Intrapulse frequency diversity performance improvement in dispersed clutter is due to:

1. The increase in signal-to-clutter ratio from the reduction in resolution volume.
2. The statistical difference between clutter and either fluctuating or nonfluctuating targets. The additional improvement from this source can be greater than that obtained from the equivalent pulse compression. This characteristic is the significant item.

1. INTRODUCTION

Intrapulse frequency diversity may be used to increase a radar's probability of detection. This will occur when targets exhibit the typical characteristic of amplitude fluctuation when carrier frequency is varied by a few percent. Even for nonfluctuating targets, intrapulse frequency diversity will increase detection probability in cases where clutter is distributed within the radar resolution cell bounding the target.

2. WAVEFORM AND PROCESSING

The radar waveform considered discontinuously shifts the transmission frequency within the pulse envelope, Figure 1. Frequency selection may be in random sequence. The pulse envelope consists of a time-contiguous set of equal subpulses, each at a different carrier frequency. The minimum frequency difference between any two subpulses within a single envelope is greater than the reciprocal of the subpulse time extent.

48 The subpulses are summed after square law (envelope) detection, noncoherent intrapulse integration. The effective signal processing bandwidth is approximately the reciprocal of subpulse length. The transmitted signal occupies a bandspan greater than the product of the number of subpulses multiplied by the signal processing bandwidth. By virtue of this wide bandspan transmitted signal, radar performance improvement is also possible in an active countermeasure environment.

Intrapulse frequency diversity results in a form of pulse compression as the time (range) resolution is determined by the subpulse length rather than that of the total pulse envelope. As will be described in this paper, the increase in the probability of detection in the presence of clutter is not dependent merely on the reduction in clutter volume. Actually, an additional significant improvement is obtained from the statistical characteristics of the target and clutter, as will be shown in Section 3.

A comparison may be made between intrapulse frequency diversity and conventional linear frequency modulation pulse compression, chirp. It is suggested that this comparison be made for equal pulse energies and widths, Figure 2a, and equal range resolutions, Figure 2b. Intrapulse frequency diversity uses noncoherent signal processing and is less efficient than the coherently processed chirp waveform, Figure 2b.

The equivalent intrapulse frequency diversity pulse compression is equal to N, the number of subpulses used. The bandspan of the transmission, Figure 2c, and spectral width, Figure 2d, is greater than N times the reciprocal range resolution.

For equal range resolution, the chirp waveform occupies a smaller bandspan or spectrum than does intrapulse frequency diversity, Figures 2e, f.

3. PROBABILITY OF DETECTION

3.1. NONFLUCTUATING TARGET IN ABSENCE OF CLUTTER

In order to isolate the effect of intrapulse frequency diversity with noncoherent subpulse integration, comparisons will be made while using constant total pulse energy. The point of interest here is to determine the effect of a number of subpulses or independent noise samples when the signal energy is maintained constant. The model used in the analysis is based on References 1 and 2.

Figure 3 presents the required signal-to-noise ratio as a function of the number of subpulses for various probabilities of detection. As expected with a nonfluctuating target model, the minimum required signal-to-noise ratio is achieved for a single subpulse or sample when using noncoherent integration. The only effect of increasing the number of subpulses within the transmitted pulse envelope is to increase the noncoherent integration loss, Figure 4.

3.2. FLUCTUATING TARGET IN ABSENCE OF CLUTTER

For a fluctuating target, independence between subpulse samples of the target is assumed. This condition is met for the majority of targets of interest when the frequency separation between subpulses is greater than the subpulse bandwidth and the target is large compared to the radar wavelength. An exponential target cross-section probability density distribution is also assumed, Reference 3.

Figure 5 presents the required signal-to-noise ratio as a function of the number of subpulses for various probabilities of detection. At probabilities of detection of 0.7 or greater, a nontrivial optimum number of subpulses exist. This is due to the opposing effects of increasing noncoherent integration loss and decreasing signal fluctuation loss as the number of independent samples is increased. As before, the total signal energy is maintained constant to isolate the effect of varying numbers of subpulses. Figure 6 presents the same information for a higher false alarm number. Similar curves are obtained for a Rayleigh target fluctuation model.

The existence of an optimum number of samples with a fluctuating target is usually masked due to the form in which the information is generally presented, signal energy being proportional to the number of samples.

Two additional methods of presenting the previous information may be useful. In Figure 7, curves of constant signal-to-noise ratio are plotted in terms of number of subpulses and probability of detection. Figure 8 may be of greater utility in determining performance when the number of subpulses is fixed.

3.3. COMPARISON OF DETECTION FOR CLUTTER-FREE FLUCTUATING AND NONFLUCTUATING TARGETS

Figure 9 indicates that the use of intrapulse frequency diversity can significantly reduce the target fluctuation loss inherent at the higher probabilities of detection.

If 30 or more subpulses are chosen for reasons based on desired pulse compression or electronic counter-countermeasures, detection performance in thermal noise is not a critical function of target fluctuation model, Figure 10. For probabilities of detection from 0.2 to 0.99 the difference in required signal-to-noise for nonfluctuating and exponentially fluctuating targets is less than one db. This indicates that, at least for the two extremes of target models considered here, a reasonable target average is obtained for 30 independent samples.

3.4. NONFLUCTUATING TARGET IN CLUTTER

The previous comparisons were conducted with constant total signal energy to isolate the singular effect of changing the number of subpulses or samples. In a clutter environment, the clutter return energy per sample will be maintained constant. The radar volumetric resolution or subpulse width will be kept constant - to again separate the effects of both target fluctuation model and the number of independent samples. The case to be considered is where the subpulse clutter return exceeds the thermal noise level, clutter limiting performance.

Major forms of clutter experienced by radar are rain, chaff, sea and ground. There is a decreasing lack of correlation and an increasing model complexity and set of assumptions for both analytic and empirical clutter data in the order previously given. Performance characteristics presented in this paper will be based on rain or dispersed chaff clutter. This performance may be considered as a boundary. Performance in a clutter environment will be reduced from this boundary as the form of clutter departs from that exhibited by rain or dispersed chaff. That is when the clutter is predominantly one or a few major specular reflectors within a radar resolution cell, rather than a large number of approximately equal-size reflectors distributed within the cell.

The clutter is assumed to be uncorrelated between the subpulse samples and to have no spectral or time foldover. These conditions will be met for rain or distributed chaff clutter if the frequency steps between the subpulses are on the order of one-and-one-half times the reciprocal of the subpulse time extent. The radar receivers are matched to the subpulse bandwidth.

The clutter is statistically equivalent to band-limited Gaussian noise. With radar resolution independent of the number of subpulses (the comparison condition previously stated), the magnitude of the subpulse signal-to-clutter ratio is independent of the number of subpulses. The case to be considered is the limiting condition of interest where the subpulse clutter return is much greater than receiver thermal noise.

The expected performance in a clutter environment with this waveform and the stated ground rules is equivalent to that of a group of noncoherently integrated pulses when total energy of the group is proportional to the number of samples.

Figures 11 and 12 present the performance achieved with intrapulse frequency diversity, noncoherent subpulse integration, and a nonfluctuating target. This performance improvement is to be compared with a radar of equal time or range resolution. If a comparison is desired with a radar using a pulsed CW transmission of time extent equal to the sum of the subpulse widths, a factor equal to the equivalent pulse compression should be added.

3.5. FLUCTUATING TARGET IN CLUTTER

The clutter model and constraints are as previously given.

Figures 13 and 14 present the performance with an exponential target fluctuating model. The considerable performance improvement over that for the nonfluctuating target model is the addition of increased sampling of the fluctuating target.

3.6. COMPARISON OF DETECTION FOR FLUCTUATING AND NONFLUCTUATING TARGETS IN CLUTTER

Figure 15 presents the signal-to-clutter ratio and probability of detection for both target models with single and with 30 frequency agile subpulses. With 30 subpulses, there is only a one-db difference in subpulse signal-to-clutter ratio for the range of probability of detection from 0.2 to 0.99 for the two extreme target fluctuation models considered.

4. APPLICATION

4.1. CHAFF CLUTTER

Signal-to-chaff clutter ratio is:

$$\frac{S}{C_c} = \frac{8\,\bar{\sigma}_t}{\sigma_c^{\,o}\,\pi\,\theta^2\,R^2\,c\,\tau\,L}$$

where:

- c = velocity of light
- L = difference between the target and clutter beamshape losses
- R = target range
- $\sigma_c^{\,o}$ = chaff cross-section density
- $\bar{\sigma}_t$ = average target cross-section
- τ = effective time resolution of waveform
- θ = antenna beamwidth

The assumptions are that the clutter fills the entire resolution cell containing the target, and there is no spectral foldover. The chaff density is 60 square meters per cubic nautical mile. The difference between the target and clutter beamshape losses is 0.5 db. For illustration, an effective time resolution of one microsecond and a beamwidth of 1.5^{o} in both dimensions was used for Figure 16.

4.2. RAIN CLUTTER

The backscattering cross-section density of rain, Reference 4, is:

$$\sigma_r^{\,o} = 1.2\,\frac{\pi^5}{\lambda^4}\,r^{1.6}\,10^{-6} \text{ square meters per cubic nautical mile}$$

where:

λ = radar wavelength in meters

r = rainfall rate in millimeters per hour

Two illustrative radar carrier frequencies are considered, 3.3 GHz and 5.6 GHz. For a rainfall rate of one mm/hour, the backscattering cross-section density at S-band is 5 square meters per cubic nautical mile; at C-band it is 44 square meters per cubic nautical mile.

An additional clutter return is received by the radar antenna sidelobes when clutter is distributed in the hemisphere surrounding the radar. This added loss factor is 0.8 db. The expression for signal-to-rain clutter ratio is similar to that for chaff, but with the additional loss factor and a substitution for rain clutter density. The signal-to-rain clutter ratio is presented in Figure 17, assuming that the rain clutter fills the entire resolution cell containing the target and that there is no spectral foldover. For illustration, the effective time resolution is one microsecond and the beamwidth is 1.5°.

4.3. WAVEFORM

The class of radar probably most constrained in instantaneous bandwidth is one using a phase-steered planar array antenna. The bandwidth limitation is required to avoid dispersion at the antenna scan angle extremes. A truly time-delay steered planar array has no similar constraint. Intrapulse frequency diversity is not even considered when frequency is used for directing the antenna beam.

The maximum instantaneous phase-steered planar array bandwidth, in percent of operating frequency, is approximately equal to the beamwidth in degrees. With a 1.5° antenna beamwidth, the approximate instantaneous bandwidth available at S-band is 50 MHz and at C-band it is 80 MHz.

Intrapulse frequency diversity clutter performance will be illustrated with 32 one-microsecond subpulses and an antenna beamwidth of 1.5°. At S-band, the minimum subpulse frequency separation is 1.5 MHz; at C-band, it is 2.5 MHz. The previous assumptions of independent subpulses and no spectral foldover are valid for this first-order analysis.

Clutter performance is limited by thermal noise unless the subpulse clutter-to-noise ratio is much greater than unity. With 32 subpulses, a probability of detection of 0.85 and a Swerling Case 2 target fluctuation model, subpulse clutter power is twice that of noise. This is at least a marginal satisfaction of the criterion.

4.4. PERFORMANCE IN CLUTTER

Performance in chaff is presented in Figure 18. For comparison, the performance of a single pulse with the same resolution volume as the 32 subpulses is shown. This single pulse could have the same 32-microsecond transmission time and could use conventional pulse compression to achieve an effective time resolution of one microsecond.

At a probability of detection of 0.85, 32 independent subpulses increase the detection range of a fluctuating target in chaff and rain clutter by a factor of eight, Figures 18, 19 and 20. The range increase for a non-fluctuating target under the same conditions is a factor of four. These increases in detection range hold for other clutter densities and radar resolution cell sizes as long as the conditions assumed in Sections 2 and 3 are maintained.

5. CONCLUSIONS

The concept of using intrapulse frequency diversity for performance improvement is not new. It is a logical extrapolation of the now ancient pulse-to-pulse frequency agility. In its more recent form, it is a closer match with phased array energy management concepts. It is equally applicable to radars with electro-mechanically positioned antennas.

Intrapulse frequency diversity is a form of noncoherent pulse compression. Its additional characteristics include performance improvement with fluctuating targets and in clutter environments, as typified by chaff and rain.

The potential increase in radar range for clutter-free fluctuating targets is presented in Figure 21.

With the constraint that subpulse clutter is much greater than noise, Figure 22 illustrates the range increase available in distributed clutter. Figure 22 combines the effects of both pulse compression and 32 independent clutter samples. The range increase available from the 32 independent clutter samples alone is illustrated in Figure 23.

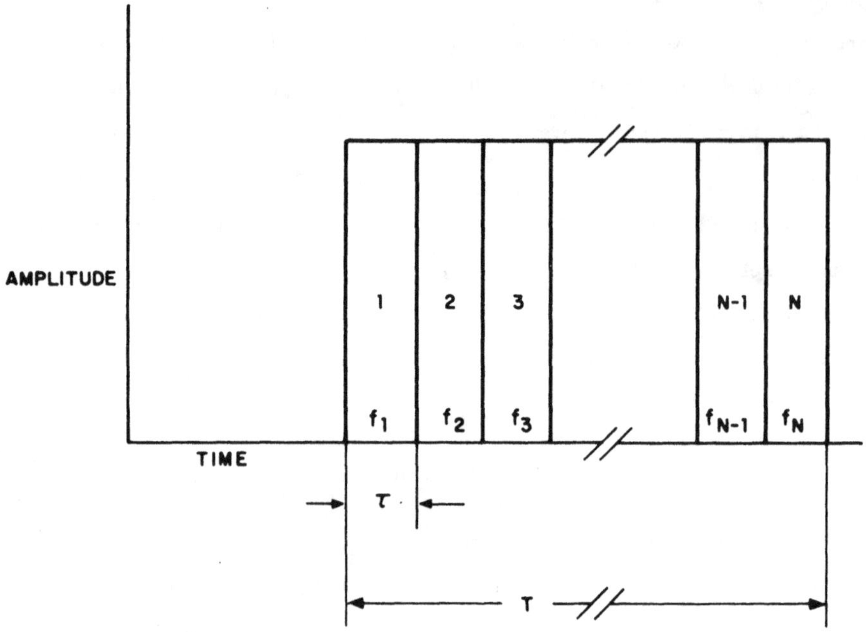

WHERE:

- f SUBPULSE FREQUENCY
- N NUMBER OF SUBPULSES
- T PULSE TIME EXTENT
- τ SUBPULSE TIME EXTENT

FIGURE 1. INTRAPULSE FREQUENCY DIVERSITY WAVEFORM.

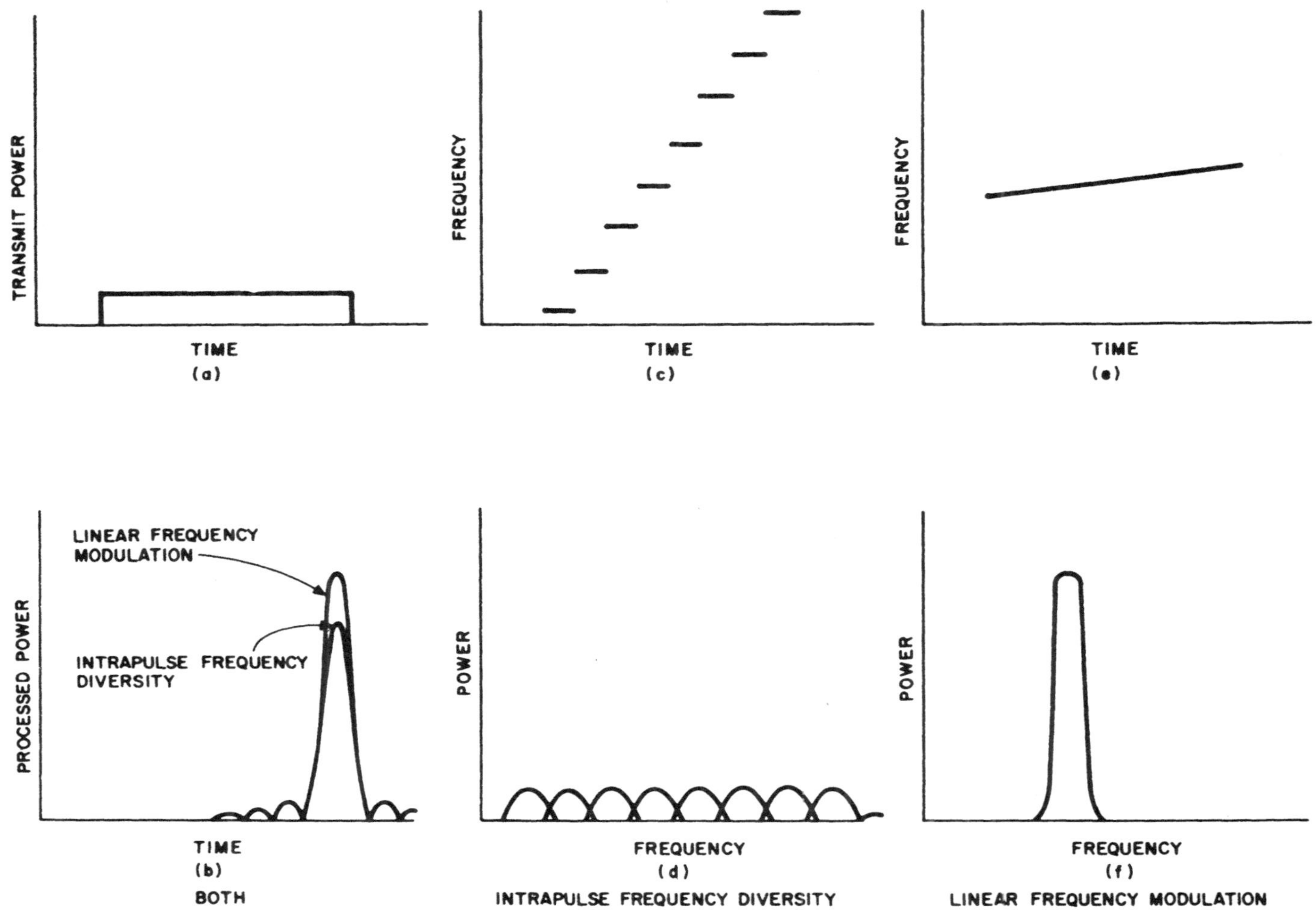

FIGURE 2. COMPARISON WITH CONVENTIONAL PULSE COMPRESSION. Time-bandwidth products equivalent.

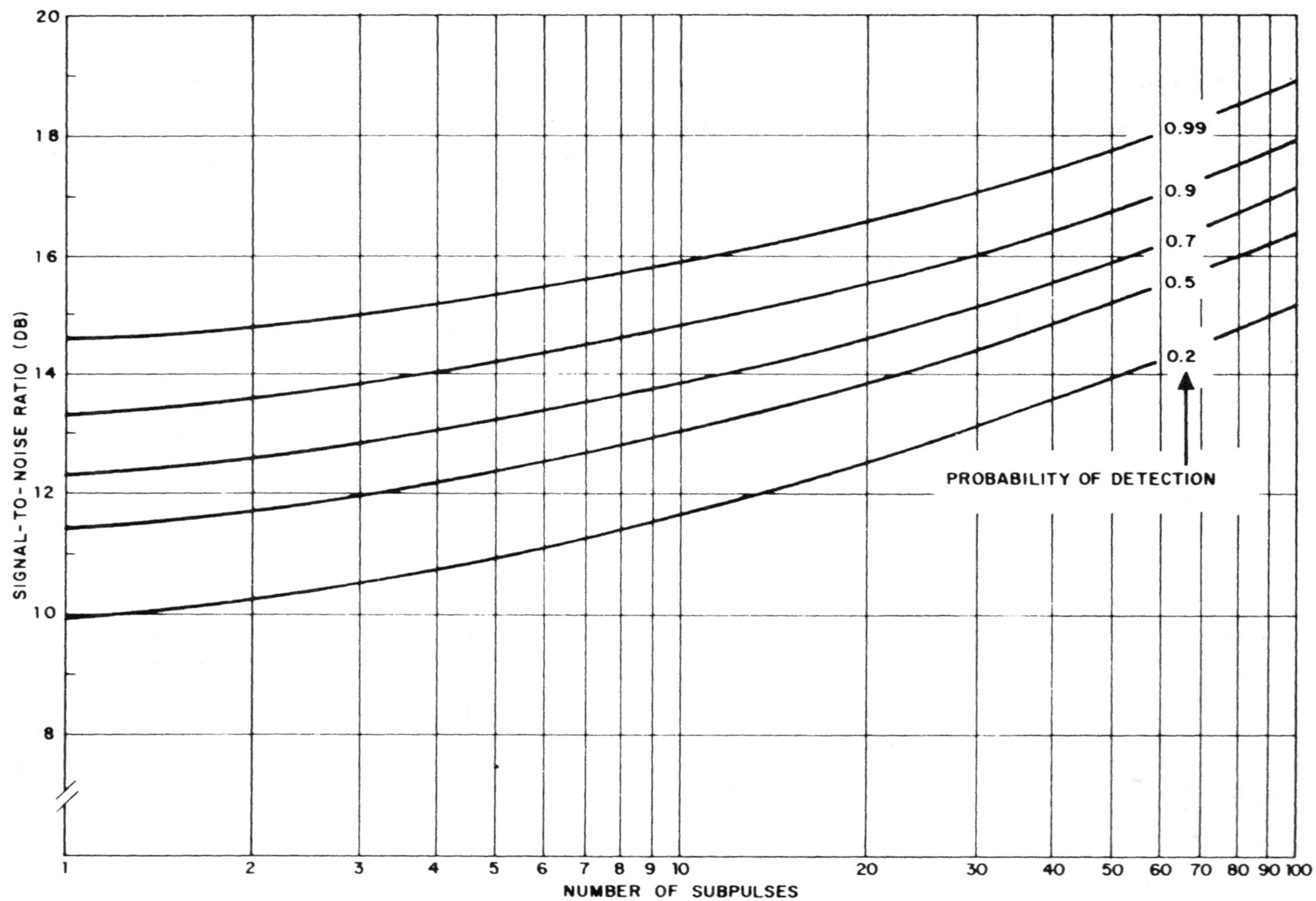

FIGURE 3. NONFLUCTUATING TARGET PROBABILITY OF DETECTION.

Independent subpulses of constant energy pulse integrated after square law detection; $n = 10^6$.

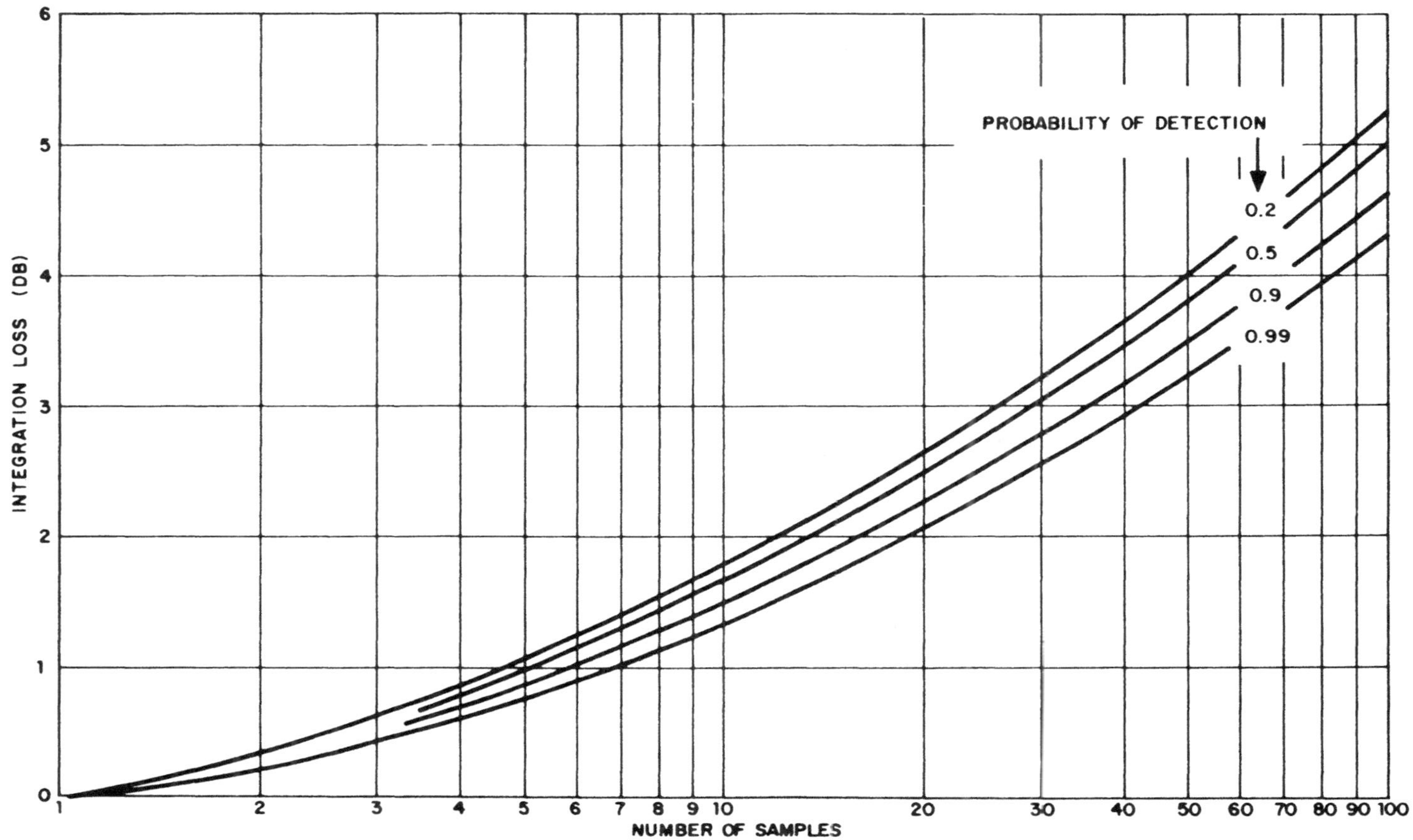

FIGURE 4. NONCOHERENT INTEGRATION LOSS.
Square law detector; $n = 10^6$.

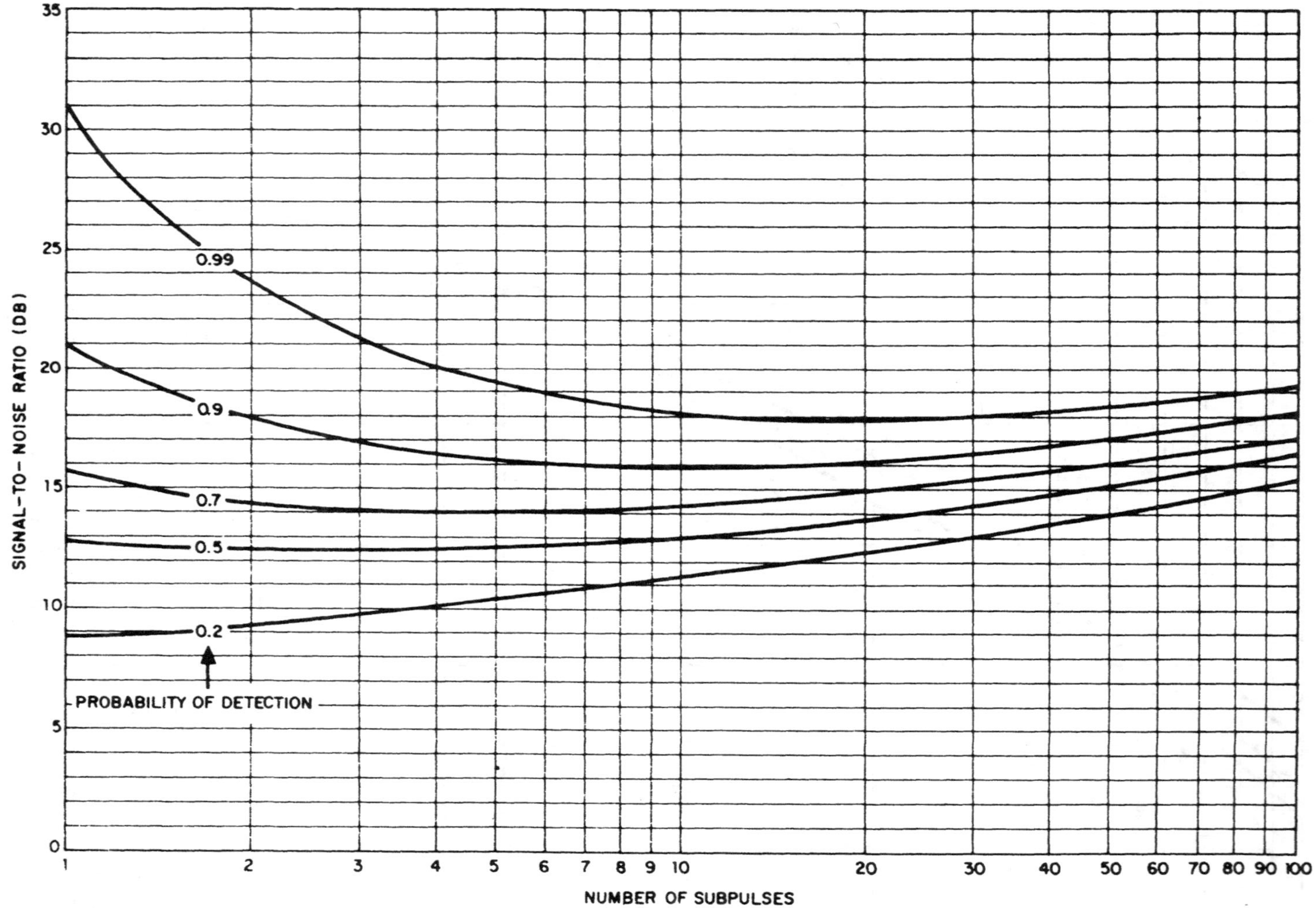

FIGURE 5. FLUCTUATING TARGET PROBABILITY OF DETECTION.

Independent subpulses of constant energy pulse integrated after square law detection. Target fluctuation model, Swerling Case 2; $n = 10^6$.

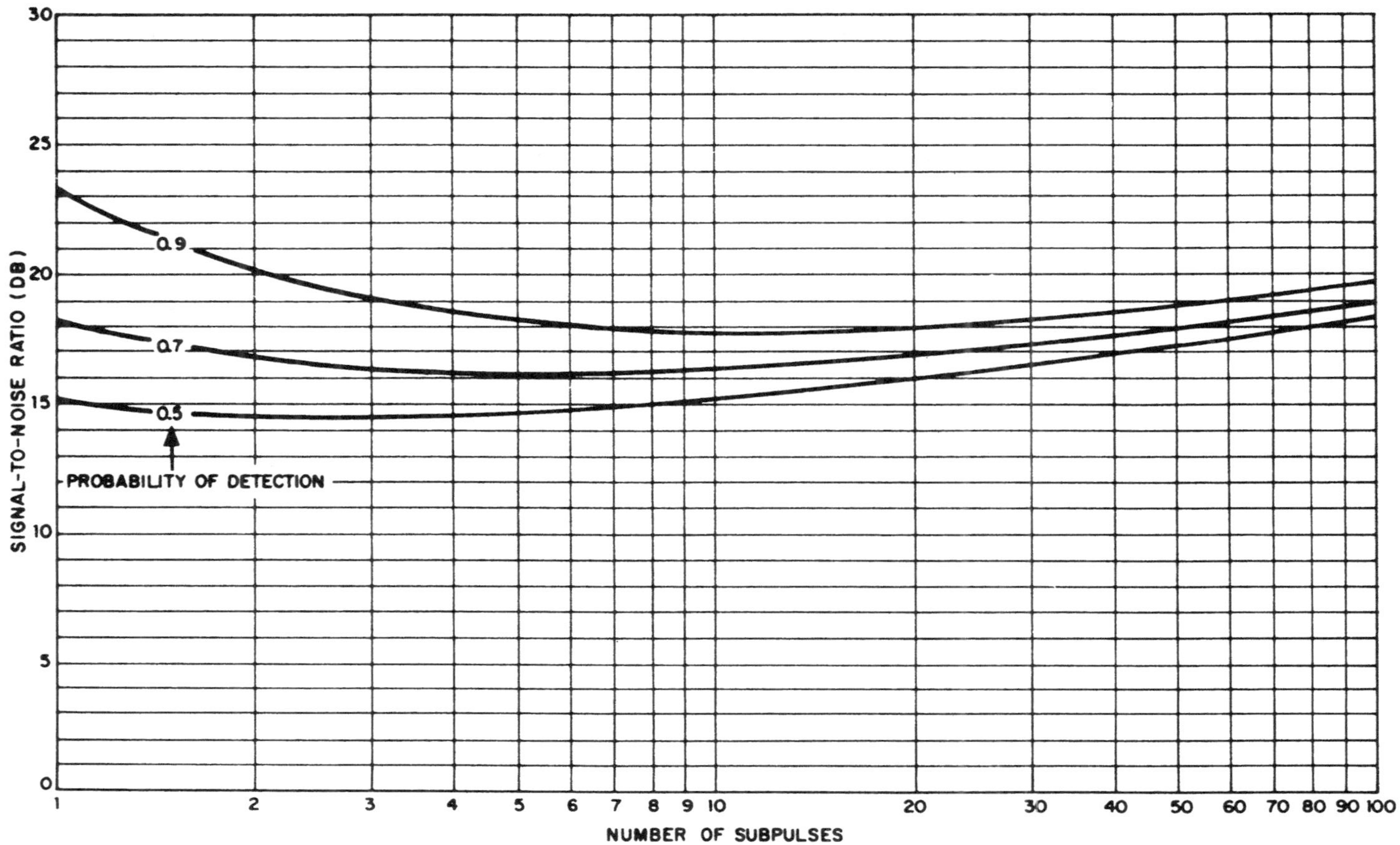

FIGURE 6. FLUCTUATING TARGET PROBABILITY OF DETECTION.

Independent subpulses of constant energy pulse integrated after square law detection. Target fluctuation model, Swerling Case 2; $n = 10^{10}$.

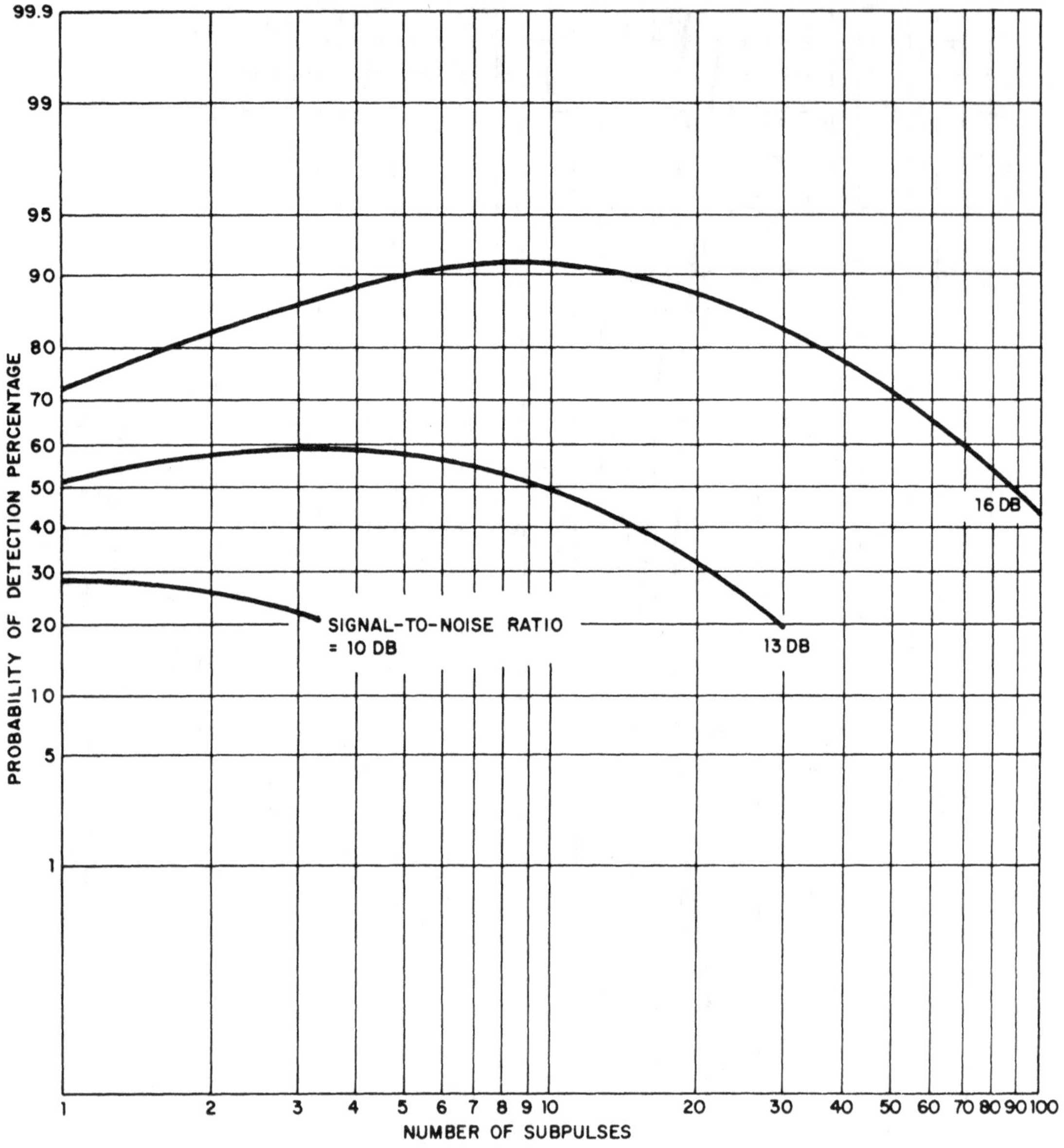

FIGURE 7. FLUCTUATING TARGET PROBABILITY OF DETECTION WITH CONSTANT SIGNAL-TO-NOISE RATIO.

Independent subpulses of constant energy pulse integrated after square law detection. Target fluctuation model, Swerling Case 2; $n = 10^6$.

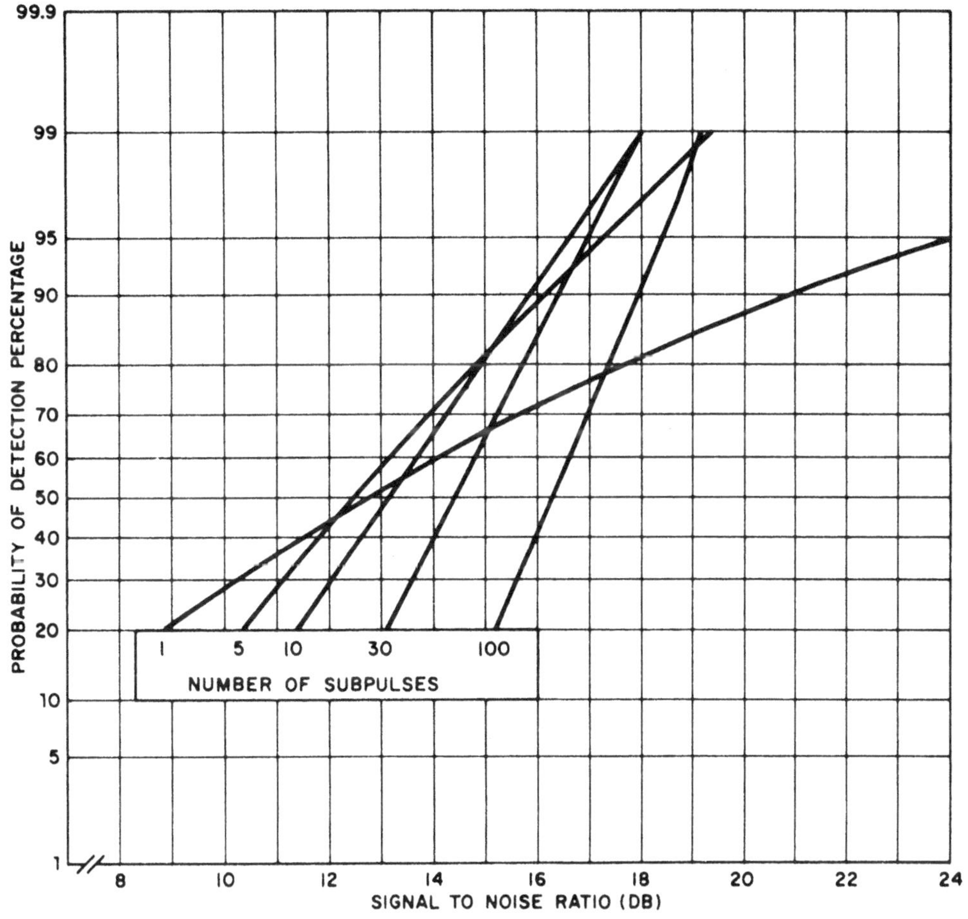

FIGURE 8. FLUCTUATING TARGET PROBABILITY OF DETECTION WITH CONSTANT NUMBER OF SUBPULSES.

Independent subpulses of constant energy pulse integrated after square law detection. Target fluctuation model, Swerling Case 2; $n = 10^6$.

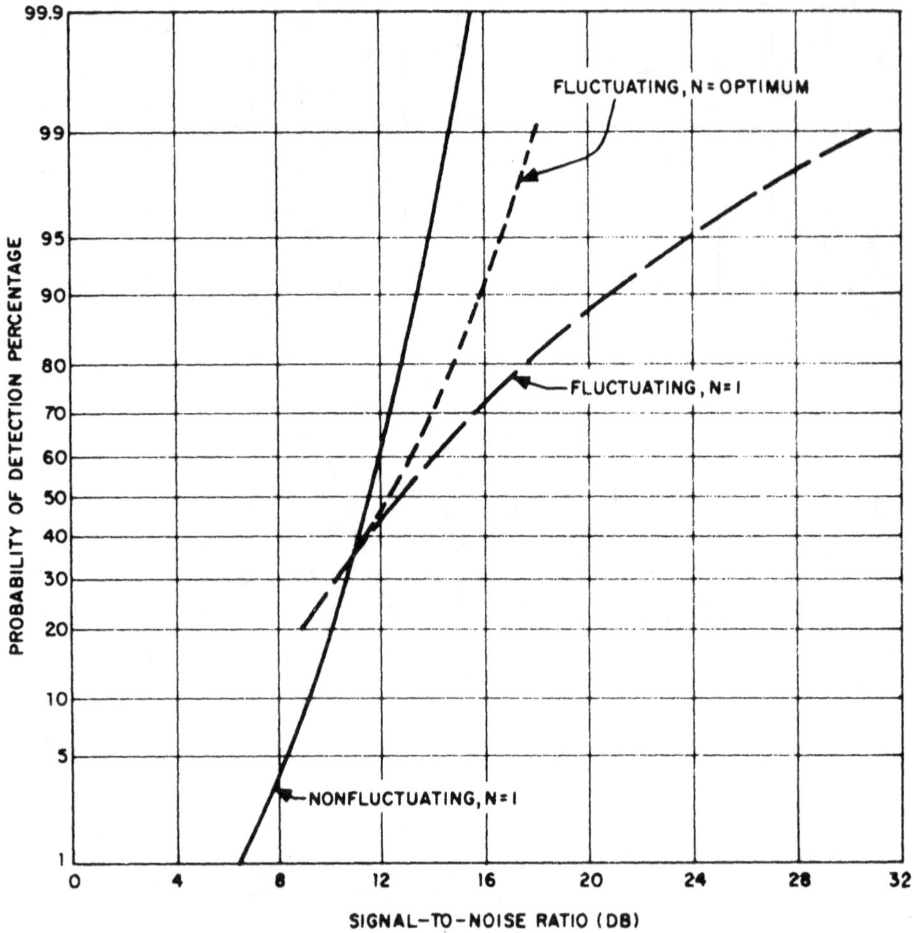

FIGURE 9. DETECTION COMPARISON WITH OPTIMUM NUMBER OF SUBPULSES.

Optimum number of independent subpulses of constant energy pulse integrated after square law detection. Fluctuating target model, Swerling Case 2; $n = 10^6$.

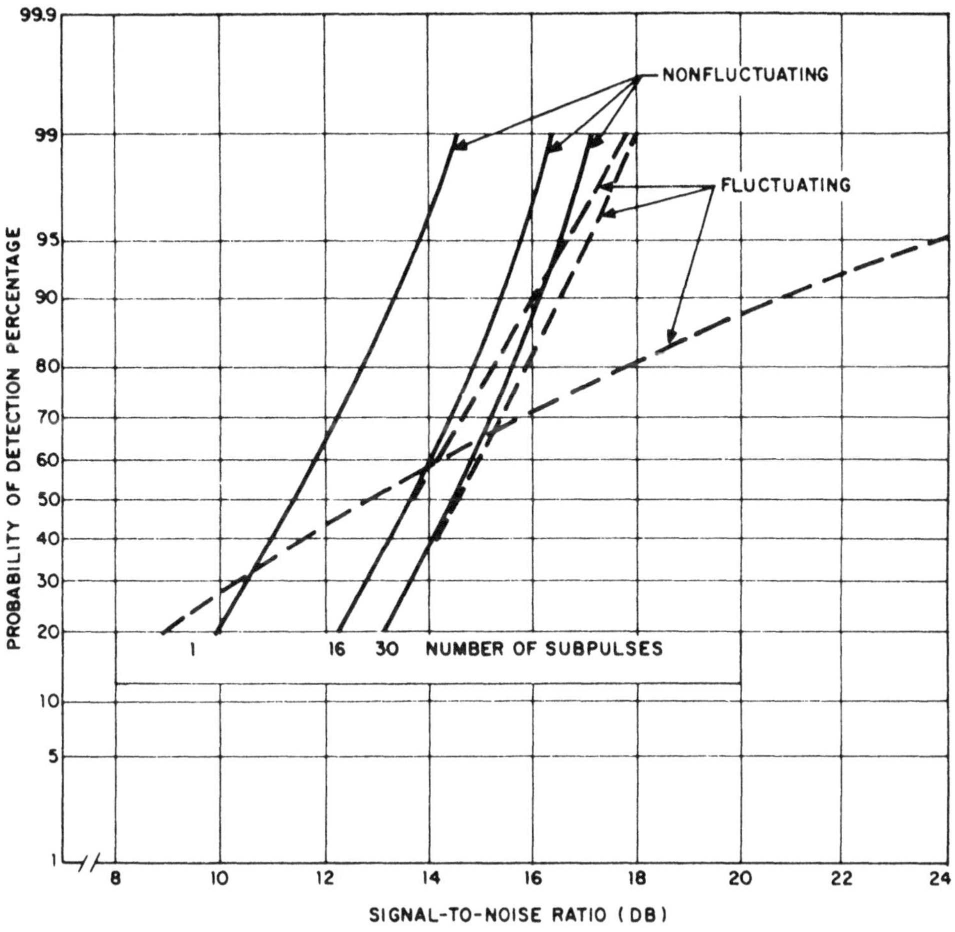

FIGURE 10. DETECTION COMPARISON BETWEEN FLUCTUATING AND NON-FLUCTUATING TARGETS.

Independent subpulses of constant energy pulse integrated after square law detection. Fluctuating target model, Swerling Case 2; $n = 10^6$.

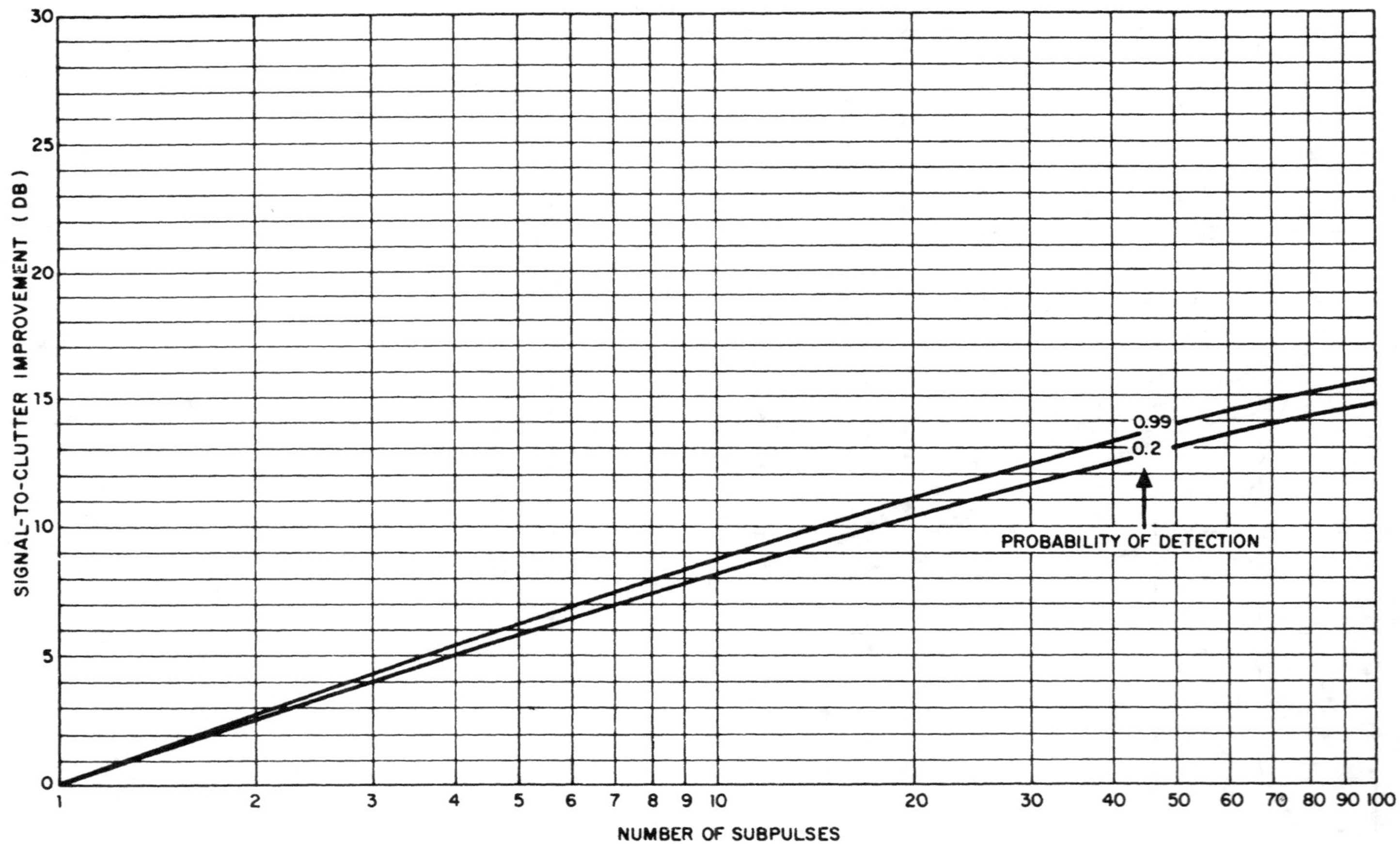

FIGURE 11. SIGNAL-TO-CLUTTER IMPROVEMENT WITH NONFLUCTUATING TARGET.

Independent, constant resolution subpulses integrated after square law detection. Subpulse clutter much greater than noise; $n = 10^6$.

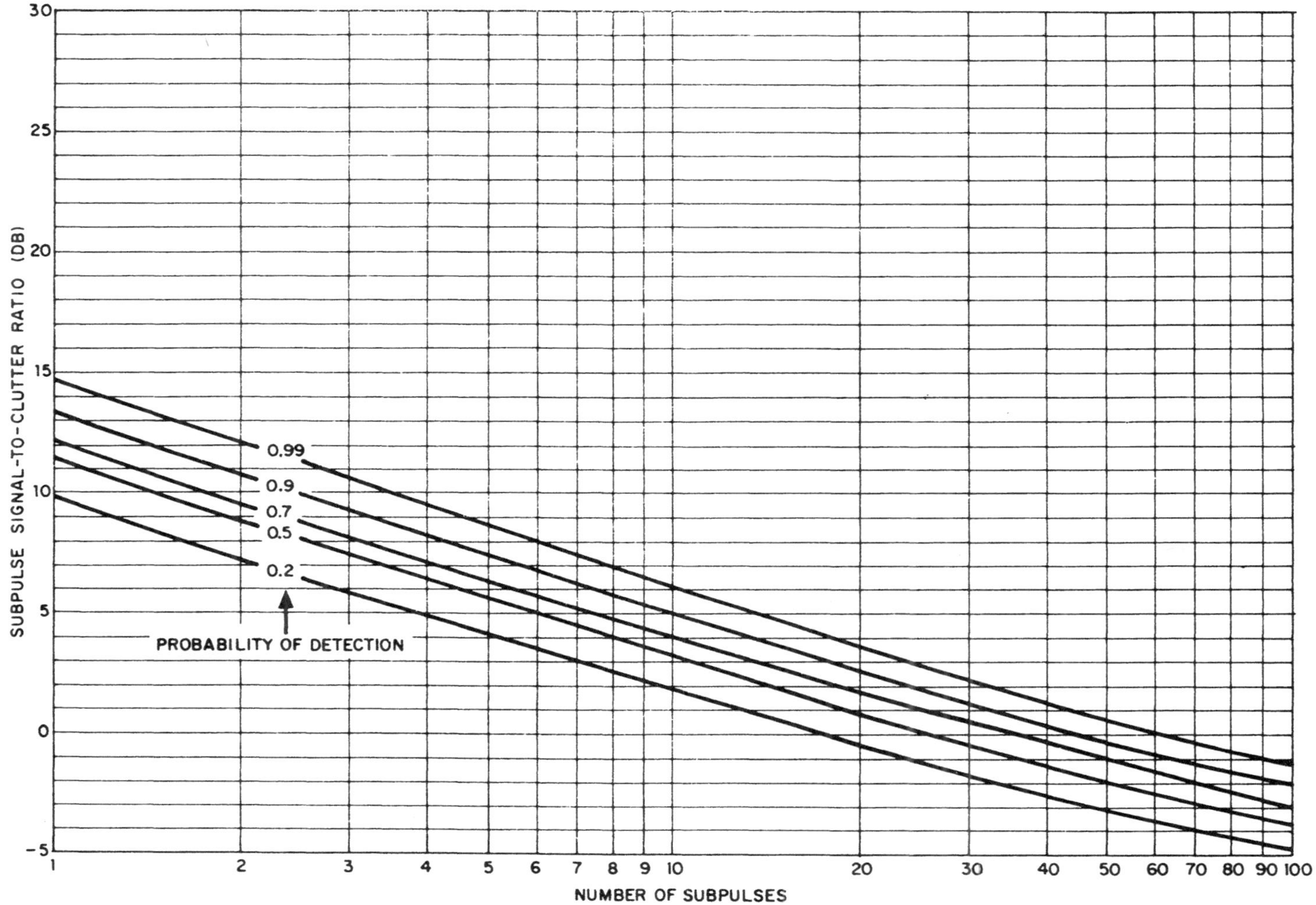

FIGURE 12. SIGNAL-TO-CLUTTER RATIO WITH NONFLUCTUATING TARGET.

Independent, constant resolution subpulses integrated after square law detection. Subpulse clutter much greater than noise; $n = 10^6$.

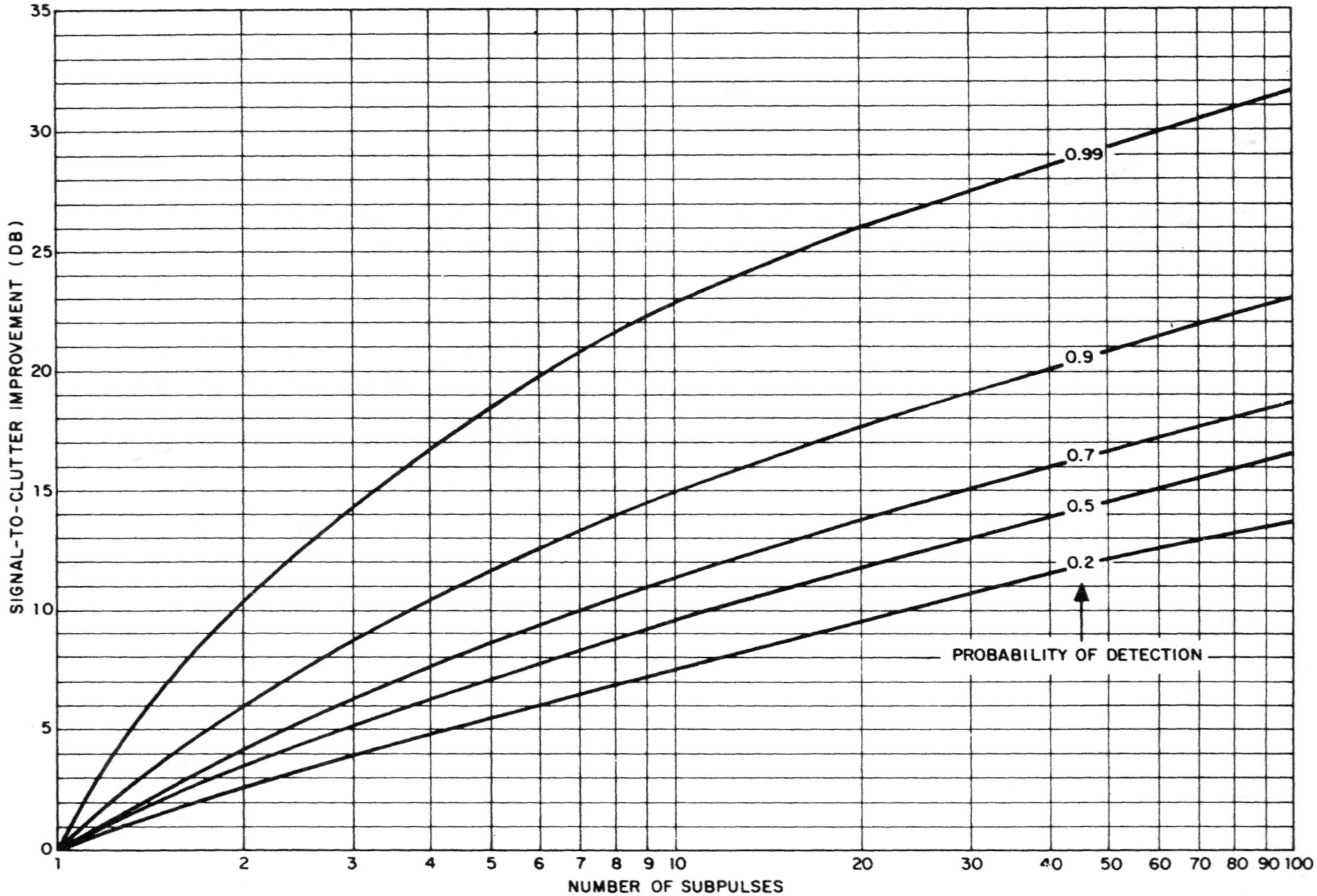

FIGURE 13. SIGNAL-TO-CLUTTER IMPROVEMENT WITH FLUCTUATING TARGET

Independent, constant resolution subpulses integrated after square law detection. Subpulse clutter much greater than noise. Target fluctuation model, Swerling Case 2; $n = 10^6$

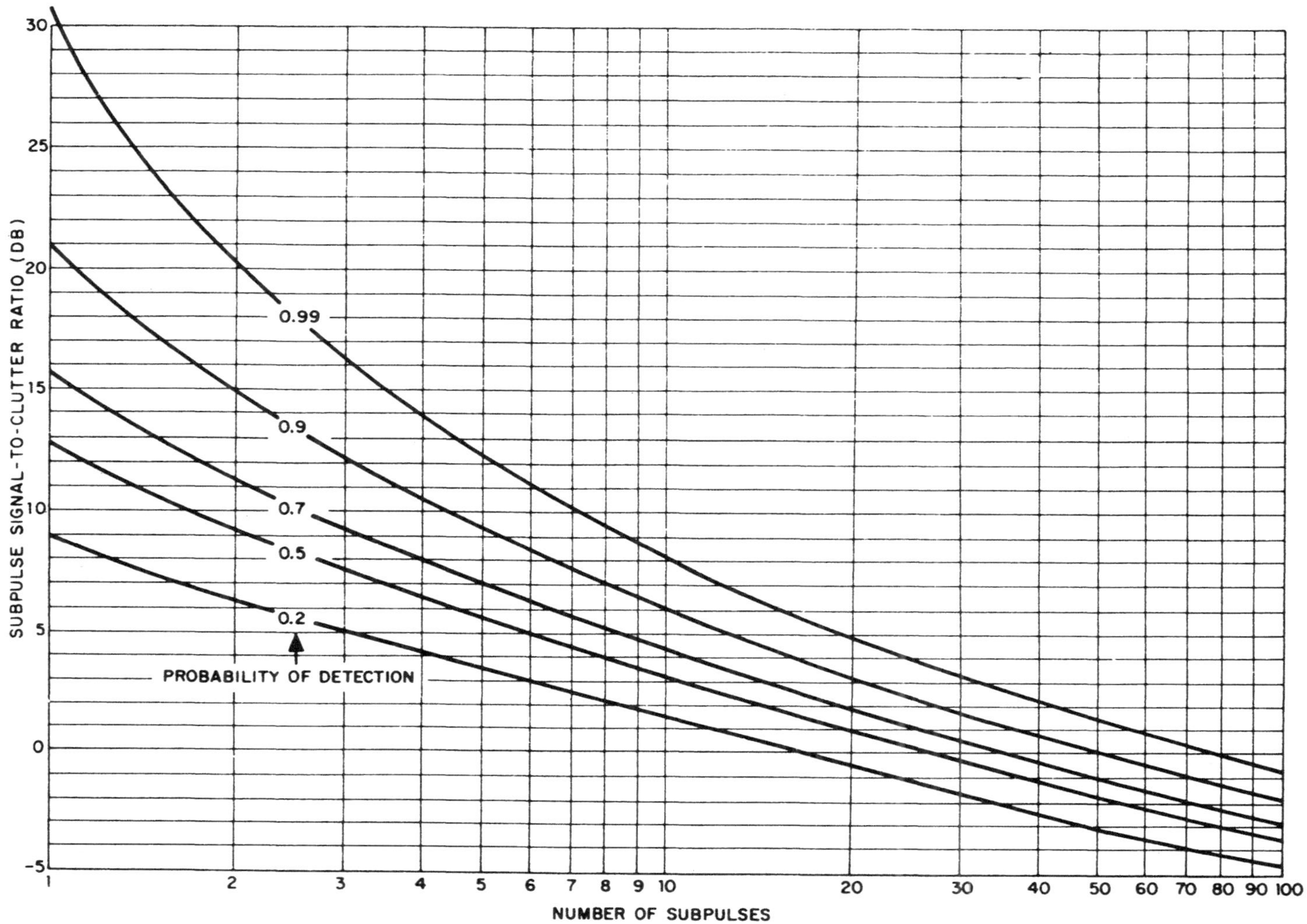

FIGURE 14. SIGNAL-TO-CLUTTER RATIO WITH FLUCTUATING TARGET.

Independent, constant resolution subpulses integrated after square law detection. Subpulse clutter much greater than noise. Target fluctuation model, Swerling Case 2; $n = 10^6$.

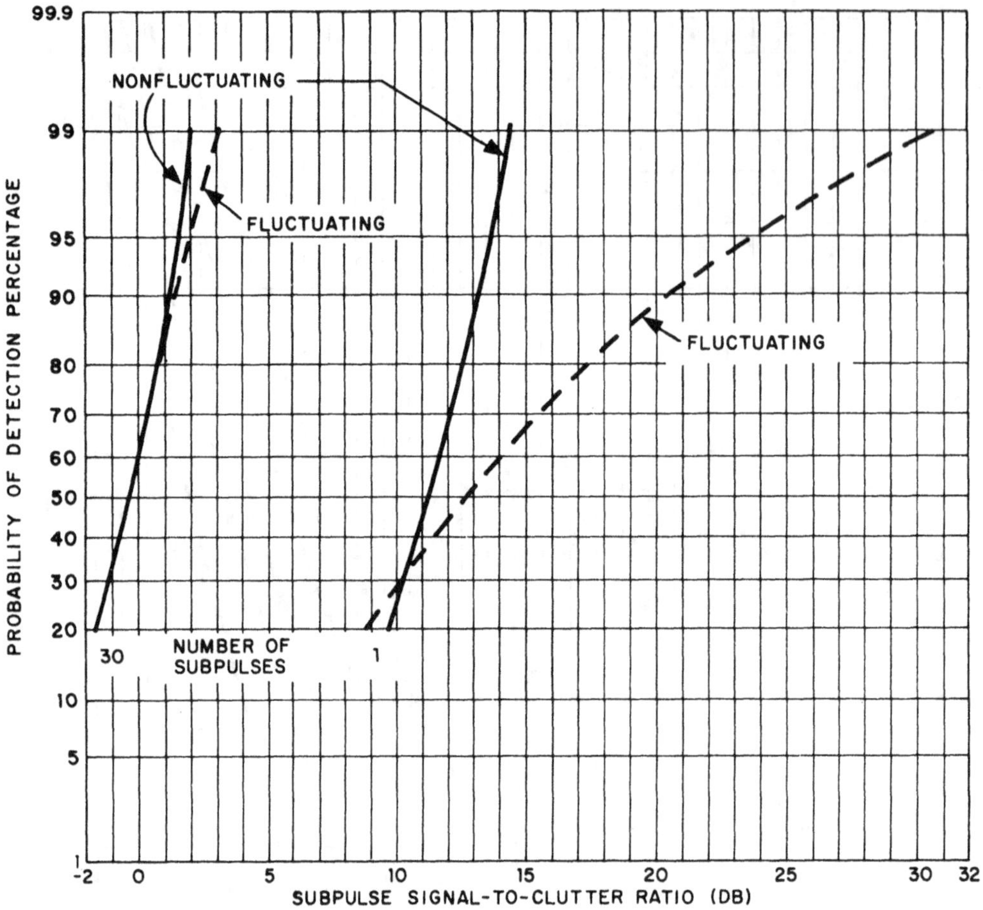

FIGURE 15. DETECTION IMPROVEMENT IN CLUTTER WITH FREQUENCY DIVERSITY.

Equal resolution waveforms. 30 independent subpulses integrated after square law detection. Subpulse clutter much greater than noise; $n = 10^6$. Fluctuating target model, Swerling Case 2.

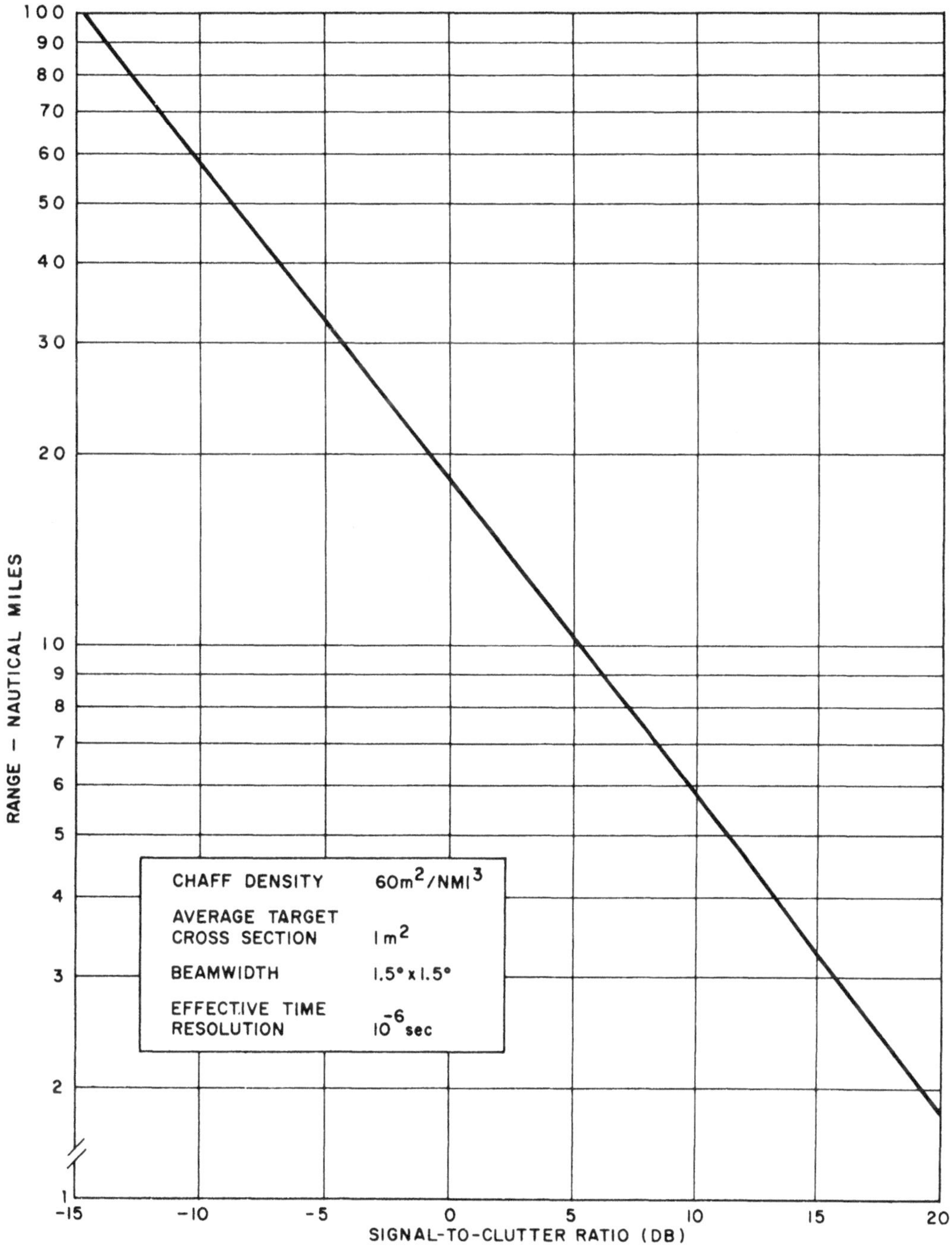

FIGURE 16. SIGNAL-TO-CHAFF CLUTTER RATIO.

69

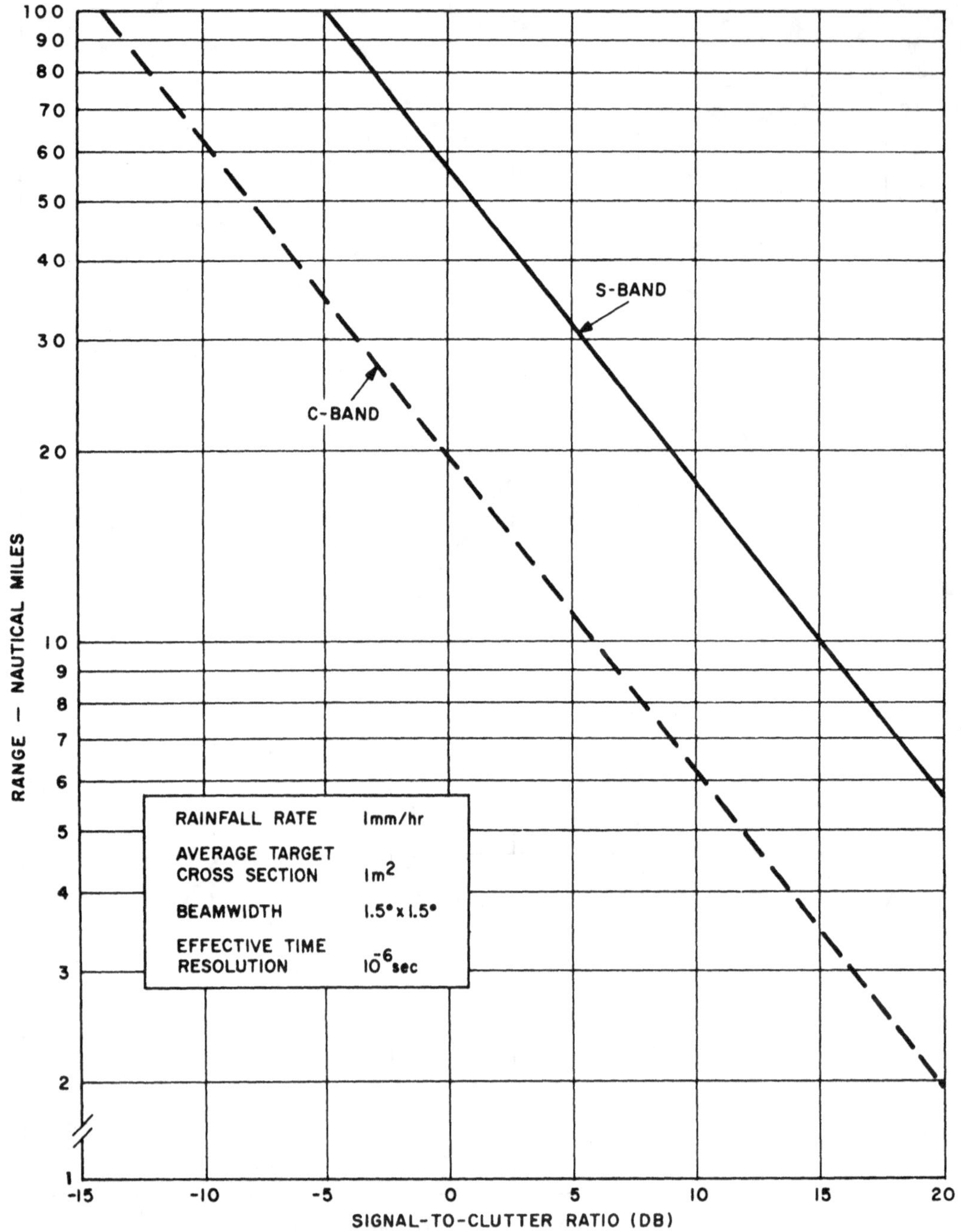

FIGURE 17. SIGNAL-TO-RAIN CLUTTER RATIO.

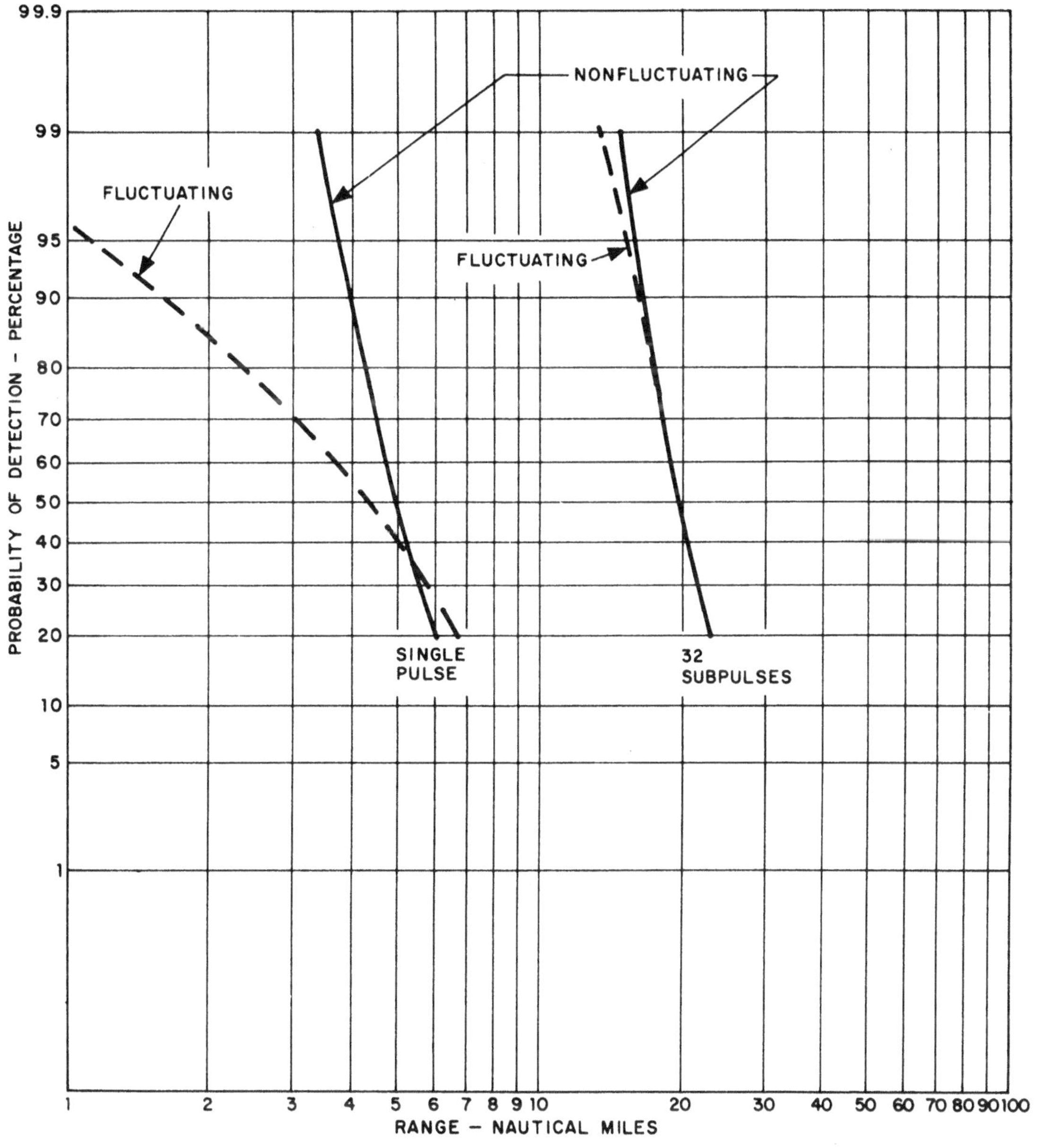

FIGURE 18. DETECTION IN CHAFF CLUTTER.

Chaff density, 60 m^2/NMI^3; target cross-section, 1 m^2; antenna beamwidth, $1.5^o \times 1.5^o$; effective time resolution, one microsecond; $n = 10^6$; Swerling Case 2 fluctuating target.

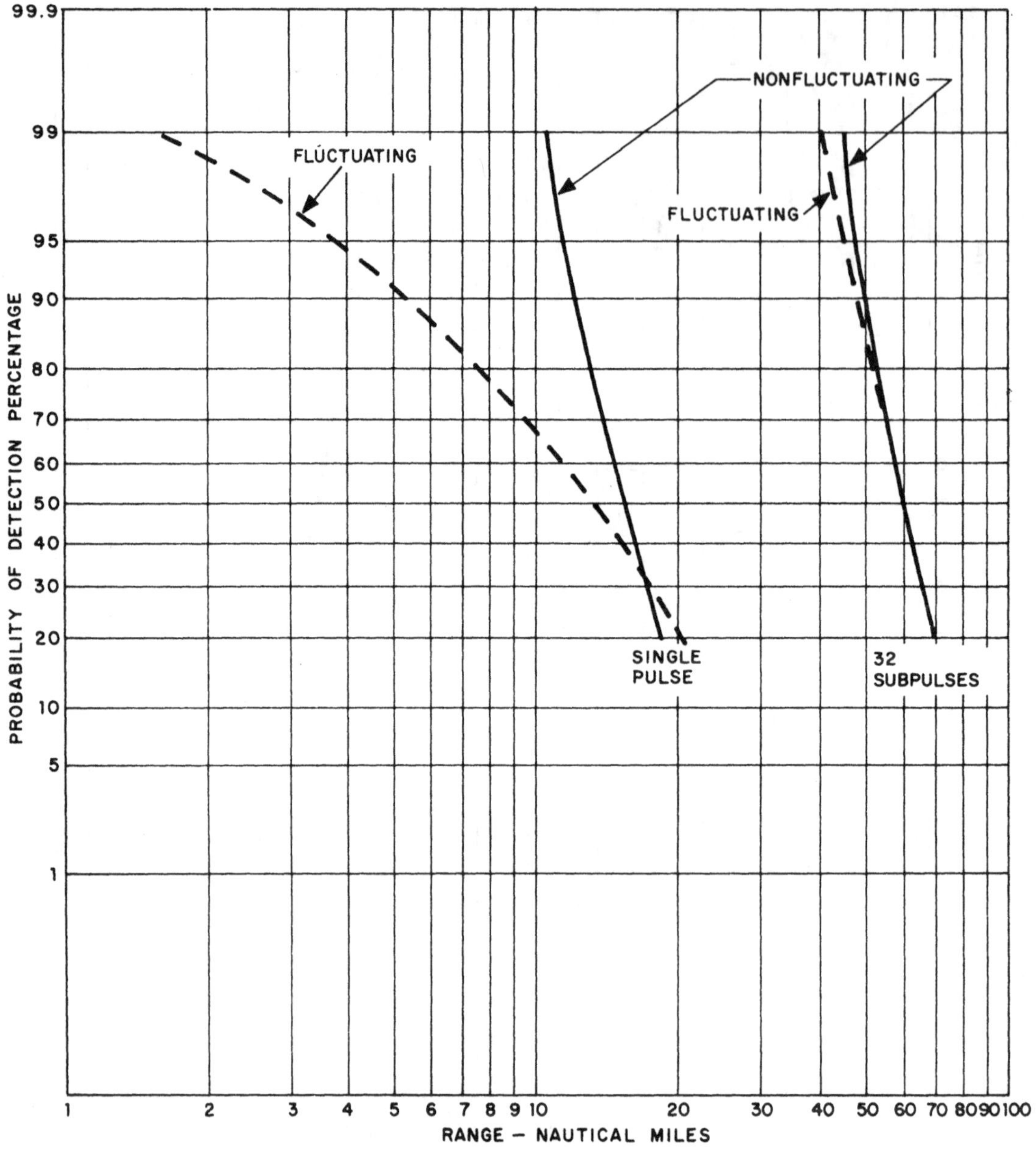

FIGURE 19. S-BAND DETECTION IN RAIN CLUTTER.

Rainfall rate 1 mm/hr; target cross-section, 1 m^2; antenna beamwidth, $1.5^o \times 1.5^o$; effective time resolution, one microsecond; $n = 10^6$; Swerling Case 2 fluctuating target; operating frequency, 3.3 GHz.

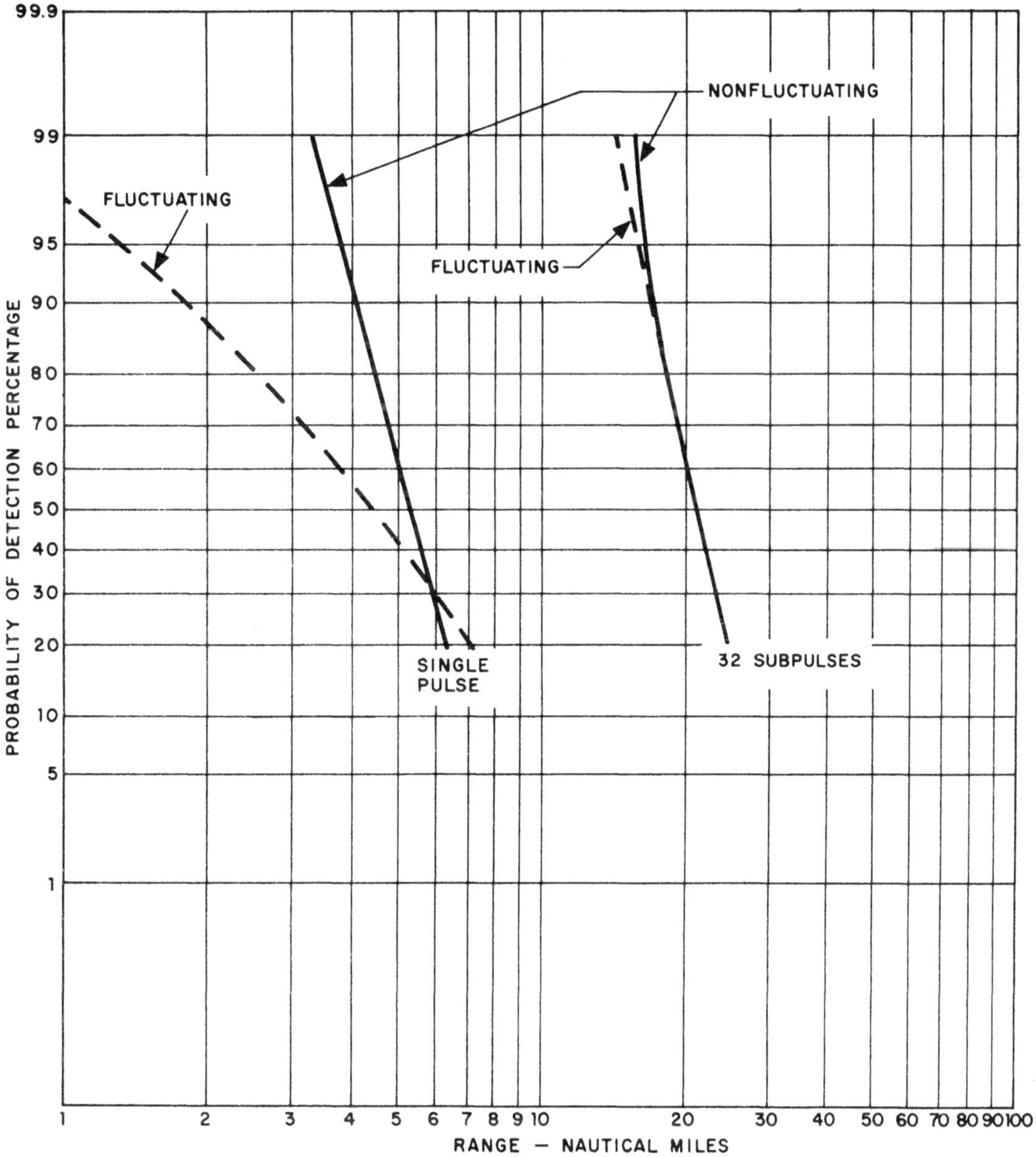

FIGURE 20. C-BAND DETECTION IN RAIN CLUTTER.

Rainfall rate, 1 mm/hr; target cross-section, 1 m^2; antenna beamwidth, $1.5^o \times 1.5^o$; effective time resolution, one microsecond; $n = 10^6$; Swerling Case 2 fluctuating target; operating frequency, 5.6 GHz.

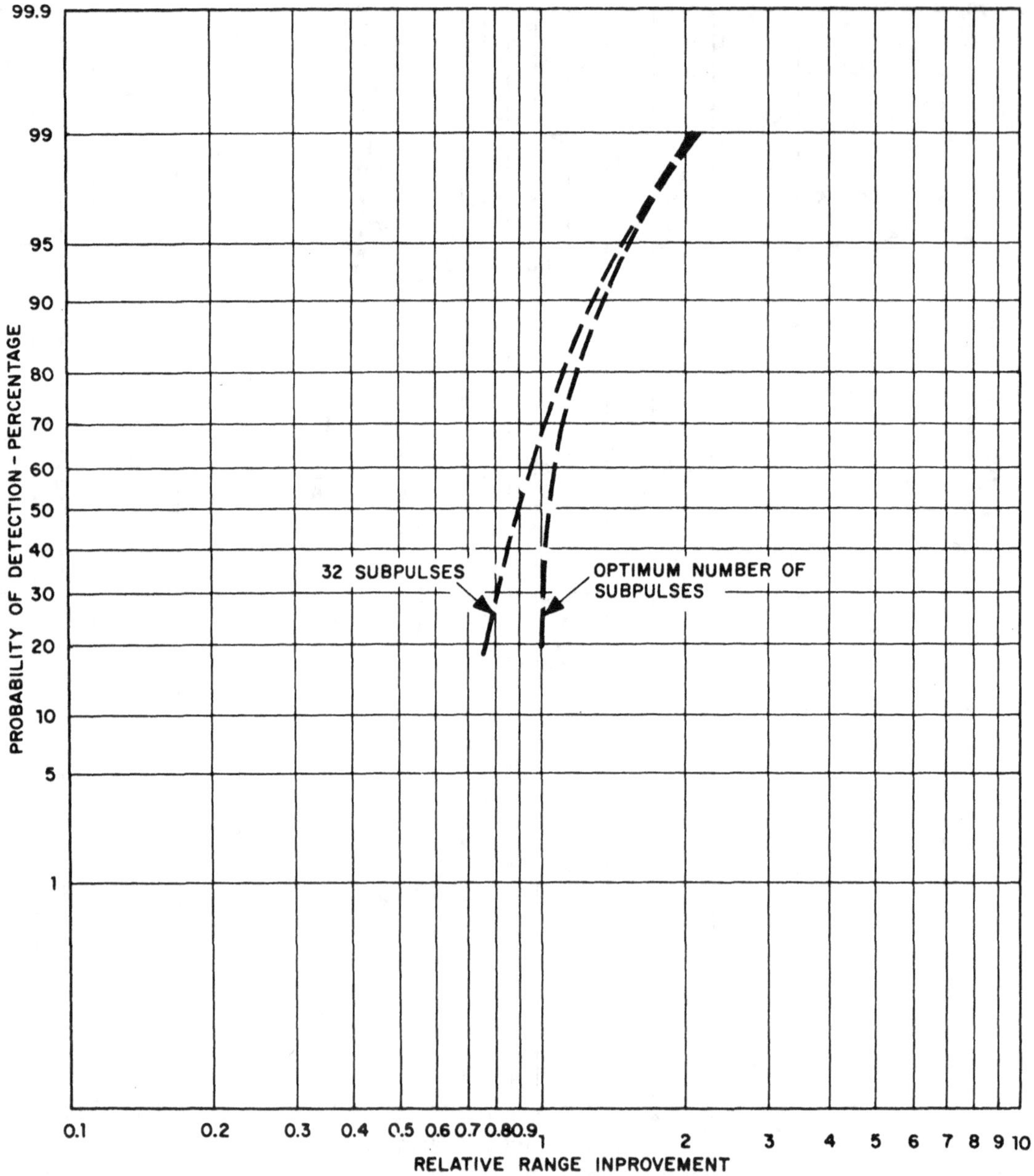

FIGURE 21. CLUTTER-FREE FLUCTUATING TARGET RANGE IMPROVEMENT.

Independent subpulses of constant energy pulse noncoherently integrated. Fluctuating target model, Swerling Case 2; $n = 10^6$.

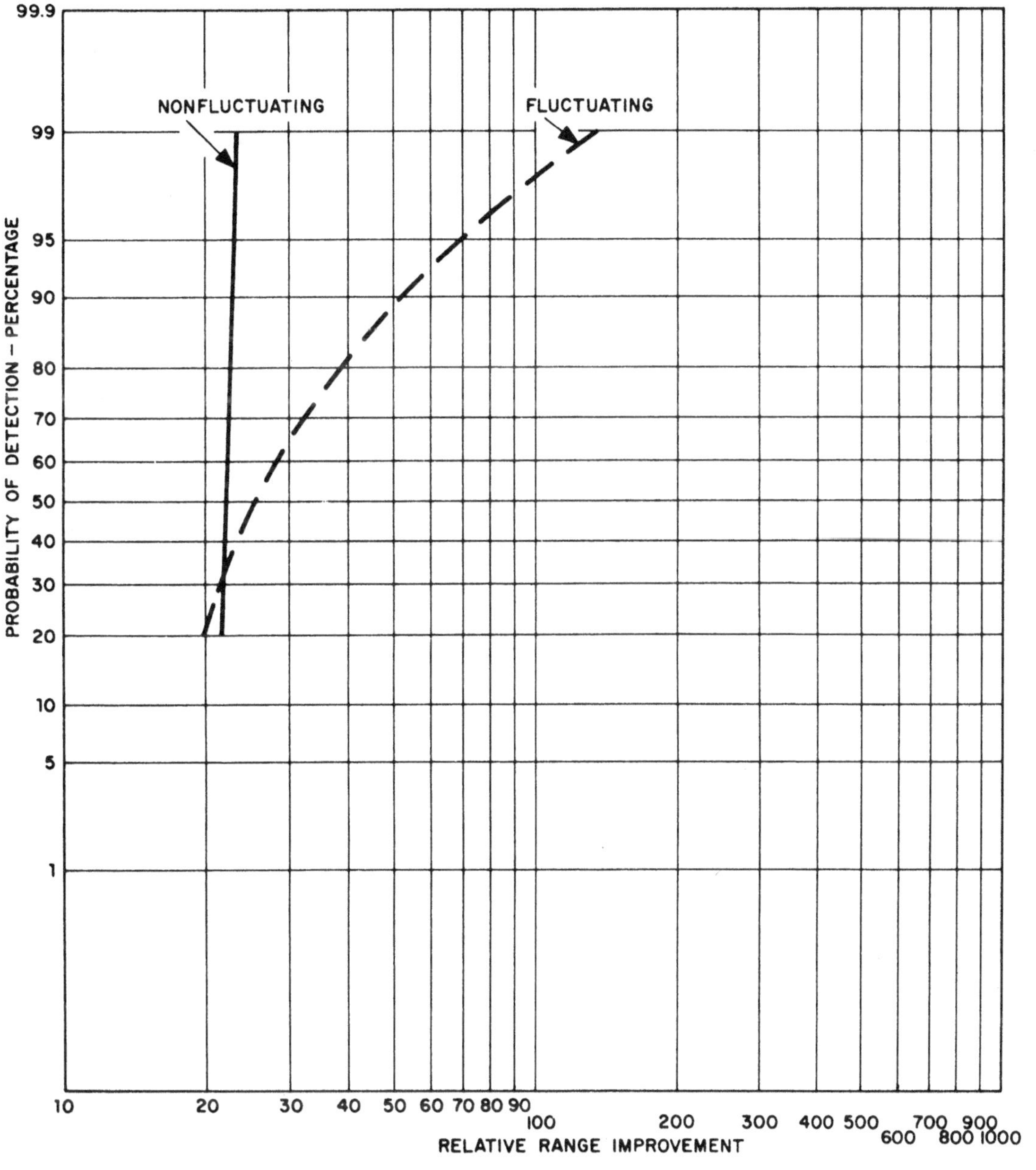

FIGURE 22. COMBINED RANGE IMPROVEMENT IN DISTRIBUTED CLUTTER.

Simultaneous effects of both pulse compression and independent clutter samples. 32 independent subpulses are noncoherently integrated. Subpulse clutter much greater than noise. Fluctuating target model, Swerling Case 2; $n = 10^6$.

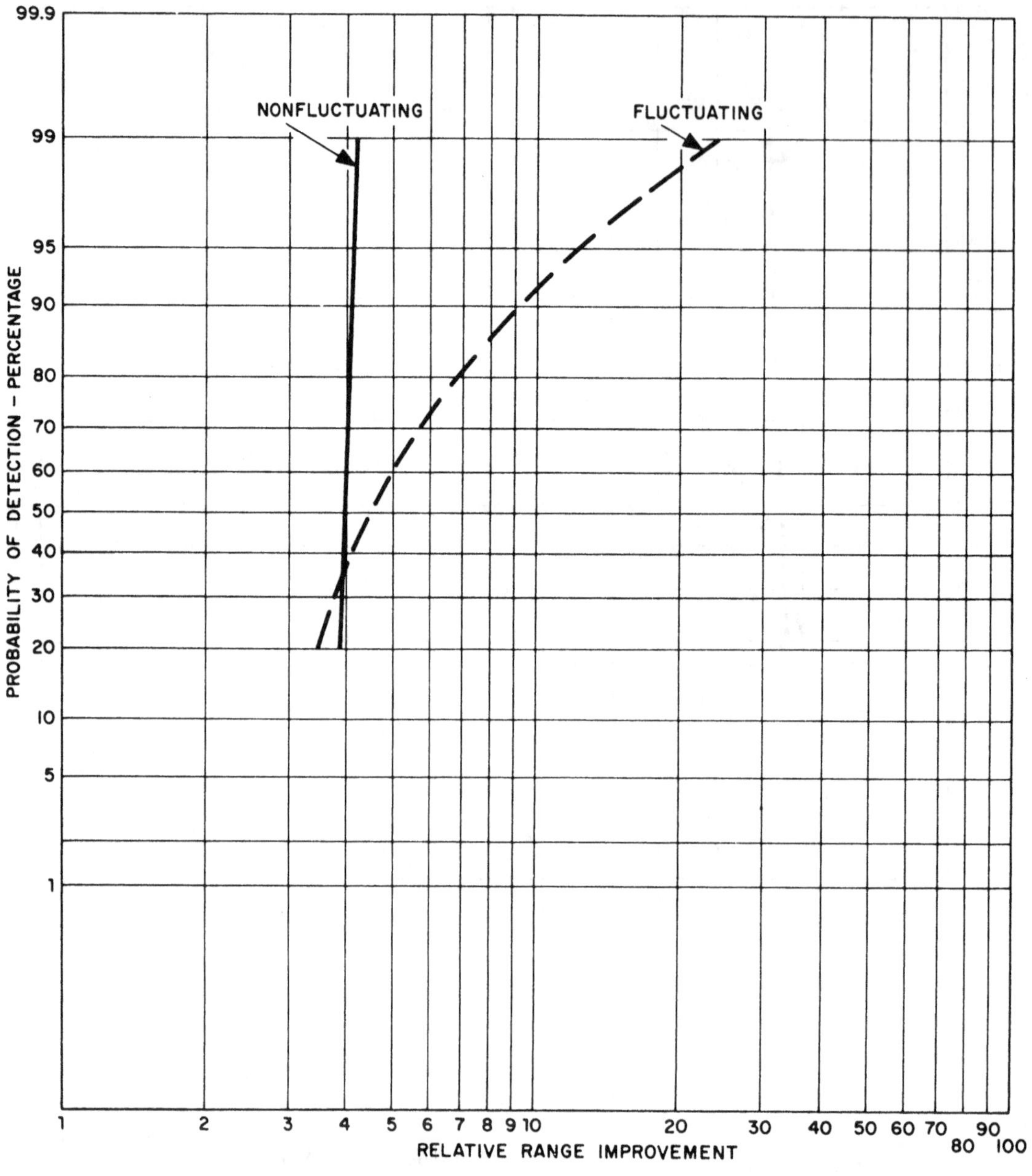

FIGURE 23. RANGE IMPROVEMENT IN DISTRIBUTED CLUTTER.

Improvement due to independent clutter samples alone. 32 independent subpulses noncoherently integrated. Subpulse clutter much greater than noise. Fluctuating target model, Swerling Case 2; $n = 10^6$.

General Detection Procedure for a Multifrequency Radar

G. Grasso and P.F. Guarguaglini

Alta Frequenza, Vol. 37, No. 2 (February 1968)
Pages 132-138

The problem of evaluating the performance of a multifrequency radar using any number of frequencies and transmitting any number of pulses on each frequency has been analytically solved. The expressions obtained are rather complex and suggest the use of a computer.

When only two frequencies are used, though, the expressions are greatly simplified and have been used for extensive calculations.

Introduction.

The use of multifrequency radars is becoming more widespread due to the advantage it offers in the detection of scan-to-scan fluctuating targets. Various schemes are possible for combining the signals relative to each frequency channel. It is in fact possible to combine the signals in RF, in IF or in video. RF combining never performs better than IF combining and is harder to implement. IF combining gives advantage over video (or postdetection) combining only if the entire system is coherent. For the great majority of cases, therefore, video combining is the most convenient. Furthermore it has the advantage that its performance is not degraded by a Doppler frequency shift.

In the literature the analysis of the Diversity Radar has been performed only when the number of pulses transmitted on each frequency is the same [1]. It is interesting to be able to evaluate the performance for any number of pulses transmitted on each frequency. The possibility to choose the number of pulses transmitted on each frequency independently gives the operator other parameters on which to act, for instance in the presence of an interference. The use of a large number of pulses transmitted on one of the frequencies, together with smaller numbers on the other frequencies, also permits the insertion of an MTI in the corresponding channel.

The purpose of this paper is to give to the designer a method of evaluating the signal-to-noise ratio necessary to obtain the desired Detection and False Alarm Probabilities for an unequal number of pulses on each frequency.

The analysis will be conducted assuming that the echoes of pulses transmitted on the same frequency are completely correlated. The frequencies used are far enough apart to consider independent the echoes of pulses transmitted on different frequencies, but are close enough to suppose that the probability distributions of all the echoes are the same.

(*) Dott. Ing. Giancarlo Grasso, Dott. Ing. PierFrancesco Guarguaglini - Advanced Studies and Research Department, SELENIA S.p.A., Roma - Institute of Electronics, University of Rome.

Mathematical Analysis.

Let v_n be the n-th raw video signal. At the output of the video integrator the total video signal will be

$$y = \sum_{n=1}^{N} v_n \tag{1}$$

where N is the number of pulses on target. If x_n is the SNR relative to the v_n signal, we may introduce

$$X = \sum_{n=1}^{N} x_n . \tag{2}$$

In reference [2] it has been proved that the characteristic function of y, $C_y(p)$, is bound to the characteristic function of X, $C_X(p)$, by the relation:

$$C_y(p) = (1 + p)^N C_X\left(\frac{p}{1 + p}\right). \tag{3}$$

The receiver studied employs M frequencies. Therefore, if K_i represents the number of pulses transmitted on each frequency, such that

$$\sum_{i=1}^{N} K_i = N ,$$

X may be expressed as

$$X = \sum_{i=1}^{M} K_i X_i \tag{4}$$

where

$$X_i = \sum_{n=1}^{K_i} x_{n,i} \tag{5}$$

if $x_{n,i}$ are the signal-to-noise ratios relative to the i-th frequency pulses.

Only the case of a different number of pulses transmitted on each frequency will be considered.

Assume that the probability distributions of the $x_{n,i}$ are the same and given by

$$p(x_{n,i}) = \frac{1}{\bar{x}} \cdot \exp\left[-\frac{x_{n,i}}{\bar{x}}\right] \tag{6}$$

where $\bar{x}$ is the average signal-to-noise ratio (SNR). Eq. (6) corresponds to target fading according to the Rayleigh distribution [3].

The characteristic function of $x_{n,i}$, obtained by transforming is:

$$C(p) = \frac{1}{1 + \bar{x}\, p} . \tag{7}$$

79

Since the K_i pulses transmitted on each frequency are completely correlated, the characteristic function of X_i may be easily evaluated from the characteristic function of the signal-to-noise ratio relative to a single pulse $C(p)$:

$$C_{X_i}(p) = C(K_i p) = \frac{1}{1 + K_i \bar{x} p}. \tag{8}$$

The characteristic function of X is the product of the characteristic functions of the M variables, being the variables independent. Therefore:

$$C_X(p) = \prod_{i=1}^{M} C_{X_i}(p) = \prod_{i=1}^{M} \frac{1}{1 + K_i \bar{x} p} \tag{9}$$

Then, using eq. (3) the characteristic function of the signal at the output of the video integrator may be easily obtained:

$$C_y(p) = \frac{1}{(1+p)^{N-M} \prod_{i=1}^{M} (1 + a_i p)} \tag{10}$$

where it has been introduced

$$a_i = 1 + K_i \bar{x}. \tag{11}$$

Eq. (10) may be expanded according to:

$$\prod_{i=1}^{M} \frac{1}{1 + a_i p} = \sum_{i=1}^{M} \frac{A_i}{1 + a_i p}. \tag{12}$$

The coefficients A_i of eq. (12) may be evaluated by solving the system:

$$B_{MM} \cdot C_{M1} = D_{M1}. \tag{13}$$

The individual terms of the matrixes are:

(13bis)

$$b_{hk} = \begin{cases} 1 & \text{for } h = 1 \\ \sum_{i=1}^{M}{}^{(1)} \sum_{j=i+1}^{M}{}^{(2)} \cdots \sum_{m}^{M}{}^{(h)} \; {}_k\delta^v{}_{i,j,\ldots,m}\, a_i a_j \ldots a_m & \text{for } h \neq 1 \end{cases}$$

$$c_{h1} = A_h$$

$$d_{h1} = \begin{cases} 1 & \text{for } h = 1 \\ 0 & \text{for } h \neq 1 \end{cases}$$

The number of the sums in each element of the matrix B_{MM} is given by the order of the corresponding row minus one. The last row terms are given by the product of $(M-1)$ a_i, excluding the K_i whose index is equal to the order of the column of which the considered term is part.

A new operator ${}_k\delta^v{}_{i,j,\ldots,m}$ has been introduced, defined according to:

$${}_k\delta^v{}_{i,j,\ldots,m} = \begin{cases} 0 & \text{for } k \text{ equal to at least one of the indexes on the right side} \\ 1 & \text{for } k \text{ different from all the indexes on the right side} \end{cases} \tag{14}$$

The characteristic function of the integrated video output may be finally written:

$$C_y(p) = \sum_{i=1}^{M} \frac{1}{(1+p)^{N-M}} \cdot \frac{A_i}{(1 + a_i p)}. \tag{15}$$

The Laplace Transform of eq. (15) will give the probability density function of y. But each term of the sum in eq. (15) may be transformed separately by couple 581.7 of Campbell-Foster [4].

Each term gives:

$$f_i(y) = \frac{A_i}{a_i} \frac{\exp\left[-\frac{y}{a_i}\right]}{\Gamma(N-M)\left[\frac{a_i - 1}{a_i}\right]^{N-M}} \cdot \gamma\left[N-M, \frac{a_i - 1}{a_i} y\right] \tag{16}$$

where

$$\gamma(v, z) = \Gamma(v) - \Gamma(v, z) \tag{17}$$

and

$$\Gamma(v, z) = \int_z^{\infty} z^{v-1} \exp[-z]\, dz. \tag{18}$$

The probability density function is therefore the sum of M terms:

$$p(y) = \sum_{i=1}^{M} f_i(y). \tag{19}$$

Correspondingly also the Detection Probability P_D will be given by the sum of M probability of detection P_{D_i}, each relative to one of the frequencies:

$$P_D = \sum_{i=1}^{M} P_{D_i}. \tag{20}$$

From eqs. (16), (17) and (18), introducing Pearson's Incomplete Function, defined as [5]:

$$I(u, p) = \frac{\int_0^{u\sqrt{p+1}} v^p \exp[-v]\, dv}{\Gamma(p+1)} \tag{21}$$

the expression of the term of the probability density function corresponding to the i-th frequency may be written:

$$f_i(y) = \frac{A_i}{a_i} \frac{\exp\left[-\frac{y}{a_i}\right]}{\left(\frac{a_i - 1}{a_i}\right)^{N-M}} \cdot I\left[\frac{a_i - 1}{a_i} \cdot \frac{y}{\sqrt{N-M}}, N - M - 1\right]. \tag{22}$$

This expression is formally analogous to one of the distributions obtained by Swerling [3] in his examination of target fluctuations. Analogously the Detection Probability can be written:

$$P_{D_i} = 1 - \int_{-\infty}^{Y_b} f(y)\, dy \tag{23}$$

where Y_b is the normalized threshold.

In the present case the complete expression for the Detection Probability is:

$$P_D = \sum_{i=1}^{M} \left\{ 1 - I\left[\frac{Y_b}{\sqrt{N-M}}, N-M-1\right] + A_i\left(1 - \frac{1}{K_i \bar{x}}\right)^{N-M} \cdot I\left[\frac{Y_b}{\left(1+\frac{1}{K_i \bar{x}}\right)\sqrt{N-M}}, N-M-1\right] \cdot \exp\left[-\frac{Y_b}{1+K_i \bar{x}}\right]\right\}. \tag{24}$$

NUMERICAL EVALUATION.

The expression (24) gives to the designer the possibility to evaluate the SNR necessary to achieve any value of Probability of Detection. The expression is though very entangled, chiefly because of the difficulty in the evaluation of the coefficients A_i. Numerical results will here be obtained only when the frequencies used are two. In this case, of great particular interest, the expression of the Probability of Detection, obtained from eqs. (24) and (13) is:

$$\begin{aligned} P_D = {} & \frac{K_1}{K_1 - K_2}\left\{1 - I\left[\frac{Y_b}{\sqrt{N-2}}, N-3\right] + \left(1 + \frac{1}{K_1 \bar{x}}\right)^{N-1} \exp\left[-\frac{Y_b}{1 + K_1 \bar{x}}\right] \cdot I\left[\frac{Y_b}{\left(1+\frac{1}{K_1 \bar{x}}\right)\sqrt{N-2}}, N-3\right]\right\} - \\ & - \frac{K_2}{K_1 - K_2}\left\{1 - I\left[\frac{Y_b}{\sqrt{N-2}}, N-3\right] + \left(1 + \frac{1}{K_2 \bar{x}}\right)^{N-1} \exp\left[-\frac{Y_b}{1 + K_2 \bar{x}}\right] \cdot I\left[\frac{Y_b}{\left(1+\frac{1}{K_2 \bar{x}}\right)\sqrt{N-2}}, N-3\right]\right\} \end{aligned} \tag{25}$$

which is valid only if $K_1 \neq K_2$.

This expression of the Detection Probability may be greatly simplified for the parameter variations of interest in radar applications. In this case, in fact, the values of the variables of the Incomplete Γ Function are such that the value of the function is very close to unity. With very good approximation therefore eq. (25) may be reduced to:

$$\begin{aligned} P_D \cong {} & \frac{K_1}{K_1 - K_2}\left(1 + \frac{1}{K_1 \bar{x}}\right)^{N-1} \exp\left[-\frac{Y_b}{1 + K_1 \bar{x}}\right] - \\ & - \frac{K_2}{K_1 - K_2}\left(1 + \frac{1}{K_2 \bar{x}}\right)^{N-1} \exp\left[-\frac{Y_b}{1 + K_2 \bar{x}}\right]. \end{aligned} \tag{26}$$

This expression has been used to obtain, by graphical interpolation, the signal-to-noise ratio necessary to achieve the desired detection conditions for several values of N, with K_1/K_2 as a parameter. The procedure is as follows: the value of Y_b corresponding to the desired Probability of False Alarm is introduced in eq. (18).

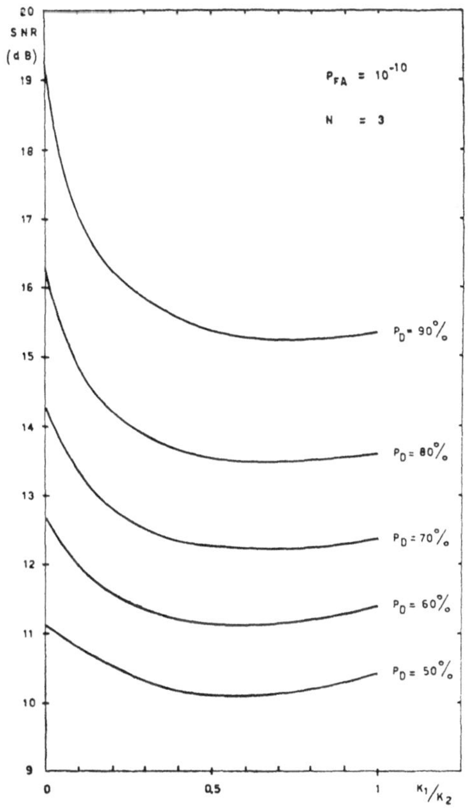

Fig. 1. — Signal-to-noise ratio vs. number of pulses on target, for $P_{FA} = 10^{-10}$ and $N = 3$.

The curve expressing the P_D vs. SNR is sketched for several values of K_1/K_2. From these curves the SNR corresponding to the desired Detection Probability may be easily extracted.

The same calculations have been performed for $P_{FA} = 10^{-10}$ and 10^{-6}, for several values of N. Figs. 1 through 6 give the SNR vs. K_1/K_2, for P_D = 50%, 60%, 70%, 80%, 90%, for N = 3, 12 and 48.

It is interesting to note that the shape of the curves is practically independent of the Probability of False Alarm and the number of pulses on target. The only parameter which affects this shape is the Probability of Detection.

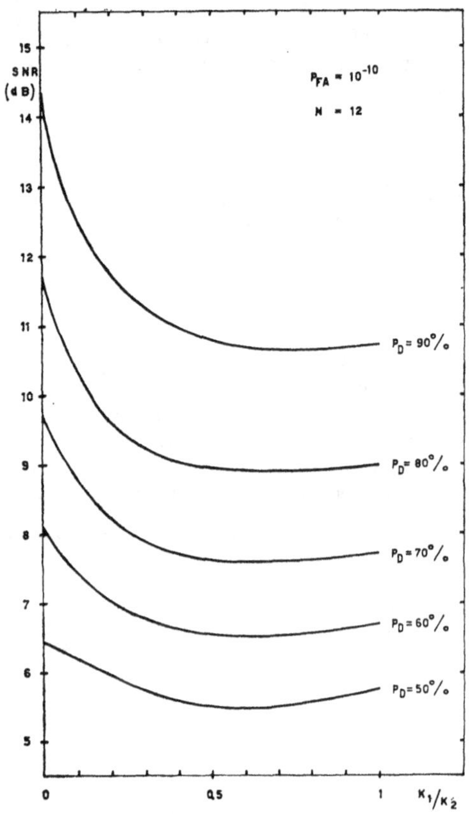

Fig. 2. — Signal-to-noise ratio vs. number of pulses on target, for $P_{FA} = 10^{-10}$ and $N = 12$.

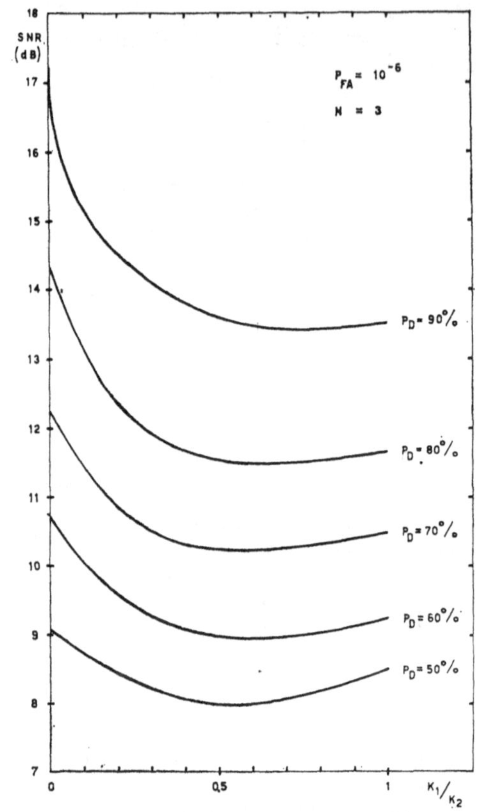

Fig. 4. — Signal-to-noise ratio vs. number of pulses on target, for $P_{FA} = 10^{-6}$ and $N = 3$.

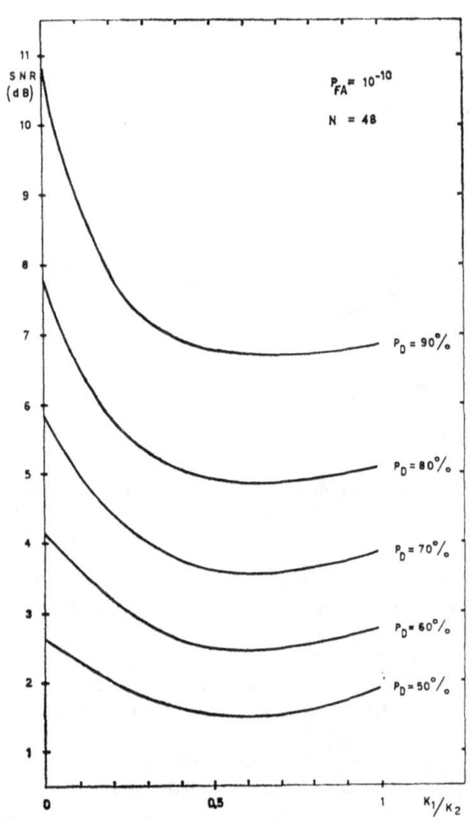

Fig. 3. — Signal-to-noise ratio vs. number of pulses on target, for $P_{FA} = 10^{-10}$ and $N = 48$.

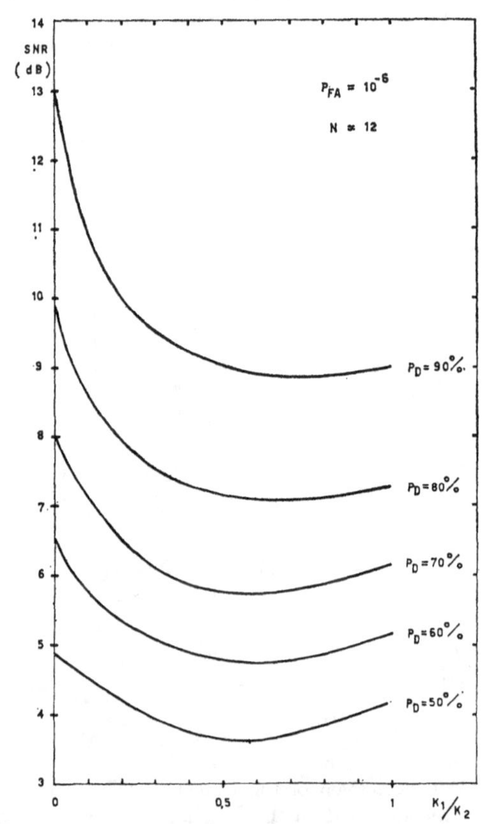

Fig. 5. — Signal-to-noise ratio vs. number of pulses on target, for $P_{FA} = 10^{-6}$ and $N = 12$.

It is possible therefore to sketch, for each value of P_D, a single curve for any value of N and P_{FA}, which shows the variation of the SNR with K_1/K_2.

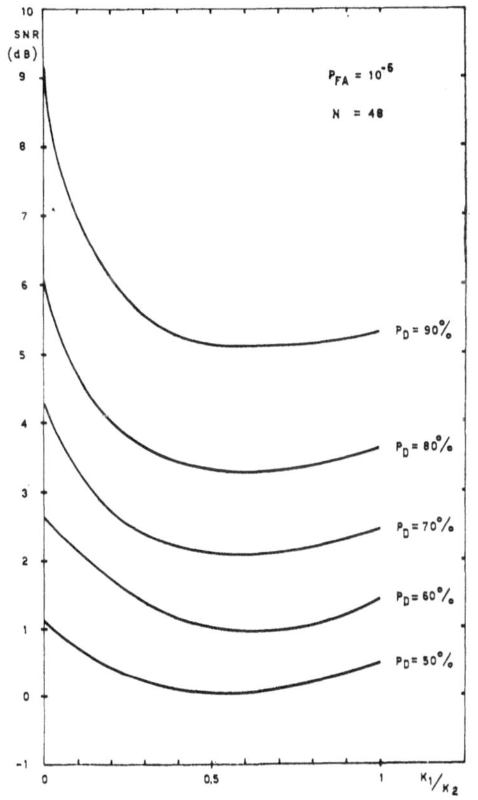

Fig. 6. — Signal-to-noise ratio vs. number of pulses on target, for $P_{FA} = 10^{-6}$ and $N = 48$.

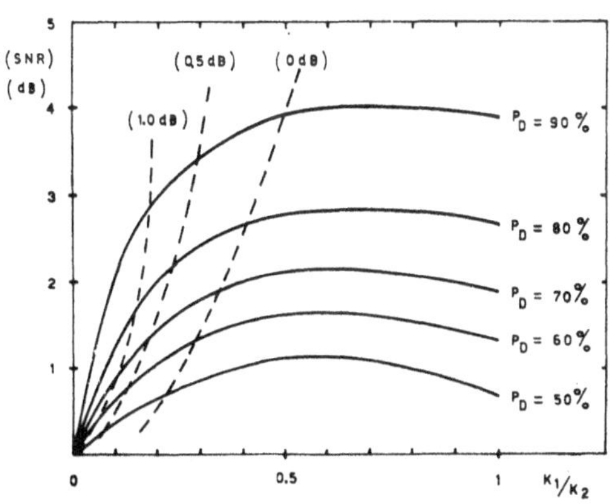

Fig. 7. — Gain on SNR vs. ratio between the number of pulses transmitted on each frequency.

The value of SNR necessary to achieve detection with a single frequency (i.e. $K_1/K_2 = 0$) is set to zero.

In this way, and by averaging the corresponding curves of figs. 1 through 6 (the approximation is never worse than 0.15 dB), the curves of fig. 7 have been sketched. A study of the figure leads to some important conclusions. The transmission of a different number of pulses on the two transmitted frequencies may cause, for low Detection Probabilities, a small improvement with re-

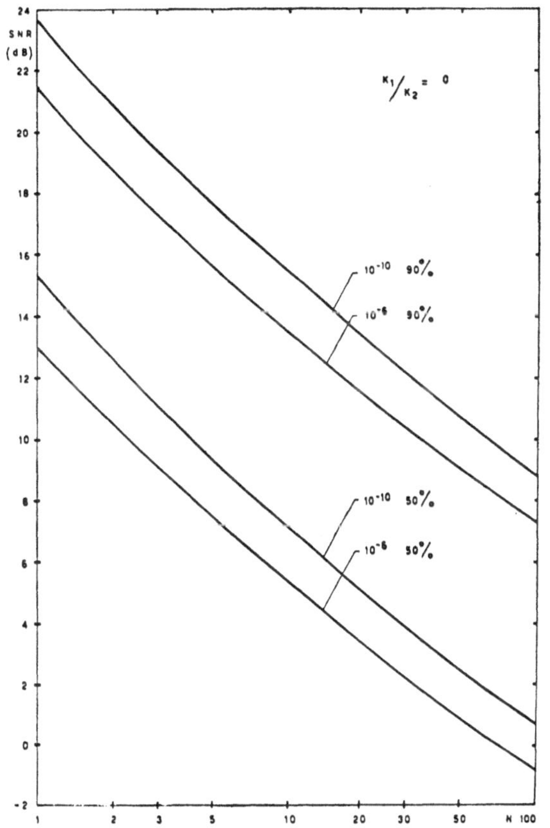

Fig. 8. - Signal-to-noise ratio vs. number of pulses on target for $K_1/K_2 = 0$.

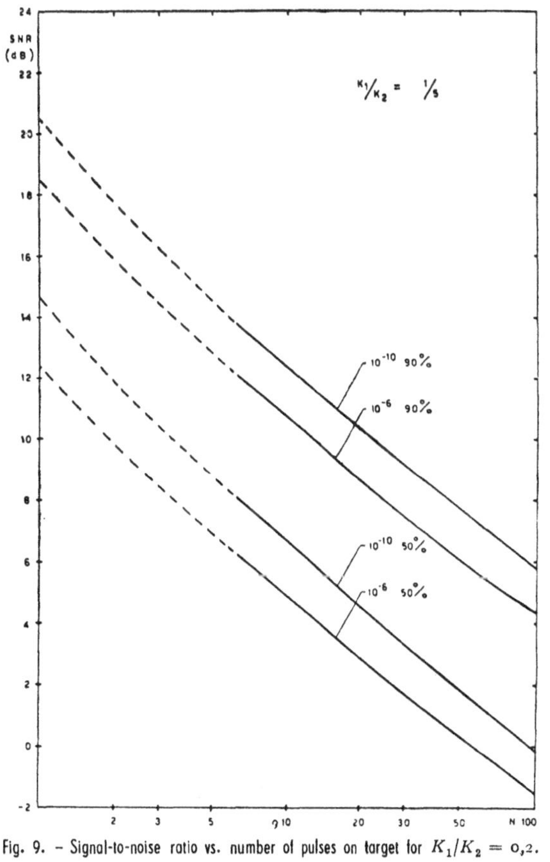

Fig. 9. - Signal-to-noise ratio vs. number of pulses on target for $K_1/K_2 = 0{,}2$.

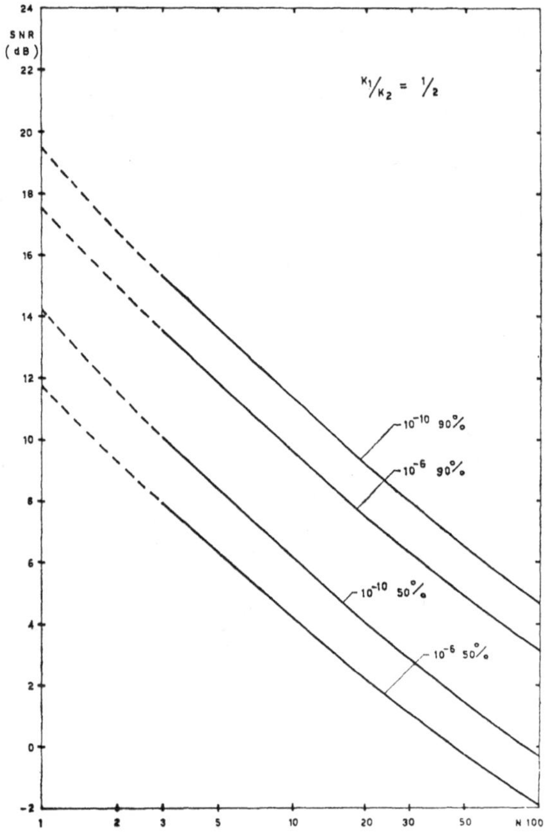

Fig. 10. – Signal-to-noise ratio vs. number of pulses on target for $K_1/K_2 = 0{,}5$.

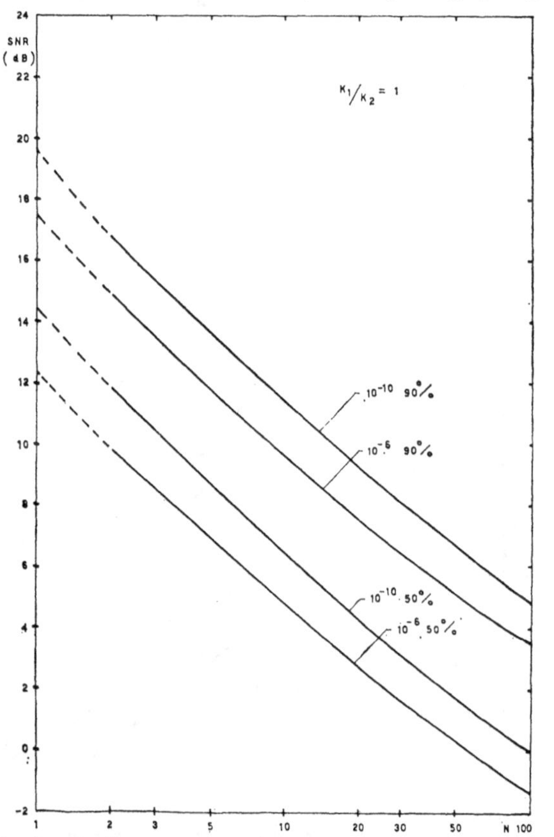

Fig. 11. - Signal-to-noise ratio vs. number of pulses on target for $K_1/K_2 = 1$.

spect to the case $K_1 = K_2$. Also, for high values of Detection Probability K_1/K_2 may differ greatly from 1, with a very small loss. The dashed lines show which values of K_1/K_2 give, in connection with several values of P_D, a loss of 0, 0.5 and 1.0 dB, compared with the classical diversity radar employing an equal number of pulses on each frequency.

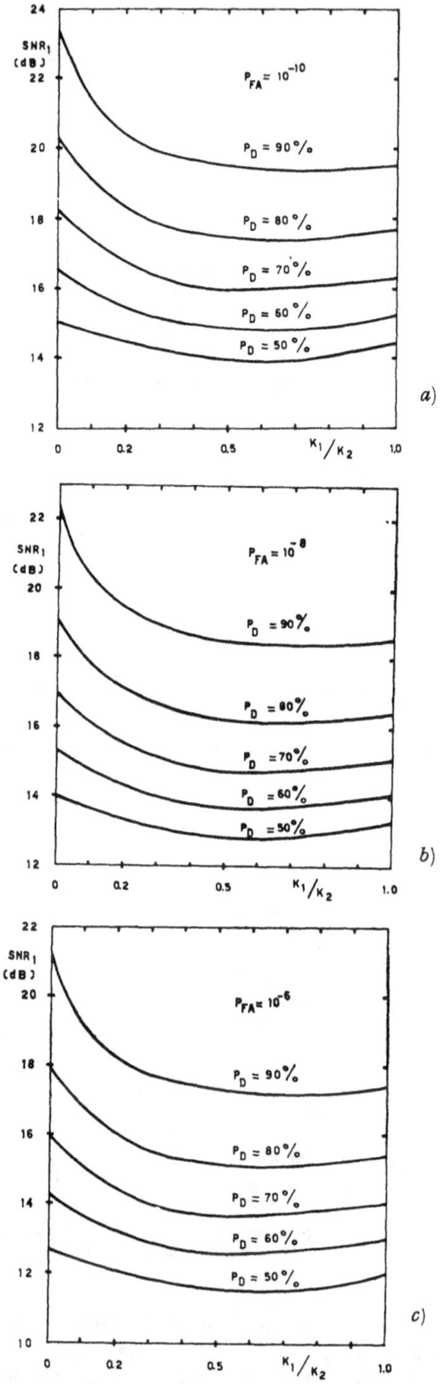

Fig. 12. — Signal-to-noise ratio necessary to achieve detection with one pulse on target, for several values of Probability of False Alarm.

The SNR for any value of the number of pulses on target has been then evaluated. Figs. 8 to 11 are examples of the SNR vs. N for typical detection situations, respectively with $K_1/K_2 = 0$, 0.2, 0.5, 1.0. The lines are dashed below the minimum number of pulses necessary to achieve physical realizability (i.e. K_1 and K_2 are both integers).

The case $K_1/K_2 = 0$ of fig. 8 corresponds to the use of a single frequency and therefore the curves are those well known for the detection, with no diversity, of a scan-to-scan fluctuating target. The case $K_1/K_2 = 1$ of fig. 11 corresponds instead to detection by a classical diversity radar, with an equal number of pulses on each frequency. The parallelism of the curves of the SNR vs. the number of pulses on target permits one to obtain simply the SNR necessary for any condition of detection. In fact the designer must only detract from the value of SNR necessary for detection of a scan-to-scan fluctuating target, the value of the improvement due to the use of diversity, which can be obtained very easily from fig. 7. This value will be dependent upon Probability of Detection and K_1/K_2, and will be independent of the Probability of False Alarm.

RESULTS AND CONCLUSION.

As a further result figures are sketched which permit a quick evaluation of the SNR necessary to achieve the desired Probabilities of Detection for any number of pulses on target.

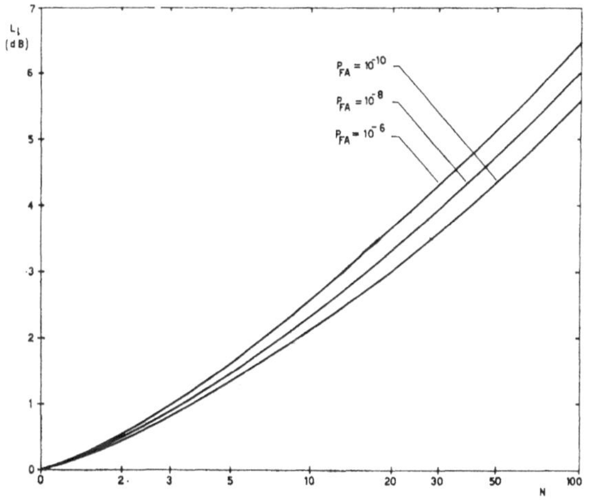

Fig. 13. — Integration loss vs. number of pulses on target.

The figures are relative to the case of two frequencies employed, but, with more complexity, analogous figures may be obtained for a larger number of frequencies.

From one of figs. 12 the signal-to-noise ratio (SNR_1), relative to the desired probabilities of detection and false alarm, corresponding to the chosen ratio K_1/K_2 may be determined. SNR_1 is the signal-to-noise ratio necessary to achieve detection in the hypothesis, not physically significant, of a single pulse on target.

To take into account also the number of pulses on target it is necessary to subtract from the value of SNR_1 found the improvement due to integration. This may be readily accomplished with the help of fig. 13 which shows the integration loss (L_i) vs. N. L_i, obtained from the curves of figs. 8 to 11, depends on the value of Probability of False Alarm, and is fairly independent of the Probability of Detection.

For further clarification an example is made. Let the values of the parameters of interest be:

$$P_D = 80\% \qquad P_{FA} = 10^{-10}$$

$$K_1/K_2 = 1/4 \qquad N = 20$$

From fig. 12*a*, in correspondence to the given value of P_D, P_{FA} and K_1/K_2, it results $SNR_1 = 18$ dB. Being 20 the pulses on target, the improvement for ideal coherent integration would be 13 dB.

From fig. 13 an integration loss of 3 dB is obtained. Therefore the net integration improvement is 10 dB. In conclusion the SNR required in the given detection conditions is 8 dB.

As a conclusion it may be said that eq. (22), in conjunction with eq. (13), gives to the designer the possibility to solve the most general detection problem for a multi-frequency radar. The case of only two frequencies used may in particular be easily solved with the help of figs. 12 and 13. In the more general case when more frequencies are used, the obtained expressions are rather complex and the use of a computer is needed.

The paper was first received on 23th *October* 1967.

BIBLIOGRAPHY

[1] G. GRASSO and P. F. GUARGUAGLINI: *Valutazioni delle prestazioni di un sistema radar a diversità di frequenza.* - « Alta Frequenza », September 1967.

[2] P. SWERLING: *More on Detection of Fluctuating Targets.* - « IEEE Trans. on I.T. », pp. 459÷460, July 1965.

[3] P. SWERLING: *Probability of Detection for Fluctuating Targets.* - « IRE Trans. on Information Theory », pp. 269÷307, April 1960.

[4] G. CAMPBELL and R. FOSTER: *Fourier Integrals for Practical Applications.* - Van Nostrand, Princeton 1961.

[5] K. PEARSON: *Tables of the Incomplete Γ-Function.* - University Press, Cambridge, 1965.

A Sequential Detection Logic for Improving Signal Detectability in Frequency-Agile Search Radars

V.G. Hansen

IEEE Transactions on Aerospace and Electronic Systems, Vol. AES-4, No. 5 (September 1968)
Pages 763-773

Abstract

Certain fading models have in the past become standard for taking into account the effects of fluctuations in target echoing area in the theoretical prediction of radar range performance. In the application of these models, fading has always been treated as an entirely random phenomenon.

In this paper it is noted that this is not necessarily true for radars which have the transmission frequency as a controlled variable, since in that case the fading characteristics of the target may be modified by properly controlling the transmission frequency.

A strategy for taking advantage of this possibility is developed and theoretical results obtained by Monte Carlo simulation for a specific logic indicate that gains in detectability of several decibels (relative to what is usually considered optimum) may be obtained when certain realistic assumptions are made for the target fading characteristics.

I. Introduction

Theoretical studies of the detectability of radar echoes in Gaussian noise have in the past been performed by many authors. Several of these have been concerned with the practical case of search radars of constant rate of scan and usually a rectangular antenna beam pattern is assumed [1]–[3]. Three different fading models have become standard for taking into account the effects of fluctuations in target echoing area resulting from specific target and radar characteristics [2], [3]. They are:

a) No fading; the echoing area σ is a constant and is determined by the target characteristics only.
b) Slow fading; the echoing area is constant during the time on target but is a random quantity from scan to scan. The echoing area is characterized by its mean value $\bar{\sigma}$ and its pdf (probability density function) $p(\sigma | \sigma)$.
c) Fast fading; the echoing area is a random quantity from sweep to sweep. It is again characterized by its mean value $\bar{\sigma}$ and pdf $p(\sigma | \bar{\sigma})$.

The physical situations which could lead to these types of fading are, respectively:

a) A "point" target (corner reflector, missile) observed with a fixed- or agile-frequency[1] radar.
b) A "complex" target (aircraft, ship) undergoing slow changes of aspect, observed with a fixed-frequency radar.
c) A "complex" target observed with a frequency-agile radar.

The pdf governing the fading of the echoing area is usually assumed to be exponential:

$$p(\sigma | \bar{\sigma}) = \frac{1}{\bar{\sigma}} \exp(-\sigma/\bar{\sigma}) \tag{1}$$

so that the Rayleigh pdf applies for the amplitude of the received echo. Swerling [2] also considers the pdf:

$$p(\sigma | \bar{\sigma}) = \frac{4\sigma}{\bar{\sigma}^2} \exp(-2\sigma/\bar{\sigma}) \tag{2}$$

which is intermediate between (1) and the no fading case. The pdf of (2) will not be considered further in this paper. Some recently published measurements support the assumption of an exponential pdf for the radar echoing area of aircrafts [4].

The theoretical signal detection curves for each of the above three fading models and post-detection integration of $M=4$, 8, and 16 hits are shown in Figs. 1, 2, and 3, respectively. These results were obtained through the use of Monte Carlo simulation techniques, as described in more detail in Section IV, and assuming:

[1] A frequency-agile radar is defined as having a transmitter of which the frequency is changed from sweep to sweep. The system properties of such radars are described in [9] and [5].

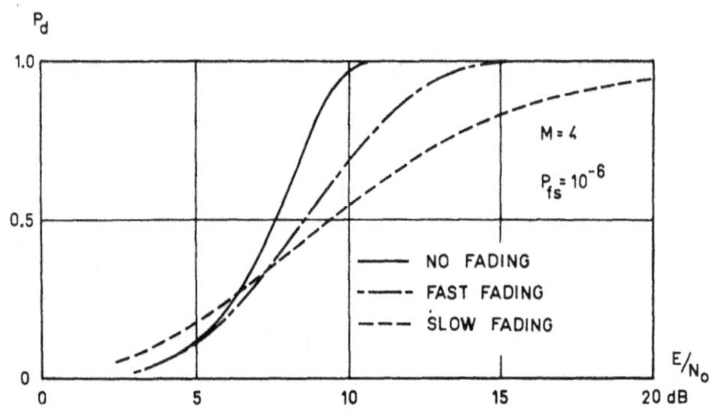

Fig. 1. Detection curves for analog post-detection integration and a Gaussian antenna beam for each of the three fading models and $M=4$.

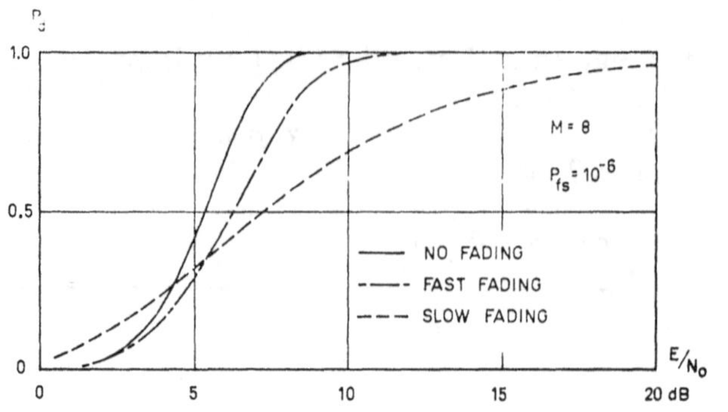

Fig. 2. Detection curves for analog post-detection integration and a Gaussian antenna beam for each of the three fading models and $M=8$.

Fig. 3. Detection curves for analog post-detection integration and a Gaussian antenna beam for each of the three fading models and $M=16$.

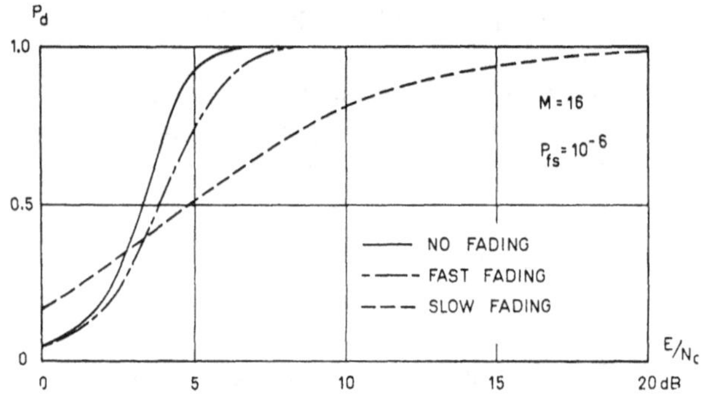

a) A Gaussian antenna beam. The number of hits is defined over the one-way 3-dB beamwidth of the antenna.
b) A matched filter for the individual signal pulses.
c) Linear envelope detection.
d) Unweighted analog post-detection integration over the one-way 3-dB width of the antenna pattern by a moving window analog integrator.[2] The threshold used is such that the false alarm probability P_{fs} applies for each range element and for each single sweep. This definition is further explained in Section IV.

These results show the effect of radar detection performance of the different types of fading and specifically, at high detection probability, the large difference ($\simeq$4 dB at $P_d=80$ percent) between the fast and slow fading cases. For some experimental results see [9].

Based on the above fading models, this paper shows that a still greater improvement may be realized with a frequency-agile transmitter if one takes advantage of the fact that the sweep-to-sweep fading of the target echoing area may be stopped at any instant by going from frequency-agile to fixed-frequency operation. A strategy for controlling the mode of radar transmission on a sweep-to-sweep basis in order to achieve such an improvement, implemented in the form of certain logical operations, is described in Section II.

II. The Sequential Logic

With a frequency-agile search radar and a complex target (i.e., fast fading) the amplitudes of the individual pulses of the received echo (without noise) might typically appear as shown by the rectangles in Fig. 4.[3] The two-way Gaussian voltage pattern of the antenna is superimposed on Fig. 4 and the number of hits per target is eight. The received pulse amplitudes correspond to randomly selected transmission frequencies and are therefore assumed to be independent Rayleigh variables modulated by the two-way voltage pattern of the antenna. It is seen that large amplitudes occur for certain frequencies (as for example sweeps 17 and 21 in Fig. 4) and it is obvious that if one of these frequencies in some way could have been determined in advance and used by the transmitter for this specific target, a significant improvement in radar performance would result. This of course is impossible; not only is the type of aircraft in the air unknown, but also the variations of echoing area with aspect angle are extreme.

The best one could hope for is, therefore, to try and find one of these "advantageous frequencies" through an actual measurement of the amplitudes of the radar echo during the time on target. During the reception of the echo shown in Fig. 4 one might for example have decided after the reception of the large amplitude in sweep 17 that the corresponding frequency should be used during the remainder of the time on target. This would have resulted in the sequence of amplitudes shown by the circles in Fig. 4.

The situation is, however, further complicated because:

a) Receiver noise is added to the radar echo, making the exact measurement of signal amplitude impossible.

[2] This detector is completely analogous to the moving window digital detector except that it works directly with the unquantized analog amplitudes.

[3] This and following video amplitude sequences were obtained from the computer simulation to be described in Section IV.

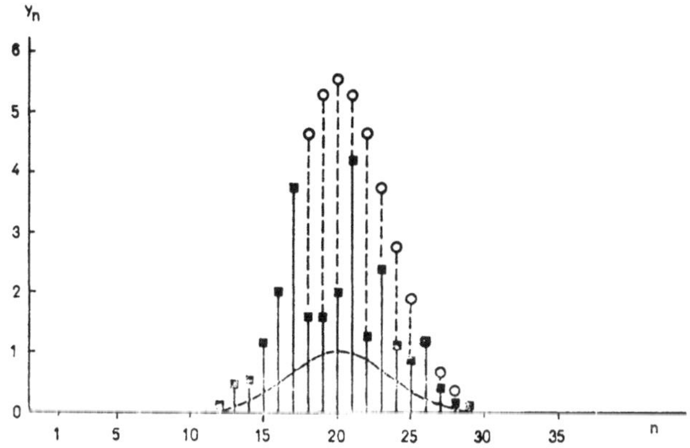

Fig. 4. Amplitudes of the individual pulses in the received echo without noise. Rectangles are for fast fading; circles are for fixed-frequency operation from sweep 17.

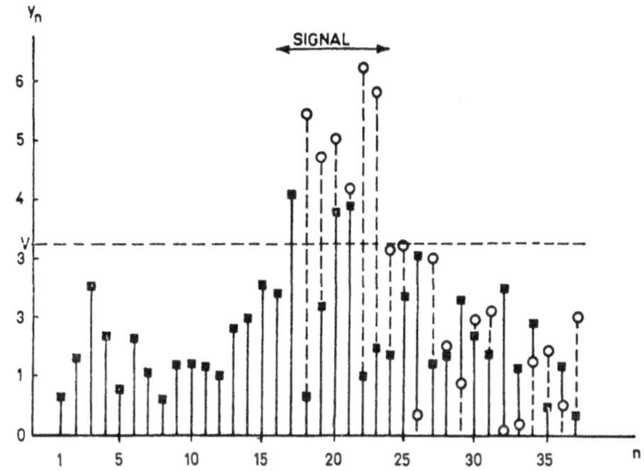

Fig. 5. Amplitudes of individual signal-plus-noise pulses in single range element at receiver output. Rectangles are for fast fading; circles are for the sequential logic with indicated threshold V and $L=2$.

Fig. 6. Flow chart of sequential logic for $L=2$.

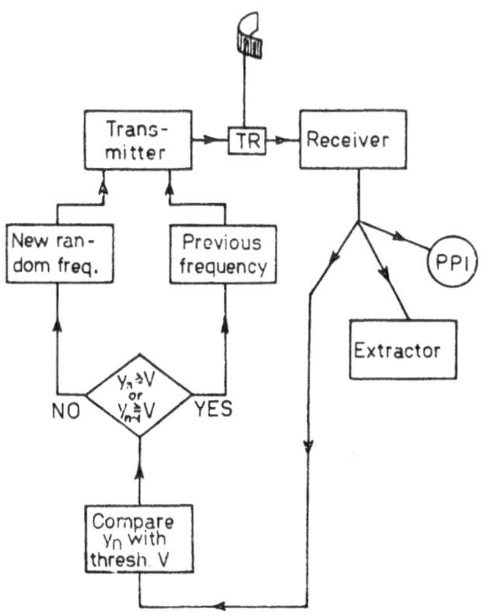

b) The azimuth of the target is unknown so that one would not know for what period of time fixed-frequency operation should be maintained when an advantageous frequency has been found.

c) A large noise amplitude may be mistaken for a signal pulse so that fixed-frequency operation would be used. In such cases it is important that the mistake be quickly discovered due to the degraded performance in the fixed-frequency mode of operation (see Figs. 1, 2, and 3, slow fading).

A logic will now be described which prior to the transmission of each new radar pulse will decide whether the same frequency as used on the previous transmission, or a new random frequency based on the principle outlined above should be used, taking into account the complications mentioned. The logic is described as it would apply to a single range element, a later section will discuss the effect of considering more than one range element simultaneously.

The amplitudes of the signal-plus-noise pulses at the receiver output in the range element considered might appear as shown by the rectangles in Fig. 5 for the case of fast fading. The amplitude received in the nth sweep is denoted y_n.

In the proposed logic each amplitude y_n is compared with a threshold V. Whenever the threshold is exceeded the transmitter is ordered to change from frequency-agile to fixed-frequency operation and the frequency to be used is the same as was used in the sweep where the threshold was exceeded. The return from fixed-frequency to frequency-agile operation is requested by the logic whenever the threshold is not exceeded in any of the L preceding sweeps. It is seen that the value chosen for L determines how rapidly frequency-agile operation will be resumed if the logic has been triggered by a noise pulse. A flow chart of this logic is shown in Fig. 6 for $L=2$.

An example of the effect of the logic for $L=2$ is shown in Fig. 5. With the logic operative the threshold is exceeded in sweep 17 and fixed-frequency operation results. The new sequence of amplitudes from sweep 18 and onwards is shown as circles in the figure. Fixed-frequency operation continues up to and including sweep 25. In sweep 26 the threshold has not been exceeded in any of the two preceding sweeps and the transmitter therefore returns to the frequency-agile mode.

With this logic there are two parameters over which performance must be optimized; the threshold level V and the value of L. Instead of the threshold level V, the corresponding probability P that noise alone will exceed this threshold is used as the parameter in the following since it is a more directly meaningful quantity.

After a few trial runs with the Monte Carlo program used for the quantitative analysis it was established that L is not a very critical parameter and the value $L=2$ was used during the optimization of P. The importance of L was later investigated in greater detail for a particular number of hits per target. The value of P was subjected to rather wide variations in the parametric study. It is obvious that in the limits $P=0$ ($V\to\infty$) and $P=100$ percent ($V=0$) the system degenerates to frequency-agile and fixed-frequency operation, respectively.

III. A Theoretical Result

General analytical results are difficult to obtain for a detection system employing the logic described in Section II. However, in the case where only 2 hits are received, and with some further simplifying assumptions, it is possible to calculate the average improvement in signal-to-noise ratio due to the logic, and to determine the threshold setting which maximizes this improvement. A similar procedure should also be applicable for more than 2 hits but the complexity of the calculations increases rapidly.

Here, it is assumed that the antenna beam shape is rectangular so that the signal-to-noise ratio for both hits is equal, and the value of L is taken as one. With $L=1$ the noise in the sweeps prior to the occurrence of the signal can have no effect on the logic and only the two signal-plus-noise hits need be considered in the analysis.

The pdf of the signal-plus-noise amplitude y when the amplitude A of the received signal is given is denoted $p(y|A)$ and the prior pdf for A is $p(A)$. From this the marginal pdf for y is

$$p(y) = \int_0^\infty p(y|A)p(A)dA. \tag{3}$$

The average signal-to-noise ratio S_1 for the first hit can be written as a constant k times the expected value of $\frac{1}{2}A^2$, i.e.,

$$S_1 = kE[\tfrac{1}{2}A^2] = \tfrac{1}{2}k\int_0^\infty A^2p(A)dA. \tag{4}$$

If on the second hit a new random frequency is used the expected signal-to-noise ratio will be the same and the average over the two hits is

$$S = \tfrac{1}{2}(S_1+S_2) = S_1. \tag{5}$$

After reception of a particular y_1 the posterior pdf for the signal amplitude A_1 is by Bayes' rule

$$p(A_1|y_1) = \frac{p(y_1|A_1)p(A_1)}{p(y_1)}\cdot \tag{6}$$

If the same frequency is used now for the second hit this posterior pdf for A_1 becomes the prior pdf for the amplitude A_2 for the second hit and the average signal-to-noise ratio for the second hit becomes

$$S_2' = k\,E[\tfrac{1}{2}A_2^2] = \tfrac{1}{2}k\int_0^\infty A_2^2p(A_2|y_1)dA_2. \tag{7}$$

The average signal-to-noise ratio of the two hits is in this case

$$S' = \tfrac{1}{2}(S_1+S_2'). \tag{8}$$

In order to obtain an improvement by using the same transmission frequency for the second hit we must have

$$S_2' \geq S_2(=S_1) \tag{9}$$

or equivalently,

$$\int_0^\infty \tfrac{1}{2}A_2^2p(A_2|y_1)dA_2 \geq \int_0^\infty \tfrac{1}{2}A_1^2p(A_1)dA_1. \tag{10}$$

If the left-hand side of this inequality is a nondecreasing function of y_1 there will exist a value V such that when y_1 exceeds V the inequality is satisfied. This value of V is therefore the threshold which should be used in the logic described in Section II. Hence, when y_1 is less than V a new random frequency is used, when y_1 exceeds V the same frequency is used on the following transmission. With this logic the average signal-to-noise ratio S_2'' obtained for the second hit is determined by averaging over all values of y_1,

$$S_2'' = S_1\int_0^V p(y_1)dy_1 + \tfrac{1}{2}k\int_V^\infty\left\{\int_0^\infty A_2^2p(A_2|y_1)dA_2\right\}p(y_1)dy_1. \tag{11}$$

The average signal-to-noise ratio for both hits is then

$$S'' = \tfrac{1}{2}(S_1+S_2'') \tag{12}$$

and the improvement due to the logic is given by $\Delta S=S''/S$.

The above calculations were carried out in Appendix I for the case where $p(A)$ follows the Rayleigh pdf. The optimum threshold value for V is found to be

$$V = \sigma_N\sqrt{2(1+S)} \tag{13}$$

where σ_N is the rms noise level and S is the signal-to-noise ratio, which for a matched filter is

$$S = \frac{E}{N_0}\cdot \tag{14}$$

Here E is the average energy of the received signal pulse and N_0 is the white-noise spectral density (one-sided). The probability P that noise alone exceeds the threshold V is, accordingly,

$$P = \int_V^\infty \frac{y}{\sigma_N^2}\exp(-y^2/2\sigma_N^2)dy = \exp\left(-\left(1+\frac{E}{N_0}\right)\right). \tag{15}$$

This relation is shown in Fig. 7. The crosses shown in this figure are related to the computer simulation as described in Section IV.

The average improvement in signal-to-noise ratio, when the threshold given by (13) is used, is

$$\Delta S = \frac{S''}{S} = 1 + \frac{S}{2e(1+S)} \tag{16}$$

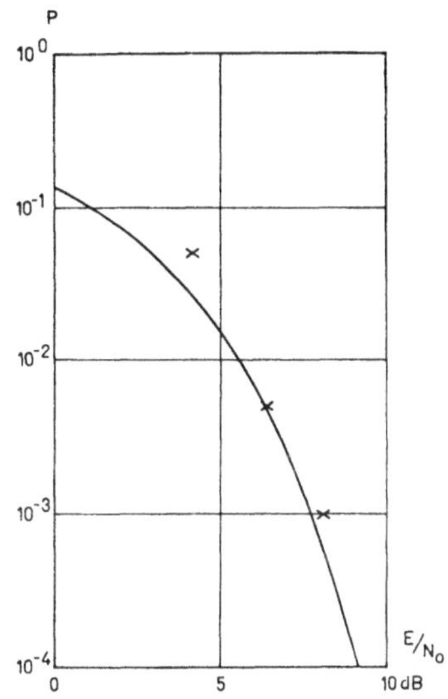

Fig. 7. Optimum value of *P* for two hits as a function of signal-to-noise ratio.

where $e=2.71828$. The theoretical improvement factor for a signal-to-noise ratio of $S=E/N_0=6$ dB amounts to $\Delta S=0.6$ dB.

IV. Monte Carlo Simulation

In order to evaluate the logic described in Section II more thoroughly, a Monte Carlo simulation was carried out on a digital computer.

In such an evaluation it is not possible, in a straightforward simulation, to perform enough repetitions of the basic Monte Carlo experiment to estimate a false alarm probablity of the order of 10^{-6}. The simplest method is therefore to determine the required threshold settings analytically and only use the simulation to determine the probability of detection and accuracy of extraction when a signal is added to the noise. Normally the analytical evaluation of a detection scheme is very much simpler for noise alone than when both signal and noise are present.

In the present investigation the logic has no effect on internal receiver noise but only modifies the signal statistics, and the detectors considered are conventional analog or digital integrators.

In order to be able to determine the angular accuracy (and also to obtain more realistic results), the simulation of the detection process had to be dynamic in azimuth, using moving window analog or digital integrators.

A detection (or false alarm if no signal is present) is in this case characterized by

1) A "target begin" when the integrated output crosses a threshold N_1 in positive direction.
2) A "target present" region as long as the integrated output stays above or equal to another threshold $N_2 \leq N_1$.
3) A "target end" when the integrated output falls below the threshold N_2.

A detection (or false alarm) in a certain sweep can, therefore, be associated with the event that the integrator output crosses the threshold N_1 in positive direction for the first time since a target end was last observed.[4] The false alarm probability P_{fs} is defined as the probability of this event taking place on any one sweep and in a single range element. With this definition the number of false alarms per scan (FAS) is

$$FAS = P_{fs} \times \text{(sweeps per scan)} \times \text{(range elements per sweep)}.$$

An expression for P_{fs} for the case of a moving window digital integrator has been derived by Hampton and McCurrach [7], and for $N_1=N_2=N$ we have

$$P_{fs} = \binom{M-1}{N-1} p^N (1-p)^{M-N+1} \tag{17}$$

where $p=\exp(-T^2/2)$ is the probability that a sample of the envelope of normalized narrow-band Gaussian noise is quantized into a 1 after the first threshold T. The thresholds which were used in the simulations were determined by (17) and are listed in Appendix II for $P_{fs}=10^{-6}$ and different values of M and N.

A similar expression was derived for the moving window analog integrator and the resulting threshold settings are listed in Appendix II for $P_{fs}=10^{-6}$.

The Monte Carlo simulation was performed in the following way: In each repetition of the basic Monte Carlo experiment a sequence of more than 4 M video amplitudes is generated where M is the number of pulses transmitted over the one-way 3-dB beamwidth of the antenna. Such a sequence of video amplitudes is shown in Fig. 5. Each video amplitude consists of a noise component plus the signal return. The signal return is maximum at the center of the sequence and is modulated by the two-way gain pattern of the antenna beam (Fig. 4) for which a Gaussian shape was assumed. The relative position of the maximum of this pattern and the signal returns is randomized over one PRF period at the start of each sequence. The logic is in operation during the generation of the video sequence.

The amplitude A_s of the received signal return is given, in the case where a new random frequency is used (fast fading), as the product of the antenna gain in the direction of the target G_θ and a reference amplitude A_s' determined by the average signal power S and the fading law. With a Rayleigh fading signal amplitude as previously assumed A_s' may be determined as

$$A_s' = \sqrt{A_1^2 + A_2^2} \tag{18}$$

[4] Or equally with the event that the integrator output falls below the threshold N_2 for the first time since a target begin was last observed.

where A_1 and A_2 are independent random variables drawn from a Gaussian distribution with zero mean and variance S, and A_s is given by

$$A_s = G_\theta A_s'. \tag{19}$$

If the frequency which was used on the previous sweep is used again, the signal amplitude is determined by (19) but using the value of A_s' from the previous sweep together with the value of G_θ for the present sweep. In either case receiver noise is added to the signal amplitude A_s. The noise is assumed Gaussian and is normalized to unity power level so that the signal-plus-noise amplitude may be determined as

$$A_{s+n} = \sqrt{(N_1 + A_s)^2 + N_2^2} \tag{20}$$

where N_1 and N_2 are independent random variables drawn from a Gaussian distribution with zero mean and unit variance.

Further, if the antenna gain pattern is normalized to unity at the beam center the signal-to-noise ratio in the receiver is equal to the signal power S. As we shall assume that a matched filter is used for the individual pulses we have

$$\frac{E}{N_0} = S \tag{21}$$

where E is the energy of the received signal (averaged over the fading distribution and not taking account of the logic) and N_0 is the (one sided) spectral density of the white noise.

When a sequence of video amplitudes has been generated it is examined by up to 6 different detectors, and detections and corresponding azimuths are stored. All the detectors considered perform a moving window integration procedure of either the analog or the quantized video, and the target-beginning–target-end beam splitting procedure was used.

For each signal-to-noise ratio $K = 300$ repetitions were made, whereafter detection probability, and bias and standard deviation of the azimuth estimate were determined.

Some estimates of the accuracy of the simulation are made in Appendix III.

V. Results

The signal detectability results of the simulation for different parameters of the sequential logic and for different numbers of hits per target, are presented in this section. Results relating to angular accuracy are found in Section VI.

Only the results for the moving window digital integrator are given since this is the case of most practical interest. Very similar results were obtained, however, for the case of the moving window analog integrator.

In the moving window digital integrator the video amplitudes are first compared with a first threshold T and quantized into two levels 0 and 1. A running sum of these digitized amplitudes is then made over the last M hits and whenever this sum exceeds a second threshold N a target detection occurs. This instant of time defines the azimuth corresponding to the target beginning. The target end occurs at the azimuth where the running sum for the first time after a target beginning drops below the same threshold (or sometimes even a somewhat lower one to prevent target splitting). The mean of target beginning and target end is taken as the estimate of target azimuth except for a constant correction due to the delay introduced by the integration process.

Results obtained for $M = 4$, 8, and 16 hits per target and a false alarm probability of $P_{fs} = 10^{-6}$ are shown in Figs. 8, 9, and 10. Each figure shows for reference the results for fast fading and slow fading, respectively.[5] The third curve on each figure is with the logic of $L = 2$ and the value of P is optimized to give the best result for detection probabilities in the range $P_d = 40$–90 percent, i.e., for threshold signals. For each curve the second threshold N was chosen to maximize detection probability. The optimum value of P is not very critical as may be seen from Fig. 11 where curves with different values of P are shown for $M = 8$ and $L = 2$.

From these results the gain in detectability offered by the logic is apparent and it is seen that the advantage increases for larger values of M. Only for large values of signal-to-noise ratio does the logic introduce a loss as compared with the fast fading case. This is not evident in the figures since this loss only occurs for detection probabilities larger than around $P_d = 99$ percent.

The effect of changing the parameter L of the logic for a given value of P is shown in Fig. 12. It is seen that for the value of $L = 4$, detection probability is improved for low signal-to-noise ratio at the expense of a degradation at high signal-to-noise ratio. For the low value of L the situation is the opposite. Some further curves were obtained in which P was varied in addition to L. These curves are not shown here but they indicated that a change in the parameter L may be partly countered by readjusting the value of P in such a way that P is decreased for larger L and increased for smaller L. In general the conclusion is that the parameter L is not very critical but that best results are obtained for small values of L. This is also desirable in the case where the logic has to operate on multiple range elements.

For the results in Figs. 8, 9, and 10, that value of E/N_0 for which the largest increase in detection probability is obtained, was determined. These values and the corresponding probabilities P used in the logic have been plotted as crosses in Fig. 7 together with the theoretical curve for two hits. This comparison shows that although the curve was derived for two hits under cetain simplifying assumptions, its applicability appears more general and it would seem to provide a useful guide for selecting the value of P optimizing detection performance for a given signal-to-noise ratio.

[5] A comparison with Figs. 1, 2, and 3 shows the losses introduced by quantizing into two levels before integration.

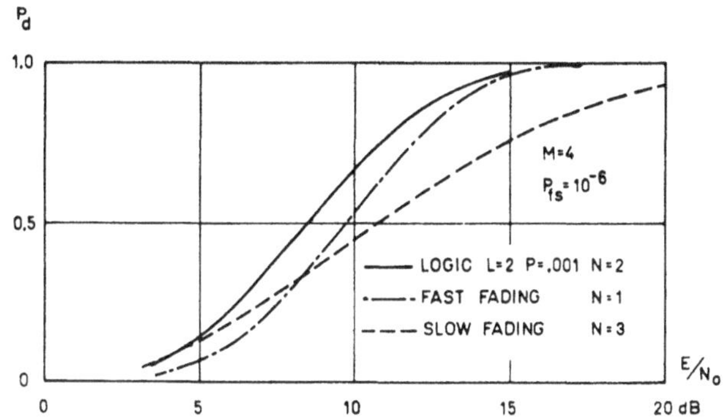

Fig. 8. Improvement in signal detectability due to sequential logic for $M=4$. Results for fast and slow fading are shown for reference.

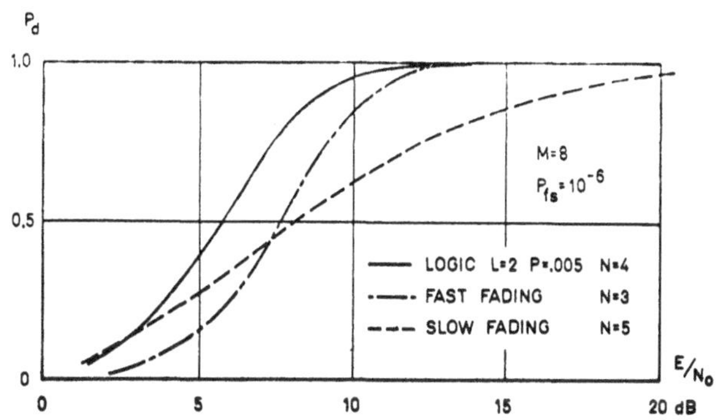

Fig. 9. Improvement in signal detectability due to sequential logic for $M=8$. Results for fast and slow fading are shown for reference.

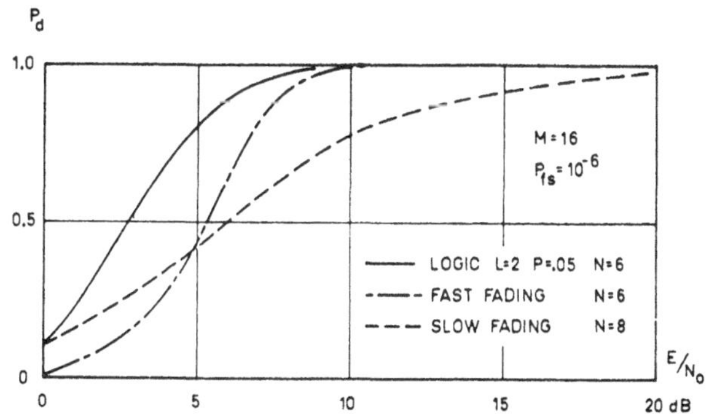

Fig. 10. Improvement in signal detectability due to sequential logic for $M=16$. Results for fast and slow fading are shown for reference.

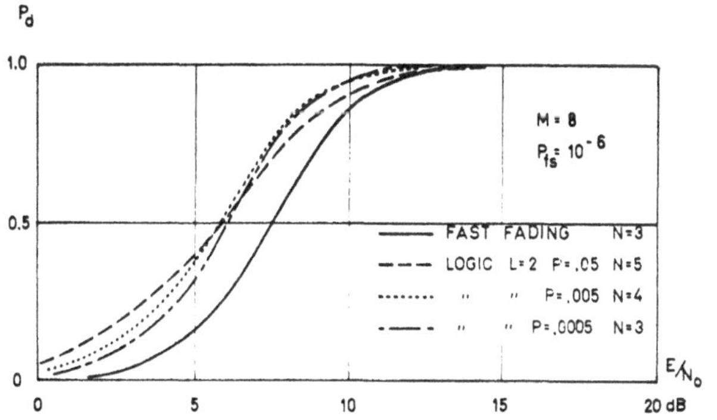

Fig. 11. Performance of sequential logic for different values of P. Result for fast fading is shown for reference.

Fig. 12. Performance of sequential logic for different values of L. The best second threshold was chosen for each case but P was kept constant. The result for fast fading is shown for reference.

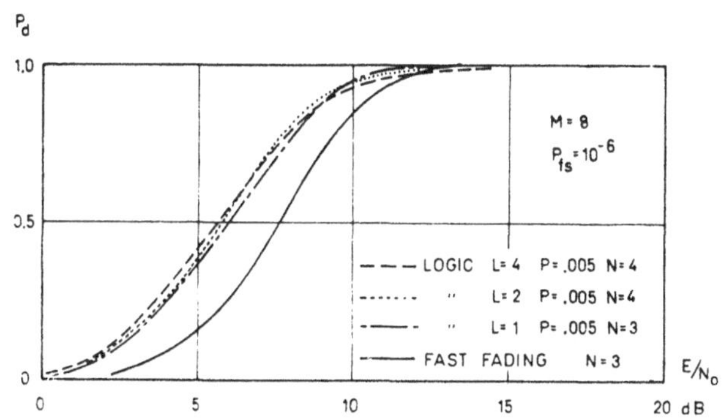

VI. Effects on Angular Accuracy

Since the sequential logic searches for an advantageous frequency during the time on target, it is obvious that whenever such a frequency is found its use will tend to upset the symmetry of the received radar echo (see Fig. 4). This in turn introduces a signal-to-noise ratio dependent bias into the estimate $\hat{\theta}$ of the target azimuth which might result in a degradation in the angular accuracy of the radar. In this section some results on the magnitude of this effect are given.

Using the beam splitting azimuth estimation method, described earlier, in the Monte Carlo simulation of the moving window digital integrator, estimates of the mean $m_{\hat{\theta}}$ and variance $\sigma_{\hat{\theta}}^2$ of the azimuth estimates $\hat{\theta}_i$, $i=1, 2, \cdots, K\,(K=300)$ may be obtained as the sample mean

$$\hat{m}_{\hat{\theta}} = \frac{1}{K} \sum_{i=1}^{K} \hat{\theta}_i$$

and the sample variance

$$\hat{\sigma}_{\hat{\theta}}^2 = \frac{1}{K-1} \sum_{i=1}^{K} (\hat{\theta}_i - \hat{m}_{\hat{\theta}})^2,$$

respectively. These expressions yield unbiased estimates of $m_{\hat{\theta}}$ and $\sigma_{\hat{\theta}}^2$ [6].

The azimuth estimates $\hat{\theta}_i$ resulting from the beam splitting operation as introduced above are assumed already to have been corrected for the known shift of half a beamwidth $\frac{1}{2}\theta_{3\,dB}$, which is introduced by beamsplitting when equal thresholds are used for the target begin and target end decisions. The estimate $\hat{m}_{\hat{\theta}}$ can therefore be interpreted as the bias in the azimuth estimate $\hat{\theta}$.

The results obtained for $\hat{m}_{\hat{\theta}}$ and $\hat{\sigma}_{\hat{\theta}} = \sqrt{\hat{\sigma}_{\hat{\theta}}^2}$ are shown in Fig. 13 as functions of signal-to-noise ratio for $K=300$.

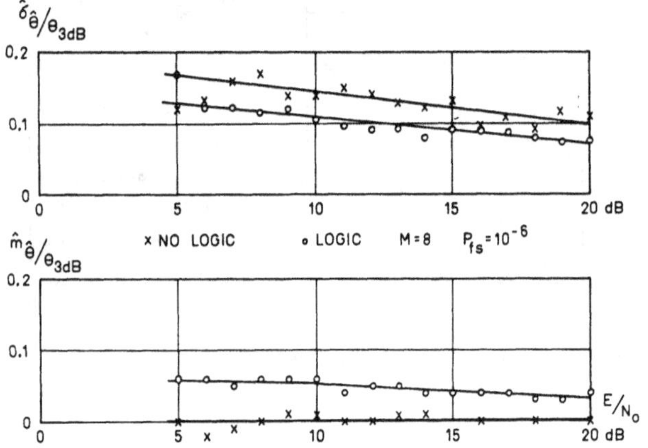

Fig. 13. Bias and standard deviation of angular estimates for fast fading (no logic) and the sequential logic (logic), respectively.

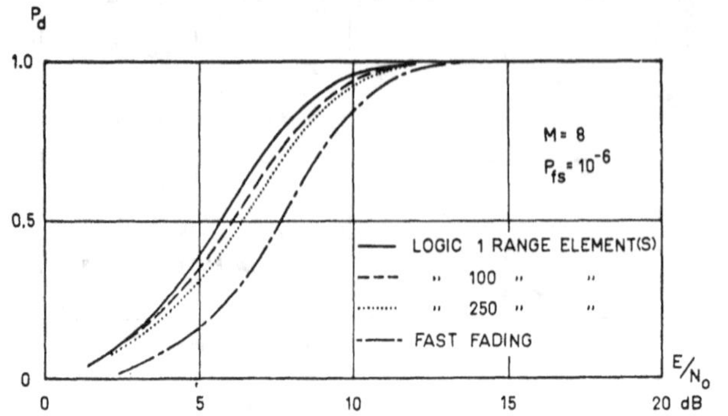

Fig. 14. Performance of sequential logic when a number of range elements are considered simultaneously. The parameters P and L of the logic were optimized in each individual case. The result for fast fading is shown for reference.

In each graph two curves are shown. The crosses apply to the fast fading case with a second threshold of $N=3$. It is seen that the bias is zero as expected. The circles represent the results obtained with the logic for $P=0.005$, $L=2$, and $N=4$. The detectability curve for this case was shown in Fig. 9.

It is seen that the logic introduces a bias of 0.05 to 0.1 of a beamwidth into the azimuth estimate at low signal-to-noise ratio where the advantage of the logic is obtained. Toward larger signal-to-noise ratio the bias decreases.

At the same time, however, the use of the logic results in a decrease in the standard deviation below that obtained for fast fading and this effect tends to compensate for the bias introduced.

It is, therefore, concluded that the degradation in angular accuracy due to the logic is negligible. It would even seem possible to estimate the magnitude of the bias by monitoring the logic and in this way improve the angular accuracy over that obtained in the fast fading case.

VII. Multiple Range Elements

All results and discussions up to this point have been concerned with the rather unrealistic case where the logic operates on only a single range element. In this section we shall consider the influence on the performance when a larger number of range elements is considered simultaneously.

The principle of the logic is unchanged in this case, only a larger number of range elements must now be considered simultaneously. This increases the probability that the logic is triggered by a noise amplitude to

$$P' = 1 - (1 - P)^J \cong JP \qquad \text{for } P \ll 1 \qquad (22)$$

where J is the number of range elements. For a fixed value of P, the increased number of range elements therefore increases the probability that the radar switches to the fixed-frequency mode of operation due to noise alone; some degradation in performance must be expected.

In order to determine this degradation, the probability that the logic is triggered by a noise amplitude in any of the J range bins is introduced into the simulation by generating a random number from a rectangular distribution (0, 1) and comparing it with P', so that if either y_n exceeds V or the random number is smaller than P' the same frequency is used on the following transmission.

Some results obtained in this way for $M=8$ are shown in Fig. 14. The number of range elements J is used as parameter and for each J the value of P was chosen as that giving the best result. It is seen that for $M=8$ more than 250 range elements can be considered simultaneously before a serious degradation occurs.

VIII. Discussion of the Fading Model

The results obtained in the previous sections depend on the applicability of the fading models discussed briefly in the Introduction. Whereas the model used in the slow fading case has been verified experimentally [4] and thus may be considered reasonably accurate, no such data are available for frequency-agile radars. The two most important considerations in this case are:

a) The frequency spacing required to produce uncorrelated target echoing areas.
b) The pdf of the target echoing area.

If the frequency spacing required for decorrelation is not small compared with the band within which the radar is frequency agile, the model becomes unrealistic and correlation between pulse-to-pulse amplitudes must be taken into account. In a recent paper by Ray [5] it is estimated that the frequency spacing required for decorrelation (to $\rho=0.5$) is 45 MHz divided by the radial target dimension in meters. This estimate indicates that for radars which are frequency agile over several hundred megahertz and for reasonable size targets (aircrafts) the decorrelation problem might not be too serious.

The exponential pdf used for the echoing area is inherently based upon the assumption that the target is made up of a large number of equally important scatterers. If this model is not valid the resulting pdf may differ from the exponential.

The main justification for the fading model which has been assumed for complex targets, as viewed by a frequency-agile radar, lies in the extrapolation of the mathematical model which has been demonstrated to account reasonably well for the fading encountered with fixed-frequency radars.

IX. Alternative Logics

The type of logic which has been described and evaluated above is primarily aimed at conventional scanning search radars. If one broadens the scope to consider also radars employing electronic scanning in which the antenna beam is programmed to scan in discrete steps, alternative logics for controlling the frequency agility might be considered. Examples are:

a) A certain threshold V is set for the individual hits. Random frequency jumping is used until a hit exceeds V. For the remainder of the time on target the corresponding frequency is used.
b) The threshold V is made a (decreasing) function of the number of the observation. Otherwise the logic is the same.
c) Combinations with the Wald sequential observer principle are made.

Some preliminary results obtained for the first of these possibilities gave improvements very similar to these found in Section V.

X. Conclusions

A sequential logic has been described for controlling the transmission frequency of a frequency-agile search radar of the type in which programming of the transmitted frequency is possible.[6] The logic is based on tentative decisions of target presence and the premise that the transmission frequency for which such a tentative decision takes place results in an above-average target echoing area for a complex target and, therefore, should be used repeatedly.

Under the assumption of the validity of the assumed fading model, simulations have shown that the logic proposed may give improvements in detectability for a single range bin of $2\frac{1}{2}$ dB or more above that usually considered the optimum. The effect of simultaneously considering a multiplicity of range elements has been investigated and it was found that the improvement in detectability may be maintained for a significant number of range elements.

Finally, it should be emphasized that with a radar of the proper type a logic as proposed may be implemented at almost negligible cost.

A disadvantage of the logic not mentioned previously is that it will tend to produce false alarms against distributed clutter, even with an ideal CFAR (constant false alarm rate) receiver, and thus, may degrade performance in such conditions toward that of a fixed-frequency radar. The reason is that if a large clutter amplitude triggers the logic and the same frequency, therefore, is used on subsequent sweeps, the clutter amplitude will keep its large value due to the correlation and a false alarm is likely to result. The only way to overcome this problem other than by coherent clutter suppression methods (MTI) would seem to be the use of a clutter recognizing circuit (as for example the output of a Dicke Fix receiver) to inhibit the logic in areas of distributed clutter.

Appendix I

Theoretical Result for Two Hits

For narrow-band Gaussian noise of variance σ_N^2 and a sinusoidal signal of amplitude A, the pdf of the envelope y of a signal-plus-noise sample is

$$p(y \mid A) = \frac{y}{\sigma_N^2} \exp\left(-\frac{y^2 + A^2}{2\sigma_N^2}\right) I_0\left(\frac{yA}{\sigma_N^2}\right). \tag{23}$$

For every new random observation frequency the amplitude A has a Rayleigh pdf as assumed for the fast fading case, i.e.,

$$p(A) = \frac{A}{\sigma_S^2} \exp\left(-\frac{A^2}{2\sigma_S^2}\right) \tag{24}$$

where σ_S^2 is the average signal power. The pdf of the amplitude of the first hit y_1 is, therefore,

$$\begin{aligned} p(y_1) &= \int_0^\infty p(y_1 \mid A) p(A) dA \\ &= \frac{y_1}{\sigma_S^2 + \sigma_N^2} \exp\left(-\frac{y_1^2}{2(\sigma_S^2 + \sigma_N^2)}\right). \end{aligned} \tag{25}$$

The average signal-to-noise ratio S_1 for the first hit is

$$S_1 = \sigma_S^2/\sigma_N^2. \tag{26}$$

After a certain amplitude y_1 has been observed, the prior pdf for the signal amplitude A_2 on the second hit, if the same frequency is used again, is from (6)

$$\begin{aligned} p(A_2 \mid y_1) = (1 + S_1) I_0\left(\frac{y_1 A_2}{\sigma_N^2}\right) \frac{A_2}{\sigma_S^2} \\ \cdot \exp\left(-\frac{y_1^2}{2\sigma_N^2} \frac{S_1}{1 + S_1}\right) \\ \cdot \exp\left(-\frac{A_2^2}{2\sigma_N^2} \frac{S_1 + 1}{S_1}\right) \end{aligned} \tag{27}$$

[6] This would, for example, be the case with a transmitter of the driven amplifier type such as is used in pulse compression radars. A reviewer has remarked that with a spin-tuned magnetron the technique would not be feasible.

where S_1 is given by (26). The average signal-to-noise ratio for the second hit is, from (7),

$$S_2' = \frac{1}{2\sigma_N^2}\int_0^\infty A_2^2 p(A_2 \mid y_1)\,dA_2. \tag{28}$$

In order to evaluate this expression the following formula [8] is required;

$$\int_0^\infty I_0(az)z^3 \exp(-b^2z^2)\,dz = \frac{1}{2b^4}\left(1+\frac{a^2}{4b^2}\right)\exp\left(\frac{a^2}{4b^2}\right). \tag{29}$$

This gives

$$S_2' = \frac{S_1}{1+S_1}\left(1+\frac{y_1^2}{2\sigma_N^2}\,\frac{S_1}{1+S_1}\right). \tag{30}$$

An improvement in the average signal-to-noise ratio is obtained if

$$S_2' = \frac{S_1}{1+S_1}\left(1+\frac{y_1^2}{2\sigma_N^2}\,\frac{S_1}{1+S_1}\right) \geq S_1 = \frac{\sigma_S^2}{\sigma_N^2} \tag{31}$$

from which the threshold of the logic is determined as

$$V = \sigma_N\sqrt{2(1+S_1)}. \tag{32}$$

Inserting into (11) to obtain the average signal-to-noise ratio for the second hit gives

$$S_2'' = S_1\int_0^V p(y_1)\,dy_1 + \int_V^\infty p(y_1)\,\frac{S_1}{1+S_1}\left(1+\frac{y_1^2}{2\sigma_N^2}\,\frac{S_1}{1+S_1}\right)dy_1.$$

With $p(y_1)$ as given by (25) and the substitution

$$z = \frac{y_1^2}{2(\sigma_S^2+\sigma_N^2)}$$

one obtains

$$S_2'' = S_1\int_0^1 e^{-z}\,dz + \int_1^\infty \frac{S_1}{1+S_1}e^{-z}\,dz + \int_1^\infty \frac{S_1^2}{1+S_1}ze^{-z}\,dz = S_1 + \frac{1}{e}\,\frac{S_1^2}{1+S_1}. \tag{33}$$

The average signal-to-noise ratio over the two hits is, therefore,

$$S'' = \frac{1}{2}(S_1+S_2'') = S_1 + \frac{1}{2e}\,\frac{S_1^2}{1+S_1} \tag{34}$$

and the improvement factor is

$$\Delta S = \frac{S''}{S} = 1 + \frac{1}{2e}\,\frac{S_1}{1+S_1}. \tag{35}$$

For $S_1 \to \infty$ this gives

$$\Delta S = 1 + \frac{1}{2e} = 1.184 \to 0.73\ \text{dB}.$$

For $S_1 = 6$ dB, ΔS is reduced to 0.60 dB.

Appendix II

Threshold Settings for $P_{fs} = 10^{-6}$

$M=$	4	8	16
Analog Integrator			
$W=$	12.4	20.0	33.5
Digital Integrator			
$N=1$ $T=$	5.26		
$N=2$ $T=$	3.86	3.97	
$N=3$ $T=$	3.16	3.36	
$N=4$ $T=$	2.62	2.93	3.20
$N=5$ $T=$		2.61	
$N=6$ $T=$		2.35	2.68
$N=8$ $T=$			2.33
$N=10$ $T=$			2.06
$N=12$ $T=$			1.83

Appendix III

Accuracy of Simulation

If K repetitions are performed of an experiment for which the probability of success is p, the total number of successes n follows a binomial distribution with mean value

$$\bar{n} = Kp$$

and standard deviation

$$\sigma_n = \sqrt{Kp(1-p)}.$$

If the probability of detection is estimated as

$$P_d = \hat{p} = \frac{n}{K}$$

the standard deviation of P_d is

$$\sigma(P_d) = \sqrt{\frac{p(1-p)}{K}}.$$

An example of a (true) detection curve and the associated limits $\pm\sigma(P_d)$ for the estimates of detection probability, is shown in Fig. 15 for $K=300$. For this detection curve the accuracy over the range of detection probabilities of main interest (0.1–0.9) may also be expressed in decibels referring to the accuracy with which the signal-to-noise ratio required to give a certain probability of detection may be determined. For a single observation the standard

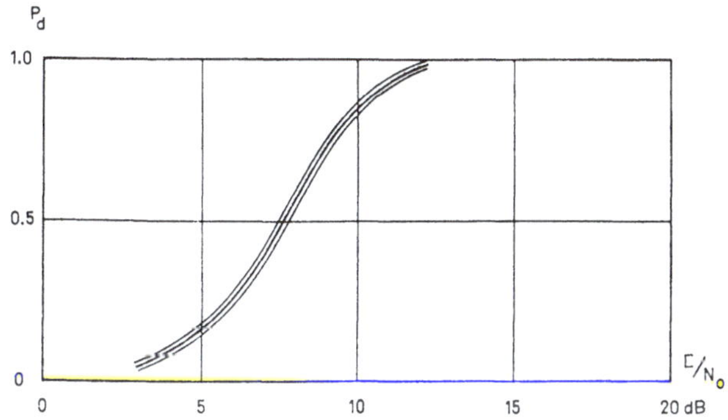

Fig. 15. Detectability curve shown for $K=300$ repetitions with $\pm\sigma(P_d)$ limits.

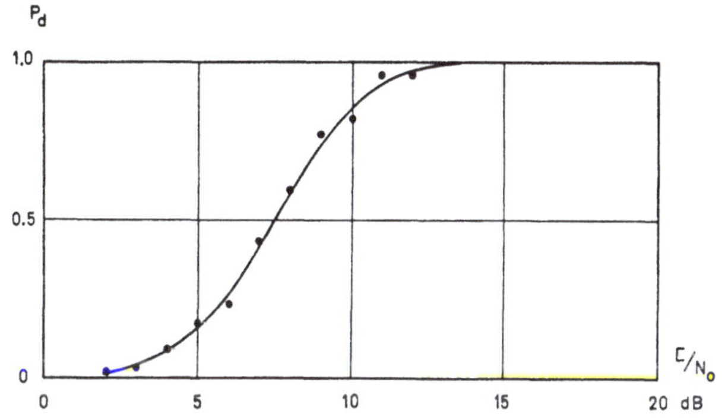

Fig. 16. Detection curve obtained by simulation. Each point represents 300 repetitions of Monte Carlo experiment.

deviation is seen to be around $\sigma_{\text{SNR}}=0.15$ dB. Since in all simulations carried out, points were determined at intervals of 1 dB in signal-to-noise ratio, some additional smoothing is obtained, so that it is estimated that the detection curves determined by simulation will be accurate within ± 0.1 dB (one-σ limits for $K=300$ repetitions).

As an illustration the result of a simulation is plotted in Fig. 16 and a detection curve is drawn by hand to give a best fit to the points. A comparison with Fig. 15 shows that most of the points fall within the $\pm\sigma(P_d)$ limits as one would expect from the theory.

ACKNOWLEDGMENT

Work on this subject commenced while the author was with the SHAPE Technical Centre, The Hague, Netherlands. In particular, the author is indebted to his former colleagues Dr. A. J. Zöttl and J. J. Jernigan, for valuable comments.

REFERENCES

[1] J. I. Marcum, "A statistical theory of target detection by pulsed radar," *IRE Trans. Information Theory*, vol. IT-6, pp. 59–267, April 1960.

[2] P. Swerling, "Probability of detection for fluctuating targets," *IRE Trans. Information Theory*, vol. IT-6, pp. 269–308, April 1960.

[3] E. L. Kaplan, "Signal-detection studies with applications," *Bell Sys. Tech. J.*, vol. 34, pp. 403–437, March 1955.

[4] T. S. Edrington, "The amplitude statistics of aircraft radar echoes," *IEEE Trans. Military Electronics*, vol. MIL-9, pp. 10–16, January 1965.

[5] H. Ray, "Improving radar range and angle detection with frequency agility," *Microwave J.*, May 1966.

[6] S. S. Wilks, *Mathematical Statistics.* New York: John Wiley, 1962.

[7] D. E. Hampton and A. A. McCurrach, "False alarm rates in a radar digital detector," Royal Aircraft Establishment, Tech. Note Rad. 789, October 1960.

[8] W. Gröbner und N. Hofreiter, *Integraltafel, Bestimmte Integrale.* Vienna, Austria: Springer, 1961.

[9] B. G. Gustavson and B-O. Ås, "System properties of jumping-frequency radars," *Philips Telecommun. Rev.*, vol. 25, no. 1, July 1964.

Vilhelm Gregers Hansen (M'60) was born in Odense, Denmark, on September 30, 1934. He received the M.Sc. degree in electrical engineering from the Royal Technical University of Denmark in 1959.

He served with the Danish Defense Research Board until 1960. From 1960 to 1966 he was employed with the SHAPE Technical Centre, The Hague, Netherlands, where he worked primarily on radar detection and antijamming techniques, and related evaluation methods. Since 1966 he has been with the Royal Technical University of Denmark, where he is teaching and doing research in the field of communication theory.

Reprinted from IEEE TRANSACTIONS
ON *AEROSPACE AND ELECTRONIC SYSTEMS*
Volume AES-4, Number 5, September, 1968
Pp. 763-773

Detection of Slow Fluctuating Targets with Frequency Diversity Channels

V. Vannicola

IEEE Transactions on Aerospace and Electronic Systems, Vol. AES-10, No. 1 (January 1974)
Pages 43-52

Abstract

In this paper an exact closed-form expression for the radar detection probability is derived and results are plotted for a frequency diversity radar receiver. The receiver model performs post-detection integration on all received pulses in all diversity channels. The target model assumed is the slow fluctuating Rayleigh-distributed (Swerling case I target) scatterer. Each of the M frequency diverse channels receives N amplitude-correlated returns to give a total of NM post square-law detection integrations. The tabulated data falls between the two extreme cases, that for which all the returns are amplitude-correlated and that for which each return is independent. The plotted results fall close to the figures obtained through simple empirical relationships.

Introduction

Radar targets generally fluctuate in amplitude and phase according to some probability density function and some fluctuation rate. If one is to determine analytically the detection probability for a particular type of target in a noise environment, it is necessary that the statistics of target and noise be known. Although it is not always possible to characterize the actual target statistics, one can make a good approximation which gives remarkable results for a wide variety of target and noise situations.

Manuscript received October 25, 1972; revised August 1, 1973.

This work sponsored by the Air Force Systems Command's Rome Air Development Center, Griffis A.F.B., N.Y. 13441.

The four target cases modeled and evaluated by Swerling [1] are well known to radar engineers and have met with great acceptance. The situations which these cases cover are of two extremes:

1) That of a slowly fluctuating target where the cross section is random and remains correlated during the scan time. The pulses reflected off such a target are correlated with each other in amplitude during that scan, although the pulse-to-pulse phase may be independent.

2) That of a rapidly fluctuating target where the cross section becomes uncorrelated from pulse to pulse. The received amplitude and phase are independent from pulse to pulse. The noise in all four models is zero-mean white Gaussian.

An excellent treatment of these four cases is given by DiFranco and Rubin [2], wherein the optimum receiver along with its performance is derived and plotted in numerous curves.

For partially correlated received pulses, several papers have been written. Swerling has developed [3] a method which is applicable to a large family of probability density functions and for fairly general correlation properties of the signal fluctuations. His method yields the characteristic function of the probability density of sum random variables, defined as Y. To obtain $p(Y)$ and P_d with this technique requires computer evaluations. Swartz [4] considered the partially correlated case for $N = 2$ and a Rayleigh amplitude density function. As to be expected, his curves fall between those of the two extremes, zero correlation and complete correlation.

In a recent paper by Barton [5], an approximate formula is given for determining the signal-to-noise ratio required for a given detection probability when there are N pulses representing n_e independent signal samples. The method is based on a very simple empirical relationship which appears to interpolate between complete correlation and zero correlation from pulse to pulse. For the examples tested, it is shown that this approach renders remarkably accurate results. A similar expression involving interpolation methods is given by Swerling [6].

In this paper, the case of M frequency diverse channels, each containing N correlated pulses, each with Rayleigh amplitude density function is evaluated. An exact closed-form solution is derived for the output probability density function and the detection probability for a given false-alarm probability and signal-to-noise ratio when all MN pulses are square-law detected and combined as shown in Fig. 1. Each of the MN pulses has independent Gaussian

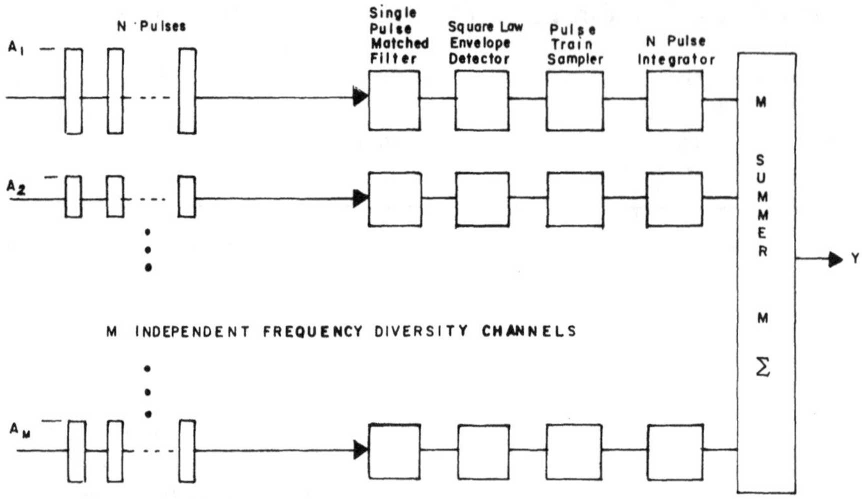

Fig. 1. *M*-channel *MN*-pulse combiner.

noise, and each group of N pulses has an independent Gaussian signal, giving a total of M independent signals in all MN pulses.

Derivation of the Probability Density Function

We may start the derivation simply by using the results of DiFranco and Rubin [2].

Determining the Threshold for False Alarm

The output probability density function for no signal present may be written from equation 10.4-17 of [2], with MN representing the total number of pulses and the variable Y representing the "convenient modified form" of equation 10.4-4 of [2]:

$$Y = \sum_{i=1}^{MN} \frac{1}{2} r_i^2 \qquad (1)$$

$$P(Y;0) = \frac{Y^{MN-1} e^{-Y}}{(MN-1)!}\,. \qquad (2)$$

From (2), we may write the false-alarm probability using equation 10.4-18a of [2]:

$$P_F = \int_{Y_b}^{\infty} P(Y;0)\,dY = \int_{Y_b}^{\infty} \frac{Y^{MN-1} e^{-Y}}{(MN-1)!}\,dY. \qquad (3)$$

This integral has been evaluated by Pachares [7] for $1 \leqslant MN \leqslant 150$ and false-alarm probabilities equal to 10^{-m}, where m is an integer ranging between 1 and 12. Using the Pachares tables, one may obtain threshold Y_b to four decimal places.

Detection Probability

Next the probability of detection is derived. We may start with the characteristic function of the signal out of the N pulse integrator, which corresponds to the Swerling case I given by DiFranco and Rubin [2] in their equation 11.2-23, and is repeated here:

$$\overline{C_{Y_i}(\xi)A_i} = \frac{(-1)^N}{(1+NA_0^2)\left(-\dfrac{1}{1+NA_0^2}+j\xi\right)(-1+j\xi)^{N-1}}\,. \qquad (4)$$

When we add M of these outputs through the adder,

$$Y = \sum_{i=1}^{M} Y_i, \qquad (5)$$

where Y_i is independent of Y_j, $i \neq j$, the result is

$$\overline{C_Y(\xi)_{A_1, A_2, \cdots, A_M}} = \frac{(-1)^{MN}}{(1+NA_0^2)^M\left(-\dfrac{1}{1+NA_0^2}+j\xi\right)^M(-1+j\xi)^{M(N-1)}}\,. \qquad (6)$$

Probability density $p(Y)$ is obtained first by factoring (6):

$$\overline{C_Y(\xi)_{A_1, A_2, \cdots, A_M}} = \frac{(-1)^{M-1}}{(1+NA_0^2)^{M-1}\left(-\dfrac{1}{1+NA_0^2}+j\xi\right)^{M-1}} \cdot \frac{(-1)^{M(N-1)+1}}{(1+NA_0^2)\left(-\dfrac{1}{1+NA_0^2}+j\xi\right)(-1+j\xi)^{M(N-1)}}\,. \qquad (7)$$

Next we use the results of equation 11.2-28 of [2],

which allows us to write the anticharacteristic function of the second factor of (7),

$$p'(Z) = \frac{\beta e^{\beta Z}}{[M(N-1)-1]!(1+\beta)^{M(N-1)}} \cdot \gamma[M(N-1), (\beta+1)Z] \tag{8}$$

where the substitution $\beta = -1/(1+NA_0^2)$ has been made.

The first factor in (7) transforms to an $(M-1)$ integration over the incomplete gamma function, $\gamma[M(N-1), (\beta+1)Z]$, of (8) and multiplier β^{M-1}. The probability density function $p(Y)$ is thus written

$$p(Y) = \frac{\beta^M e^{\beta Y}}{(1+\beta)^{M(N-1)}} \underbrace{\int_0^Y \int_0^{Z_{M-1}}}_{(M-1)} \cdots \int_0^{Z_2} \frac{\gamma[M(N-1), (\beta+1)Z_1]}{[M(N-1)-1]!} \cdot dZ_1, dZ_2, \cdots, dZ_{M-1} \, . \tag{9}$$

This $(M-1)$-fold integral is evaluated by expanding the integrand [8] to the form

$$\frac{\gamma(n, x)}{(n-1)!} = 1 - \left[1 + x + \frac{x^2}{2!} + \frac{x^3}{3!} + \cdots + \frac{x^{n-1}}{(n-1)!}\right] e^{-x}, \tag{10}$$

performing the integrations, and collecting terms. The result is

$$p(Y) = \frac{(-\beta)^M e^{\beta Y}}{(1+\beta)^{MN-1}} \left\{ \frac{[(1+\beta)Y]^{M-1}}{(M-1)!} + \sum_{i=0}^{M-2} (-1)^{M-1-i} \frac{(MN-2-i)!}{(MN-M-1)!(M-1-i)!} \cdot \frac{[(1+\beta)Y]^i}{i!} - (-1)^{M-1} e^{-(1+\beta)Y} \cdot \sum_{j=0}^{MN-M-1} \frac{(MN-2-j)!}{(MN-M-1-j)!(M-1)!} \cdot \frac{[(1+\beta)Y]^j}{j!} \right\}. \tag{11}$$

The probability of detection

$$P_D = \int_{Y_b}^{\infty} p(Y)\, dY = 1 - \int_0^{Y_b} p(Y)\, dY \tag{12}$$

can be found in a similar manner by performing the integration and collecting terms. The expression which results can be contracted using the incomplete gamma functions, according to (10), to yield

$$P_D = 1 - \frac{\beta^{M-1}}{(1+\beta)^{MN-1}} \left\{ \sum_{i=0}^{M-1} \left(\frac{1+\beta}{\beta}\right)^i \cdot \frac{(MN-2-i)!}{(MN-M-1)!(M-1-i)!} \frac{\gamma(1+i, -\beta Y_b)}{i!} + \beta \sum_{j=0}^{MN-M-1} (1+\beta)^j \frac{(MN-2-j)!}{(MN-M-1-j)!(M-1)!} \cdot \frac{\gamma(1+j, Y_b)}{j!} \right\}, \tag{13}$$

from which curves may be constructed after choosing the appropriate threshold Y_b from Pachares' [7] tables. In using Pachares' tables, the total number of pulses m corresponds to MN.

The expression in (13) resembles an expression derived by James [9] for target detection in noise when the target fading period is less than the post-detection integration and greater than the interpulse period. The expression is in terms of target, post-correlator, and noise bandwidths, and, with a degree of effort, can be related to the topic treated in this paper.

Computational Results and Checking With Other Results

Equation (13) is used to compute the curves in Figs. 2 through 21. These figures present the probability of detection versus single-pulse rms signal-to-noise ratio for different combinations of correlated pulses per channel N and number of diversity channels. Each figure shows a family of curves for different false-alarm probabilities. Note that these curves fall between the two extreme cases, i.e., when $M = 1$, Swerling case I, and when $N = 1$, Swerling case II, as presented in DiFranco and Rubin [2]. The curves for case I, $M = 1$, and case II, $N = 1$, have been repeated here and, as expected, they check with those of [2]. The 3-dB discrepancy between the curves presented here and those of DiFranco and Rubin [2] is due to the fact that we are plotting P_D against the average signal-to-noise ratio A_0^2, rather than the average of the peak, $\mathscr{R}_p = 2A_0^2$ of [2]. The crosses in Figs. 5 and 15 are computed using Barton's paper [5]. Note the agreement of these points with the curves which were plotted using (13). Regions of departure occur at high probabilities of detection, as well as low probabilities.

The data presented in these figures fall within the capability limits of the computer, and are accurate to within a probability of detection of ±0.10 percent for up to a total of $MN = 100$ pulses. In order to extend the evaluation of (13) to more than $MN = 100$ pulses, a computer having higher precision is required. The Appendix contains a discussion of the computer programming, as well as the program used to compute the data presented in the figures.

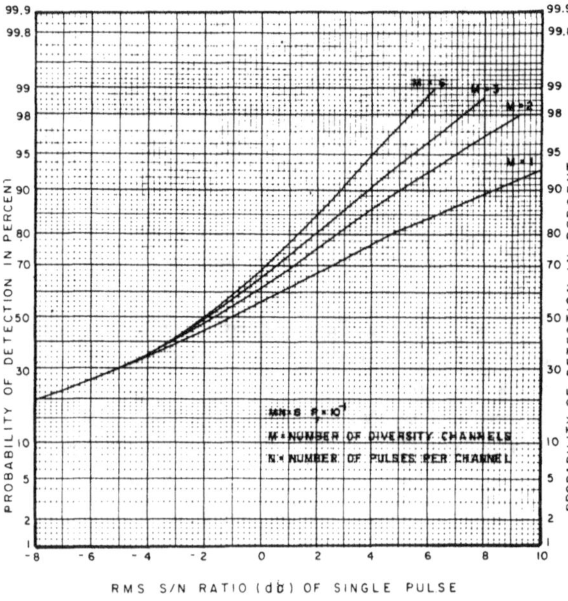

Fig. 2.

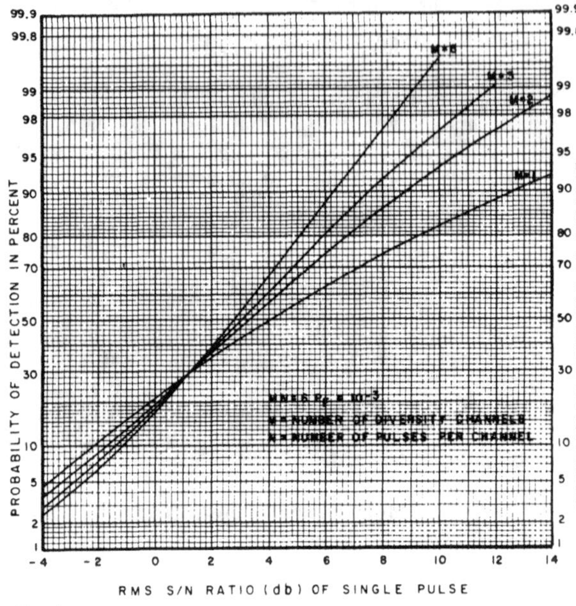

Fig. 3.

Fig. 4.

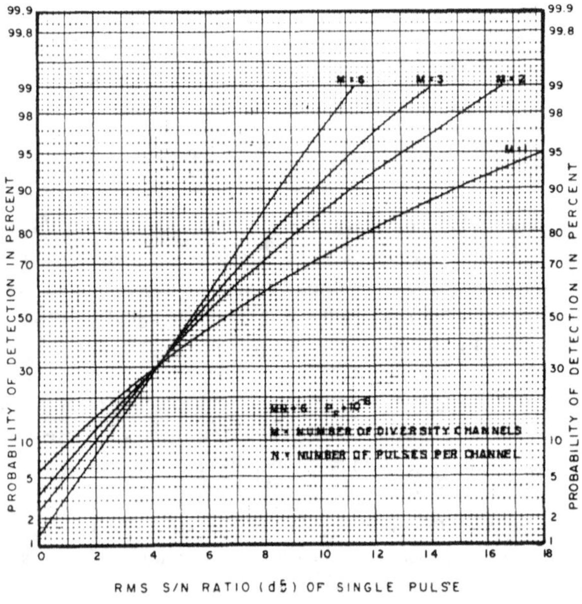

Fig. 5.

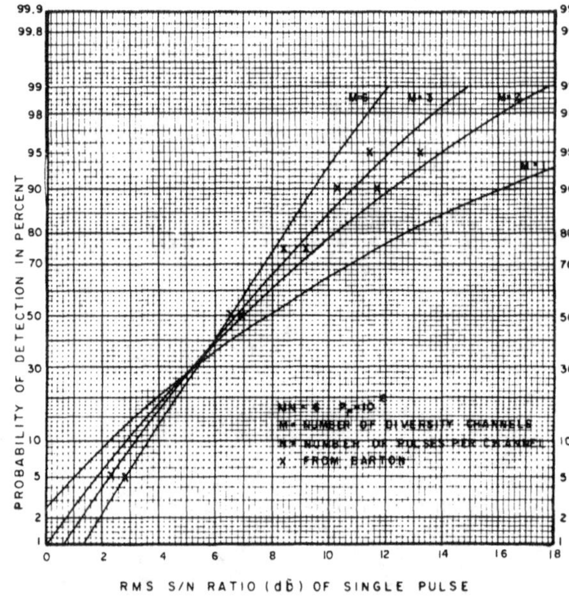

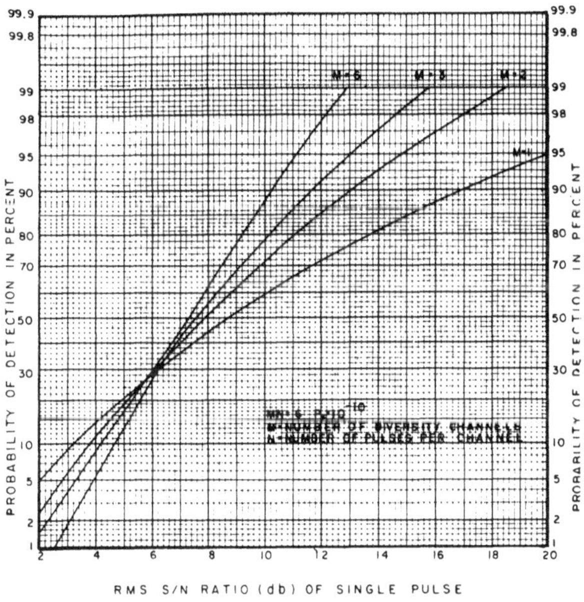

Fig. 6.

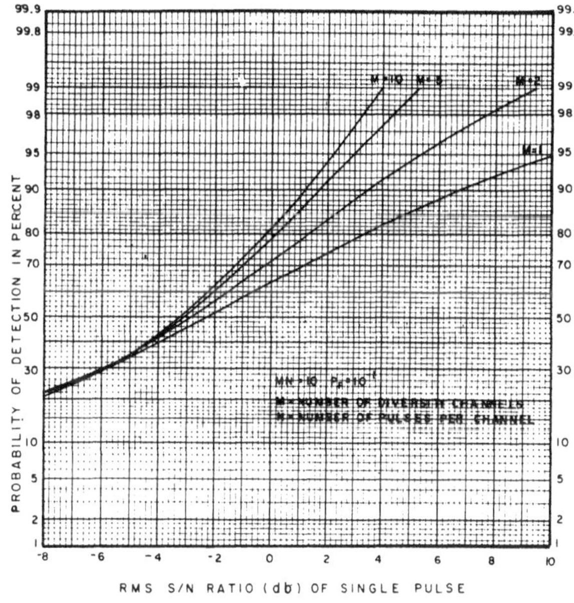

Fig. 7.

Fig. 8.

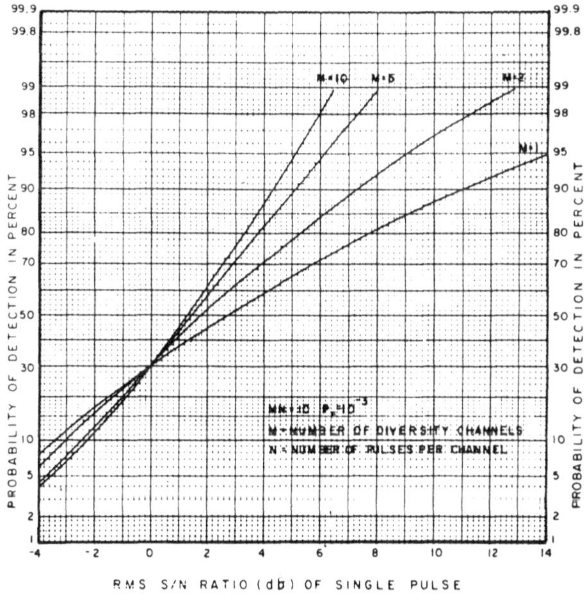

Fig. 9.

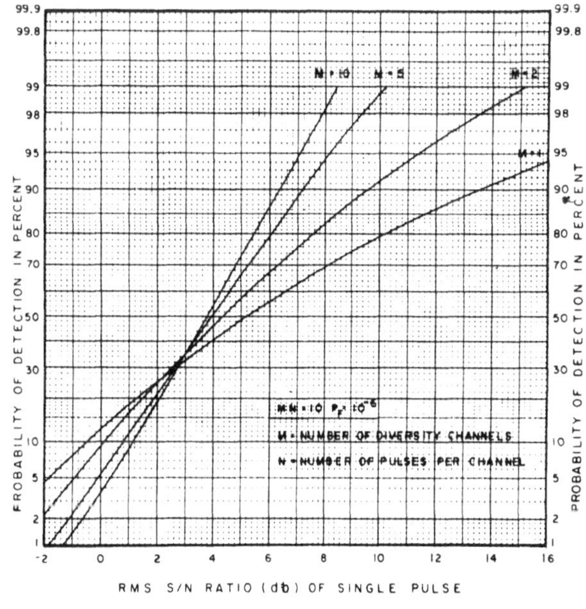

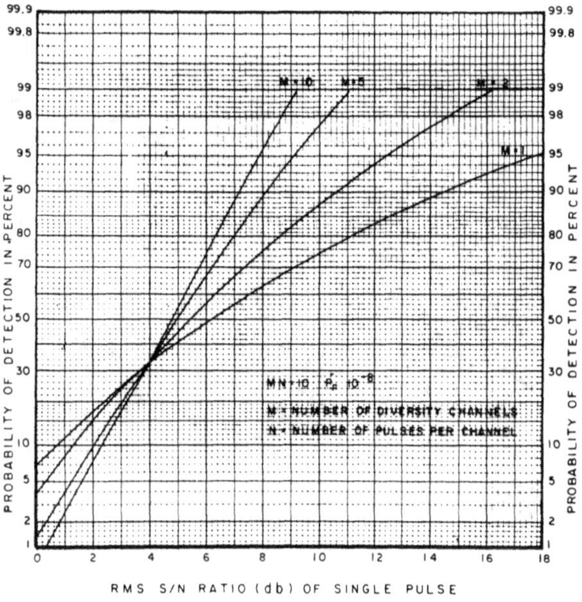

Fig. 10.

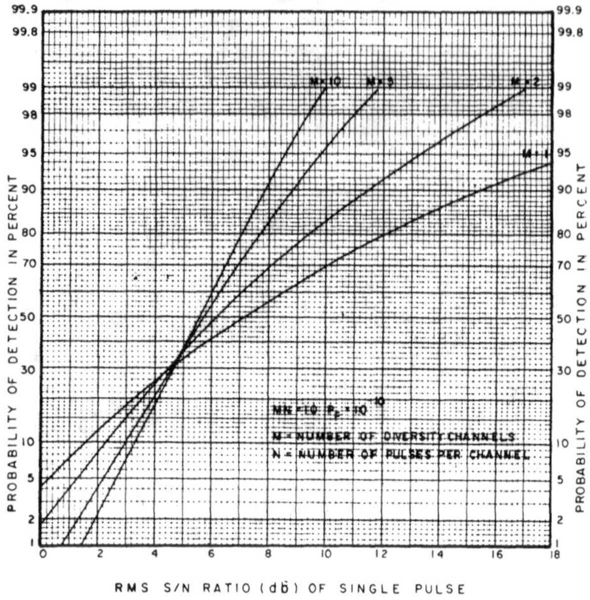

Fig. 11.

Fig. 12.

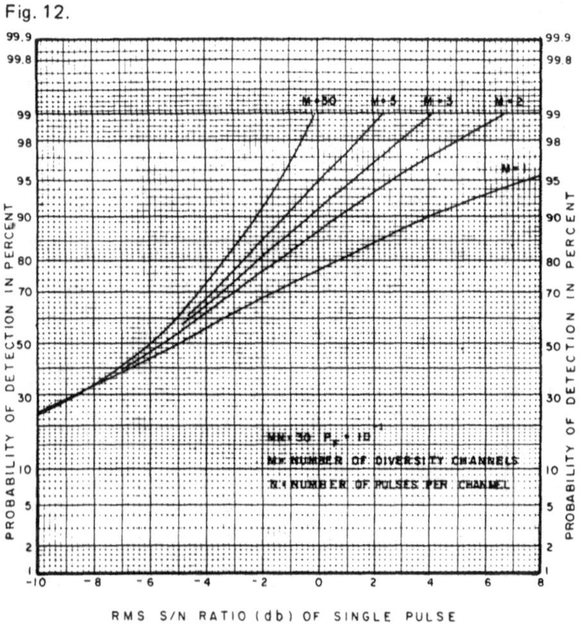

Fig. 13.

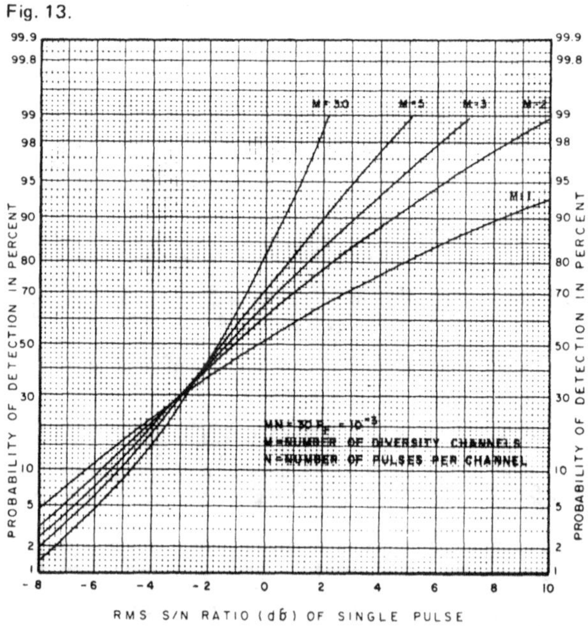

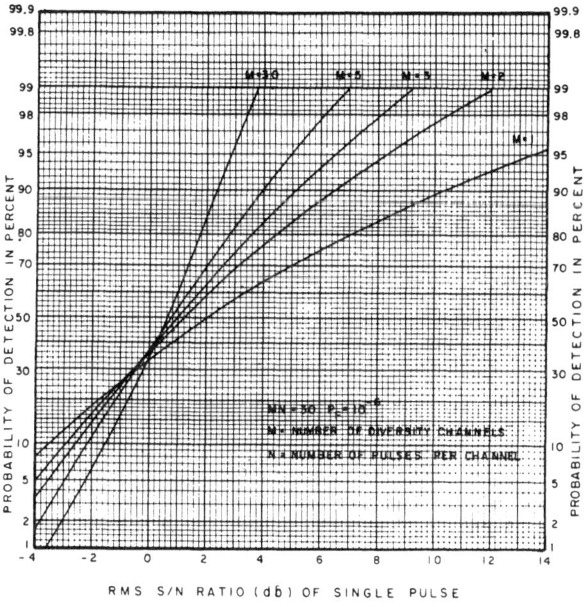

Fig. 14.

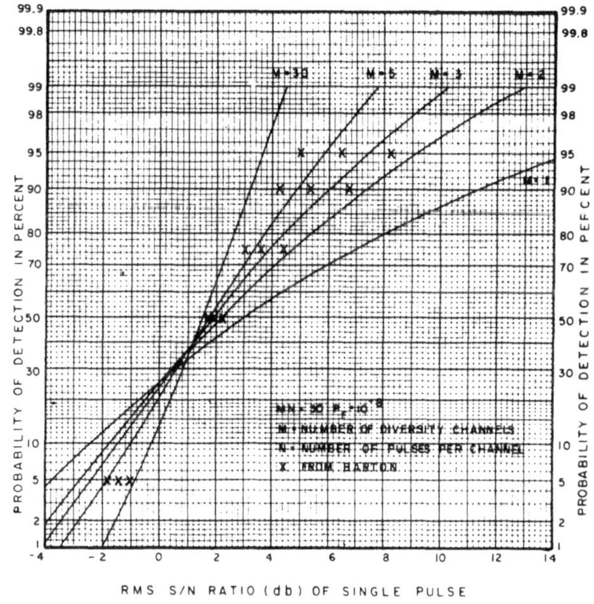

Fig. 15.

Fig. 16.

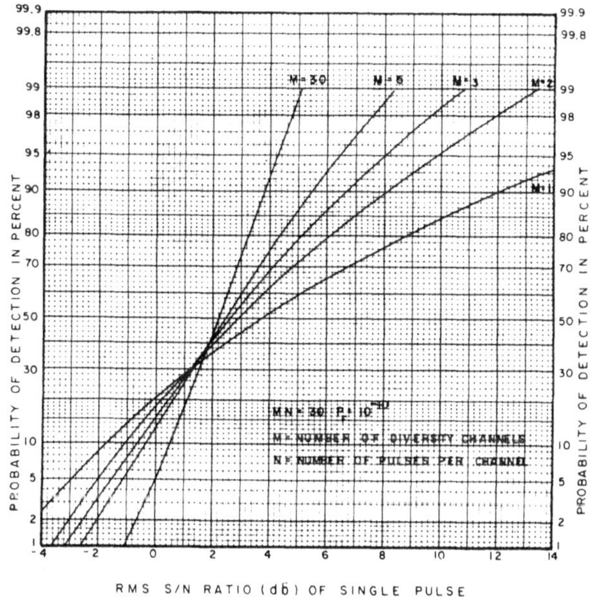

Fig. 17.

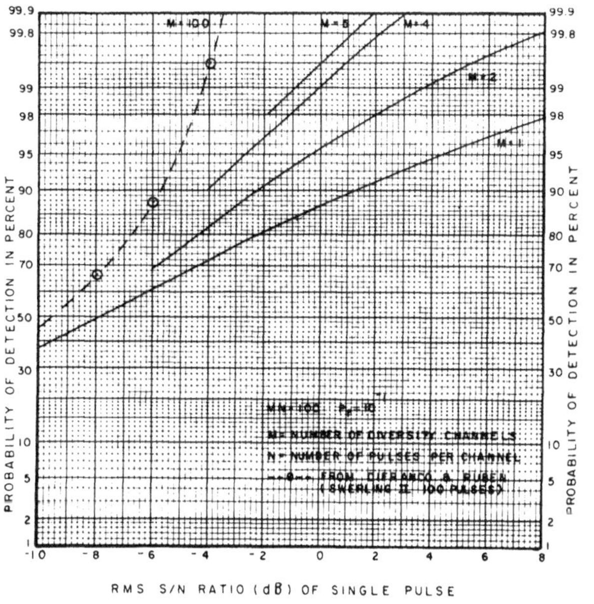

106

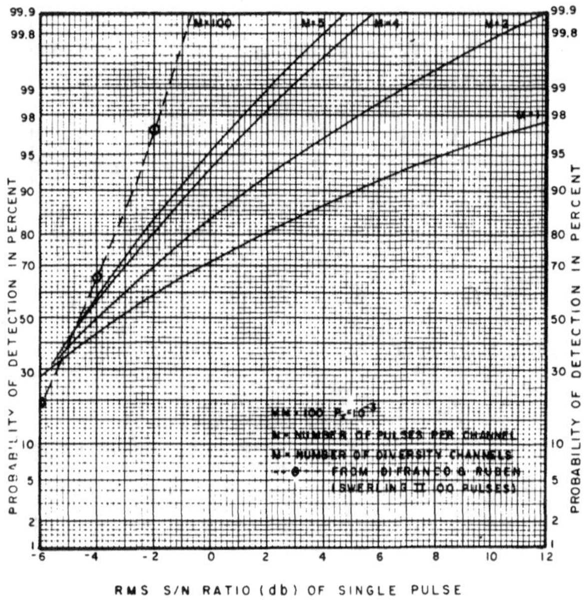

Fig. 18.

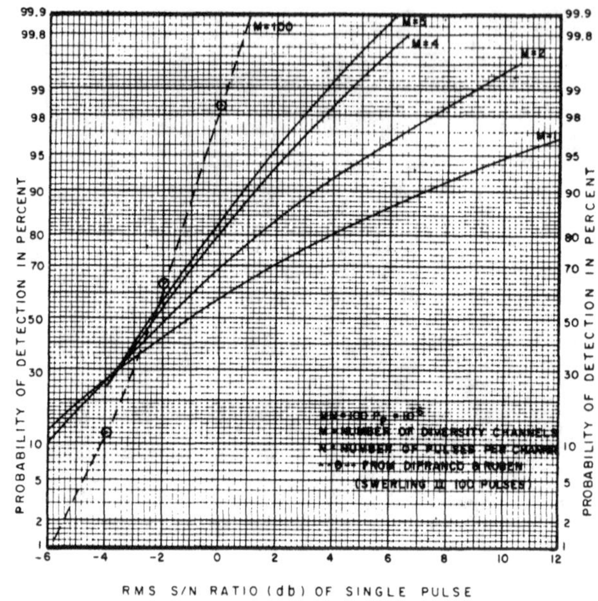

Fig. 19.

Fig. 20.

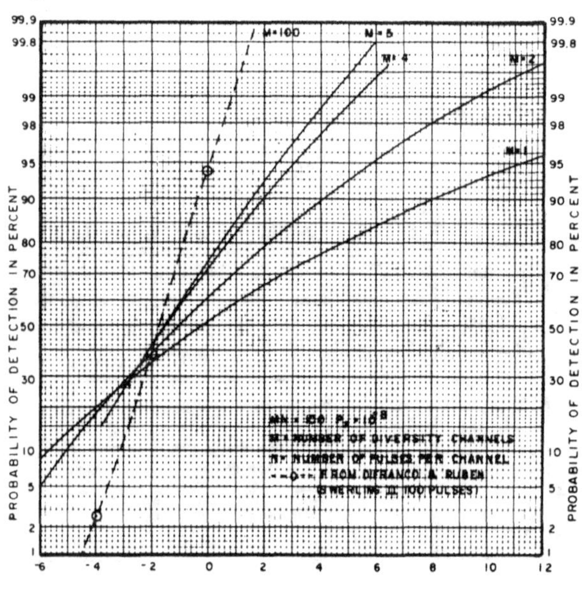

Fig. 21.

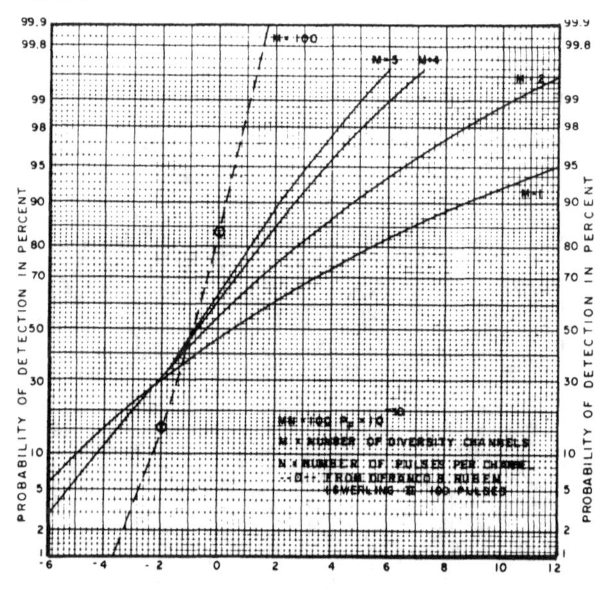

Appendix

Computational Procedure for Probabilities of Detection

The expression for the probability of detection is given by

$$P_D = 1 - \frac{\beta^m}{\rho(Nm + N - 1)} \left[\sum_{i=0}^{m} \left(\frac{\rho}{\beta}\right)^i \cdot \frac{(n + m - i - 1)!}{(n-1)!(m-i)!} \, \frac{\gamma(i + 1, -\beta Y_0)}{i!} + \beta \sum_{j=0}^{n-1} \rho^j \, \frac{(n + m - j - 1)!}{(n - j - 1)!m!} \, \frac{\gamma(j + 1, Y_0)}{j!} \right] \quad (14)$$

where

$$m = M - 1$$
$$n - M(N - 1)$$
$$\beta = -1/(1 + NA_0{}^2)$$
$$\rho = \beta + 1$$
$$\gamma(k, x) = \text{incomplete gamma function.}$$

The evaluation of this function presents some computational difficulties which were not entirely resolved in the time available. For typical values of the parameters, many of the quantities [the factorials, β^m, or $\gamma(k, x)$] become greater than 10^{38} or less than 10^{-38} and exceed the capacity of the computer. This problem was overcome very simply by taking logarithms. However, for cases involving relatively large m and low signal-to-noise ratio, the numbers comprising the summations become very large and opposite in sign. For example, for $m = 14$ and $N = 2$, the largest terms are on the order of 10^{20}. Since the final answer is a probability, it must lie between 0 and 1. As such, more than 20-place accuracy is required in the terms of the summations, whereas single precision on a computer offers only 8 places. Also, much greater accuracy can be required for other combinations of the parameters.

For the present, we have not included results over the entire range of interest of all the variables. It is expected, however, that the expression could be rewritten or re-expressed to be compatible with the limitations of the computer. The following is a listing of the actual Fortran program used to calculate the results:

```
10   COMMON FAC(400),H(200),GAM2(2,200),ISIG(200)
20 COMMON PDT(30),YO(30,5),NPR(30,36),NP(30),A(20)
60  READ: NT
70  READ: NC
80  DO 8 I=1,NC;8 READ: (YO(I,J),J=1,5)
90  READ: (NP(I),I=1,NC)
100  DO 10 I=1,NC;NPAIR=2*NP(I)
110  10  READ: (NPR(I,K),K=1,NPAIR)
120 DO 55 I=1,9;55 A(I)=2.*I-10.
130  ISIG(1)=1
140  DO 81 I=2,NT;J=I-1;81 ISIG(I)=-1*ISIG(J)
150  2 FORMAT (3X,F12.4,6X,I2,6X,I3,6X,F7.2,6X,F12.8)
160   FAC(1) = 0.
170  FAC(2)=0.
180  DO 80 I=3,NT
190  J=I-1
200  80  FAC(I)=FAC(J)+ALOG(FLOAT(J))
210  DO 6 I1=1,NC;NPAIR=2*NP(I1)
220  DO 6 K=2,NPAIR,2;K1=K-1;M=NPR(I1,K1);N=NPR(I1,K);XN=N
222  PRINT 102,N,M;102 FORMAT ("- FOR N =",I3," AND M =",I3)
229  PRINT 104;PRINT 103,(A(I),I=1,9);PRINT 120;120 FORMAT ("0")
230  DO 6 J=1,5;Y=YO(I1,J)
233  104 FORMAT ("-                        < S/N RATIO (DB) >")
236  103 FORMAT ("   Y   ",9F6.0)
240  M=NPR(I1,K1);N=NPR(I1,K);XN=N
244  DO 5 M1=1,9;A2=0.
246 A0=10.**(A(M1)/20.)
248  BETA=-1./(1.+XN*A0*A0)
250  P=-BETA*Y
252  MS=M+1;CALL GA(2,MS,P)
255  B1=ALOG(-BETA)
258  NS=(M+1)*(N-1);ROA=BETA+1.;R1=ALOG(ROA)
260  N6=N*M+N-1
270  IF(M1.EQ.1) CALL GA(1,NS,Y)
272  DO 4 I=1,NS;IA=I-1;M12=M+2;M13=ISIG(M12)
274  H2=ALOG(FAC2(NS,M,IA))
276  H3=(M+1)*B1;H4=(N6-IA)*R1
278  A7=H3-H4+GAM2(1,I)+H2;IF(ABS(A7).GT. 50.) GOTO 4
280  A2=A2+M13*EXP(A7);4 CONTINUE
330  MS=M+1;DO 3 I=1,MS;IA=I-1;M10=M-IA+1
344  G2=ALOG(FAC1(NS,M,IA))
345  G3=(N-IA)*B1;G4=(N6-IA)*R1
350  A6=G3-G4+GAM2(2,I)+G2;IF(ABS(A6).GT.50.) GOTO 3
360  A2=A2+ISIG(M10)*EXP(A6); 3 CONTINUE
620  IF(ABS(A2).GT. 50.) A2=-50.
630  PD=1.0-A2
640  PDT(M1)=PD;5 CONTINUE;PRINT 89,Y,(PDT(J8),J8=1,9)
650  89 FORMAT (F7.2,1X,9F6.3)
660  6 CONTINUE
670  STOP;END
680   FUNCTION FAC1(NS,M,IA)
690   N = M - IA
700   P = 1.
710   DO 10 I = 1,N
720   10 P = P*(NS-1+I)/I
730   FAC1 = P
740   IF(M .EQ. IA)FAC1 = 1.
750   RETURN
760   END
770   FUNCTION FAC2(NS,M,IA)
780   N = M
790   P = 1.
800   DO 10 I = 1,N
810   10 P = P*(NS-IA-1+I)/I
820   FAC2 = P
830   IF(M .EQ. 0) FAC2 = 1.
840   RETURN
850   END
1000  SUBROUTINE GA(I3,NS,X)
1005  COMMON FAC(400),H(200),GAM2(2,200)
1010  C=ALOG(X)
1020   NMAX=2.*X+10.;IF(NMAX.GT.30) NMAX=30
1030  IMAX=NS+NMAX;XN=1.5*X
1040  DO 10 I=1,IMAX;IA=I-1
1050  10 H(I)=IA*C-X-FAC(I);H5=1.0
1060  DO 20 I=1,NS;GAM2(I3,I)=0.
1070  G3=I
1080  IF(XN.GT.G3) GOTO 80
1090  J=I+1;J1=J+NMAX-1;H8=H(J);GAM=0.
1100  DO 30 K=J,J1
1110  H1=H(K)-H8;IF(ABS(H1).GT.50.) GOTO 30
1120  GAM=GAM+EXP(H1)
1130  GAM2(I3,I)=ALOG(GAM)+H8
1140  30 CONTINUE;GO TO 20
1150  80 IF(ABS(H(I)).GT.50.) GOTO 20
1160  H5=H5-EXP(H(I))
1170  GAM2(I3,I)=ALOG(H5)
1180  20 CONTINUE;RETURN;END

READY
```


Acknowledgment

The author wishes to acknowledge J.V. DiFranco of Sperry Rand for his comments on the original manuscript of this paper. He also acknowledges E. Bromley of the Rome Air Development Center for overcoming the programing difficulties in obtaining the data presented in the figures, and for preparing the Appendix.

The author also wishes to acknowledge Manuel Ares, Charles B. Clarkson, and Judd J. Gosten of General Electric Company for their treatment of the same subject. The author became aware of their work after IEEE acceptance of this article.

References

[1] P. Swerling, "Probability of detection for fluctuating targets," *IEEE Trans. Information Theory*, vol. IT-6, pp. 269-308, April 1960.
[2] J.V. DiFranco and W.L. Rubin, *Radar Detection*. Englewood Cliffs, N.J.: Prentice-Hall, 1968, chs. 9-11.
[3] P. Swerling, "Detection of fluctuating pulsed signals in the presence of noise," *IRE Trans. Information Theory*, vol. IT-3, pp. 175-178, September 1957.
[4] M. Swartz, "Effects of signal fluctuation on the detection of pulse signals in noise." *IRE Trans. Information Theory*, vol. IT-2, pp. 66-71, June 1956.
[5] D.K. Barton, "Simple procedures for radar detection calculations," *IEEE Trans. Aerospace and Electronic Systems*, vol. AES-5, pp. 837-846, September 1969.
[6] P. Swerling, "Recent developments in target models for radar detection analysis," presented at the AGARD Avionics Technical Symp. on Advanced Radar Systems, Istanbul, Turkey, May 25-29, 1970.
[7] J. Pachares, "A table of bias levels useful in radar detection problems," *IRE Trans. Information Theory*, vol. IT-4, pp. 38-45, March 1958.
[8] M. Abramowitz and I.A. Stegun, *Handbooks of Mathematical Functions*. Washington D.C.: National Bureau of Standards, 1965, pp. 260-262.
[9] P.W. James, "Detection of fading targets in noise: The case in which the fading is neither fast nor slow," Royal Aircraft Establishment, U.K., Tech. Note RAD 828, January 1963, ASTIA Doc. 406158.

Vincent C. Vannicola (M'67) was born in New Haven, Conn., on September 8, 1932. He received the B.A. degree from the University of Connecticut at Storrs in 1956, and the M.S. degree in electrical engineering and the E.E. degree from Syracuse University, Syracuse, N.Y., in 1969 and 1973, respectively.

From 1956 to 1960 he was employed by the Light Military Electronics Department of General Electric Company, Utica, N.Y., where he was engaged as a product design engineer. He has been with the Rome Air Development Center, Griffiss Air Force Base, Rome, N.Y., since August 1960, where he has worked on high-power microwave components in oversized transmission line configurations, broadband microwave networks, and pulse compression systems. He is currently working in signal processing, statistical detection of radar signals, and adaptive radar.

Frequency Agility Radar Range Calculation Using Number of Independent Pulses

G. Lind

IEEE Transactions on Aerospace and Electronic Systems, Vol. AES-12, No. 6 (November 1976)
Pages 811-815

Abstract

A method for accurately calculating the radar range improvement introduced by frequency agility is presented. This calculation is made by using the concept of independent pulses. The number of independent target or clutter pulses is calculated by a simple method based on such basic combinatorial probability theory as drawing differently colored balls from an urn.

Manuscript received October 30, 1975. Copyright 1976 by The Institute of Electrical and Electronics Engineers, Inc.

I. Introduction

Frequency agility radar, where the frequency has a random variation from pulse to pulse, exhibits improved performance compared to a fixed-frequency radar in several ways. Radar range, glint reduction, and electronic counter-countermeasures (ECCM) properties are among them. This correspondence discusses the influence of frequency agility on radar range. A method is given whereby frequency agility range improvement can be calculated in great detail. It will account for the effects on the target signal (target decorrelation) as well as the effect on clutter signals (clutter decorrelation). The effect of the frequency agility bandwidth, pulse repetition frequency, time-on-target, and target (or clutter) depths are included as well as the possible decorrelation the fixed-frequency radar may have, in order to make the comparison correct.

Range improvements using random agile frequency are a function of the number of independent pulses, both target echo pulses, and clutter pulses. The important thing is the change in the number of independent pulses from fixed frequency to frequency agility. In the case of fixed frequency a certain natural decorrelation is obtained, caused by the movement of the target, internal clutter motion, or motion of the radar platform. This gives rise to a certain number of independent pulses, which must be compared with the number that can be attained with frequency agility.

With frequency agility the number of independent pulses can obviously never exceed the number of pulses obtained during an illumination time. It may, however, be less, particularly if the target (or clutter) has a limited depth or if the bandwidth of the frequency agility is limited. Several pulses can then come so near each other in frequency that their frequency spacing is insufficient for decorrelation.

One method of calculating the number of independent pulses when using frequency agility is given below. The method gives the mean value and also the statistical distribution. The former may be used for rough estimates and the latter when particularly high accuracy is wanted. The method is based on the theory of correlation diameter and on the reasoning from the classical probability theory, with balls picked out of urns.

Two cases are dealt with, with decorrelation only from frequency agility (no fixed frequency decorrelation) and with the combined effect of frequency agility and natural decorrelation. The first case is dealt with in Section III and the latter in Section IV.

The influence on range of a certain number of independent pulses is considered briefly in Sections V and VI. Section V deals with the influence of the decorrelation of the target echo, while Section VI deals with the decorrelation of clutter and its influence on the range.

II. Number of Frequency Channels

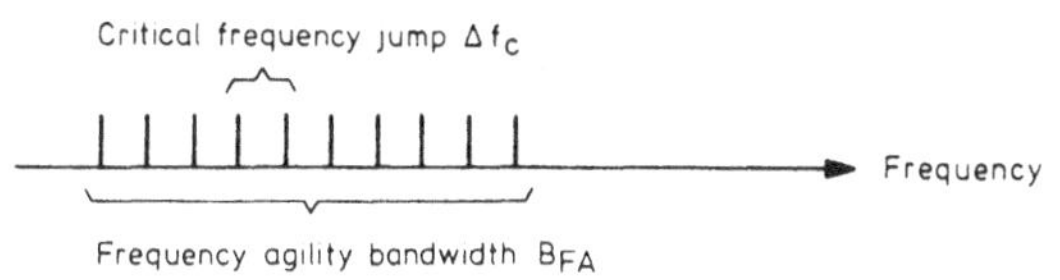

Fig. 1. Diagram of frequency channels.

As has been mentioned above, the number of independent pulses cannot be greater than the number of pulses n during the illumination time. For the case without natural decorrelation at fixed frequency, it cannot be greater than the number of frequency channels ν, a concept that is defined below. In actual fact, in this case the mean value of the number of independent pulses must be less than both n and ν. The relationships resulting from the combined effect of natural fixed frequency decorrelation and decorrelation through frequency agility are dealt with in Section V.

The number of frequency channels is defined in accordance with Fig. 1. This gives the relationship

$$\nu = 1 + B_{\mathrm{FA}}/\Delta f_c \tag{1}$$

where B_{FA} is the frequency agility bandwidth and Δf_c is the critical frequency jump, which, because of the definition used, is identical with the concept correlation diameter used by statisticians. The correlation diameter is the smallest distance which with certainty gives independence. The critical frequency jump or correlation diameter is obtained from the formula

$$\Delta f_c = c/2D \tag{2}$$

where c is the velocity of light and D is the depth of the target. The target or clutter model is a large number of equally large reflectors evenly distributed over the depth D. See [1] and [2].

As examples, the number of frequency channels has been calculated for two cases.

Case 1: Target, depth $D = 5$ m (aircraft). Frequency agility bandwidth $B_{FA} = 450$ MHz.

$$\Delta f_c = (3 \times 10^8)/(2 \times 5) = 30 \text{ MHz}$$

$$\nu = 1 + 450/30 = 16.$$

Case 2: Clutter, depth $D = 75$ m, corresponds to distributed clutter at the pulse duration 0.5 μs. Frequency agility bandwidth $B_{FA} = 50$ MHz.

$$\Delta f_c = (3 \times 10^8)/(2 \times 75) = 2 \text{ MHz}$$

$$\nu = 1 + 50/2 = 26.$$

III. Average Number of Independent Pulses, with Decorrelation Only Through Frequency Agility

If the number of pulses n per illumination time and the number of frequency channels ν are known, the average number of independent pulses can be obtained using classical probability theory. The problem is reformulated to the drawing of n balls out of an urn. The urn contains an infinite number of balls of different colors in equal proportion. We are then interested in the number of *different* colors k obtained as the result of a draw. As a mean value we obtain

$$E\{k\} = \nu[1 - (1 - 1/\nu)^n] \tag{3}$$

where k is the number of independent pulses (colors) and $E\{k\}$ is the average value of k.

Proof of (3) is given in the Appendix, which also includes the formula material for the probability of getting exactly k independent pulses (different colors), i.e., the distribution is found for which $E\{k\}$ is the mean value. The latter may be exploited on the occasions when exceptionally high accuracy is necessary. The use of the mean value $E\{k\}$ for calculating something that has a nonlinear dependence on k, for example, detection probability, will of course give an approximate result.

Example: Number of pulses $n = 5$. Number of channels $\nu = 5$. Average number of independent pulses $E\{k\} = 3.4$. The exact distribution is given in Table I.

The formula in (3) can be approximated for large numbers of frequency channels ($\nu \gg 1$) in the following way:

$$E\{k\} \approx \nu\,[1 - e^{-n/\nu}]. \tag{4}$$

IV. Average Number of Independent Pulses, with Decorrelation Through Frequency Agility and Natural Decorrelation

As before, the number of pulses per illumination time is n. The illumination time is τ_0. The correlation time of the clutter or target is τ_c. With fixed frequency this gives the number of independent pulses m as follows:

$$m = 1 + \tau_0/\tau_c \leqslant n. \tag{5}$$

The number of independent pulses can never be greater than the number of pulses n even if the correlation time is very short.

The pulses are now divided up into m groups, each of n/m pulses. Decorrelation with frequency agility is now applied for each group in accordance with (3). The number of independent pulses for a group of n/m pulses is then given by

$$\nu[1 - (1 - 1/\nu)^{n/m}].$$

The total number of independent pulses is

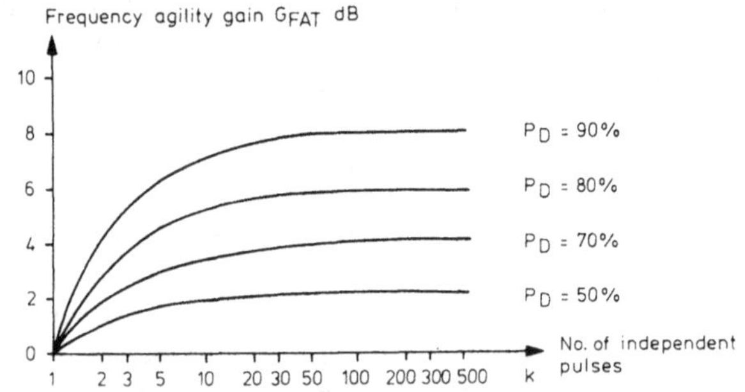

Fig. 2. Frequency agility gain versus number of independent pulses, for detection probabilities from 50 to 90 percent.

TABLE I

k	1	2	3	4	5
Probability	0.0016	0.0960	0.4800	0.3840	0.0384

$$E\{k\} = m\nu\,[1 - (1 - 1/\nu)^{n/m}]. \tag{6}$$

The approximation for $\nu \gg 1$ applies here:

$$E\{k\} \approx m\nu\,[1 - e^{-n/m\nu}]. \tag{7}$$

The reason for the above treatment is that the two decorrelation mechanisms are normally independent and are dependent on quite different parameters of a target or clutter. For example, frequency agility decorrelation of a target depends on its physical extension in depth, while fixed-frequency decorrelation depends mainly on the physical extension in width and its real or apparent augular velocity. Because the mechanisms are different they have been considered as leading to completely new draws with completely new colors each time the natural decorrelation has given rise to a new position, a new game.

V. Frequency Agility Range: Effect of the Target

The number of independent echo pulses affects the frequency agility gain, which is that part of what is known as the fluctuation loss obtained with fixed frequency that is eliminated by frequency agility. This has been calculated, among others, by Guarguaglini [3] and is shown in Fig. 2 for a detection probability P_D over the range from 50 to 90 percent.

Another way of expressing G_{FAT} is the following equation given by Barton [5], which is empirical but surprisingly accurate:

$$G_{FAT} = (1 - 1/R)\, G_{MAX} \tag{8}$$

where G_{FAT} and G_{MAX} are in dB's and G_{MAX} is the well known fluctuation loss for $k = 1$ on the target of interest.

Example: Detection probability P_D = 90 percent; mean number of independent pulses $E\{k\}$ = 6.5, which gives G_{FAT} = 6.8 dB.

VI. Frequency Agility Range: Effect of Clutter

Frequency agility affects the range in the presence of clutter through the integration gain. When calculating the range in the presence of noise, the range is improved by more than one pulse during the illumination time through integration, and this is expressed in terms of what is known as the integration gain. However, this presupposes decorrelation, which is in fact obtained in noise.

No integration gain is obtained for clutter that is completely correlated during the whole illumination time. For clutter that is partially correlated an integration gain is obtained which is a function of the equivalent number of uncorrelated pulses. This applies irrespective of the reason for the decorrelation. With fixed frequency there is usually some decorrelation of clutter, which is insignificant for ground clutter, but is often quite appreciable for rain and sea clutter. Platform motion can result in a certain decorrelation even with ground clutter, particularly with airborne systems. In general, frequency agility gives a very much more complete decorrelation. In the case of point clutter, which unlike distributed clutter consists of individual reflectors resolved by the radar, there is sometimes no decorrelation at all, neither with fixed frequency, platform motion, nor frequency agility. This is especially the case with ground clutter and sea clutter, low radar site, low grazing angle, and extremely high resolution of the radar.

The frequency agility gain may then be given by

$$G_{FAC} = G_{IFA} - G_{IFF} \tag{9}$$

where G_{FAC} is the frequency agility gain for clutter, G_{IFA} is the integration gain for frequency agility, G_{IFF} is the integration gain for fixed frequency, all expressed in dB.

The integration gain with noncoherent integration has been dealt with, among others, by Hall [4], who divides it up into parts in accordance with the following equation:

$$G_I = 10^{10} \log (n) - L_I(n) \tag{10}$$

where G_I is the integration gain in dB, n is the number of pulses for noise and the number of independent pulses for clutter, and $L_I(n)$ is the integration loss in dB.

Barton [5] has pointed out that the integration loss with noncoherent integration is actually a detector loss that is obtained because the weaker of two signals that are obtained at the same time is suppressed in an envelope detector (small signal suppression). The integration loss is determined by the total integrated signal-to-noise ratio

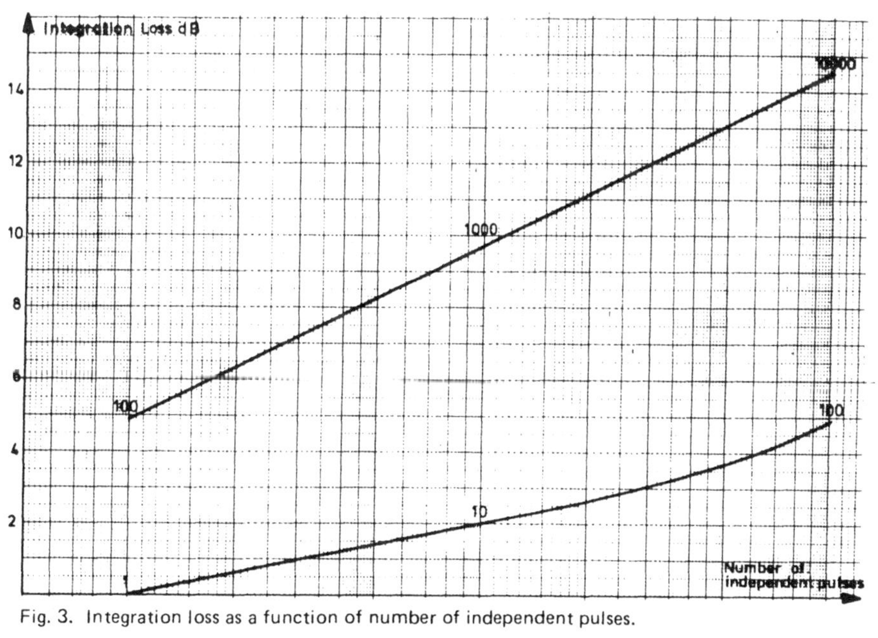

Fig. 3. Integration loss as a function of number of independent pulses.

required for detecting and only varies a little with respect to the remaining parameters, such as the false alarm ratio. It is shown in Fig. 3.

The number of independent clutter pulses with fixed frequency, if there is any decorrelation, is obtained from (5). For frequency agility the method described in Section III or IV is used.

Example: Illumination time τ_0 = 10 ms. PRF = 2000 Hz. Number of pulses n = 20. Clutter is in accordance with case 2, Section II. Number of frequency channels ν = 26. Clutter correlation time τ_0 = 4.5 ms (which corresponds to, for example, sea clutter in the X band with sea condition 5 of [6, Sec. 3.6 and Fig. 7-6].

In the case of fixed frequency the number of independent pulses will be given by $m = 1 + \tau_0/\tau_c = 3.2$. In accordance with (10) the integration gain in the case of fixed frequency will then be $G_{IFF} = 5.0 - 1.0 = 4.0$ dB.

The number of pulses n = 20, the number of independent pulses in the case of fixed frequency m = 3.2, and the number of frequency channels ν = 26 gives the number of independent pulses in the frequency agility case from (6): $E\{k\} = 18.1$.

In the case of frequency agility the integration gain will then be in accordance with (10): $G_{IFA} = 12.6 - 2.5 = 10.1$ dB. The frequency agility gain for clutter in this case will then be in accordance with (9): $G_{FAC} = 10.1 - 4.0 = 6.1$ dB.

Appendix

Proof of (3); Formula Material for Exact Calculation

Assume that we have an infinite number of balls in an urn, divided up among ν colors, with equal numbers of each color. We draw n balls out of the urn. For each color the probability that at least one of the balls drawn has that color is given by

$$1 - (1 - 1/\nu)^n.$$

The mean number of different colors is then the number of colors times the probability that a certain color is represented, or

$$E\{k\} = \nu\,[1 - (1 - 1/\nu)^n]. \tag{3}$$

The mean value is insufficient for an exact calculation, it being necessary to know $P_{k,n}$, the probability of obtaining exactly k different colors if we draw n balls and the number of colors is ν, and we now obtain the following recurrence formula:

$$P_{k,n} = P_{k,n-1} \cdot k/\nu + P_{k-1,n-1}\,[1 - (k-1)/\nu] \tag{11}$$

for $k \leqslant n$ with starting points $P_{1,1} = 1$, $P_{k,1} = 0, k \neq 1$.

Equation (11) may be understood from the following reasoning. When $n - 1$ balls have been drawn, another one ball is drawn. This can have a color that has already been drawn, and the number of colors will then not be changed, with the probability k/ν. It can also have a new color so that the number of colors drawn increases by 1 from $k - 1$ to k, with the probability $[1 - (k - 1)/\nu]$.

GÖRAN LIND
Philips Elektronikindustrier AB
175 20 Järfälla, Sweden

References

[1] W.P. Birkemeier and N.B. Wallace, "Radar tracking accuracy improvement by means of pulse-to-pulse frequency modulation," *AIEE Trans. (Commun. Electron.)*, vol. 81, pp. 571-575, 1962.

[2] J.C. Simon, "Quelques problemes de fluctuations en radioélectricité," *Ann. Radioélectricité*, Jan. 1955.

[3] P.F. Guarguaglini, "A unified analysis of diversity radar systems," *IEEE Trans. Aerosp. Electron. Syst.*, pp. 318-320, Mar. 1968.

[4] W.M. Hall. "General radar equation," Space/Aeronautics Research and Development 1962-1963.

[5] D.K. Barton, "Simple procedures for radar detection calculations," *IEEE Trans. Aerosp. Electron. Syst.*, pp. 837-846, Sept. 1969.

[6] F. Nathanson, *Radar Design Principles.* New York: McGraw-Hill, 1969.

Section III Studies by SHAPE Technical Centre

During the late 1960s, the SHAPE Technical Centre at The Hague, Netherlands, performed extensive analyses of performance of various search radars. To aid in this effort, "standard methods for predicting and specifying performance of air surveillance radar systems" were developed, and published as a NATO unclassified document by J.A. Bijvoet, TR-50-U, April 1969, AD860025. The methods were largely based on work by Hall and by Blake (see Volume 2 of *Radars*), and gave coverage contours for selected probabilities of detection (P_d = 0.5 and 0.8).

Because the detection performance of frequency diversity radars had not been covered adequately in prior literature and reports, a major analytical effort was devoted to this area, and three supplementary reports associated with TR-50-U represent the results. Since these reports are not generally available to the radar community, they are reprinted here in their entirety. Basically, the data used as an input for predicting range of the radar in TR-50-U are the detectability factors, $D(n, n_e)$, for stated probabilities of detection and false alarm (see Section 2). Figures 1 through 8 of Reprint 3-1 provide these factors for P_d = 0.5 and 0.8, $P_f = 10^{-6}$ and 10^{-10}, and Swerling Case I/II and Case III/IV targets, when linear video integration is used with one to four diversity channels. Figures 2 through 29 of Reprint 3-2 provide similar data for P_d = 0.5 and 0.8, $P_f = 10^{-4}$ and 10^{-6}, and the same Swerling cases, using triple diversity in either of two modes:

Mode 0 — linear addition of channels.

Mode 1 — logic-product combination according to which the output [(a+b) (b+c) (c+a)] is tested against the threshold. Operation with no jamming and with one or two channels jammed are considered. Details of analysis for these cases are contained in Reprint 3-3.

It is interesting to test for consistency the several methods represented by these SHAPE reports, by Guarguaglini, by Vannicola, by Barton, and by Bakut. Let us take as an example a system providing a total of 30 pulses, received in one, two, or three channels, and find the detectability factor for P_d = 0.8 and $P_f = 10^{-6}$ on a Rayleigh target.

(a) Poelman (Reprint 3-1). Figure 1, for $m \times n_h$ = 30, gives:

$m = 1,\ D(30,1) = +7.0$ dB

$m = 2,\ D(30,2) = +4.5$ dB

$m = 3,\ D(30,3) = +3.7$ dB

(b) Guarguaglini (1968). Figure 1, for all P_f and n, gives:

$$\left.\begin{array}{l}\mu = 1,\ \Delta(snr) = 5.8\text{ dB}\\ \mu = 2,\ \Delta(snr) = 3.2\text{ dB}\\ \mu = 3,\ \Delta(snr) = 2.3\text{ dB}\end{array}\right\}\begin{array}{c}\text{Add to}\\ D_o(30)\\ = +1.5\end{array}\left\{\begin{array}{l}D(30,1) = +7.3\text{ dB}\\ D(30,2) = +4.7\text{ dB}\\ D(30,3) = +3.8\text{ dB}\end{array}\right.$$

(c) Vannicola (Reprint 2-4). Figure 14, for MN = 30, gives:

$M = 1,\ D(30,1) = +7.1$ dB

$M = 2,\ D(30,2) = +4.5$ dB

$M = 3,\ D(30,3) = +3.6$ dB

(d) Barton (1969). Starting from single-pulse detectability factor on steady target and using Fig. 1 of Section 2:

$$\begin{array}{rl} D_o(1) = +12.6\text{ dB} & \text{Add} \\ L_i(30) = +\ \ 3.7\text{ dB} & \left\{\begin{array}{ll} L_f(1) = +5.6\text{ dB} & D(30,1) = +7.1\text{ dB} \\ L_f(2) = +2.8\text{ dB} & D(30,2) = +4.3\text{ dB} \\ L_f(3) = +1.9\text{ dB} & D(30,3) = +3.4\text{ dB}\end{array}\right. \\ -10\log 30 = \underline{-14.8\text{ dB}} & \\ D_o(30) = +\ \ 1.5\text{ dB} & \end{array}$$

(e) Bakut (1963). See general equation for $q_o(n_o, m_o)$, with $n_o = n = 30$ and m_o = 1, 2, or 3.

$$\left.\begin{array}{l}K_{2n}^{-1}(10^{-6}) = 127\\ K_{2n}^{-1}(0.8) = 50.6\end{array}\right\}\quad K_{2n}^{-1}(10^{-6}) - K_{2n}^{-1}(0.8) = 76.4$$

$$K_2^{-1}(0.8) = 2\ln(1/0.8) = 0.446,\quad D(30,1) = \frac{q_o}{n} = 5.78 \text{ or } +7.6\text{ dB}$$

$$K_4^{-1}(0.8) = 1.67,\quad D(30,2) = \frac{q_o}{n} = 3.09 \text{ or } +4.9\text{ dB}$$

$$K_6^{-1}(0.8) = 3.1,\quad D(30,3) = \frac{q_o}{n} = 2.5 \text{ or } +4.0\text{ dB}$$

(f) Bakut (1969). Approximation from Table 2-3 with $n_o = \gg 1$.

$$\begin{array}{l}\phi^{-1}(1 - 10^{-6}) = 4.75\\ \phi^{-1}(0.80) = 0.84\end{array}\quad 2[\phi^{-1}(1 - F) + \phi^{-1}(D)] = 11.18$$

For $m = 1,\ K_2^{-1}(0.8) = 0.446,\quad D(30,1) = 4.58$ or +6.6 dB

$m = 2,\ K_4^{-1}(0.8) = 1.67,\quad D(30,2) = 2.44$ or +3.9 dB

$m = 3,\ K_6^{-1}(0.8) = 3.1,\quad D(30,3) = 1.98$ or +3.0 dB

(g) Bakut (1969). Eq. (5.3.6) from Table 2-1 with $n = 30$, and $m_o = 1$.

$$D(30,1) = \frac{q_o}{n} = \frac{127 - 2(29)}{0.446} - 1 \div 30 = 5.12 \text{ or } +7.1 \text{ dB}$$

Summary of D(30,m)	m = 1	m = 2	m = 3	Worst Error (dB)
(a) Poelman	+7.0	+4.5	+3.7	0.1
(b) Guarguaglini	+7.3	+4.7	+3.8	0.2
(c) Vannicola	+7.1	+4.5	+3.6	0.1
(d) Barton	+7.1	+4.3	+3.4	0.2
(e) Bakut	+7.6	+4.9	+4.0	0.5
(f) Bakut	+6.6	+3.9	+3.0	0.6
(g) Bakut	+7.1	–	–	–
Exact Value	+7.1	–	–	–

It is apparent that all methods are in close agreement, with the approximate procedures (b), (d), (e), and (f) having errors of up to 0.6 dB.

A second area of Poelman's study concerns relative advantages of diversity and single-frequency systems. The problem here is to characterize accurately the options available. Poelman (Table I of Reprint 3-1) estimates 1.7 and 2.9 dB respective loss in dual and triple diversity systems from RF multiplexing and "video recombination loss." Then, considering the option of simply increasing peak power in a single channel or distributing the power over m channels, he finds the "diversity efficiency" to be small or nonexistent at $P_d = 0.8$. This result can be explained in terms of integration loss, circuit loss, and diversity gain as follows:

	Degree of Diversity		
	Single	*Double*	*Triple*
Fluctuation loss, $L_f(m)$	5.6	2.8	1.7
Diversity gain	0.0	2.8	3.9
Integration loss (for $n_h = 3$)	-0.7	-1.4	-1.9
Circuit loss	0.0	-1.7	-2.9
Relative gain	-0.7	-0.3	-0.9
Diversity efficiency	0.0	+0.4	-0.2

Slightly different values actually are derived from the precise data, but the result confirms that diversity offers no net advantage under these conditions, and is disadvantageous for the Case III/IV target. Poelman notes that radars which are limited in peak power gain by multiplexing output frequencies with time offsets to avoid waveguide breakdown.

Other assumptions lead to different results. For example, if the Deutch intrapulse diversity were used with a single transmitter tube, a broadband receiver front end, and efficient digital or analog processing to eliminate "video recombination loss," the comparison at $P_d = 0.9$ would be as follows:

	Degree of Diversity		
	Single	*Double*	*Triple*
Fluctuation loss	8.4	4.2	2.8
Diversity gain	0.0	4.2	5.6
Integration loss	-0.7	-1.4	-1.9
Circuit loss	0.0	0.0	0.0
Relative gain	-0.7	+2.8	+3.7
Diversity efficiency	0.0	+3.5	+4.4
Output pulse width	τ	$\tau/2$	$\tau/3$

Even greater gain is realized if the prf and output pulse width are held constant, thereby providing increasing average power with increasing diversity. While it can be argued that pulse compression would offer increased power in the single-channel case, the minimum practical pulse compression ratio is usually on the order of 20, and there is no convenient way to use a 2:1 or 3:1 increase in available transmitter duty factor. Also, if implemented with dual magnetron transmitters (e.g., AN/TPN-19 airport surveillance radar; see Ward et al., bibliography of Section 6), the available peak power and allowable pulse width may be constant, thereby giving a 3 dB average power advantage in the dual-diversity system. The potential diversity advantages in clutter are noted in Section 2.

With these cautions as to application of their conclusions to other systems, the SHAPE results are reprinted here for general use by the radar community.

Performance Evaluation of Frequency Diversity Radars

A.J. Poelman

Supplement 2 to Technical Report, TR-50-U, SHAPE Technical Centre, The Hague, Netherlands (April 1970)

SUMMARY

STC Technical Report TR-50-U describes procedures for evaluating the performance of radar systems which have a single transmitter-receiver channel. This Supplement contains corresponding information relating to frequency diversity radars, and includes graphs of the detectability factor.

Frequency diversity radars are radars which use two or more transmitters, each operating on a different frequency, in a multiplex arrangement on the same RF transmission channel. Examples of practical systems using this technique are the NATO Early Warning Radars and several types of air traffic control radars.

The reasons for undertaking the work reported here are twofold; first, currently available data is limited to a few cases only and, second, there is some uncertainty concerning performance quotations and the validity of approximations in previous work.

Herscovici and Detape (Ref. 1) have analyzed the performance of the various types of frequency diversity radar, including their PPI-display characteristics and the associated responses on the part of the operator. However, the only detectability graphs given in Ref. 1 relate to a three-channel system, to a single value of the number of hits per channel, and to a specific PPI display. Also, the study is restricted to Swerling case I/II type scintillating targets.

Grasso and Guargualini (Ref. 2) were also concerned with Swerling case I/II type scintillating targets; they produced graphs of the linear video addition type of diversity for 50% and 90% probabilities of detection. TR-50-U, however, bases performance parameters on probabilities of detection of 50% and 80%.

This Supplement contains graphs of the detectability factor for the two sorts of Swerling target cross-section fluctuation model (known as cases I and II and cases III and IV, respectively), consideration being restricted to the linear video addition type of recombination. These were prepared by extending the procedures given by Marcum (Ref. 3) and Swerling (Ref. 4). Attention is also paid to the efficiency of this type of frequency diversity radar, taking into account the inevitable losses in the RF multiplexing and video recombination unit.

It is concluded, from the results shown in Fig. 9, that the complexities inherent in the use of frequency diversity can be justified only when there are power-handling limitations in the tube or waveguide, when a spread of frequencies is deemed necessary in the jamming context, or when continuity of detection is highly important.

It appears therefore that, where the adoption of frequency diversity is contemplated because of pulse peak power limitations, it would be more efficient to increase the pulse length and, if necessary and possible, apply pulse compression. If, on the other hand, the limitation is the average power, then obviously the best answer is to use pulse-to-pulse frequency agility.

This work was performed under Task IV.b and V.d of the STC Programme of Work for 1969/70.

1. INTRODUCTION

There are several ways in which transmitter power can be increased so as to improve the coverage of a radar system.

(1) <u>Increase the average power on target in a single channel system</u>

Means of doing this, and the possible difficulties which might be encountered, are as follows:

(a) Increase the peak power; limited by the power-handling capabilities of the waveguides, and by costs.

(b) Increase the pulsewidth; limited by the desired range resolution. Or increase the pulsewidth and apply pulse compression.

(c) Increase the p.r.f; limited by the maximum unambiguous range.

(d) Increase the number of hits per target; limited by the desired data rate.

(2) <u>Employ more than one transmitter (each on a different frequency but using a common antenna)</u>

In such a system the pulses from the different transmitters are transmitted sequentially.

The received target-returns in the different channels are rephased in time and then suitably recombined at video level before transmission to automatic or manual extractors.

Radars using this method of operation are known as "Frequency Diversity Radars".

This supplement relates to the performance of frequency diversity radars using common linear video addition type of recombination.

Advantages of the application of frequency diversity radars are:

(1) As the pulses of the different transmitters are transmitted sequentially, no increase is required in the power-handling capability of the waveguide system.

(2) In addition to having more total transmitted power, by reason of the extra transmitter, frequency diversity radars also have improved detection performance if the groups of echoes on the different channels are uncorrelated, and the group of echoes on each channel is correlated. Basically this means that the greater the number of uncorrelated target returns, the less is the influence of the amplitude scintillation.

(3) If one of the transmitters fails, coverage will be maintained and, as a result, continuity of detection preserved.

(4) A better ECCM performance will be obtained because of the spread of frequencies. In addition, special types of diversity recombination can be used with the possibility of further improving the ECCM performance, depending on the particular type of jamming encountered.

Unfortunately, the use of RF multiplexing and video recombination in frequency diversity radars introduces losses which limit their practical efficiency.

There are several ways in which the video signals in the different channels can be recombined (see Ref. 1). For example, in a three-channel system, where a, b and c are the channels, the following combinations exist:

$$\left.\begin{aligned} A_3 &= (a + b + c) \\ M_3 &= (a \times b \times c) \\ D_1 &= (ab + bc + ca) \\ D_2 &= (a + b)\ (b + c)\ (c + a) \end{aligned}\right\} \quad (1.1)$$

Herscovici and Detape proved (in Ref. 1) that the linear video addition type of recombination, A_3 in (1.1), is fundamentally the most efficient. They further showed that the performances of the mixed combinations, D_1 and D_2 in (1.1), are not much worse than that of the A_3 type. Because D_1 and D_2 are more efficient against certain types of jamming, these combinations are of importance.

This Supplement is concerned only with deriving the performance of the linear video addition type of frequency diversity radar, taking account of both the Swerling target cross-section fluctuation models. The following assumptions have been made:-

- the transmitters all have the same peak power and pulsewidth
- there is complete correlation of the group of echoes on each of the individual channels
- there is adequate spacing between the frequencies of the transmitters, so that there is no correlation between the groups of echoes on the different channels.

At present, insufficient measured data is available to show whether these conditions apply to existing radars and targets.

The derivation of the performance of the interesting D_1 type of diversity may be covered in a separate Supplement to TR-50-U at a later date.

126

2. FORMULAE FOR THE PROBABILITY OF DETECTION

The formulae for the probability of detection, which appear below, were used to calculate the detectability curves and these in turn provided the relevant detectability factors. (The derivations of the formulae appear in the Appendix.)

An m-channel system was considered, with transmitters of equal peak power and pulsewidth, n_h hits per channel, square-law detection in each of the channels, linear video addition recombination, and a threshold after integration. It was further assumed that additive internal narrow-band gaussian noise of equal power was present in the channels, and that there was complete correlation between echoes on the same channel and no correlation between echoes on different channels. The threshold levels, Y_b, are related to the probability of false alarm, P_f, by

$$P_f = 1 - I\left\{\frac{Y_b}{\sqrt{m\, n_h}}\, ,\; m\, n_h - 1\right\} \tag{2.1}$$

where I is the incomplete gamma function as defined by Pearson.

The formulae for the probability of detection, P_d, were obtained from the derived expressions for the probability density function of v, $\phi(v)$, the signal after recombination and integration, via

$$P_d = \int_{Y_b}^{\infty} \phi(v)dv = 1 - \int_{0}^{Y_b} \phi(v)dv \tag{2.2}$$

Both the Swerling target scintillation models, (see Ref. 4) were considered and resulted in the following probability density functions:

<u>type 1</u>

$$\phi(x,\bar{x}) = \frac{1}{\bar{x}} \exp\left(-\frac{x}{\bar{x}}\right), \quad x \geq o \tag{2.3}$$

<u>type 2</u>

$$\phi(x,x) = \frac{4x}{(\bar{x})^2} \exp\left(-\frac{2x}{\bar{x}}\right), \quad x \geq o \tag{2.4}$$

x = signal-to-noise power ratio

$\bar{x}$ = average of x over all target fluctuations

<u>Case 1</u> : Swerling type 1 target scintillation model

$$P_d = 1 - \sum_{k=o}^{\infty} \Bigg[(mn_h + k - 1)^{C}(k) \cdot \frac{(m)_k}{(mn_h)_k} \cdot \frac{(n_h\bar{x})^k}{(1 + n_h\bar{x})^{m+k}} \cdot I\left\{ \frac{Y_b}{\sqrt{mn_h + k}}, \; mn_h + k - 1 \right\} \Bigg], \tag{2.5}$$

where,

$$(p)^{C}(q) = \frac{p!}{(p-q)!(q)!}$$

and

$$(n)_k = \begin{cases} 1, \text{ if } k = o \\ (n) \times (n + 1) \times \ldots\ldots.(n + k - 1) \\ \qquad\qquad \text{otherwise.} \end{cases}$$

Other expressions for P_d are

$\underline{n_h = 1}$

$$P_d = \sum_{k=1}^{m} \left[{}_{(m)}C_{(k)} \cdot (1 + \bar{x})^k \cdot \phi^{(k-1)}(v = Y_b) \right], \tag{2.6}$$

$\underline{n_h > 1}$

$$P_d = 1 - I\left\{\frac{Y_b}{\sqrt{mn_h - m}}, mn_h - m - 1\right\} + \sum_{k=1}^{m}\left[{}_{(m)}C_{(k)} \cdot (1 + n_h\bar{x})^k \cdot \phi^{(k-1)}(v = Y_b)\right],$$

where,

$$\phi^{(k-1)}(v=Y_b) = \left[\left(\frac{d}{dv}\right)^{k-1}\left\{\phi(v)\right\}\right]_{v=Y_b}$$

$$\phi(v) = \frac{1}{(1 + n_h\bar{x})^m} \cdot \frac{1}{(mn_h - 1)!} \cdot v^{mn_h - 1} \cdot \exp(-v) \cdot {}_1F_1\left\{m, mn_h, \frac{v}{1 + \frac{1}{n_h\bar{x}}}\right\},$$

and ${}_1F_1$ is a confluent hypergeometric function, called the Kummer function,

$${}_1F_1(\alpha, \beta, z) = 1 + \frac{\alpha}{\beta} \cdot \frac{z}{1!} + \frac{\alpha(\alpha+1)}{\beta(\beta+1)} \cdot \frac{z^2}{2!} + \ldots\ldots = \sum_{k=o}^{\infty}\left[\frac{(\alpha)_k}{(\beta)_k} \cdot \frac{z^k}{k!}\right].$$

<u>Case 2</u> : Swerling type 3 target scintillation model

$$P_d = 1 - \sum_{k=o}^{\infty} \left[{}_{(mn_h+k-1)}C_{(k)} \cdot \frac{(2m)_k}{(mn_h)_k} \cdot \frac{\left(\frac{n_h\bar{x}}{2}\right)^k}{\left(1+\frac{n_h\bar{x}}{2}\right)^{2m+k}} \cdot I\left\{ \frac{Y_b}{\sqrt{mn_h+k}} , mn_h+k-1 \right\} \right] . \tag{2.7}$$

Other expressions for P_d are

<u>$n_h = 2$</u>

$$P_d = \sum_{k=1}^{2m} \left[{}_{(2m)}C_{(k)} \cdot (1+\bar{x})^k \cdot \phi^{(k-1)}(v = Y_b) \right] , \tag{2.8}$$

<u>$n_h > 2$</u>

$$P_d = 1 - I\left\{ \frac{Y_b}{\sqrt{mn_h-2m}} , mn_h-2m-1 \right\} + \sum_{k=1}^{2m} \left[{}_{(2m)}C_{(k)} \left(1 + \frac{n_h\bar{x}}{2}\right)^k \cdot \phi^{(k-1)}(v = Y_b) \right] ,$$

where,

$$\phi(v) = \frac{1}{\left(1+\frac{n_h\bar{x}}{2}\right)^{2m}} \cdot \frac{v^{mn_h-1}}{(mn_h-1)!} \exp\left(-\frac{v}{1+\frac{n_h\bar{x}}{2}}\right) \cdot {}_1F_1\left\{ mn_h - 2m, mn_h, \frac{-v}{1+\frac{2}{n_h\bar{x}}} \right\} .$$

3. PRESENTATION OF RESULTS AND PERFORMANCE EVALUATION PROCEDURE

Curves (corresponding to the format used in TR-50-U) of the detectability factor, D_s, versus the total number of hits, mn_h, for both Swerling target scintillation models, and with parameters practical values for P_d, P_f and number of channels m, are given in Fig. 1 - 8.

For comparison purposes, curves for a one-channel system are included in these graphs.

<u>How to use the results for performance evaluation purposes</u>

The radar range equations given in TR-50-U can be used in respect of frequency diversity radars.

It is to be noted that:

D_s : The detectability factor, D_s, as given in Fig. 1 to 8 must be used. The detectability factor is given as a function of the total number of hits received by all channels.

P_t : The effective transmitter peak power, P_t, is the peak power of one channel (all channels are assumed to have the same peak power and pulsewidth).

L_r,L_t : The waveguide loss on transmission, L_t, and reception, L_r, must include the losses in the RF multiplexer.

L_x : The receiving system loss, L_x, must include the loss due to video recombination and the increased collapsing loss.

Note that, unlike the procedure given in TR-50-U, the diversity gain is not included in L_x.

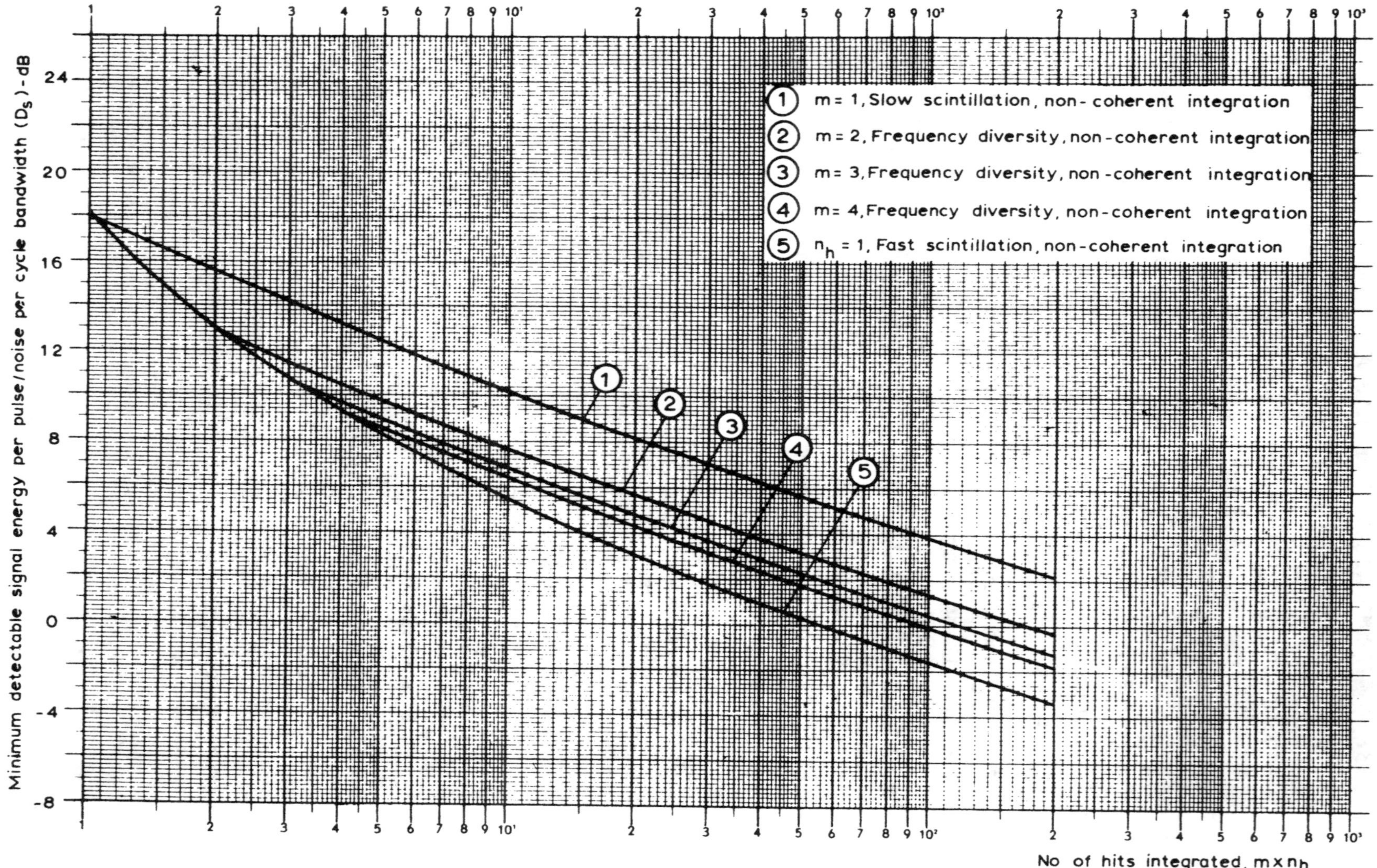

Fig. 1 Required signal-to-noise ratio, for P_d=0.8 and P_f=10^{-6}, for an m-channel frequency diversity radar and a Swerling case I/II type scintillating target.

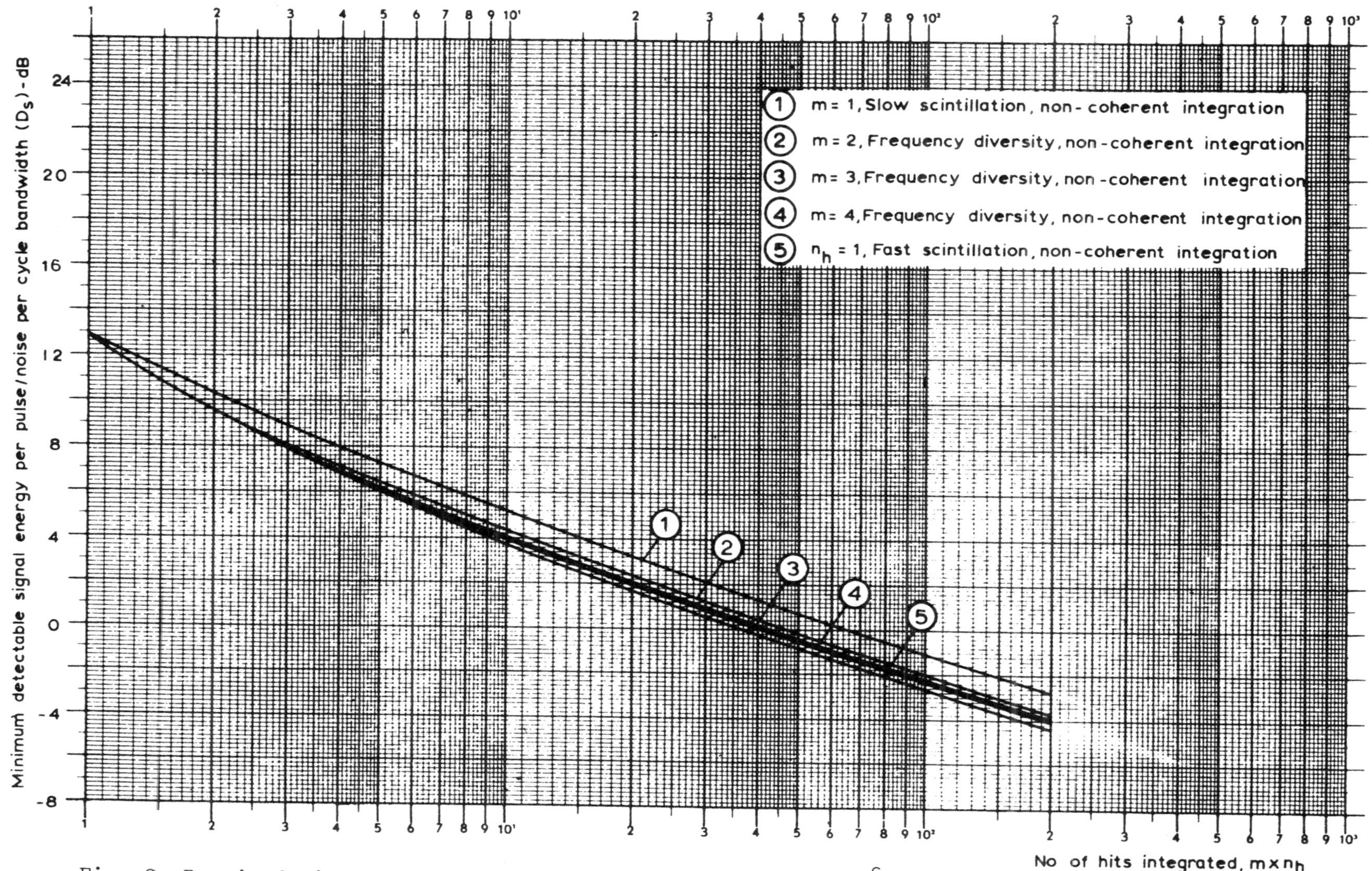

Fig. 2 Required signal-to-noise ratio, for P_d=0.5 and P_f=10^{-6}, for an m-channel frequency diversity radar and a Swerling case I/II type scintillating target.

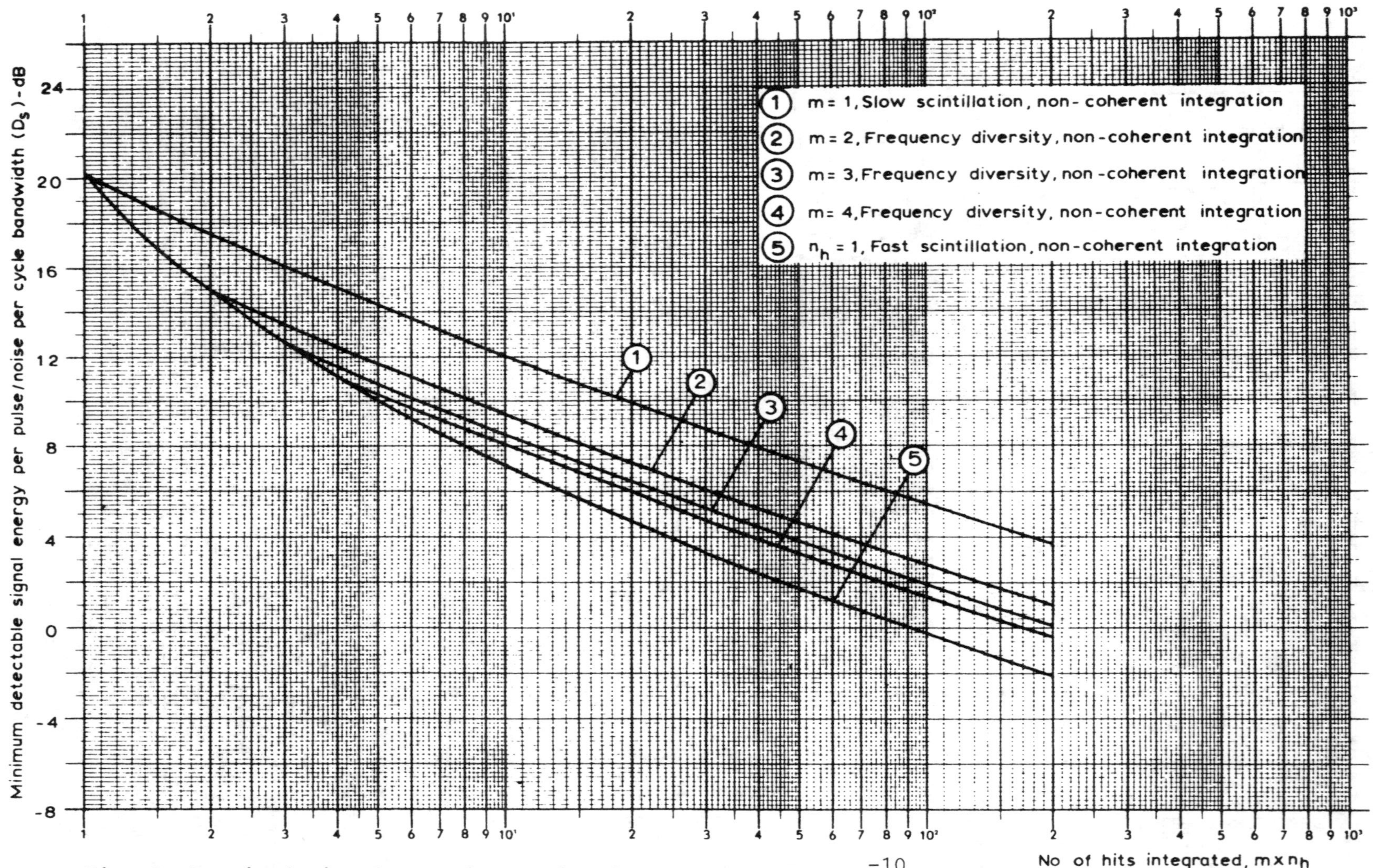

Fig. 3 Required signal-to-noise ratio, for P_d=0.8 and P_f=10^{-10}, for an m-channel frequency diversity radar and a Swerling case I/II type scintillating target.

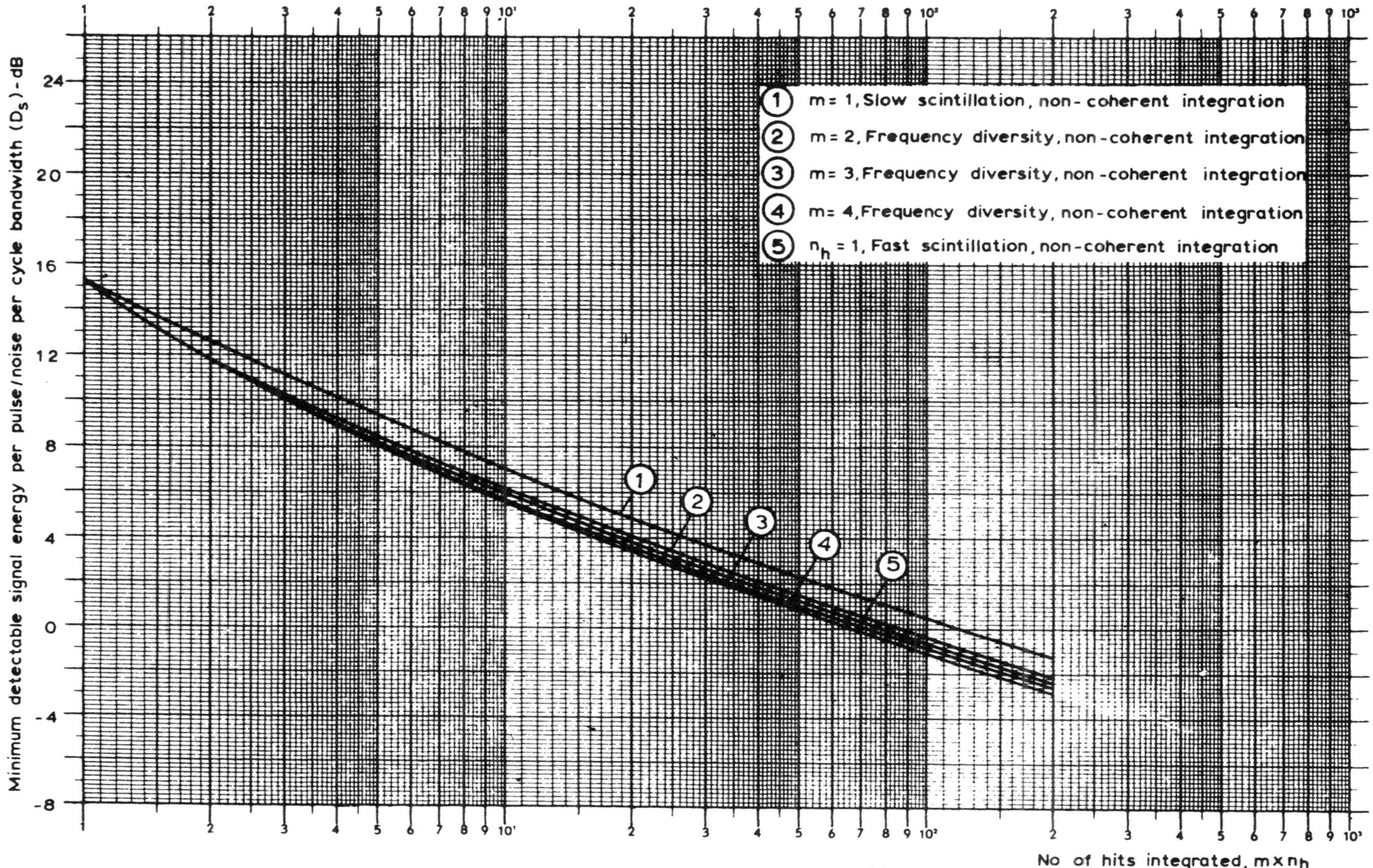

Fig. 4 Required signal-to-noise ratio for P_d=0.5 and P_f=10^{-10}, for an m-channel frequency diversity radar and a Swerling case I/II type scintillating target.

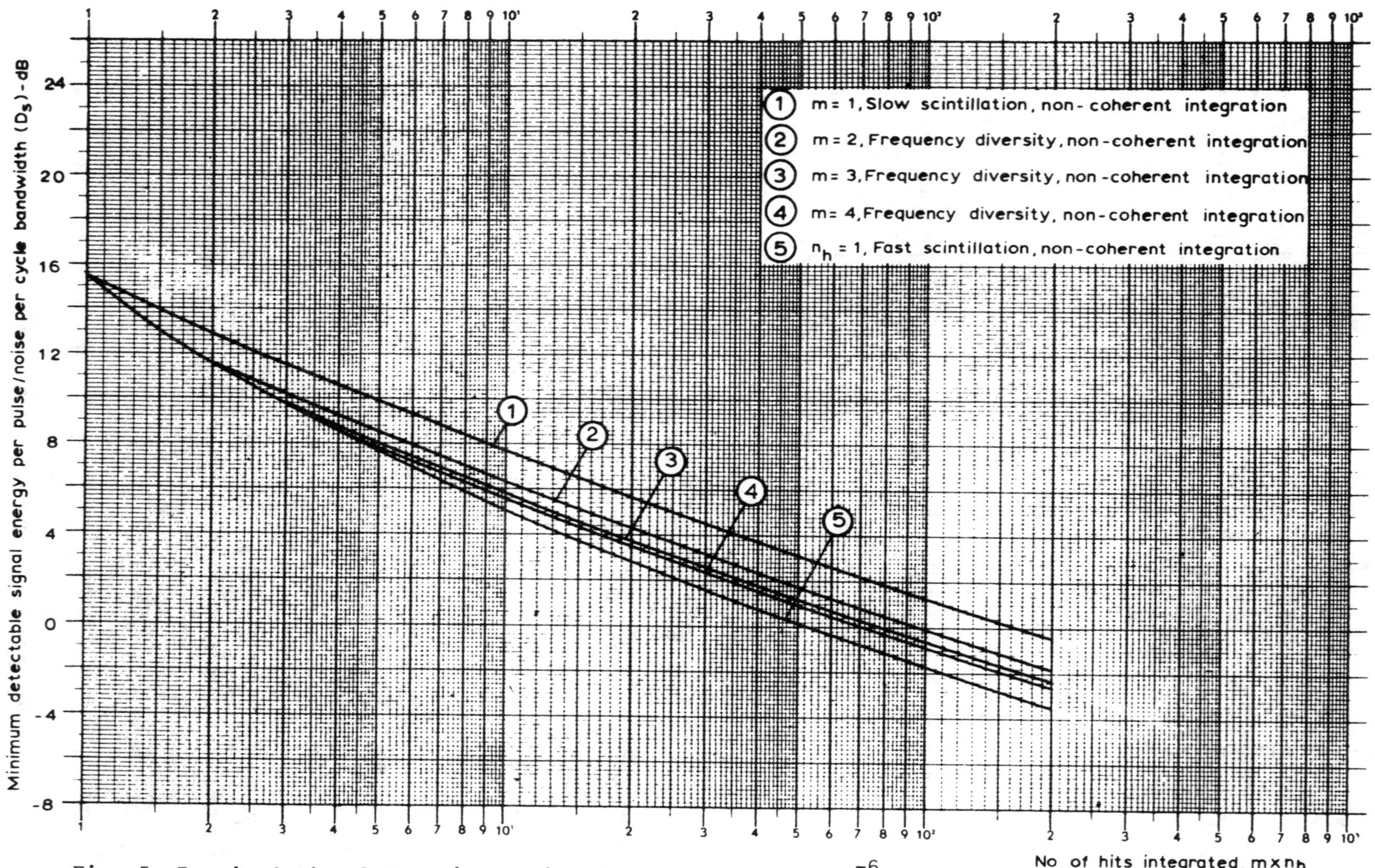

Fig. 5 Required signal-to-noise ratio, for P_d=0.8 and P_f=10^{-6}, for an m-channel frequency diversity radar and a Swerling case III/IV type scintillating target.

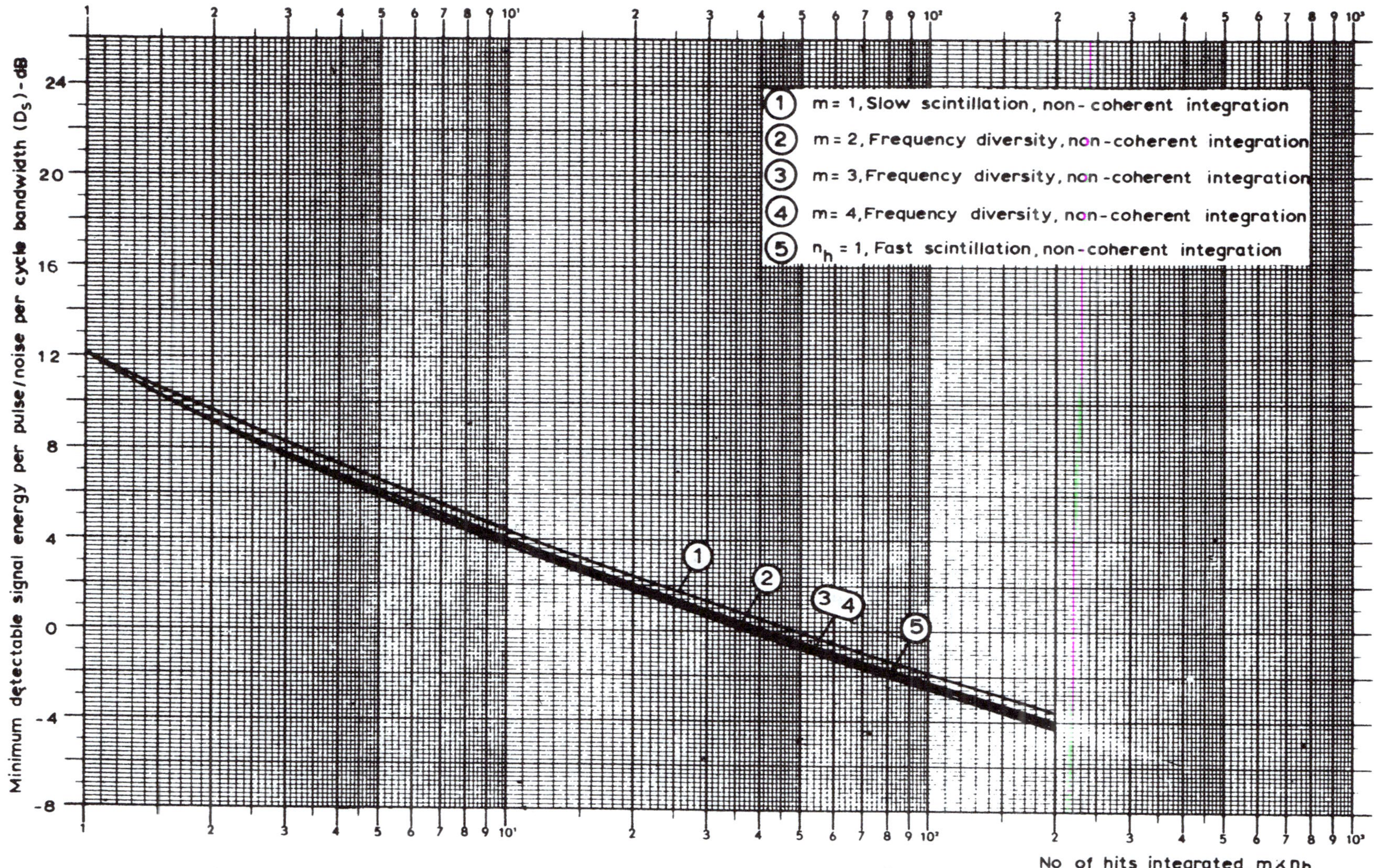

Fig. 6 Required signal-to-noise ratio, for P_d=0.5 and P_f=10^{-6}, for an m-channel frequency diversity radar and a Swerling case III/IV type scintillating target.

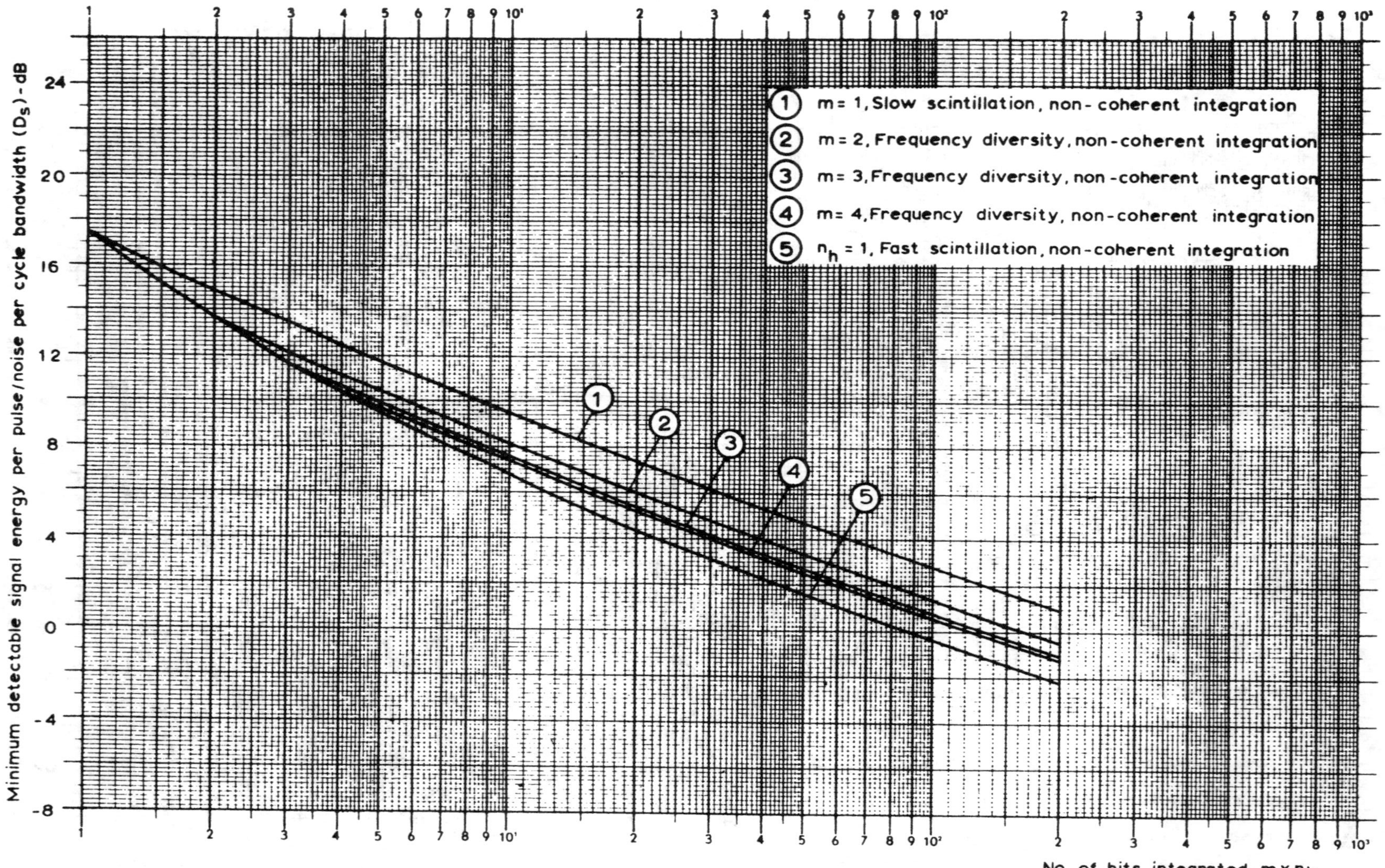

Fig. 7 Required signal-to-noise ratio, for P_d=0.8 and P_f=10^{-10}, for an m-channel frequency diversity radar and a Swerling case III/IV type scintillating target.

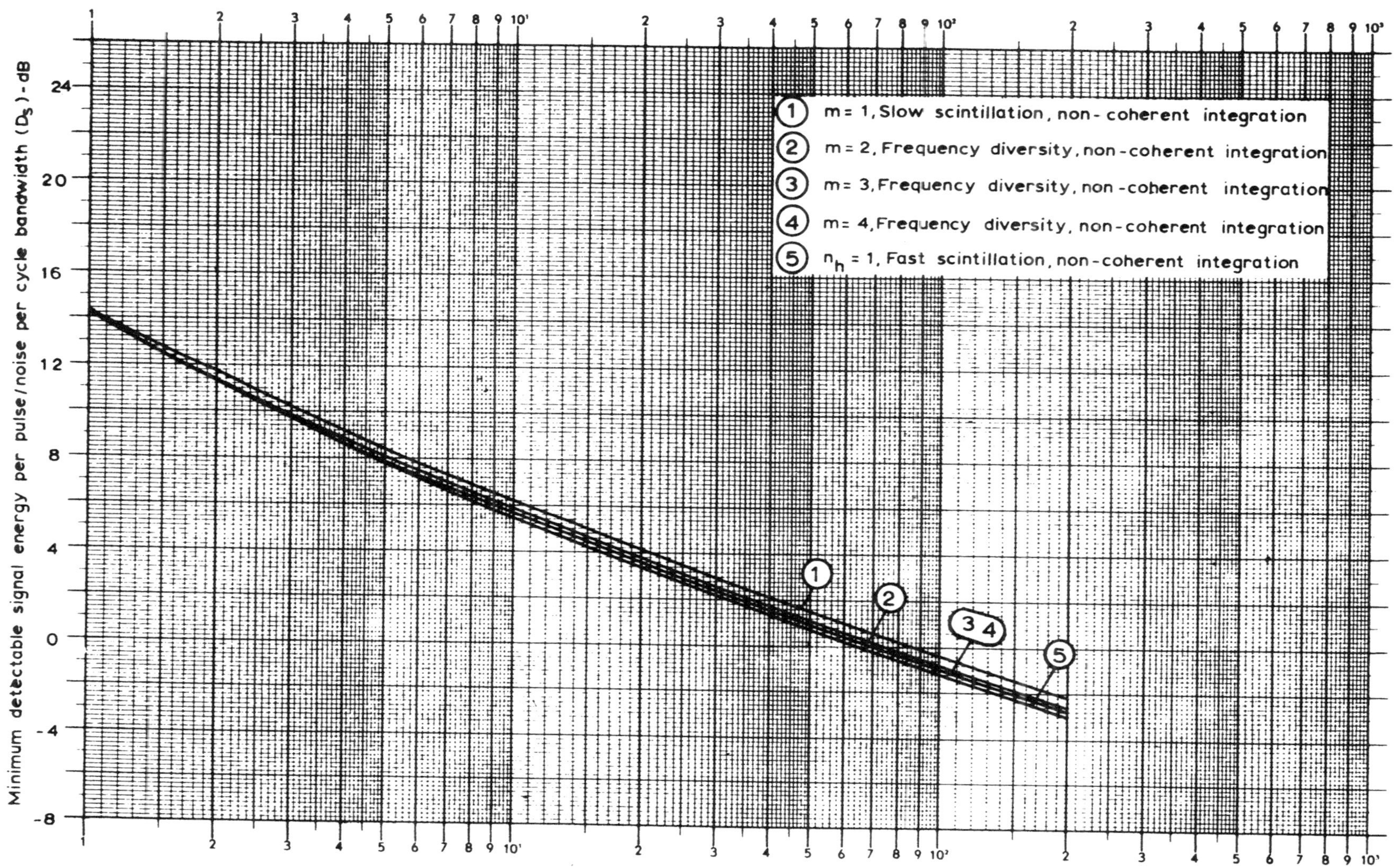

Fig. 3 Required signal-to-noise ratio, for P_d=0.5 and P_f=10^{-10}, for an m-channel frequency diversity radar and a Swerling case III/IV type scintillating target.

4. DISCUSSION OF DIVERS TY GAIN, EFFICIENCY AND LOSSES

The diversity gain, ΔD_s, is defined as the gain in required signal-to-noise ratio resulting from increasing the number of used frequencies and increasing the total power radiated, as compared with a single-channel system.

The diversity efficiency is the gain in required signal-to-noise ratio of the actual system as compared with the case in which additional power was used to increase the single channel pulse power.

In Fig. 9, the diversity gain is plotted against the number of channels for a limited number of values for the different parameters. The coherent integration line is drawn in these graphs to enable the diversity efficiency to be easily measured.

It can be seen from graphs that the diversity efficiency reduces when the number of hits per channel or the number of channels increases. It also follows that for missile-like targets, or to some extent also for modern, short wing, aircraft, as represented by the Swerling case III/IV type scintillating model, the diversity efficiency is non-existent, even for $P_d = 0.8$.

The diversity arrangement is of course achieved at the expense of an RF multiplexer, which introduces an appreciable waveguide loss on both transmission and reception, and a video recombination unit, which also introduces an appreciable loss.

The figures for these losses given in Table 1 were obtained by averaging the losses quoted in various contractors' specifications relating to radar systems of this type. The figures are presumed to be typical.

TABLE 1

Extra Losses Inherent in Frequency Diversity Radars

Type of loss	2-channel system per channel	3-channel system per channel
RF multiplexing loss	2 x 0.35 dB	2 x 0.7 dB
Video recombination loss	1 dB	1.5 dB
Total	1.7 dB	2.9 dB

Subtracting these losses shows (Fig. 9) that, even for Swerling case I/II type scintillating targets and $P_d = 0.8$, the diversity efficiency largely vanishes.

It should be noted that the gains shown here are only valid if the different transmitter frequencies used are sufficiently far apart to avoid correlation occurring between the groups of echoes on the different channels. This may become difficult in the case of many modern types of aircraft and on missile-type targets. Correlation between the groups of echoes on the different channels results in less diversity gain.

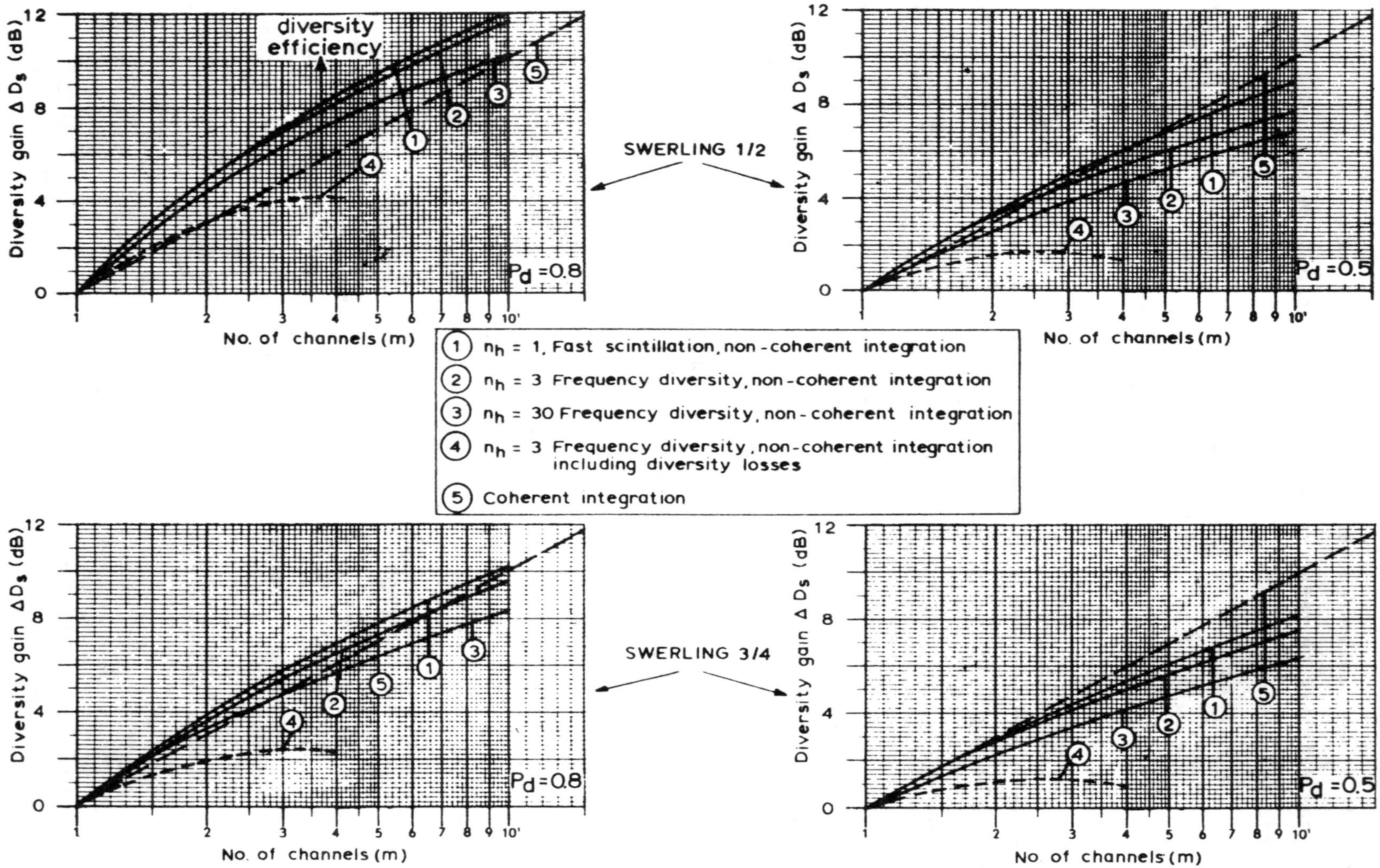

Fig. 9 Diversity gain versus number of channels (linear addition), $P_f=10^{-6}$.

70196

5. CONCLUSIONS

It is concluded, from the results shown in Fig. 9, that the complexities inherent in the use of frequency diversity can be justified only when there are power-handling limitations in the tube or waveguide, when a spread of frequencies is deemed necessary in the jamming context, or when continuity of detection is highly important.

It appears therefore that, where the adoption of frequency diversity is contemplated because of pulse peak power limitations, it would be more efficient to increase the pulse length and, if necessary and possible, apply pulse compression. If, on the other hand, the limitation is the average power, then obviously the best answer is to use pulse-to pulse frequency agility.

This Supplement contains curves of the detectability factor for the common linear video addition type of frequency diversity radar system. The procedure for using the results for the purpose of performance evaluation, using the standard methods given in STC Technical Report TR-50-U, is explained.

APPENDIX

DERIVATION OF FORMULAE FOR THE PROBABILITY OF DETECTION

This Appendix gives the derivation of formulae for the probability of detection of a frequency diversity system which uses m channels, ν equal transmitter peak powers (and pulsewidth), and n_h hits per channel, when additive internal narrow-band gaussian noise of equal power is present in the channels.

Consideration is restricted to the case of the linear video addition type of recombination, it is assumed that the returns on each channel are completely correlated, while the groups of returns on the different channels are not correlated.

Square-law detection and linear-law integration are considered because the linear envelope detector is more difficult to deal with and it may be expected (see Ref.3) that the results from both types of detectors will be approximately equal.

Finally, two well-known types of target cross-section fluctuation models are considered (see Ref. 1) which result in the following probability density functions for the input signal-to-noise power ratios.

(a) Swerling cases I/II type model,

$$\phi(x,\bar{x}) = \frac{1}{\bar{x}} \exp\left(-\frac{x}{\bar{x}}\right), \quad x \geqslant 0, \tag{A.1}$$

(b) Swerling case III/IV type model,

$$\phi(x,\bar{x}) = \frac{4x}{(\bar{x})^2} \exp\left(-\frac{2x}{\bar{x}}\right), \quad x \geqslant 0, \tag{A.2}$$

where

x = signal-to-noise power ratio on the different channels

and

$\bar{x}$ = average of x over all target fluctuations

When $\phi(v)$ is the probability density function of the recombined video, v, it follows that the probability of detection,

$$P_d = \int_{Y_b}^{\infty} \phi(v)dv = 1 - \int_{0}^{Y_b} \phi(v)dV, \tag{A.3}$$

where Y_b, the threshold level, is related to the probability of false alarm, P_f, by, (see Ref. 3).

$$P_f = 1 - I\left\{\frac{Y_b}{\sqrt{m.n_h}}, m.n_h - 1\right\}, \tag{A.4}$$

where

$m.n_h$ = total number of integrated independent variates of noise

and

I = the incomplete gamma-function as defined by Pearson.

For the determination of Y_b, which corresponds with desired values of P_f, we refer to Ref. 5.

(a) <u>Swerling type 1 target scintillation model</u>

From Ref. 3, it follows for the characteristic function for the sum of n_h pulses of signal plus noise,

$$\bar{C}_{n_h}(p) = \frac{1}{(1+p)^{n_h - 1}\left\{1+p(1+n_h\bar{x})\right\}}$$

Then the characteristic function of the sum, v, of m uncorrelated groups of each n_h pulses is,

$$\bar{C}^{m}_{n_h}(p) = \left[\bar{C}_{n_h}(p)\right]^{m} = \frac{1}{(1+p)^{mn_h-m}\left\{1+p(1+n_h\bar{x})\right\}^{m}} \tag{A.5}$$

The inverse Laplace transform of (A.5) can be found with pair (581.1) of Ref. 6. It follows,

$$\phi(v) = \frac{1}{(1+n_h\bar{x})^{m}} \cdot \frac{1}{(mn_h-1)!} \cdot v^{\,mn_h-1}\exp(-v) \cdot {}_1F_1\left\{m, mn_h, \frac{v}{1+\frac{1}{n_h\bar{x}}}\right\}, \tag{A.6}$$

where

$${}_1F_1(\alpha,\beta,z) = 1 + \frac{\alpha}{\beta}\,\frac{z}{1!} + \frac{\alpha(\alpha+1)}{\beta(\beta+1)}\cdot\frac{z^2}{2!} + \cdots\cdots = \sum_{k=0}^{\infty}\left[\frac{(\alpha)_k}{(\beta)_k}\,\frac{z^k}{k!}\right], \tag{A.7}$$

is a confluent hypergeometric function, the so-called Kummer function.

There are two methods, at least, by which expressions for the probability of detection, P_d, may be obtained.

(i) Method 1

Substitution of (A.6) in (A.3) and using (A.7) results in,

$$P_d = 1 - \frac{1}{(1+n_h\bar{x})^m (mn_h-1)!} \sum_{k=o}^{\infty} \left[\frac{(m)_k}{(mn_h)_k} \cdot \frac{1}{k!} \cdot \right.$$

$$\left. \left\{ \frac{1}{1+\frac{1}{n_h\bar{x}}} \right\}^{n_h} \int_o^{Y_b} v^{mn_h+k-1} e^{-v} dv \right] \tag{A.8}$$

The incomplete gamma function as defined by Pearson, is,

$$I(u,p) = \int_o^{u\sqrt{p+1}} \frac{e^{-v} \cdot v^p}{p!} \, dv \tag{A.9}$$

With (A.9) it follows for (A.8),

$$P_d = 1 - \sum_{k=o}^{\infty} \left[{}_{(mn_h+k-1)}C_{(k)} \cdot \frac{(m)_k}{(mn_h)_k} \cdot \frac{(n_h\bar{x})^k}{(1+n_h\bar{x})^{m+k}} \cdot \right.$$

$$\left. I\left\{ \frac{Y_b}{\sqrt{mn_h+k}} , mn_h+k-1 \right\} \right], \tag{A.10}$$

where

$${}_pC_q = \frac{p!}{(p-q)!.q!} \tag{A.11}$$

Equation (A.10) is a general expression for P_d; if $m = 1$, we have Swerling case 1 and if $n_h = 1$, we have Swerling case 2.

(ii) Method 2

From (A.5) it follows,

$$\bar{C}^{m}_{n_h}(p) - \frac{1}{(1+p)^{mn_h-m}} = \frac{1}{(1+p)^{mn_h-m}} \cdot \left[\frac{1}{\left\{1+p(1+n_h\bar{x})\right\}^{m}} - 1 \right],$$

or,

$$\bar{C}^{m}_{n_h}(p) = \frac{1}{(1+p)^{mn_h-m}} + \left[1 - \left\{ 1+p(1+ n_h\bar{x}) \right\}^{m} \right] . \bar{C}^{m}_{n_h}(p) .$$

Hence,

$$\bar{C}^{m}_{n_h}(p) = \frac{1}{(1+p)^{mn_h-m}} - \sum_{k=1}^{m} \left[{}_{(m)}C_{(k)} \cdot (1+n_h\bar{x})^{k} . p^{k} . \bar{C}^{m}_{n_h}(p) \right], \tag{A.12}$$

where ${}_{(m)}C_{(k)}$ is defined with (A.11).

Now,

$$\frac{v^{k-1} e^{-v}}{(k-1)!} \longleftrightarrow \frac{1}{(1+p)^{k}}, \quad k > o \tag{A.13}$$

and if

$$\phi(v) \longleftrightarrow \bar{C}^{m}_{n_h}(p) ,$$

then

$$\left\{\left(\frac{d}{dv}\right)^k \phi(v)\right\} = \phi^{(k)}(v) \longleftrightarrow p^k \bar{C}^m_{n_h}(p) - p^{k-1}\phi(v=0)$$

$$-p^{k-2}\phi^{(1)}(v=0) \ldots\ldots\ldots\ldots - \phi^{(k-1)}(v=0) \qquad \text{(A.14)}$$

From (A.6) it follows,

$$\phi^{(k)}(v=0) = 0, \quad \text{if } k=0,1,2,\ldots\ldots, mn_h-2$$

and (A.15)

$$\phi^{(k)}(v=0) = \frac{1}{(1+n_h\bar{x})^m}, \quad \text{if } k=mn_h-1$$

Hence, with (A.13), (A.14) and (A.15, it follows that the inverse Laplace transform of (A.12),

$\underline{n_h = 1}$

$$\phi(v) = -\sum_{k=1}^{m}\left[{}_{(m)}C_{(k)} \cdot (1+\bar{x})^k \cdot \phi^{(k)}(v)\right],$$

(A.16)

$\underline{n_h > 1}$

$$\phi(v) = \frac{v^{mn_h-m-1} \cdot e^{-v}}{(mn_h-m-1)!} - \sum_{k=1}^{m}\left[{}_{(m)}C_{(k)} \cdot (1+n_h\bar{x})^k \cdot \phi^{(k)}(v)\right].$$

With (A.3) and (A.16) it follows that the probability of detection,

<u>$n_h = 1$</u>

$$P_d = \sum_{k=1}^{m} \left[{}_{(m)}C_{(k)} \cdot (1 + \bar{x})^k \cdot \phi^{(k-1)}(v = Y_b) \right]$$

(A.17)

<u>$n_h > 1$</u>

$$P_d = 1 - I \left\{ \frac{Y_b}{\sqrt{mn_h - m}}, \; mn_h - m - 1 \right\} + \sum_{k=1}^{m} \left[{}_{(m)}C_{(k)} \cdot (1 + n_h \bar{x})^k \cdot \phi^{(k-1)}(v = Y_b) \right]$$

(b) <u>Swerling type 3 target scintillation model</u>

From Ref. 3 it follows for the characteristic function for the sum of n_h pulses of signal plus noise,

$$\bar{C}_{n_h}(p) = \frac{1}{(1+p)^{n_h - 2} \cdot \left\{ 1 + p\left(1 + \frac{n_h \bar{x}}{2}\right) \right\}^2}$$

Then the characteristic function of the sum, v, of m uncorrelated groups of each n_h pulses is,

$$\bar{C}^m_{n_h}(p) = \frac{1}{(1+p)^{mn_h - 2m} \left\{ 1 + p\left(1 + \frac{n_h \bar{x}}{2}\right) \right\}^{2m}} \qquad (A.18)$$

With pair (581.1) of Ref. 6 it follows for the inverse Laplace transform of (A.18),

$$\phi(v) = \frac{1}{(1+\frac{n_h\bar{x}}{2})^{2m}} \cdot \frac{v^{mn_h-1}}{(mn_h-1)!} \cdot e^{-v} \cdot {}_1F_1\left\{2m, mn_h, \frac{v}{1+\frac{2}{n_h\bar{x}}}\right\}, \qquad (A.19)$$

where ${}_1F_1$ is defined with (A.7).

Application of the relation

$${}_1F_1\,(\alpha,\beta,z) = e^{z} \cdot {}_1F_1(\beta-\alpha,\beta,-z)$$

in (A.19) results in a sometimes more convenient expression for $\phi(v)$. Then we get,

$$\phi(v) = \frac{1}{\left(1+\frac{n_h\bar{x}}{2}\right)^{2m}} \cdot \frac{v^{mn_h-1}}{(mn_h-1)!} \cdot \exp\left\{-\frac{v}{1+\frac{n_h\bar{x}}{2}}\right\} \cdot$$

$${}_1F_1\left\{mn_h-2m, mn_h, \frac{-v}{1+\frac{2}{n_h\bar{x}}}\right\}, \qquad (A.20)$$

where now the Kummer series is finite.

Again we shall use two methods to obtain formulae for P_d.

(1) <u>Method 1</u>

Substitution of (A.19) in (A.3) and using (A.7) and (A.9), results in,

$$P_d = 1 - \sum_{k=o}^{\infty}\left[(mn_h+k-1)^{C}(k) \cdot \frac{(2m)_k}{(mn_h)_k} \cdot \frac{\left(\frac{n_h\bar{x}}{2}\right)^k}{\left(1+\frac{n_h\bar{x}}{2}\right)^{2m+k}}\right.$$

$$\left. \cdot\, I\left\{\frac{Y_b}{\sqrt{mn_h+k}},\; mn_h+k-1\right\}\right]. \qquad (A.21)$$

If m = 1 we have Swerling case 3 and if $n_h = 1$ we have Swerling case 4.

(ii) Method 2

From (A.18) it follows that

$$\bar{C}^{m}_{n_h}(p) - \frac{1}{(1+p)^{mn_h-2m}} = \frac{1}{(1+p)^{mn_h-2m}}\left[\frac{1}{\left\{1+p\left(1+\frac{n_h\bar{x}}{2}\right)\right\}^{2m}} - 1\right]$$

Hence,

$$\bar{C}^{m}_{n_h}(p) = \frac{1}{(1+p)^{mn_h-2m}} - \sum_{k=1}^{2m}\left[{}^{(2m)}C_{(k)} \cdot \left(1+\frac{n_h\bar{x}}{2}\right)^{k} \cdot p^{k} \cdot \bar{C}^{m}_{n_h}(p)\right] \qquad (A.22)$$

From (A.19) it follows that

$$\phi^{(k)}(v=0) = 0, \text{ if } k=0,1,2\ldots\ldots, mn_h-2$$

and (A.23)

$$\phi^{(k)}(v=0) = \frac{1}{\left(1+\frac{n_h\bar{x}}{2}\right)^{2m}}, \text{ if } k = mn_h-1$$

Because it is rather complicated to derive the inverse Laplace transform of (A.22) if $n_h = 1$, we restrict ourselves to the cases $n_h \geq 2$.

With (A.13), (A.14) and (A.23) we obtain for the inverse Laplace transform of (A.22),

$\underline{n_h = 2}$

$$\phi(v) = -\sum_{k=1}^{2m} \left[{}_{(2m)}C_{(k)} \cdot (1 + \bar{x})^k \cdot \phi^{(k)}(v) \right],$$

(A.24)

$\underline{n_h > 2}$

$$\phi(v) = \frac{v^{mn_h - 2m - 1} \cdot e^{-v}}{(mn_h - 2m - 1)!} - \sum_{k=1}^{2m} \left[{}_{(2m)}C_{(k)} \cdot \left(1 + \frac{n_h \bar{x}}{2}\right)^k \cdot \phi^{(k)}(v) \right]$$

Hence with (A.3) it follows that the probability of detection,

$\underline{n_h = 2}$

$$P_d = \sum_{k=1}^{2m} \left[{}_{(2m)}C_{(k)} \cdot (1+\bar{x})^k \cdot \phi^{(k-1)}(v = Y_b) \right]$$

$\underline{n_h > 2}$

$$P_d = 1 - I\left\{ \frac{Y_b}{\sqrt{mn_h - 2m}},\ mn_h - 2m - 1 \right\} + \sum_{k=1}^{2m} \left[{}_{(2m)}C_{(k)} \right.$$

(A.25)

$$\left. \left(1 + \frac{n_h \bar{x}}{2}\right)^k \cdot \phi^{(k-1)}(v = Y_b) \right]$$

REFERENCES

1. S. Herscovici et A. Detape, "Performances d'un Radar Diversité", Annales de Radioelectricité, Tome 12, No. 50, Octobre 1957, pp. 285-298.

2. G. Grasso, P.F. Guarguaglini, "Valutazione delle Prestazioni di un Sistema Radar a Diversita di Frequenza", Alta Frequenza, Vol. 36, N.9, Settembre 1967, pp 799-805.

3. J.I. Marcum, "A Statistical Theory of Target Detection by Pulsed Radar", Mathematical Appendix.

   IRE Trans. on Information Theory, April 1960, pp. 145-169.

   Special Monograph Issue.

4. P. Swerling, "Probability of Detection for Fluctuating Targets",

   IRE Trans. on Information Theory, April 1960, pp. 269-308.

   Special Monograph Issue.

5. J. Eendebak, "Detectability Curves for Swerling Cases III and IV Target Models".

   SHAPE Technical Centre, Technical Report TR-50-U, Supplement I, (December 1969), (NATO Unclassified).

6. G.A. Campbell and R.M. Foster, "Fourier Integrals for Practical Applications", D. van Nostrand Company, Inc., 1948.

Further Notes on Performance Evaluation of Frequency Diversity Radar

A.J. Poelman and E.J. Bennée

Supplement 3 to Technical Report TR-50-U, SHAPE Technical Centre, The Hague, Netherlands (May 1973)

SUMMARY

STC Technical Report TR-50-U (Ref. 1) describes procedures for evaluating the performance of radar systems which have a single transmitter-receiver channel. Supplement 1 describes a computer-based method of evaluating the performance of single-channel system against missile-type targets. Supplement 2 deals with the performance evaluation of multi-channel frequency diversity radar systems operating in the linear-addition mode of video recombination in a quiet and clear environment. The present supplement contains corresponding information relating to triple-channel frequency diversity radar systems operating in the linear-addition mode and the D_1 logic-product mode of video recombination in an interference or jamming environment.

The D_1 logic-product mode was first mentioned in the work of Herscovici and Detape (Ref. 2) and is characterized by the recombination of twice the smallest video and the intermediate video. This means that the channel with the largest video is cancelled; thus this mode has an electronic counter-countermeasure (ECCM) feature when only one channel is jammed.

Modern frequency diversity radar systems usually employ three channels and usually operate in the linear-addition mode and the D_1 logic-product mode of video recombination. There is thus a need for procedures for evaluating the performance of such systems against the main target types both in clear and in jamming environments. This supplement meets that need: the work described was performed under Task V.d of the STC Programme of Work for 1971/72.

Graphs are provided of the detectability factor for the two types of Swerling target cross-section fluctuation model (known as Cases I/II and Cases III/IV respectively), and for both modes of video recombination. Distinction is made between the cases where no channel is jammed, one channel is jammed, and two channels are simultaneously jammed with equal power. (Jamming noise is assumed to have the same type of statistical distribution as the internal noise.) The graphs were prepared by extending the procedures given by Marcum (Ref. 3) and Swerling (Ref. 4).

The Supplement describes the performance evaluation procedure and also examines the efficiency of the mixed video recombination mode D_1.

The derivations of the formulae for the probabilities of false alarm and of detection are not included in the Supplement. They will be presented in an STC Technical Memorandum (Ref. 5), to be published in mid-1973, together with a discussion on related computer studies. This Memorandum will also examine the efficiency of some other mixed video recombination modes.

The following conclusions can be drawn from the results given in the present Supplement. Both in a quiet environment and when two channels are jammed, the performance (min. required average signal power - to - average internal noise power ratio) of the logic-product mode is not much different from that of the linear-addition mode. The degradation is 1 to 3 dB in a quiet environment, and the gain is -2 to 3 dB when two channels are jammed. When only one channel is jammed, the logic-product mode is far preferable to the linear-addition mode, as might be expected. Since the performance of both modes of operation is very poor when two channels are jammed, Ref. 5 will examine a mode which has an ECCM in such a jamming environment.

1. INTRODUCTION

Supplement 2 of TR-50-U (Ref. 1) deals with the performance evaluation of frequency diversity radar systems operating in the linear-addition mode of video recombination, and in a clear and quiet environment. In its introduction the Supplement lists some alternative modes of video recombinations mentioned in the original work of Herscovici and Detape (Ref. 2).

These authors defined two mixed modes of video recombination D_1 and D_2. In the case of a triple-channel system

$$D_1 = (a.b + b.c + c.a) \tag{1a}$$

and

$$D_2 = (a + b).(b + c).(c + a) \tag{1b}$$

where a, b, and c are the three channels. They concluded that:

- in a clear and quiet environment the performances of D_1 and D_2 are not much worse than those of the most efficient linear-addition mode of recombination, known as A_3, where

$$A_3 = (a + b + c), \tag{2}$$

- they are more efficient than the A_3 mode against certain types of jamming, and are therefore of importance.

Reference 2 does not state explicitly whether the products in (1) are meant to be of the arithmetic type or logical type. However, when logical products are considered in (1), the ECCM feature of the D_1 and D_2 modes becomes quite obvious. We then have, for the recombined videos,

$$D_1 \text{ mode}: \quad u_1 = 2y_s + y_i \tag{3a}$$

$$D_2 \text{ mode}: \quad u_2 = y_s + y_i, \tag{3b}$$

where y_s and y_i are the smallest and the intermediate respectively of the videos on the three channels a, b, and c. This means that when only one channel is jammed, e.g. in the case of a spot-jammer or repeater-jammer, in general the jammed channel will be cancelled.

On the other hand, the linear-addition mode of video recombination A_3 produces

$$u_o = y_s + y_i + y_1, \tag{3c}$$

where y_1 is the largest video. This obviously does not provide an ECCM feature.

Modern frequency diversity radar systems usually employ three channels, and usually operate in the A_3 mode and D_1 logic-product mode of video recombination. Therefore, in this Supplement, of the D_1 and D_2 logic-product modes only the D_1 mode in a triple-channel system is considered.

This Supplement is concerned with deriving the performance of triple-channel frequency diversity radar systems operating in:

- the D_1 logic-product mode, here called Mode 1, and
- the A_3 linear-addition mode, here called Mode 0,

(see equations (3a) and (3c), and Fig. 1), when stationary additive narrow-band white Gaussian noise is present in the channels.

The three following environmental conditions are studied:

- no channel jammed; here called Condition III
- one channel jammed; here called Condition IIE
- two channels simultaneously jammed; here called Condition IEE.

In this context a parameter J is defined for each video channel by

$$J = \frac{\text{average external noise power}}{\text{average internal noise power}} = \frac{\sigma_j^2}{\sigma_n^2}$$

We can now combine the above notations to give the following cases

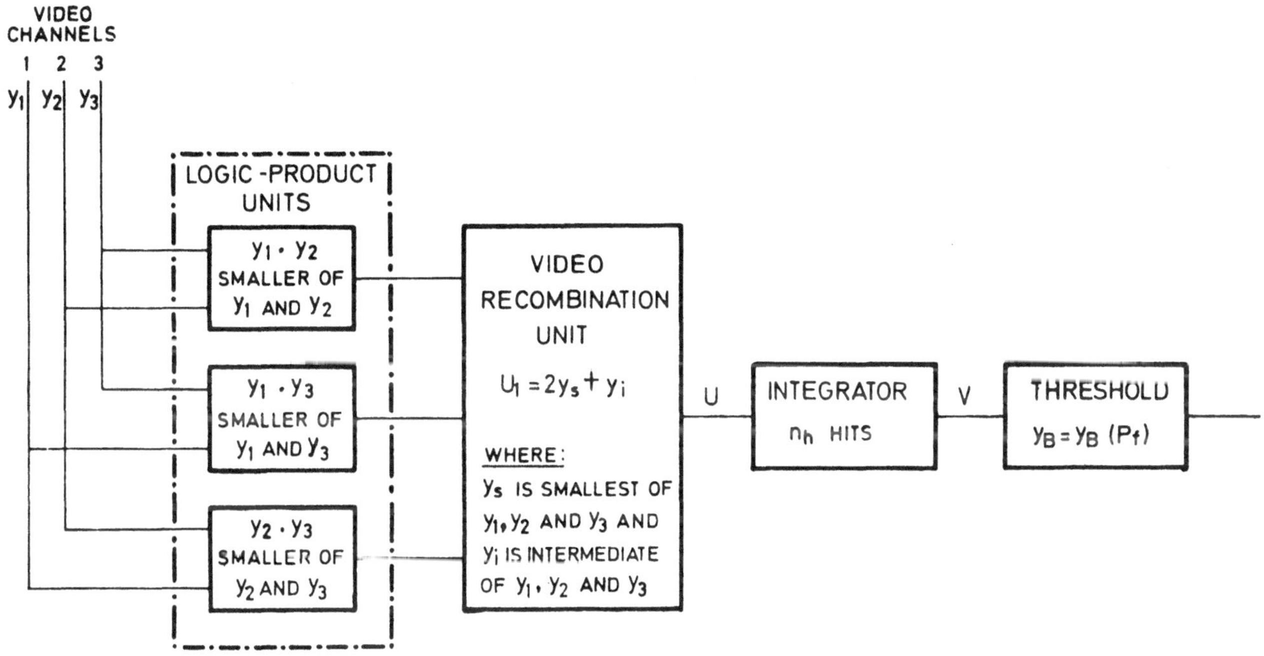

(a) Logic-product mode of video recombination (Mode 1)

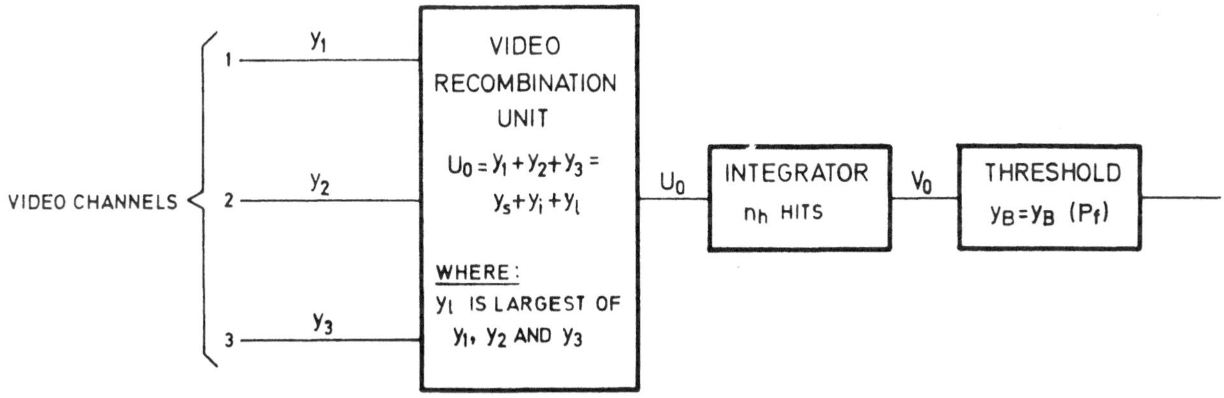

730348

(b) Linear-addition mode of video recombination (Mode 0)

Fig. 1 Triple-channel frequency diversity operation: video recombination techniques considered for evaluation of:

Environmental condition	Mode of operation	Case
III	0	III-0
III	1	III-1
IIE	0	IIE-0
IIE	1	IIE-1
IEE	0	IEE-0
IEE	1	IEE-1

Here capitals I and E denote respectively a channel with only internal noise and a channel with external noise also, due to interference or jamming.

It is noted that when all three channels are simultaneously jammed with equal power, the cases EEE-0 and EEE-1 can be considered as special cases of III-0 and III-1, at least when the jamming noise has the same type of statistical distribution as the internal noise.

The performance of Mode 0 for $J \neq 0$ is also evaluated so that attention can be given to the efficiency (relative gain) of Mode 1.

The required information on the detectability factor is obtained by extending some procedures given by Marcum (Ref. 3) and Swerling (Ref. 4), and applying them to both Swerling target cross-section fluctuation models, i.e. Cases I/II and III/IV.

For simplicity, the following assumptions are made in the analysis:

- the transmitters all have the same peak power and pulse width
- there is complete correlation of the group of target echoes on each of the individual channels
- there is adequate spacing between the frequencies of the transmitters, so that there is no correlation between the groups of target echoes on the three channels
- the target average cross-sections for the three transmitter frequencies are identical
- the average internal noise powers on the channels are equal
- the external noise or jamming has the same type of statistical distribution as the internal noise
- when two channels are jammed the average jamming noise powers on these channels are equal.

The derivations of the formulae for the probabilities of false alarm and of detection are not included in this Supplement: they will appear in Ref. 5, together with a discussion of the computer studies. Reference 5 will also examine the efficiency of the following logic-product modes of video recombination for a triple-channel frequency diversity radar system:

- Mode 2 : $u_2 = y_s + y_i$, $(a + b).(b + c).(c + a)$, i.e. D_2 of Ref. 2
- Mode 3 : $u_3 = y_s + 2y_i = 2u_2 - u_4 = 2u_1 - 3u_4$
- Mode 4 : $u_4 = y_s$, $(a.b.c)$, i.e. M_3 of Ref. 2.

2. FORMULAE FOR THE PROBABILITIES OF FALSE ALARM AND OF DETECTION

2.1 GENERAL

The formulae for the probability of false alarm, which are given in Section 2.2, are used to calculate the normalized threshold levels. The formulae for the probability of detection, which are given in Section 2.3, are used to calculate the detectability curves, and these in turn provide the desired detectability factors. Detectability curves are curves of the probability of detection versus the average input signal power - to - average internal noise power ratio, the parameters being the probability of false alarm, the number of hits per channel, and the relative jamming strength. The detectability factors are the required minimum values for the average input signal power - to - average internal noise power ratio for particular values of the probabilities of detection and false alarm and the parameters involved. The derivations of the formulae presented in Sections 2.2 and 2.3 will be presented in Ref. 5.

Because of mathematical complexities in treating the logic-product Mode 1 of frequency diversity operation, only analytical expressions for the probability of detection are obtained for the case where the target is scintillating according to the Swerling Case I/II model and the number of video recombinations n_h equals one. For the Swerling Case I/II scintillating target with $n_h > 1$ and for the Swerling Case III/IV scintillating target the detectability curves are determined by Monte Carlo simulation.

A triple-channel system is considered, with transmitters of equal peak power and pulse width, n_h hits per channel, square-law detection in each of the channels, linear-addition (Mode 0) and logic-product (Mode 1) video recombination, and a threshold after integration. It is also assumed that:

- stationary additive internal narrow-band white Gaussian noise with zero mean and of equal (average) power are present in the channels

- the noises from external origin (jamming or interference), if present in the channels, are of equal (average) power and have the same type of statistical distribution as the internal noise
- there is complete correlation between the n_h target echoes on the same channel (slow scintillation) and no correlation between the echoes on different channels (fast scintillation).
- the receivers operate linearly over the signal-strength range considered.

Square-law detection is considered because the linear envelope detector is more difficult to deal with, and it can be assumed (see Ref. 3) that both types of detector produce very similar results.

2.2 PROBABILITY OF FALSE ALARM

The normalized threshold levels y_B are related to the probability of false alarm P_f by the equation

$$P_f = \int_{y_B}^{\infty} f_v(v)dv = 1 - \int_0^{y_B} f_v(v)dv \tag{4}$$

where $f_v(v)$ is the probability density function of the normalized* signal v after video recombination and integration. Here only the noises from internal and external origin contribute to the signal v.

2.2.1 Mode 0; $u_o = y_1 + y_2 + y_3 = y_s + y_i + y_l$

For this mode, equation (4) becomes

$$P_f = 1 - (J+1)^{-mn_h} \sum_{k=0}^{\infty} \left[C_k^{mn_h+k-1} \left\{\frac{J}{J+1}\right\}^k I\left\{\frac{y_B}{\sqrt{(3n_h+k)}}, 3n_h+k-1\right\}\right] \tag{5}$$

where $J = \dfrac{\text{average external noise power}}{\text{average internal noise power}} = \dfrac{\sigma_J^2}{\sigma_n^2}$

$$C_k^{mn_h+k-1} = \frac{(mn_h+k-1)!}{k!(mn_h-1)!},$$

---

* *With respect to the average internal noise power*

and I is Pearson's expression for the incomplete gamma function, given by

$$I(u,p) = \frac{1}{\Gamma(p+1)} \int_0^{u\sqrt{(p+1)}} e^{-t} \, t^p \, dt, \qquad (5a)$$

and where

$m = 1$ for Case IIE - 0

and

$m = 2$ for Case IEE - 0

2.2.2 Mode 1; $u_1 = 2y_s + y_i$

For this mode, equation (4) takes the form

$$P_f = 1 - \sum_{k=1}^{3} \sum_{l=1}^{n_h} \left[\frac{\Phi_{kl}(p=-a_k)}{(l-1)! \, (a_k)^{n_h-l+1}} \, I\left\{ \frac{a_k y_B}{\sqrt{(n_h-l+1)}}, \, n_h-1 \right\} \right] \qquad (6)$$

where

$$\Phi_{kl}(p) = \frac{d^{l-1}}{dp^{l-1}} \left\{ \frac{Q(p)}{P_k(p)} \right\},$$

$$P_k(p) = \frac{P(p)}{(p+a_k)^{n_h}},$$

$$P(p) = \prod_{k=1}^{3} \left\{ (p+a_k)^{n_h} \right\},$$

$$Q(p) = (b_1 p + b_o)^{n_h}$$

and I is Pearson's expression for the incomplete gamma function.

In the above, the following values are used for a_k, b_1, and b_o:

Case	a_1	a_2	a_3	b_1	b_o
IIE-1	$\frac{2J+3}{3(J+1)}$	$\frac{J+2}{J+1}$	2	$-\frac{2(J+3)}{3(J+1)}$	$\frac{2(J+2)(2J+3)}{3(J+1)^2}$
IEE-1	$\frac{2}{J+1}$	$\frac{J+2}{J+1}$	$\frac{J+3}{3(J+1)}$	$-\frac{2(2J+3)}{3(J+1)^2}$	$\frac{2(J+2)(J+3)}{3(J+1)^3}$

If J approaches infinity, then in Case IIE-1 y_s and y_i (see equation (3a) and Fig. 1) will be selected from the two channels which have internal noise only. For this particular case the following relationship between P_f and Y_B has been derived:

$$P_f = 1 - 2^{n_h} \sum_{l=1}^{n_h} \left[(-3)^{l-1} C_{n_h-1}^{n_h+l-2} \left\{ \frac{I\left\{\frac{2y_B}{3\sqrt{(n_h-l+1)}}, n_h-1\right\}}{(1.5)^{-(n_h-l+1)}} + (-1)^{-(n_h+l-1)} I\left\{\frac{y_B}{\sqrt{(n_h-l+1)}}, n_h-1\right\} \right\} \right] \qquad (7)$$

where

$$C_{n_h-1}^{n_h+l-2} = \frac{(n_h+l-2)!}{(n_h-1)!(l-1)!},$$

and I is Pearson's expression for the incomplete gamma function.

2.2.3 Approximations when J>>0

Approximate relationships between P_f and y_B for J>>0 have been derived for the Cases IIE-0, IEE-0, and IEE-1; the validity of these relationships increases with increasing J. The results are:

Case IIE-0

If J>>0,

$$u_o \approx y_1,$$

where y_1 is the video of the jammed channel.

$$P_f \approx 1 - I\left\{\frac{y_B}{(J+1)\sqrt{n_h}},\ n_h - 1\right\} \tag{8}$$

<u>Case IEE-0</u>

If J>>0,

$$u_o \approx y_i + y_1,$$

or u_o is the linear addition of the videos of the two jammed channels.

$$P_f \approx 1 - I\left\{\frac{y_B}{(J+1)\sqrt{(2n_h)}},\ 2n_h - 1\right\} \tag{9}$$

<u>Case IEE-1</u>

If J>>0,

$$u_1 \approx y_i,$$

where y_i is the smallest video of the two jammed channels.

$$P_f \approx 1 - I\left\{\frac{2y_B}{(J+1)\sqrt{n_h}},\ n_h - 1\right\} \tag{10}$$

In equations (8), (9), and (10) I is again Pearson's expression for the incomplete gamma function, given in (5a).

It is clear that for Case IIE-1 no approximate relation between P_f and y_B for J>>0 can be derived, other than equation (7), which in fact is exact for $J = \infty$.

2.3 PROBABILITY OF DETECTION

The formulae for the probability of detection P_d are obtained from the expressions derived for the probability density function $f_v(v)$ of the normalized signal v after video recombination and integration via the equation

$$P_d = \int_{y_B}^{\infty} f_v(v)\, dv = 1 - \int_0^{y_B} f_v(v)\, dv. \tag{11}$$

Here y_B is determined by the desired P_f, and target echoes now contribute to the signal v.

Both Swerling target cross-section fluctuation models (see Ref. 4) are considered, which result in the following probability density functions for the input signal power - to - average internal noise power ratios in the channels:

Swerling Case I/II target model

$$f_x(x) = \frac{1}{\bar{x}} \exp(-\frac{x}{\bar{x}}), \quad x \geq 0 \tag{12a}$$

Swerling Case III/IV target model

$$f_x(x) = \frac{4x}{(\bar{x})^2} \exp(-\frac{2x}{\bar{x}}), \quad x \geq 0 \tag{12b}$$

where

x = signal power - to - average internal noise power ratio on the different channels, and $\bar{x}$ = average of x over all target fluctuations; assumed to be the same for all channels.

2.3.1 Mode 0; $u_o = y_1 + y_2 + y_3 = y_s + y_i + y_l$

For this mode we have

$$P_d = 1 - \left\{ \prod_{k=1}^{4} (a_k)^{m_k} \right\} \cdot \sum_{k=1}^{4} \sum_{l=1}^{m_k} \left[\frac{\Phi_{kl}(p=-a_k)}{(l-1)!\,(a_k)^{m_k-l+1}} \, I\left\{ \frac{a_k y_B}{\sqrt{(m_k-l+1)}}, m_k-1 \right\} \right] \tag{13}$$

where

$$\Phi_{kl}(p) = \frac{d^{l-1}}{dp^{l-1}} \left\{ \frac{1}{P_k(p)} \right\}$$

$$P_k(p) = \frac{P(p)}{(p+a_k)^{m_k}}$$

$$P(p) = \prod_{k=1}^{4} \left\{ (p+a_k)^{m_k} \right\}$$

and I is Pearson's expression for the incomplete gamma function, given at (5a).

The values to be taken for a_k and m_k depend on the target model and on the cases, as follows:

a_k and m_k	Swerling Case I/II		Swerling Case III/IV	
	Case IIE-0	Case IEE-0	Case IIE-0	Case IEE-0
a_1	$\frac{1}{J+1+n_h\bar{x}}$	$\frac{1}{1+n_h\bar{x}}$	$\frac{1}{J+1+n_h\frac{\bar{x}}{2}}$	$\frac{1}{1+n_h\frac{\bar{x}}{2}}$
a_2	$\frac{1}{1+n_h\bar{x}}$	$\frac{1}{J+1+n_h\bar{x}}$	$\frac{1}{1+n_h\frac{\bar{x}}{2}}$	$\frac{1}{J+1+n_h\frac{\bar{x}}{2}}$
a_3	$\frac{1}{J+1}$	1	$\frac{1}{J+1}$	1
a_4	1	$\frac{1}{J+1}$	1	$\frac{1}{J+1}$
m_1	1	1	2	2
m_2	2	2	4	4
m_3	n_h-1	n_h-1	n_h-2	n_h-2
m_4	$2(n_h-1)$	$2(n_h-1)$	$2(n_h-2)$	$2(n_h-2)$

2.3.2 Mode 1; $u_1 = 2y_s + y_i$

Because of mathematical complexities, expressions for P_d have been obtained only for Swerling Case I/II target and n_h=1. We have

$$P_d = \sum_{k=1}^{3} \frac{a_k}{\alpha_k} \exp(-\alpha_k y_B) \tag{14}$$

where a_k and α_k depend on the cases, as follows:

a_k and α_k	Case III-1	Case IIE-1	Case IEE-1
a_1	$-\frac{4}{3(1+\bar{x})}$	$-\frac{2(J+2+2\bar{x})}{(1+\bar{x})(J+3+3\bar{x})}$	$-\frac{2(J+2+2\bar{x})}{(J+1+\bar{x})(2J+3+3\bar{x})}$
a_2	$\frac{2}{1+\bar{x}}$	$-a_1 + \frac{2}{4J+3+3\bar{x}}$	$-a_1 - \frac{2}{J-3-3\bar{x}}$
a_3	$-\frac{2}{3(1+\bar{x})}$	$-(a_1+a_2)$	$-(a_1+a_2)$
α_1	$\frac{2}{1+\bar{x}}$	$\frac{1}{1+\bar{x}} + \frac{1}{J+1+\bar{x}}$	$\frac{1}{1+\bar{x}} + \frac{1}{J+1+\bar{x}}$
α_2	$\frac{1}{1+\bar{x}}$	$\frac{1}{3}\left\{\frac{2}{1+\bar{x}} + \frac{1}{J+1+\bar{x}}\right\}$	$\frac{1}{3}\left\{\frac{1}{1+\bar{x}} + \frac{2}{J+1+\bar{x}}\right\}$
α_3	$\frac{2}{1+\bar{x}}$	$\frac{2}{1+\bar{x}}$	$\frac{2}{J+1+\bar{x}}$

3. PRESENTATION OF RESULTS

3.1 GENERAL

Extremely difficult computation problems arise in the accurate determination of the normalized threshold levels y_B. The degree of difficulty of these computations increases with increasing number of hits per channel n_h, with increasing values for the average external noise power - to - average internal noise power ratio J, and with decreasing values for the probability of false alarm P_f.

Unfortunately, despite the use of the standard double precision feature on the CDC 6400 computer, it was found to be impossible to obtain threshold values for all cases without rewriting all the programs, and using arithmetic scaling techniques.

In order to fill the gap in threshold level information, some approximate relations between P_f and y_B have been determined (see Section 2.2.3). It is believed that in both single-channel (Ref. 6) and frequency diversity systems the average signal power - to - average noise power ratio required for a certain detection quality is not a very sensitive function of P_f and consequently of y_B.

Since the only y_B values of practical interest are those which correspond with very low values for P_f ($P_f < 10^{-3}$), in this context there is no point in trying to obtain results by Monte Carlo simulation. The determination of accurate threshold levels would require millions of samples, which is quite impracticable.

Again, equation (13) for the probability of detection P_d for Mode 0 is rather cumbersome to deal with for larger values for n_h unless J=0. Therefore this equation is used to determine the detectability curves only for $n_h = 1$. For $n_h > 1$, Monte Carlo simulation is applied to obtain the desired information, and also for Mode 1 (except for $n_h = 1$ and Swerling Case I/II).

Except for the determination of y_B, Monte Carlo simulation is a very suitable technique for obtaining the detectability curves and relevant detectability factors, since the only detectability factors of practical interest are those that correspond with P_d's larger than or equal to 0.5.

The number of hits per channel n_h or number of video recombinations considered in this study is limited to a maximum of 20. This is because of the difficulties in determining threshold levels for larger values of n_h (even for low J values), and the computer time required for the determination of the detectability curves by Monte Carlo simulation, which increases with increasing value for n_h. Moreover, it is considered that information for $n_h>20$ is of limited practical interest.

The problems that arise in the computation of threshold levels, and the simulation study will be discussed in Ref. 5.

3.2 THRESHOLD LEVELS

The threshold levels y_B corresponding with two practical values for P_f, 10^{-6} ($\pm$ 10^{-9}) and 10^{-4}($\pm$ 10^{-7}), for various values of the parameters n_h and J, and for the various cases considered, are presented in Tables 1 to 4 (page 27 -30). Due to computation problems, threshold levels could not be given for all parameter combinations.

Table 3 shows that for Case IIE-1 y_B approaches an upper limit as J tends to infinity. These upper limits are determined by n_h and the internal noise signals in the unjammed channels, since the probability that the jammed channel has the largest video increases to unity when J increases to infinity.

Approximate threshold levels based on the formulae in Section 2.2.3 are presented in Tables 5 to 8*. Where possible, corresponding figures for the accurate P_f are included: these are obtained by inserting the approximate y_B values in the accurate equations for $P_f = P_f(y_B)$. It is believed that for J $\geqslant$18 dB, and for all cases and n_h values up to 20, the approximate y_B values correspond reasonably well with the P_f

---

*See pages 188-91

values considered. This supposition is supported by the information given in Tables 7 and 8.

3.3 DETECTABILITY FACTORS

The detectability factor D_s is defined as the average signal power - to - average internal noise power ratio that is required in the video channels (or at the inputs to logic-product or video-recombination units) for a given P_d and P_f (see Ref. 1, Fig. 13). Expressed analytically

$$D_s = \frac{\text{minimum detectable signal energy per pulse}}{\text{internal noise per cycle bandwidth}} \text{ (dB)}$$

Curves showing the variation of D_s with J and with the number of hits per channel n_h are presented in Fig. 2 to 29 (page 35). They cover both Swerling cases, values of J up to 30dB, P_f of 10^{-6} and 10^{-4} , and P_d of 0.8 and 0.5. The curves relating to Cases IIE-0, IEE-0, and IEE-1 are in the format used in TR-50-U: for Case IIE-1, covered by Fig. 10 to 13, a more suitable format is used.

For ease of reference, the coverages of Fig. 2 to 29 are listed below:

Fig.	Case	Swerling model	P_d	P_f
2	IIE-0	I/II	0.8	10^{-6}
3			0.5	
4			0.8	10^{-4}
5			0.5	
6		III/IV	0.8	10^{-6}
7			0.5	
8			0.8	10^{-4}
9			0.5	
10	IIE-1	I/II	0.8, 0.5	10^{-6}
11				10^{-4}
12		III/IV		10^{-6}
13				10^{-4}
14	IEE-0	I/II	0.8	10^{-6}
15			0.5	
16			0.8	10^{-4}
17			0.5	
18		III/IV	0.8	10^{-6}

Fig.	Case	Swerling model	P_d	P_f
19			0.5	
20			0.8	10^{-4}
21			0.5	
22	IEE-1	I/II	0.8	10^{-6}
23			0.5	
24			0.8	10^{-4}
25			0.5	
26		III/IV	0.8	10^{-6}
27			0.5	
28			0.8	10^{-4}
29			0.5	

3.3.1 Cases IIE-0, IEE-0, and IEE-1

It seems that in Mode 0, with J = 12 dB and $n_h \leqslant 5$, or with $J \geqslant 18$ dB and $n_h \leqslant 20$, use of the approximate threshold levels gives reasonably accurate information on D_s. The same is true in Mode 1 with J = 12 dB and $n \leqslant 3$, or $J \geqslant 18$ dB and $n_h \leqslant 20$.

This conclusion is supported by the fact that in these ranges the linear relationship between D_s and J (which one expects for larger values for J) becomes clearly visible. It means that for J > 30 dB information on D_s can be obtained by simple linear extrapolation in the relevant Figures.

In the Cases where the information on D_s obtained by using the approximate y_B values was apparently not accurate enough over the full n_h range considered (Mode 0: J = 6 dB and to a less extent J = 12 dB, and Mode 1: J = 12 dB and to a less extent J = 18 dB), the presented curves were obtained by interpolation in the J field. The same technique was applied in those cases where not even approximate information on D_s could be obtained because no approximate y_B values were available.

For general information the curves $D_s = D_s(n_h)$ for J = 6 dB and 12 dB (Mode 0) and J = 12 dB and 18 dB (Mode 1), stemming from use of the approximate y_B values without correction by interpolation in the J field, are represented by dotted lines in the Figures concerned.

3.3.2 Case IIE-1

For this case no approximate relationship between P_f and y_B for J>>0 can be derived. Due to computational difficulties, even when $J=\infty$ (leading to the simplified equation $P_f = P_f(y_B)$), it appeared only possible to obtain y_B values for n_h up to 5.

Therefore the presented curves $D_s = D_s(J\cdot)$ for $n_h = 5, 10, 20$, where D_s is known only for $J = 0$ ($-\infty$dB) and for $n_h = 5$ also for $J = \infty$(∞dB), were obtained by extrapolation in the n_h field.

3.3.3 Case III-0

The curves for Case III-0, of which those for $P_f = 10^{-6}$ appear also in Supplement 2 of TR-50-U (Ref. 1), are included for comparative purposes.

4. PERFORMANCE EVALUATION PROCEDURE

The radar range equation for a clear environment, as quoted in TR-50-U, can be used for frequency diversity radars operating in the linear-addition mode (Mode 0) or the logic-product mode (Mode 1) of video recombination, to determine the clear-environment coverage and (as a function of J) the coverage in a jamming environment.

The following facts should be noted about the parameters appearing in that equation:

D_s : The detectability factor D_s as given in Fig. 2 to 29 must be used. Except for Case IIE-1, the detectability factor is given as a function of the number of hits per channel n_h (which equals the number of video recombinations integrated), with parameters J, P_d, and P_f. For Case IIE-1, D_s is presented as a function of J because this gives a more suitable format.

P_t : The effective transmitter peak power P_t is the peak power of one channel (all three channels are assumed to have the same peak power and pulse width).

L_t, L_r: The waveguide loss on transmission L_t and on reception L_r must include losses in the RF multiplexer. Only the loss per channel must be taken into account.

L_x : The receiving system loss L_x must include:

(1) the loss in a logic-product unit; only the loss per channel must be taken into account (this loss is assumed to be negligible)

(2) the video recombination loss (see Note)

(3) the increased collapsing loss (see Note)

(4) the loss in signal power - to - average internal noise power ratio due to ECCM and MTI fixes, pulse compression processing loss,

etc.; only the loss per channel must be taken into account (it is supposed that the three channels apply the same type of receiver simultaneously, which in practice is not the case).

Note

Since D_s is defined in the video channels, or at the inputs to the video recombination unit, both the video recombination loss and collapsing loss must be transformed to the inputs of this unit, and be equally divided over the channels. This means that for a triple-channel system operating in Mode 0 or Mode 1, L_x must include the video recombination loss and the collapsing loss each divided by three.

The diversity gain or degradation is not included in L_x as in the procedure given in TR-50-U: instead, it is included in D_s.

The procedures given here for determining countermeasures-environment coverages such as:

- the self-screening and mutual-screening coverage for one jammer only
- the quiet target coverage in a mixed jamming raid environment

differ from those given in TR-50-U in that knowledge of the parameter J is required since D_s depends on J.

The parameter J is introduced because it is a convenient way of dealing with evaluation studies of the present kind. Indeed its use in unavoidable in multi-channel system evaluations where it is desired:

- to consider any jamming strength
- to study the different types of video recombination
- to compare the performance of the different types of video recombination.

In the case of one jammer only and one or more channels jammed, J in each channel can be determined from the following formula*, provided the average internal noise power in the video channels is known:

$$J = \frac{\sigma_j^2}{\sigma_n^2} = \frac{1}{\sigma_n^2} \cdot \frac{E_{rp} G_{rj} c^2}{(4\pi)^2 f^2 R_j^2 L_r L_p L_{aj}} \qquad (15)$$

*For the meaning of symbols, see end of this section.

where

σ_j^2 = average jamming noise power

σ_n^2 = average internal noise power.

When σ_n^2 is not known the simplest way to obtain J is to consider the self-screening range for a single-channel system. We then have (see TR-50-U)

$$S_{min} = \frac{P_t \tau_t G_t \sigma}{4\pi R_{ss}^2 E_{rp} L_t L_p L_a} = S_{min}(G_{rj} = G_r, J). \quad (16)$$

It follows that

$$S_{min}(G_{rj} = G_r,\ J) = S_{min}(G_{rj} = G_r,\ J = 0) \cdot \frac{1}{1 + J(G_{rj} = G_r)} \quad (17)$$

and when $J(G_{rj} = G_r)$ is determined $J(G_{rj})$ is known.

It is noted that it is assumed that the external noise and internal noise are both narrow-band Gaussian processes, and that the receivers operate linearly over the signal strength range considered here. As a consequence, J in each of the video channels equals J at the input to the corresponding receiver.

When σ_n^2 is, known equation (15) can also be used to determine J in a jamming raid environment. However, when σ_n^2 is not known the determination of J becomes more complicated, although in principle equations (16) and (17) could be used here also (see TR-50-U).

Finally, it should be noted that:

- the graphs of D_s are obtained under a number of assumptions (see Section 2.1).
- if all three channels are jammed with equal and known strength the graphs for J=0 can be used. It follows that

  $D_s(J) = D_s(J=0) + J$ in dB.
- in the case of special ECCM and MTI fixes, pulse compression processing, etc., the presented information on D_s can only be used when the probability density functions of the noise signals

and target echoes (in isolation, after square-law detection in the channels) have not been changed too much,

- parameter J is defined in the video channels (or at the input to the receivers). In the case of special ECCM fixes the improvement in J must be known in order to be able to perform the radar coverage calculations.

List of Symbols

c = velocity of light

E_{rp} = effective jammer power density in watts per hertz

f = radar frequency in megahertz

G_r = maximum power gain of receiving antenna

G_{rj} = power gain of receiving antenna in direction of jammer

G_t = maximum power gain of transmitting antenna

L_a = one-way atmospheric loss factor for target return

L_{aj} = atmospheric loss factor for jammer signals

L_p = one-way scanning loss factor due to beam shape

R_j = jammer range

R_{ss} = self-screening range

S_{min} = minimum detectable signal-to-noise ratio at the input of the receiver

σ = mean equivalent target echoing area in square meters

5. DISCUSSION OF MIXED VIDEO RECOMBINATION GAIN

The mixed video recombination gain ΔD_s is defined here as the gain in the required average signal power - to - average internal noise power ratio D_s in the video channels as compared with a system operating in the linear-addition mode (Mode 0).

In Fig. 30 and 31 the logic-product Mode 1 gain for condition IIE where only one channel is jammed is plotted against the average external noise power - to - average internal noise power ratio J(dB) for both Swerling models, $10^{-6}<P_f<10^{-4}$ and with parameters n_h and P_d. (ΔD_s appears rather insensitive to P_f).

In Fig. 32 the logic-product Mode 1 gain for condition IEE where two channels are simultaneously jammed is plotted against J for both Swerling models, $10^{-6}\leqslant P_f\leqslant 10^{-4}$, and with parameters n_h and P_d (ΔD_s again appears rather insensitive to P_f).

From these graphs the following characteristics can be seen:

Quiet environment ($J=0$ or $-\infty$dB)

The degradation of Mode 1 is relatively small - between 1 and 3 dB for the parameter combinations (n_h, P_f, P_d) and the target scintillation models considered.

One channel jammed

As was expected, for J>6 dB Mode 1 is far superior to Mode 0. The gain of Mode 1 as a function of J does not vary very much with the parameter values (n_h, P_f, P_d) and the target scintillation models considered.

Two channels jammed (with equal average power)

The difference in the performance of the two modes is relatively small. Mode 1 shows a gain of -2 to 3 dB for the parameter combinations (n_h, J, P_f, P_d) and the target scintillation models considered.

In this Supplement the diversity gain and efficiency, as defined in Supplement 2 of TR-50-U, are not evaluated because the authors consider that the defined mixed video recombination gain (or degradation) is a more meaningful parameter in this study area. However, information on diversity gain and efficiency for the cases IIE-0 and IIE-1 can easily be obtained by comparing the results presented in Fig. 2 to 13 with those for a jammed single-channel system.

Since the losses in logic-product units can be assumed negligibly small the discussion in Supplement 2 on extra losses in a triple-channel diversity system operating in the linear-addition mode also holds for such a system operating in the logic-product Mode 1.

6. CONCLUSIONS

When only one channel is jammed (condition IIE), the logic-product Mode 1 type of video recombination is far superior to the linear-addition Mode 0. However, it is doubted whether in such circumstances this mode of logic-product video recombination is necessarily the most efficient one. This problem will be considered in Ref. 5.

When two channels are simultaneously jammed with equal (average) power the performances of both Modes 0 and 1 are very poor. Therefore in Ref. 5 attention will be paid to a mixed video recombination mode which has a ECCM feature in such a jamming environment.

Case	Noise power ratio $J = \frac{(P_n)_E}{(P_n)_I}$ in dB	Number of hits per channel, n_h, or number of video recombinations integrated						
		1	2	3	4	5	10	20
III-0	$-\infty$	19.1292	25.4126	30.9571	36.1144	41.0221	63.5482	104.252
	0	30.4036	38.7689	46.1870	53.1102	59.7116	90.0821	144.100
IIE-0	6	70.8983	87.8183	102.247	115.505	128.101	185.221	-
external noise	12	-	-	-	-	-	-	-
present in one	18	-	-	-	-	-	-	-
channel	24	-	-	-	-	-	-	-
	30	-	-	-	-	-	-	-
	0	34.7260	45.3090	54.6510	63.3419	71.6116	109.550	178.009
IEE-0	6	84.5258	109.006	130.315	150.153	169.036	255.025	-
external noise	12	-	-	-	-	-	-	-
present in two	18	-	-	-	-	-	-	-
channels	24	-	-	-	-	-	-	-
	30	-	-	-	-	-	-	-

Table 1

Threshold levels for Mode 0 and $P_f = 10^{-6}$

Case	Noise power ratio $J = \frac{(P_n)_E}{(P_n)_I}$ in dB	Number of hits per channel, n_h, or number of video recombinations integrated						
		1	2	3	4	5	10	20
III-0	$-\infty$	13.9282	19.5672	24.5947	29.3065	33.8163	54.7515	93.1630
	0	21.1930	28.8363	35.6395	42.0137	48.1132	76.4045	128.216
IIE-0	6	48.2584	63.1789	76.1981	88.2637	99.6827	152.340	257.553
external noise	12	–	–	–	–	–	–	–
present in one	18	–	–	–	–	–	–	–
channel	24	–	–	–	–	–	–	–
	30	–	–	–	–	–	–	–
	0	24.8461	34.3833	42.8716	50.8196	58.4218	93.6646	158.203
IEE-0	6	59.7851	81.6656	101.095	119.122	136.269	216.046	–
external noise	12	–	–	–	–	–	–	–
present in two	18	–	–	–	–	–	–	–
channels	24	–	–	–	–	–	–	–
	30	–	–	–	–	–	–	–

Table 2

Threshold levels for Mode 0 and $P_f = 10^{-4}$

Case	Noise power ratio $J = \frac{(P_n)_E}{(P_n)_I}$ in dB	Number of hits per channel, n_h, or number of video recombinations integrated						
		1	2	3	4	5	10	20
III-I	$-\infty$	14.5087	18.0372	21.1135	23.9567	26.6495	38.8927	60.7026
	0	17.4932	21.7959	25.5494	29.0193	-	-	-
IIE-I	6	20.0785	25.1324	29.5472	33.6303	-	-	-
external noise	12	21.6335	27.1772	32.0242	36.5076	-	-	-
present in one	18	22.1720	27.8920	32.8946	37.5230	-	-	-
channel	24	22.3200	28.0890	33.1349	37.8034	-	-	-
	30	22.3579	28.1395	33.1966	37.8762	-	-	-
	∞	22.3714	28.1563	33.2174	37.9003	42.3336	-	-
	0	22.0361	27.5323	32.3247	36.7524	-	-	-
IEE-I	6	38.4104	48.4633	57.1927	64.4218	-	-	-
external noise	12	119.712	146.350	169.278	190.316	-	-	-
present in two	18	451.068	546.490	627.859	-	-	-	-
channels	24	-	-	-	-	-	-	-
	30	-	-	-	-	-	-	-

Table 3

Threshold levels for Mode 1 and $P_f = 10^{-6}$

Case	Noise power ratio $J = \frac{(P_n)_E}{(P_n)_I}$ in dB	Number of hits per channel, n_h, or number of video recombinations integrated						
		1	2	3	4	5	10	20
III-I	$-\infty$	9.9035	13.0889	15.8778	18.4665	20.9279	32.2182	52.6114
	0	11.9668	15.8515	19.2542	22.4133	–	–	–
IIE-I	6	13.7975	18.3611	22.3632	26.0805	–	–	–
external noise	12	14.9198	19.9250	24.3184	28.4009	–	–	–
present in one	18	15.3123	20.4761	25.0103	29.2243	–	–	–
channel	24	15.4206	20.6284	25.2017	29.4523	–	–	–
	30	15.4483	20.6675	25.2508	29.5109	–	–	–
	∞	15.4584	20.6803	25.2676	29.5302	33.5877	–	–
	0	15.1247	20.0781	24.4159	28.4420	–	–	–
IEE-I	6	26.7049	35.6662	43.4990	50.7475	–	–	–
external noise	12	80.5681	104.401	125.011	144.012	–	–	–
present in two	18	301.400	386.193	458.815	525.380	–	–	–
channels	24	–	–	–	–	–	–	–
	30	–	–	–	–	–	–	–

Table 4

Threshold levels for Mode 1 and $P_f = 10^{-4}$

Case	Noise power ratio $J = \frac{(P_n)_E}{(P_n)_I}$ in dB	Number of hits per channel, n_h, or number of video recombinations integrated: 1	2	3	4	5	10	20
	6	69.0801	83.4422	95.6459	106.752	117.158	163.552	244.132
IIE-0	$P_f \rightarrow$	1.56×10^{-6}	2.29×10^{-6}	3.25×10^{-6}	4.48×10^{-6}	6.11×10^{-6}	21.6×10^{-6}	145×10^{-6}
external noise	12	234.872	283.703	325.196	362.958	398.336	556.076	830.050
present in one	18	898.041	1,084.75	1,243.40	1,387.78	1,523.05	2,126.17	3,173.72
channel	24	3,550.71	4,288.93	4,916.20	5,487.07	6,021.90	8,406.56	12,548.4
	30	14,161.4	17,105.6	19,607.4	21,884.2	24,017.3	33,528.1	50,047.1
	6	83.4422	106.752	127.063	145.811	163.552	244.132	387.701
IEE-0	$P_f \rightarrow$	1.23×10^{-6}	1.45×10^{-6}	1.71×10^{-6}	1.92×10^{-6}	2.18×10^{-6}	3.72×10^{-6}	-
external noise	12	283.703	362.958	432.015	495.757	556.076	830.050	1,318.18
present in two	18	1,084.75	1,387.78	1,651.82	1,895.54	2,126.17	3,173.72	5,040.12
channels	24	4,288.93	5,487.07	6,531.04	7,494.68	8,406.56	12,548.4	19,927.8
	30	17,105.6	21,884.2	26,047.9	29,891.2	33,528.1	50,047.1	79,478.8

Table 5

Approximate threshold levels for Mode 0 and $P_f = 10^{-6}$,
and the corresponding accurate P_f values

Case	Noise power ratio $J = \frac{(P_n)_E}{(P_n)_I}$ in dB	Number of hits per channel, n_h, or number of video recombinations integrated						
		1	2	3	4	5	10	20
IIE-0 external noise present in one channel	6	46.0542	58.7820	69.6409	79.5691	88.9100	130.965	205.156
	$P_f \rightarrow$	1.56×10^{-4}	2.21×10^{-4}	3.05×10^{-4}	4.07×10^{-4}	4.59×10^{-4}	19.7×10^{-4}	68.4×10^{-4}
	12	156.584	199.859	236.779	270.535	302.294	445.281	697.530
	18	598.705	764.166	905.331	1,034.40	1,155.83	1,702.54	2,667.02
	24	2,311.92	3,021.39	3,579.54	4,089.85	4,569.98	6,731.60	10,545.0
	30	9,441.11	12,050.3	14,276.4	16,311.7	18,226.5	26,847.8	42,056.9
IEE-0 external noise present in two channels	6	58.7820	79.5691	97.8360	114.812	130.965	205.156	339.456
	$P_f \rightarrow$	1.20×10^{-4}	1.41×10^{-4}	1.58×10^{-4}	1.78×10^{-4}	2.45×10^{-4}	3.15×10^{-4}	-
	12	199.859	270.535	332.642	390.362	445.281	697.530	1,154.15
	18	764.165	1,034.40	1,271.87	1,492.56	1,702.54	2,667.02	4,412.93
	24	3,021.39	4,089.85	5,028.77	5,901.35	6,731.60	10,545.0	17,448.0
	30	12,050.3	16,311.7	20,056.4	23,536.5	26,847.8	42,056.9	69,588.5

Table 6

Approximate threshold levels for Mode 0 and $P_f = 10^{-4}$, and the corresponding accurate P_f values

Case	Noise power ratio $J = \frac{(P_n)_E}{(P_n)_I}$ in dB	Number of hits per channel, n_h, or number of video recombinations integrated						
		1	2	3	4	5	10	20
IEE-1 external noise present in two channels	6	35.5400	41.7211	47.8229	53.3761	58.5788	81.7758	122.066
	$P_f \rightarrow$	4.62×10^{-6}	11.6×10^{-6}	24.1×10^{-6}	44.3×10^{-6}	-	-	-
	12	117.436	141.852	162.598	181.479	199.168	278.038	415.025
	$P_f \rightarrow$	1.31×10^{-6}	1.65×10^{-6}	2.03×10^{-6}	2.45×10^{-6}	-	-	-
	18	449.020	542.374	621.698	693.890	761.524	1,063.09	1,586.86
	$P_f \rightarrow$	1.07×10^{-6}	1.13×10^{-6}	1.19×10^{-6}	-	-	-	-
	24	1,775.36	2,144.46	2,458.10	2,743.53	3,010.95	4,203.28	6,274.20
	30	7,080.71	8,552.83	9,803.70	10,942.1	12,008.6	16,764.0	25,023.6

Table 7

Approximate threshold levels for Mode 1 and $P_f = 10^{-6}$,
and the corresponding accurate P_f values

Case	Noise power ratio $J = \frac{(P_n)_E}{(P_n)_I}$ in dB	Number of hits per channel, n_h, or number of video recombinations integrated						
		1	2	3	4	5	10	20
IEE-1 external noise present in two channels	6	23.0271	29.3910	34.8204	39.7845	44.4550	65.4825	102.578
	P_f →	4.18×10^{-4}	8.76×10^{-4}	15.6×10^{-4}	25.1×10^{-4}	-	-	-
	12	78.2921	99.9293	118.389	135.267	151.147	222.640	348.765
	P_f →	1.31×10^{-4}	1.62×10^{-4}	1.96×10^{-4}	2.32×10^{-4}	-	-	-
	18	299.352	382.083	452.665	517.199	577.915	851.272	1,333.51
	P_f →	1.07×10^{-4}	1.13×10^{-4}	1.18×10^{-4}	1.23×10^{-4}	1.92×10^{-4}	-	-
	24	1,183.59	1,510.70	1,789.77	2,044.92	2,284.99	3,365.80	5,272.50
	30	4,720.56	6,025.15	7,138.19	8,155.83	9,113.28	13,423.9	21,028.5

Table 8

Approximate threshold levels for Mode 1 and $P_f = 10^{-4}$, and the corresponding accurate P_f values

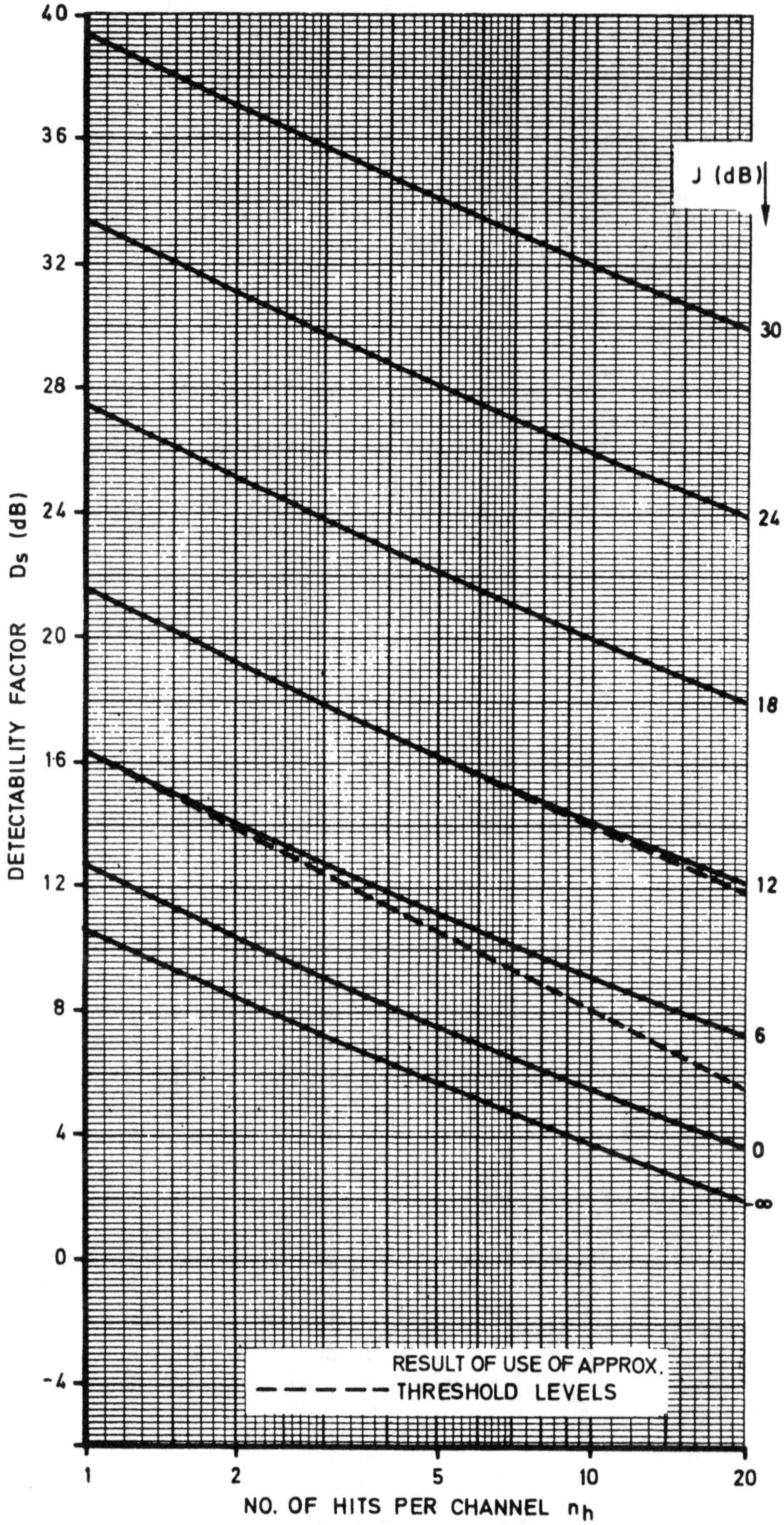

J (dB)
30
24
18
12
6
0
∞
DETECTABILITY FACTOR Ds (dB)
40
36
32
28
24
20
1·6
12
8
4
0
-4
RESULT OF USE OF APPROX.
THRESHOLD LEVELS
1
2
5
10
20
NO. OF HITS PER CHANNEL nh

Fig. 2

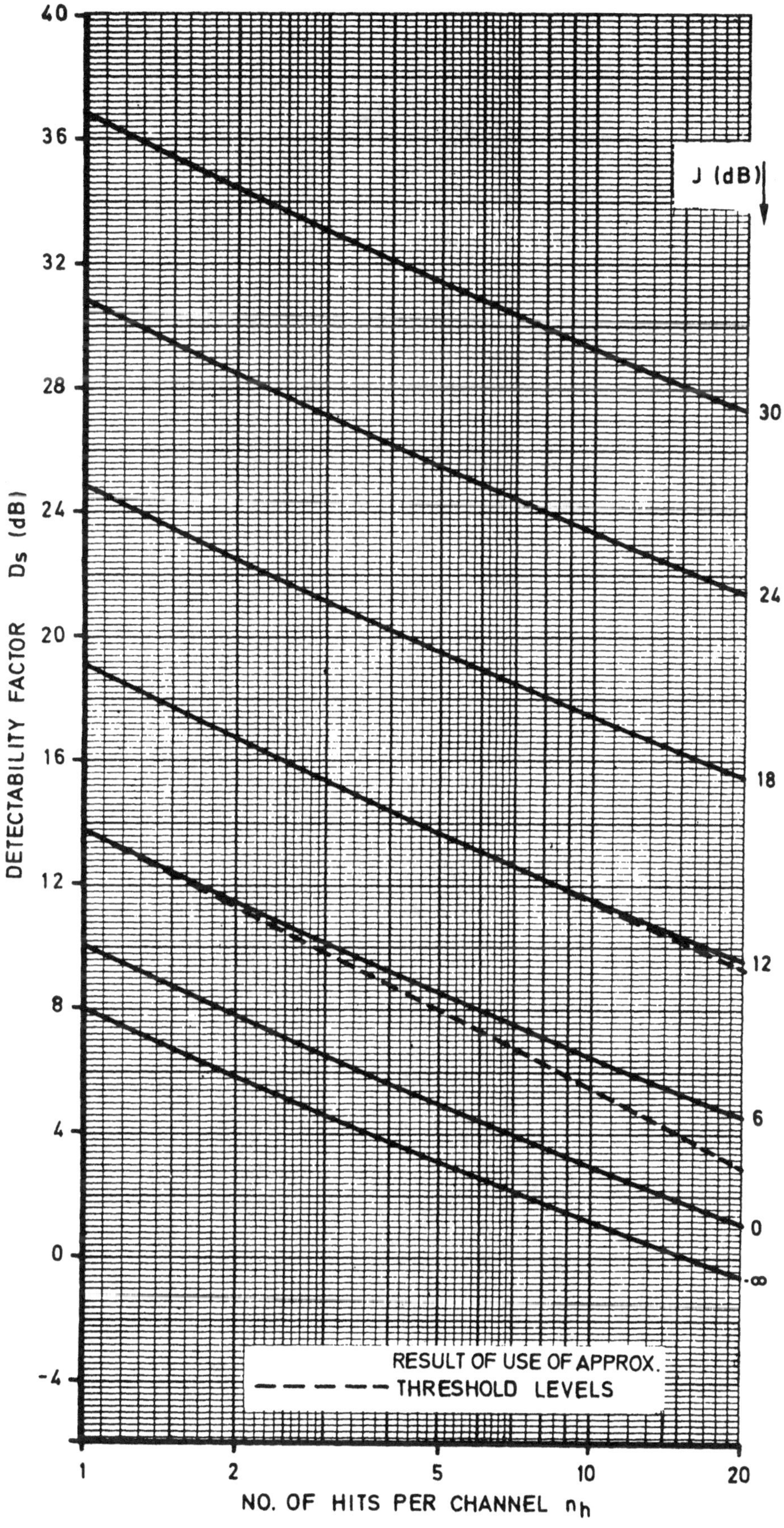

J (dB)
30
24
18
12
6
0
-∞
DETECTABILITY FACTOR Ds (dB)
40
36
32
28
24
20
16
12
8
4
0
-4
1
2
5
10
20
NO. OF HITS PER CHANNEL n_h
RESULT OF USE OF APPROX.
THRESHOLD LEVELS

Fig. 3

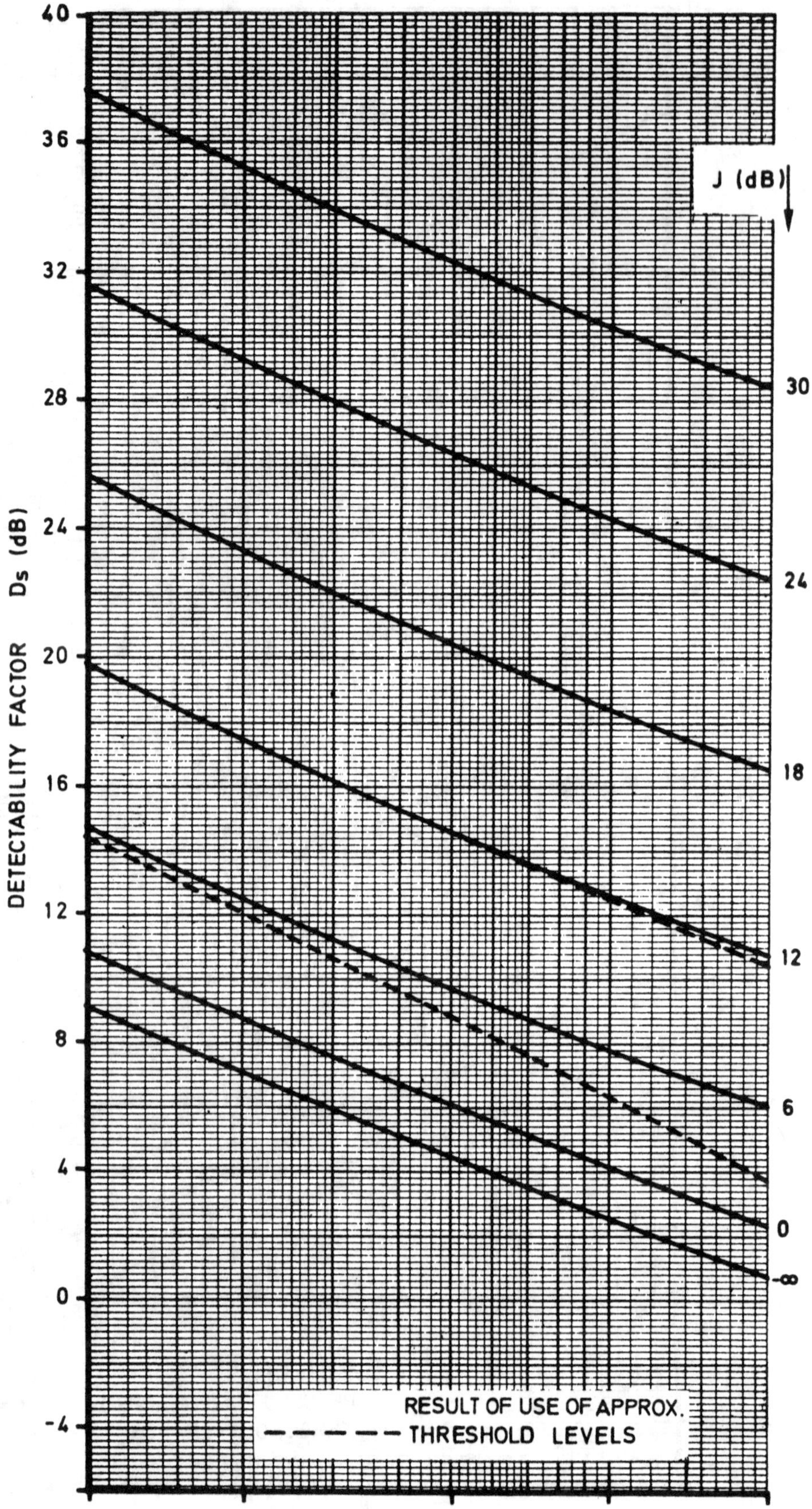
DETECTABILITY FACTOR Ds (dB)
J (dB)
30
24
18
12
6
0
-∞
RESULT OF USE OF APPROX.
THRESHOLD LEVELS
NO. OF HITS PER CHANNEL n_h

Fig. 4

194

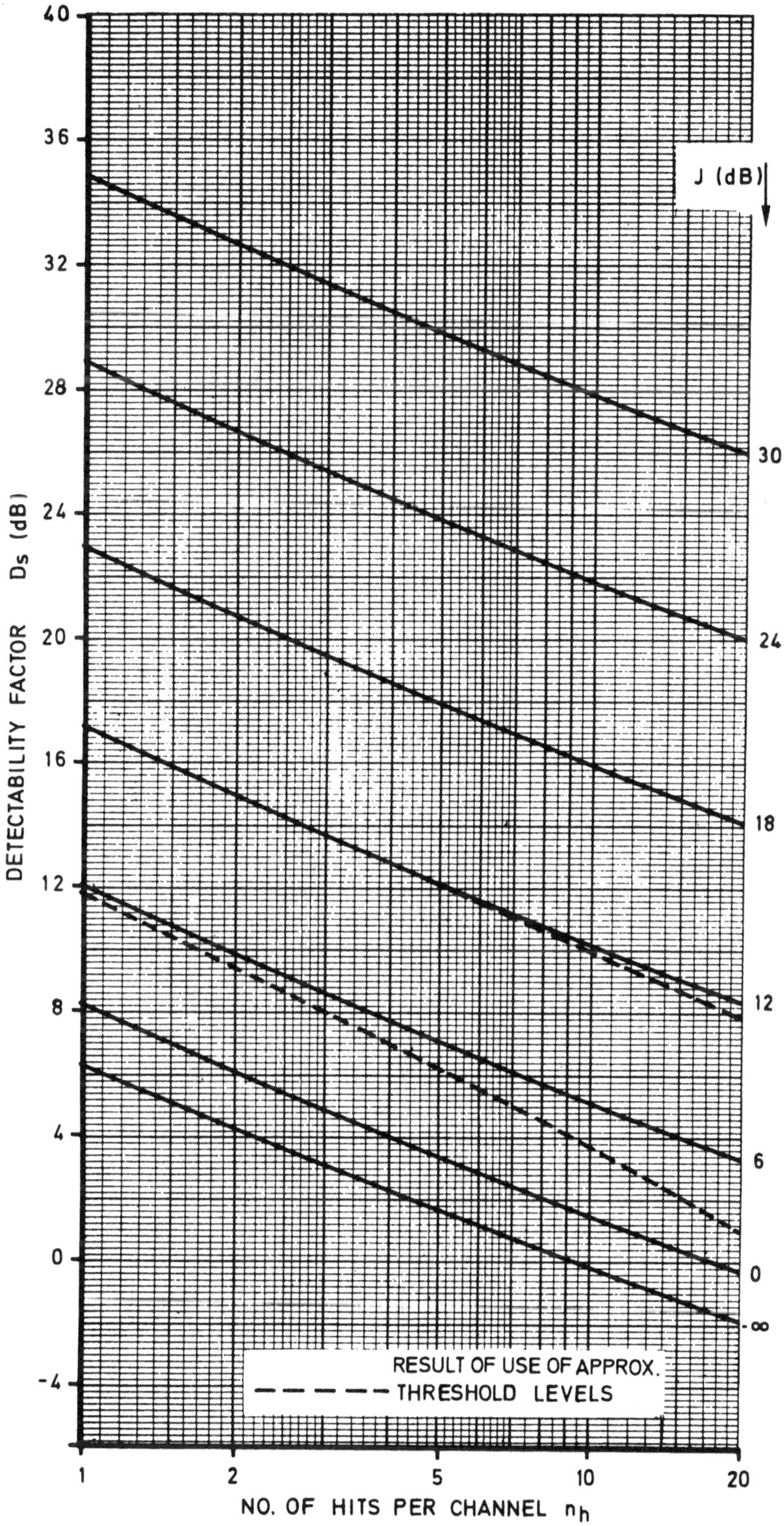
J (dB)
30
24
18
12
6
0
-∞
RESULT OF USE OF APPROX. THRESHOLD LEVELS
DETECTABILITY FACTOR Ds (dB)
NO. OF HITS PER CHANNEL nh

Fig. 5

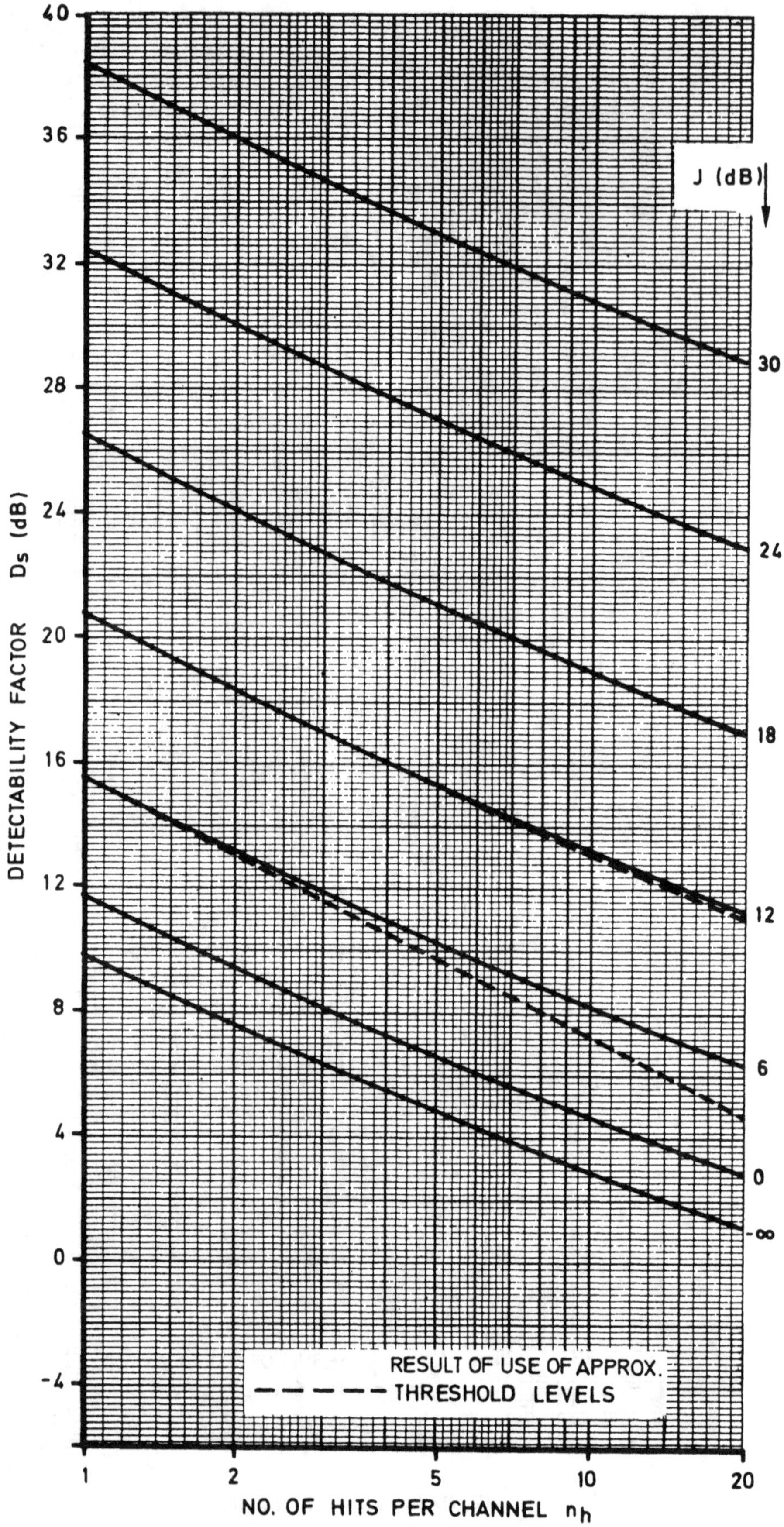
40
36
32
28
24
20
16
12
8
4
0
-4
DETECTABILITY FACTOR Ds (dB)
J (dB)
30
24
18
12
6
0
-∞
RESULT OF USE OF APPROX.
THRESHOLD LEVELS
1
2
5
10
20
NO. OF HITS PER CHANNEL nh

Fig. 6

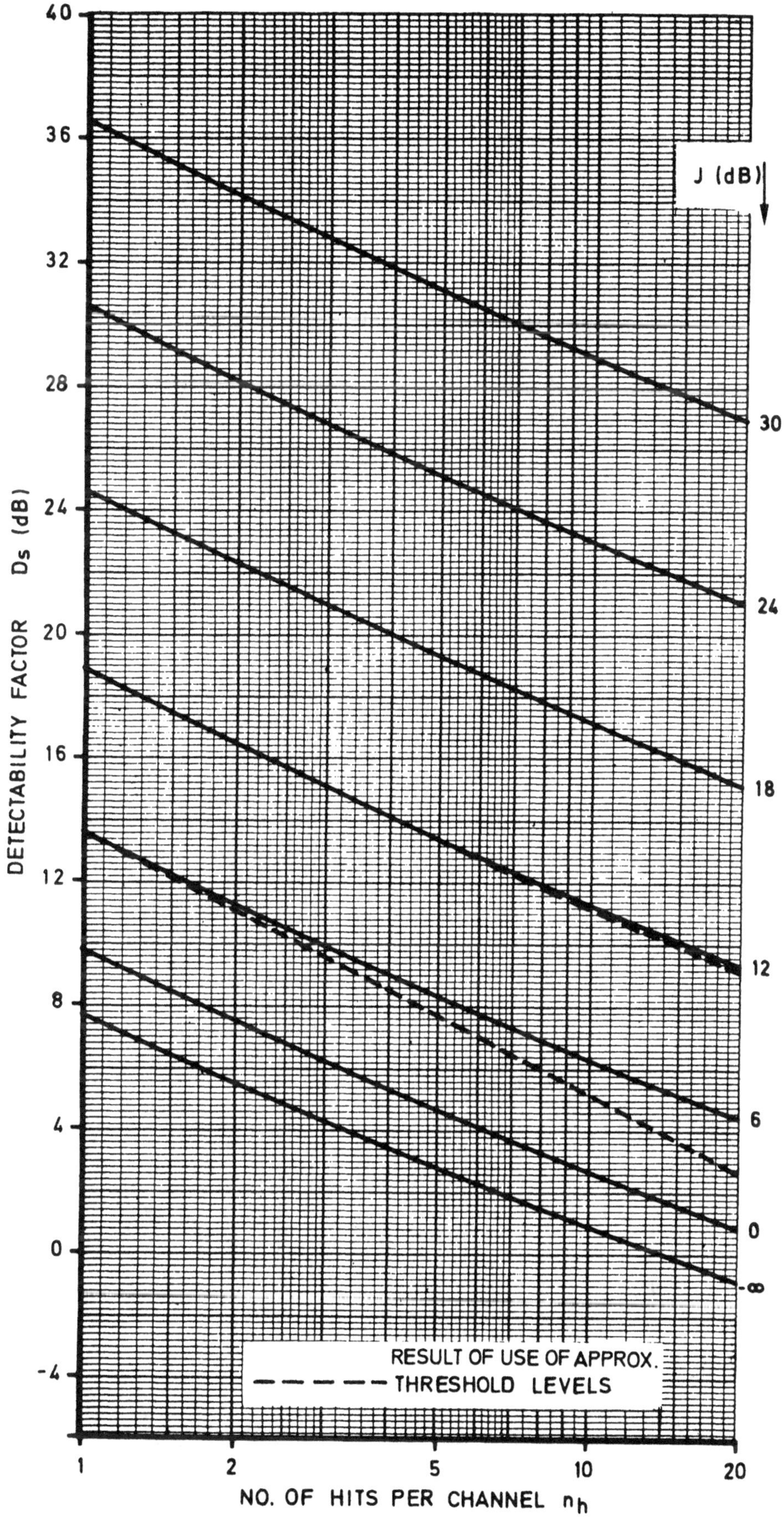

DETECTABILITY FACTOR Ds (dB)
40
36
32
28
24
20
16
12
8
4
0
-4
J (dB)
30
24
18
12
6
0
-∞
RESULT OF USE OF APPROX. THRESHOLD LEVELS
1
2
5
10
20
NO. OF HITS PER CHANNEL n_h

Fig. 7

197

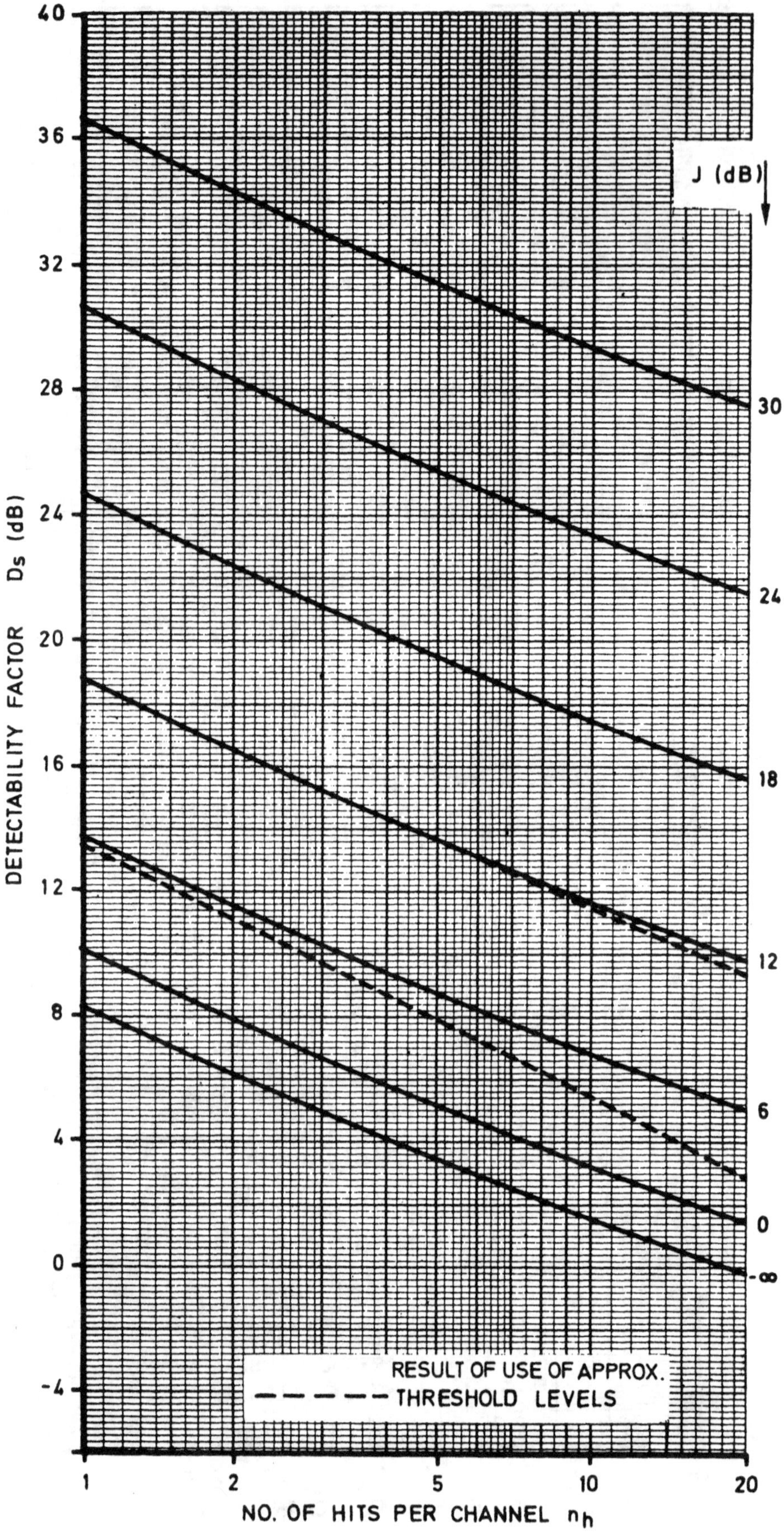

DETECTABILITY FACTOR Ds (dB)
40
36
32
28
24
20
16
12
8
4
0
-4
J (dB)
30
24
18
12
6
0
-∞
RESULT OF USE OF APPROX.
THRESHOLD LEVELS
1
2
5
10
20
NO. OF HITS PER CHANNEL n_h

Fig. 8

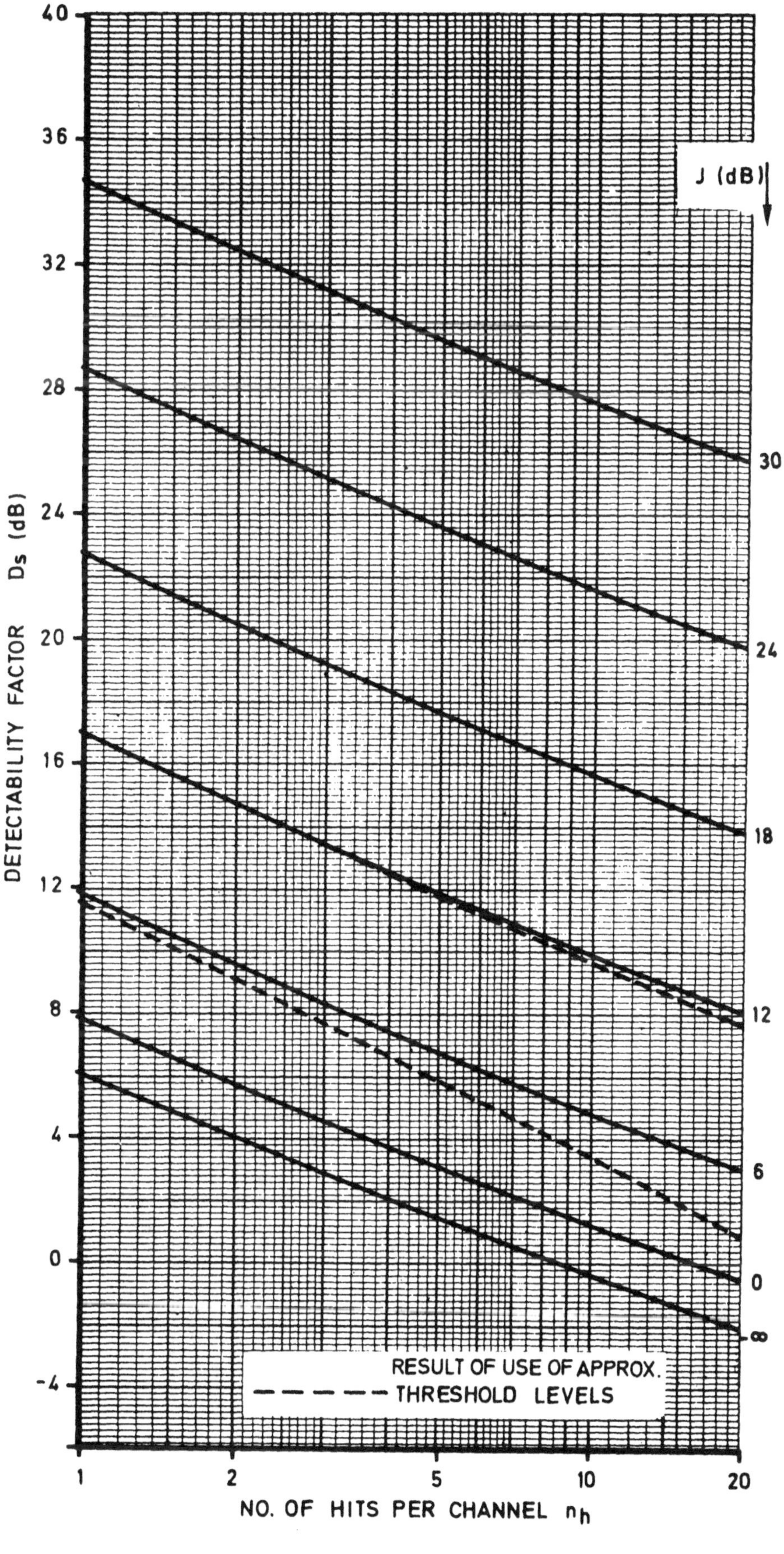

DETECTABILITY FACTOR Ds (dB)
40
36
32
28
24
20
16
12
8
4
0
-4
J (dB)
30
24
18
12
6
0
∞
RESULT OF USE OF APPROX.
THRESHOLD LEVELS
1
2
5
10
20
NO. OF HITS PER CHANNEL n_h

Fig. 9

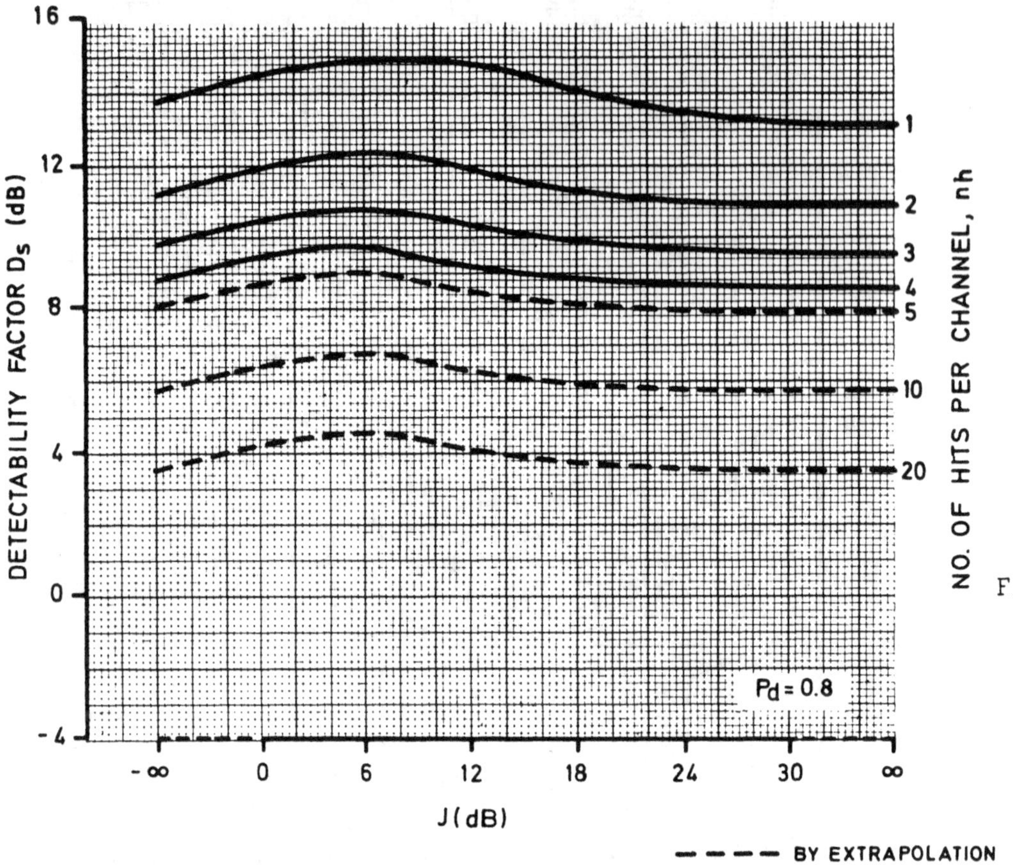

DETECTABILITY FACTOR D_s (dB)
NO. OF HITS PER CHANNEL, nh
J(dB)
P_d = 0.8
---- BY EXTRAPOLATION

(a)

Fig. 10

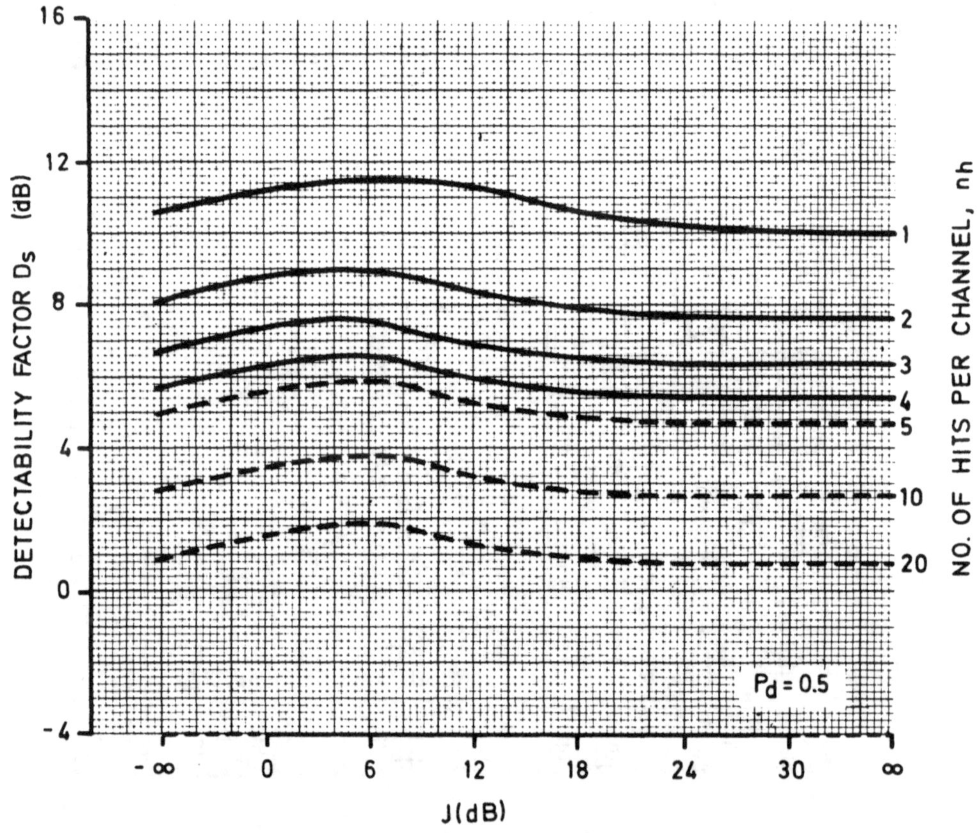

DETECTABILITY FACTOR D_s (dB)
NO. OF HITS PER CHANNEL, nh
J(dB)
P_d = 0.5

(b)

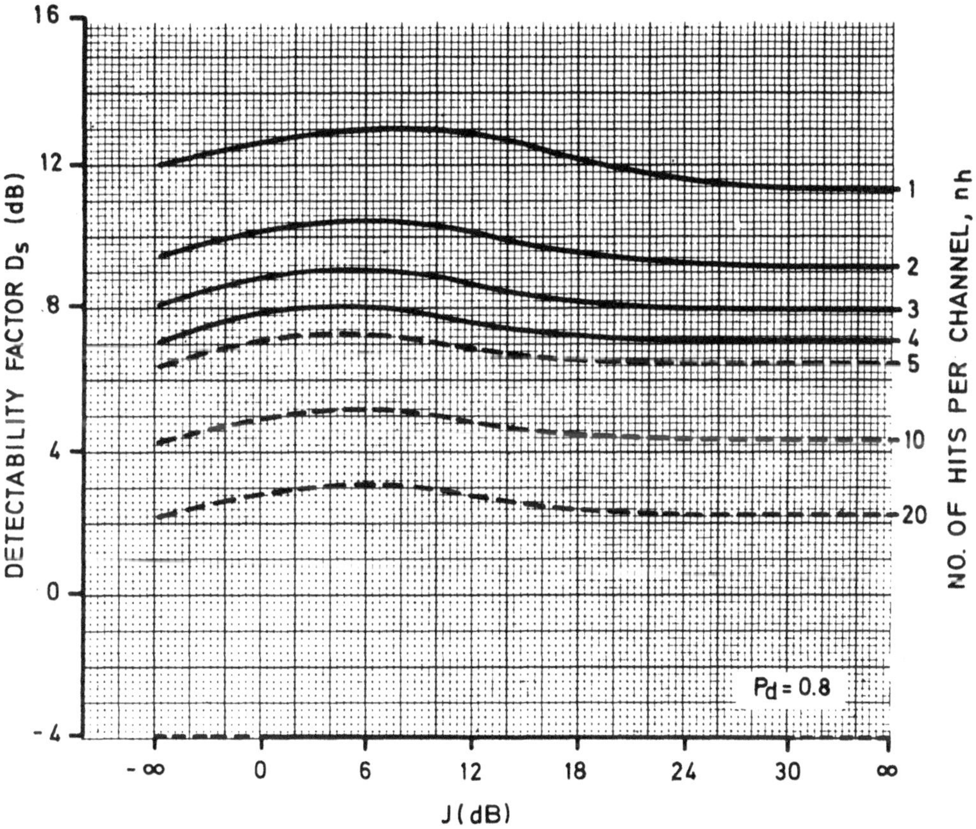

(a)

Fig. 11

— — — — BY EXTRAPOLATION

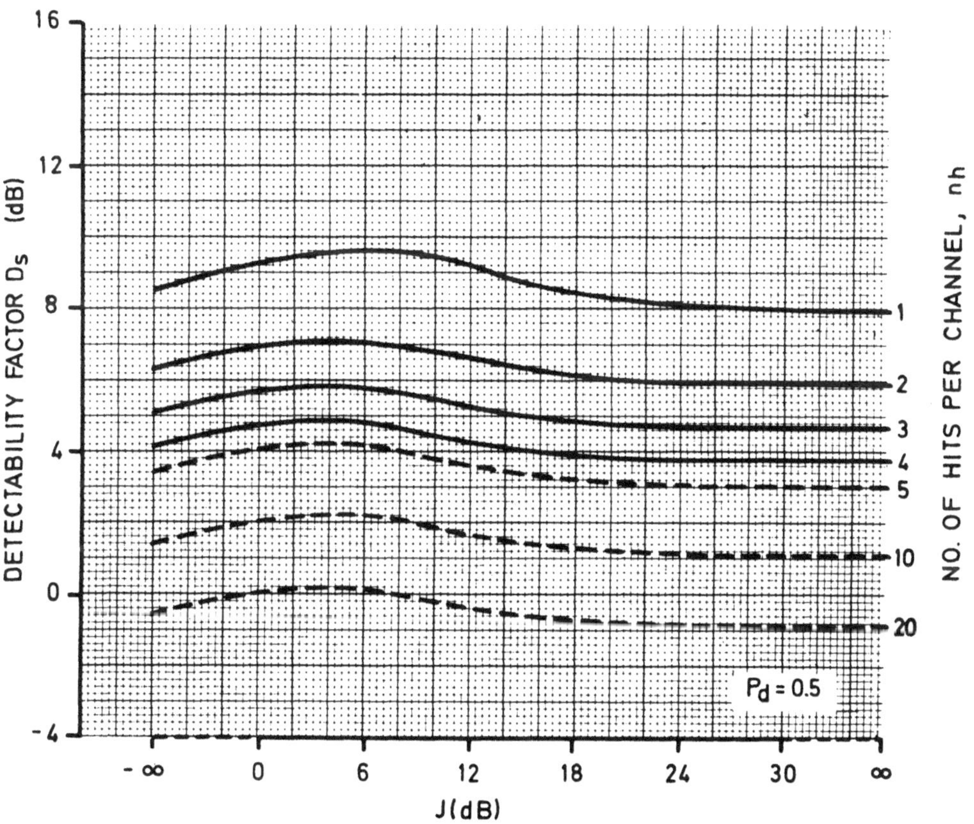

(b)

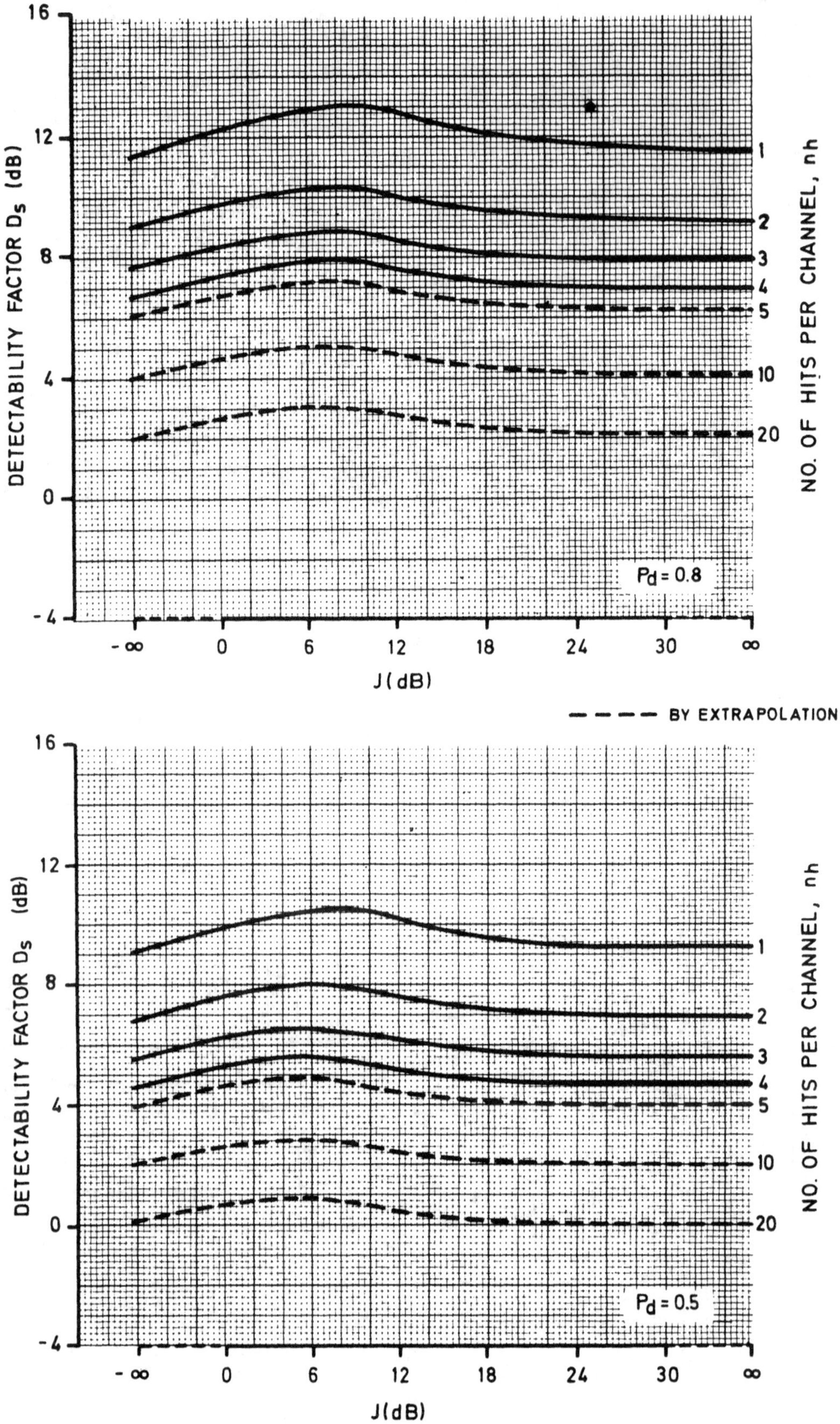

DETECTABILITY FACTOR D_S (dB)
NO. OF HITS PER CHANNEL, nh
1
2
3
4
5
10
20
P_d = 0.8
J(dB)
-∞ 0 6 12 18 24 30 ∞
16 12 8 4 0 -4
BY EXTRAPOLATION
P_d = 0.5

(a)

(b)

Fig. 12

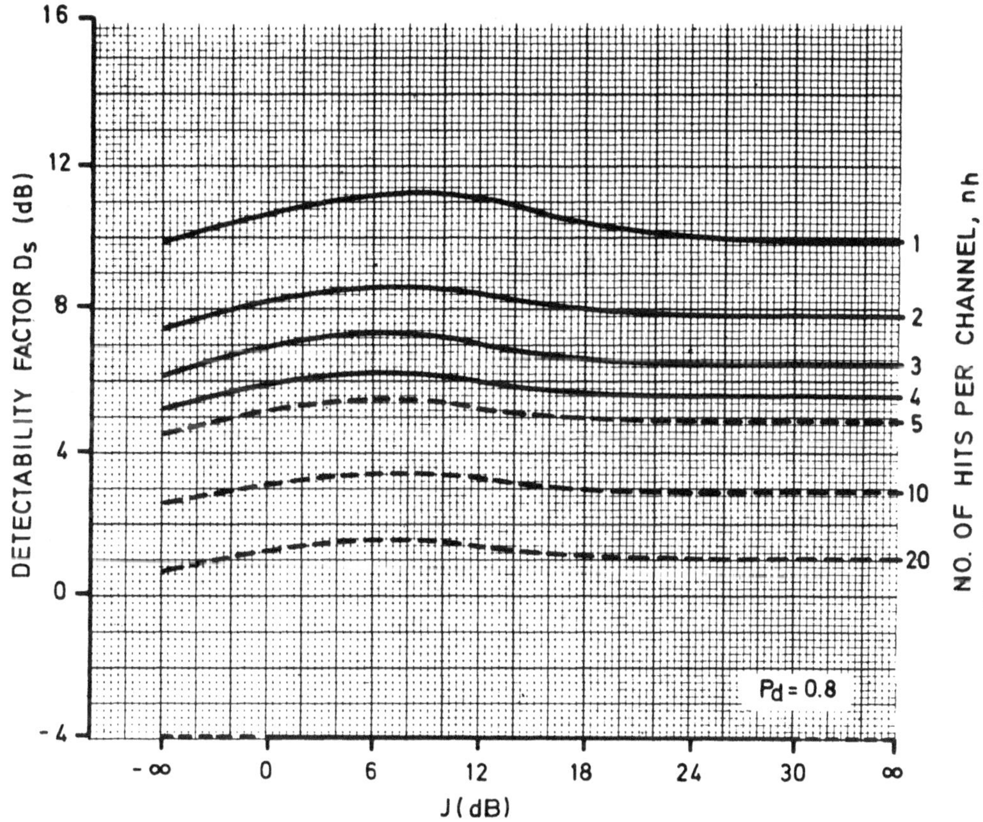
DETECTABILITY FACTOR D_s (dB)
NO. OF HITS PER CHANNEL, nh
P_d = 0.8
J(dB)

(a)

Fig. 13

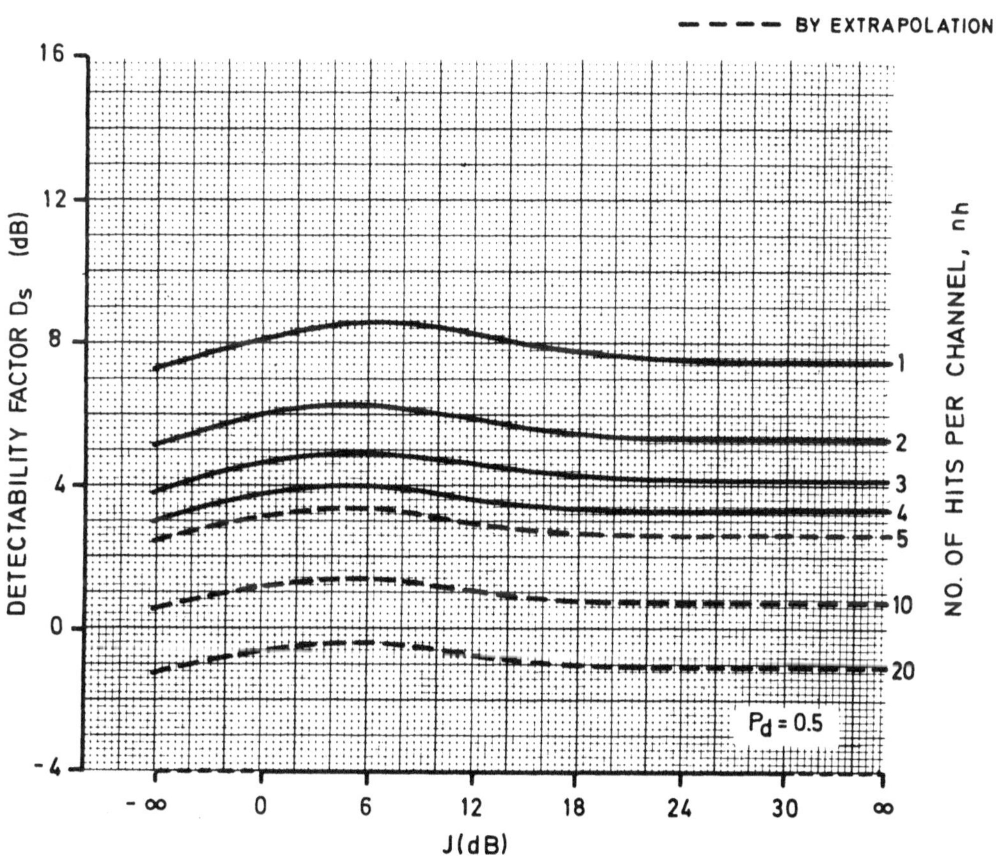
BY EXTRAPOLATION
DETECTABILITY FACTOR D_s (dB)
NO. OF HITS PER CHANNEL, nh
P_d = 0.5
J(dB)

(b)

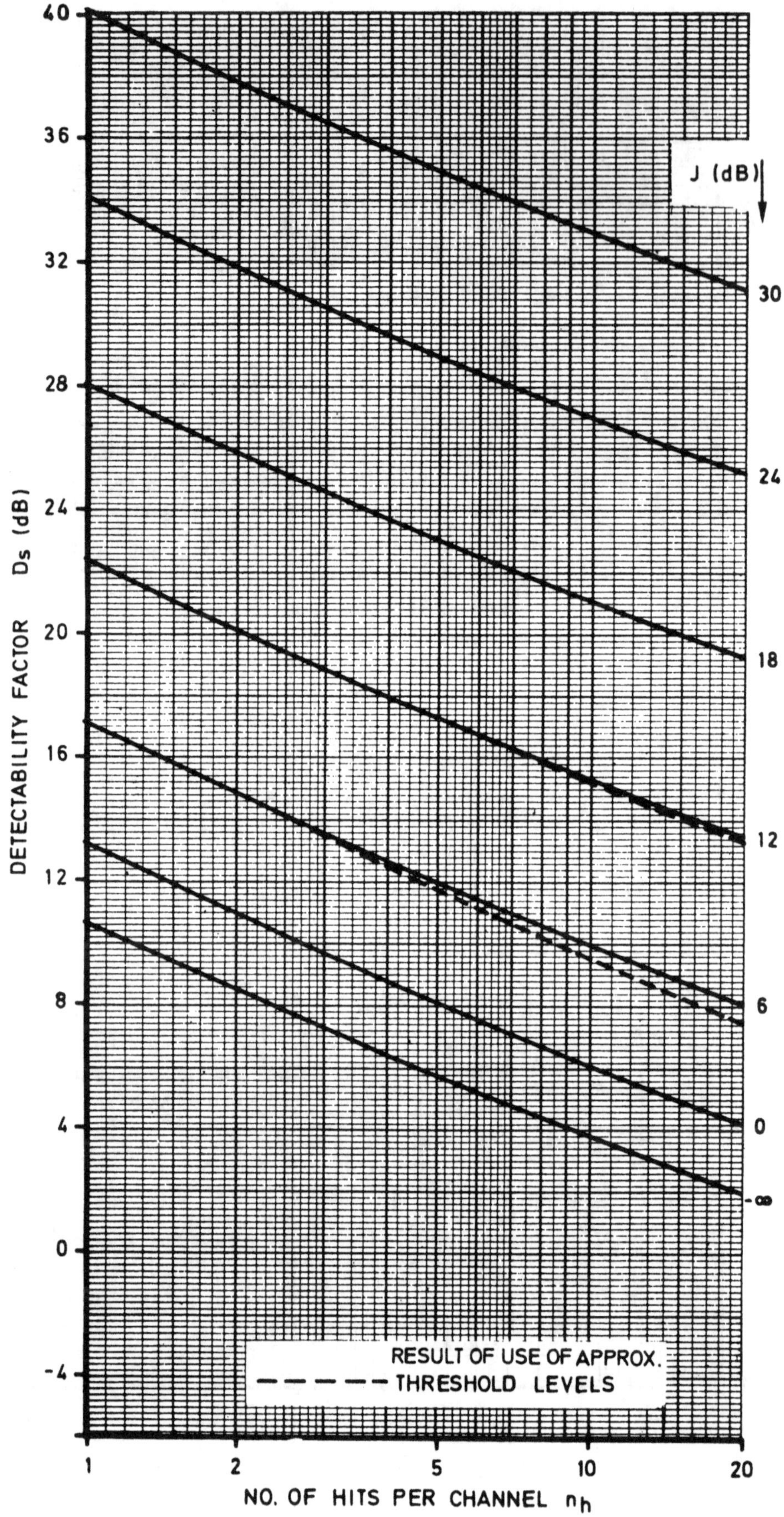
DETECTABILITY FACTOR Ds (dB)
NO. OF HITS PER CHANNEL nh
J (dB)
30
24
18
12
6
0
-∞
RESULT OF USE OF APPROX.
THRESHOLD LEVELS

Fig. 14

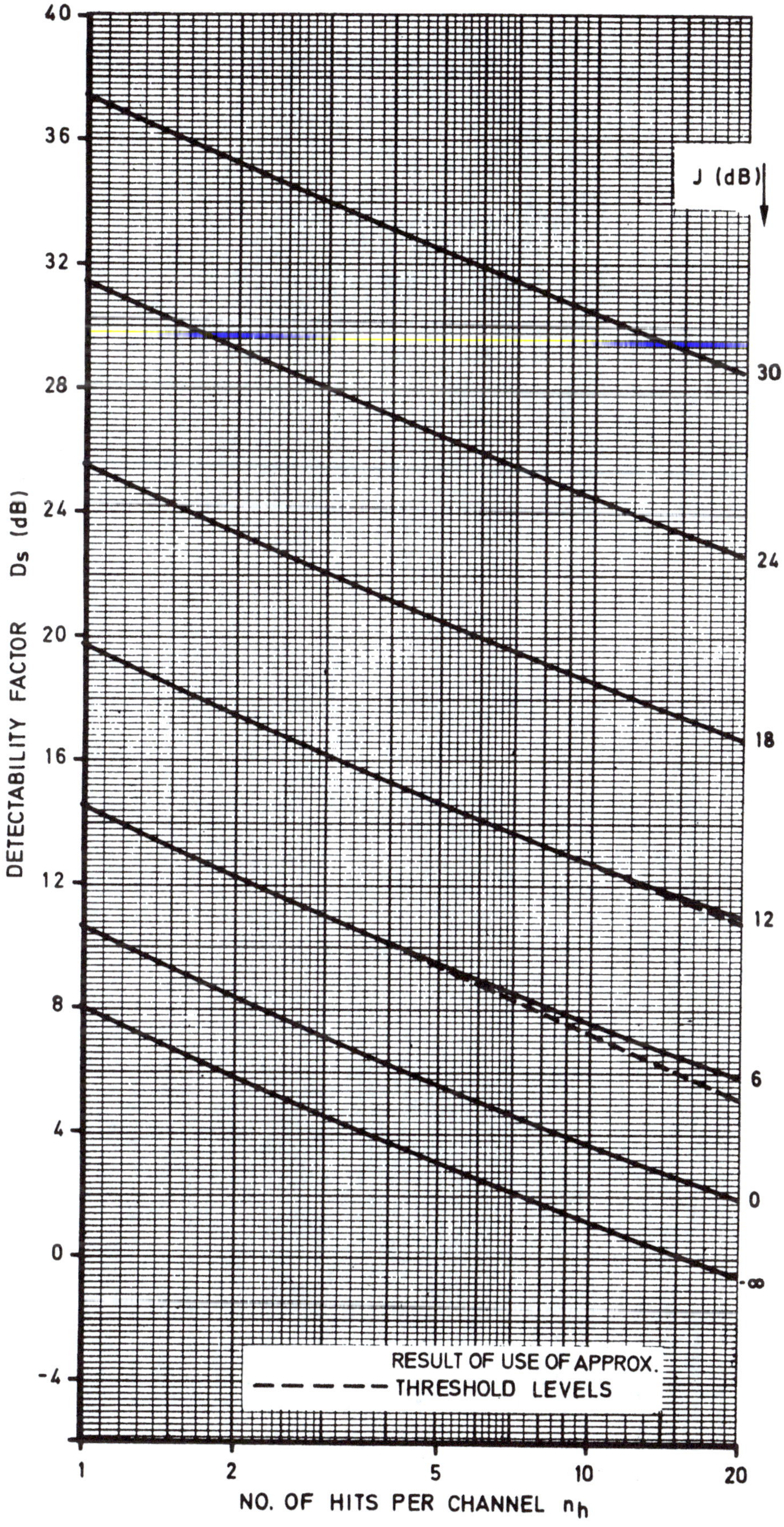

40
36
32
28
24
20
16
12
8
4
0
-4
DETECTABILITY FACTOR Ds (dB)
J (dB)
30
24
18
12
6
0
-∞
RESULT OF USE OF APPROX.
THRESHOLD LEVELS
1
2
5
10
20
NO. OF HITS PER CHANNEL n_h

Fig. 15

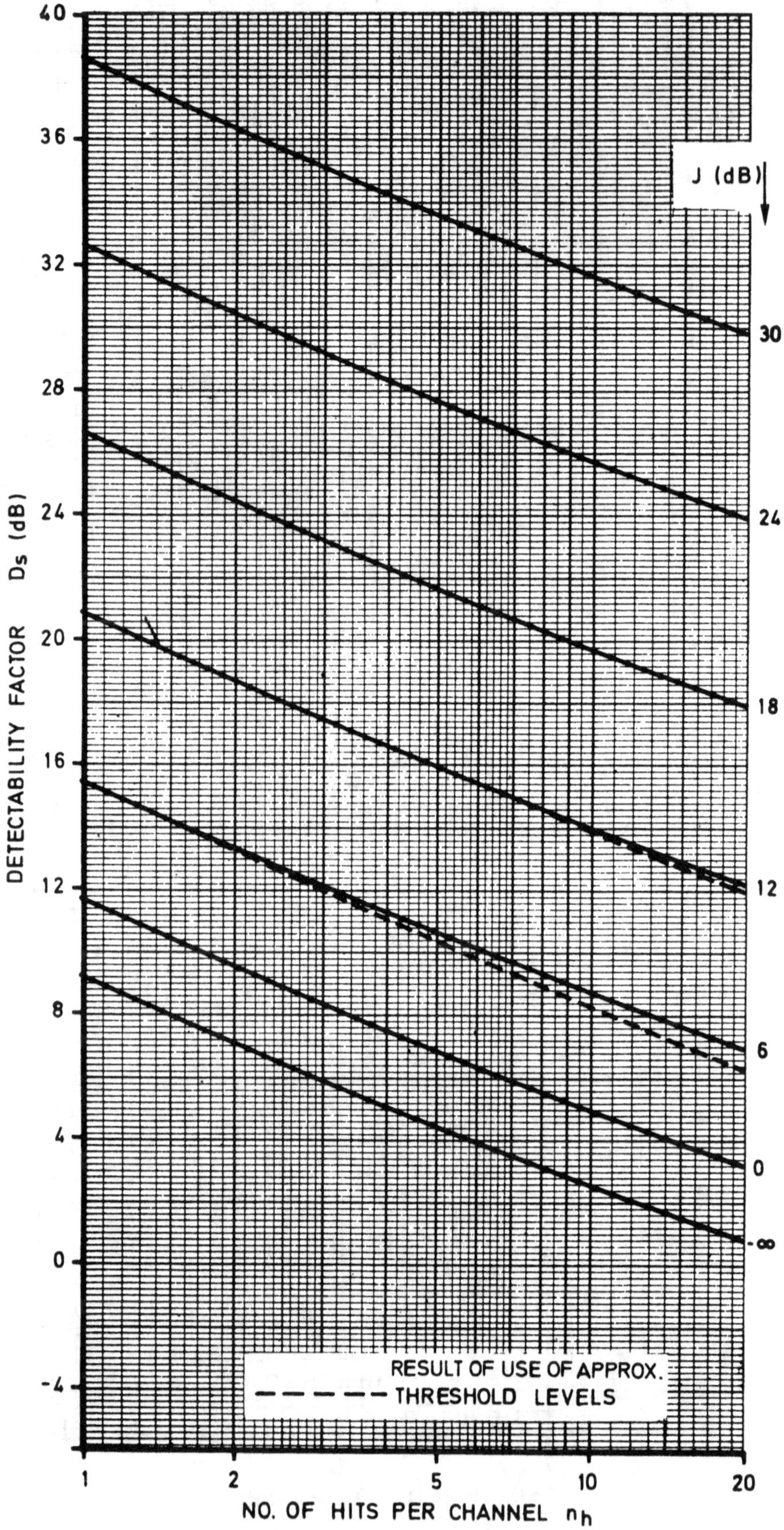

J (dB)
30
24
18
12
6
0
-∞
40
36
32
28
24
20
16
12
8
4
0
-4
DETECTABILITY FACTOR Ds (dB)
RESULT OF USE OF APPROX.
THRESHOLD LEVELS
1
2
5
10
20
NO. OF HITS PER CHANNEL nh

Fig. 16

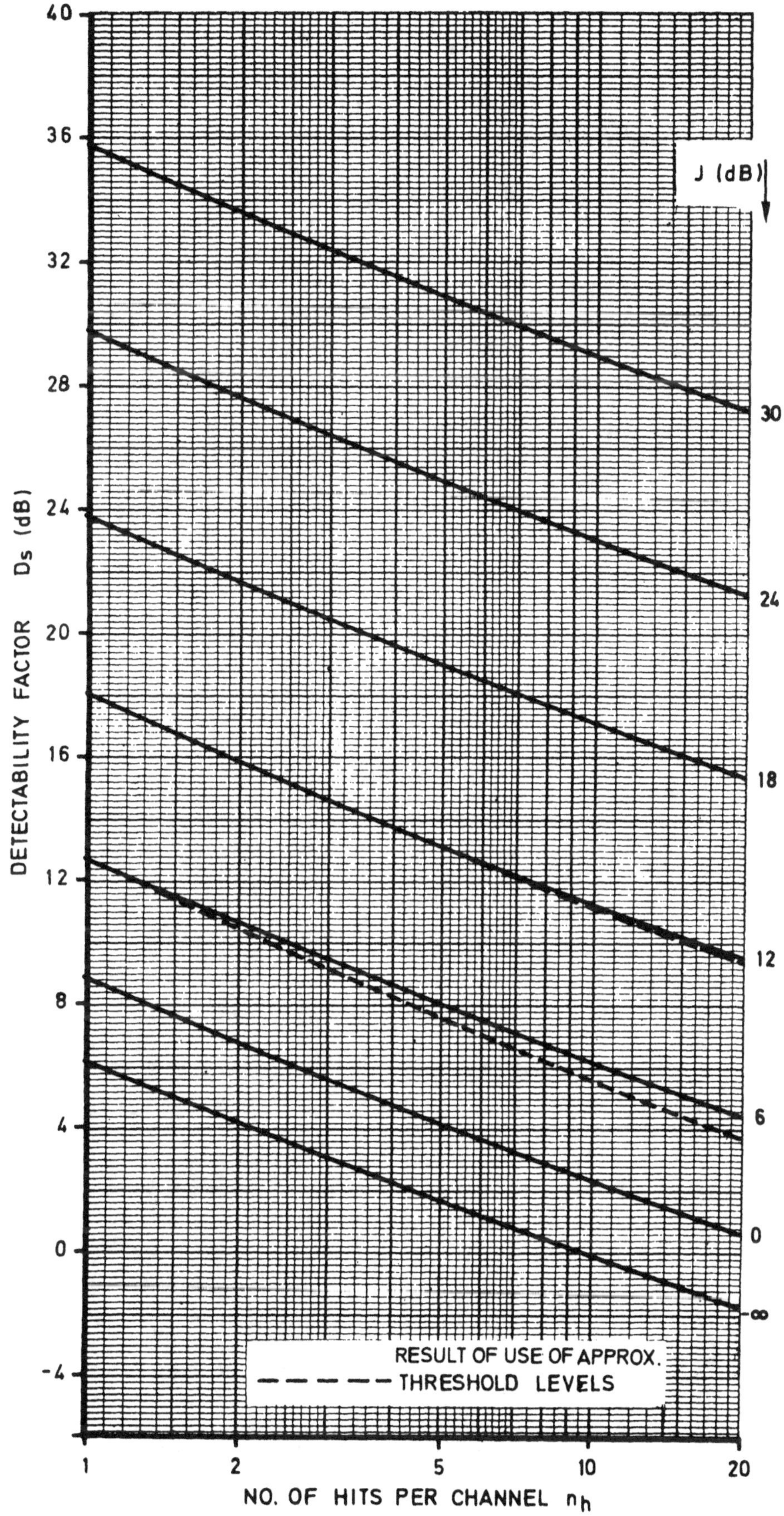

J (dB)
30
24
18
12
6
0
-∞
DETECTABILITY FACTOR Ds (dB)
NO. OF HITS PER CHANNEL n_h
RESULT OF USE OF APPROX.
THRESHOLD LEVELS

Fig. 17

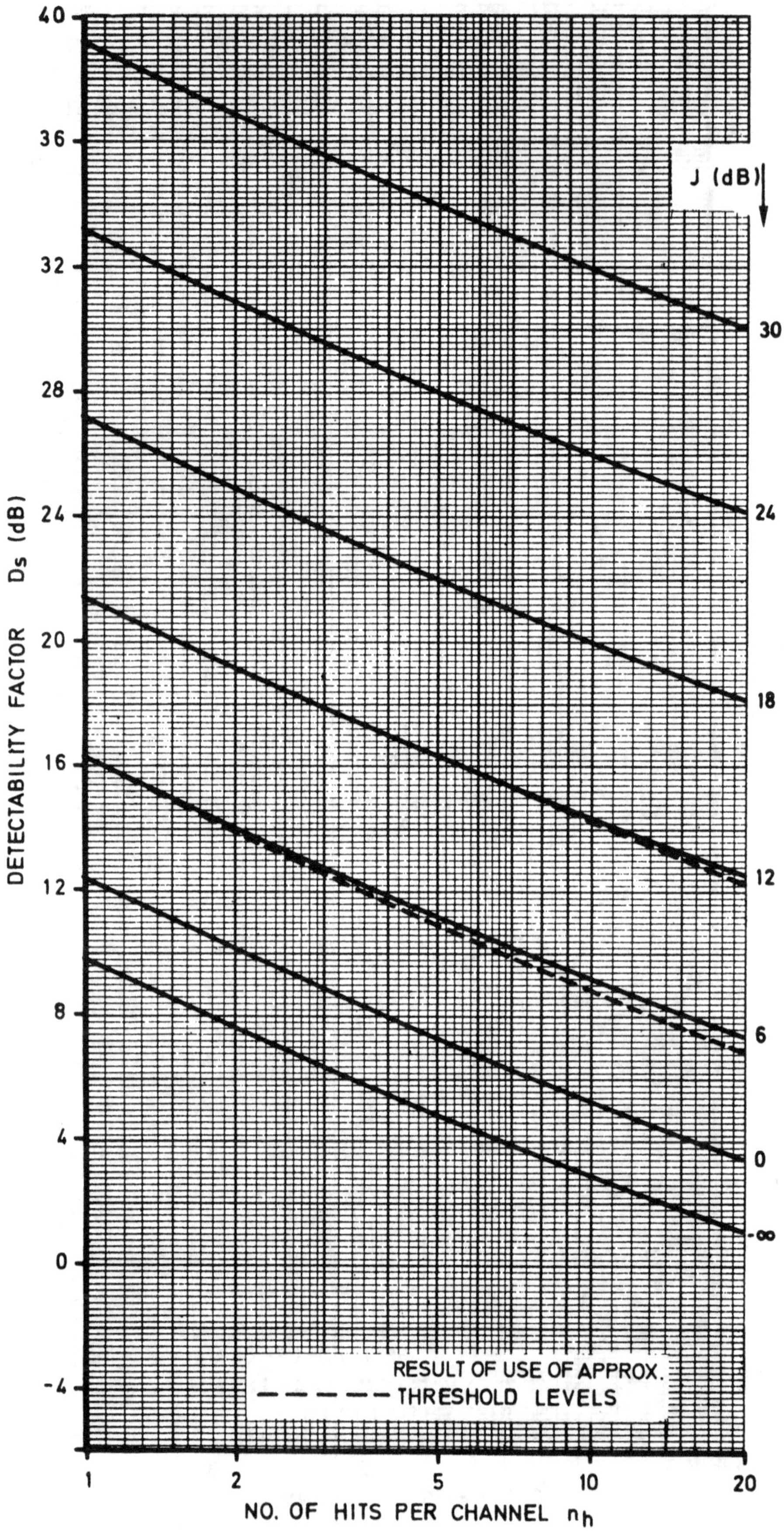
DETECTABILITY FACTOR Ds (dB)
40
36
32
28
24
20
16
12
8
4
0
-4
J (dB)
30
24
18
12
6
0
-∞
RESULT OF USE OF APPROX.
THRESHOLD LEVELS
1
2
5
10
20
NO. OF HITS PER CHANNEL n_h

Fig. 18

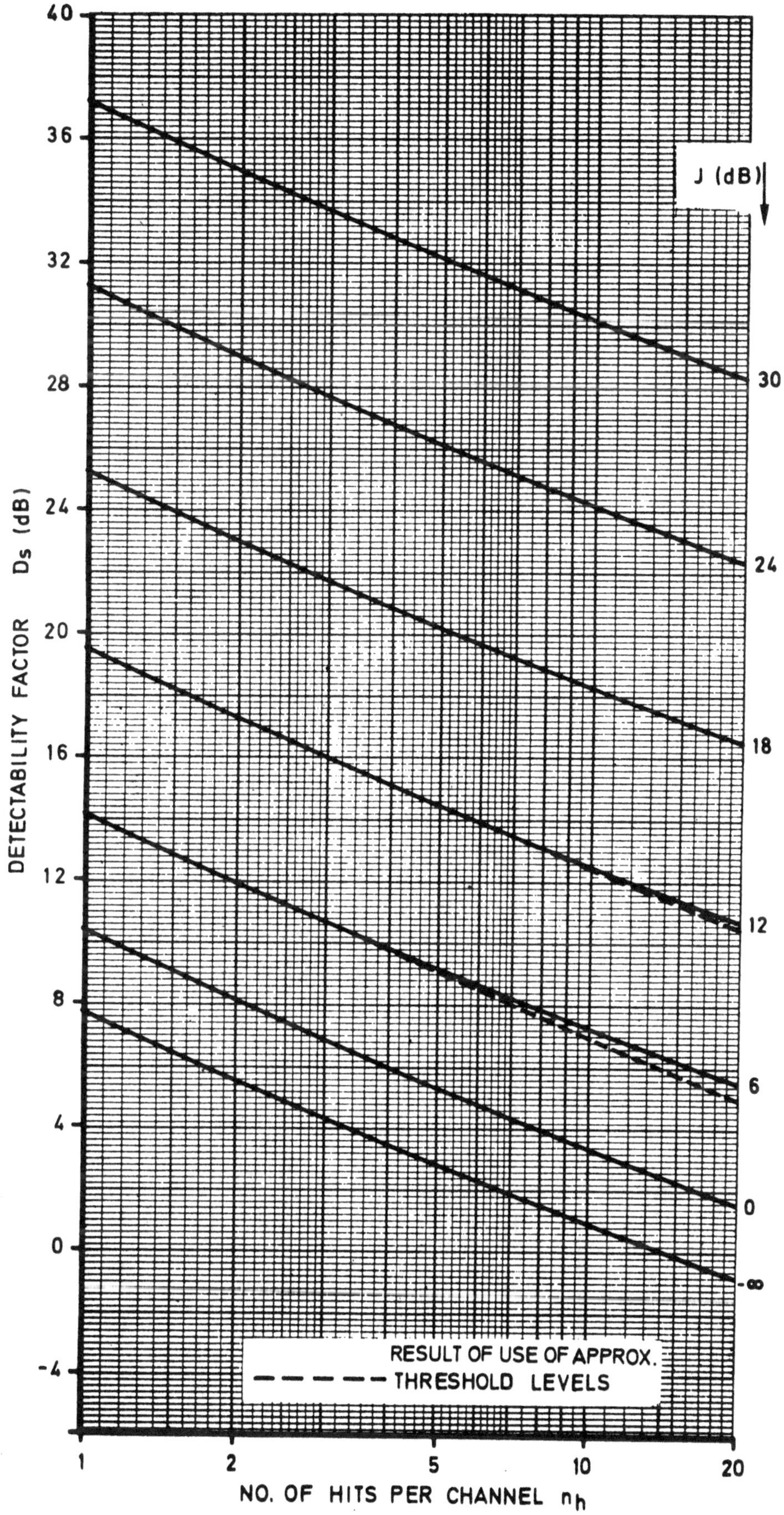
40
36
32
28
24
20
16
12
8
4
0
-4
DETECTABILITY FACTOR Ds (dB)
J (dB)
30
24
18
12
6
0
-∞
RESULT OF USE OF APPROX.
THRESHOLD LEVELS
1
2
5
10
20
NO. OF HITS PER CHANNEL nh

Fig. 19

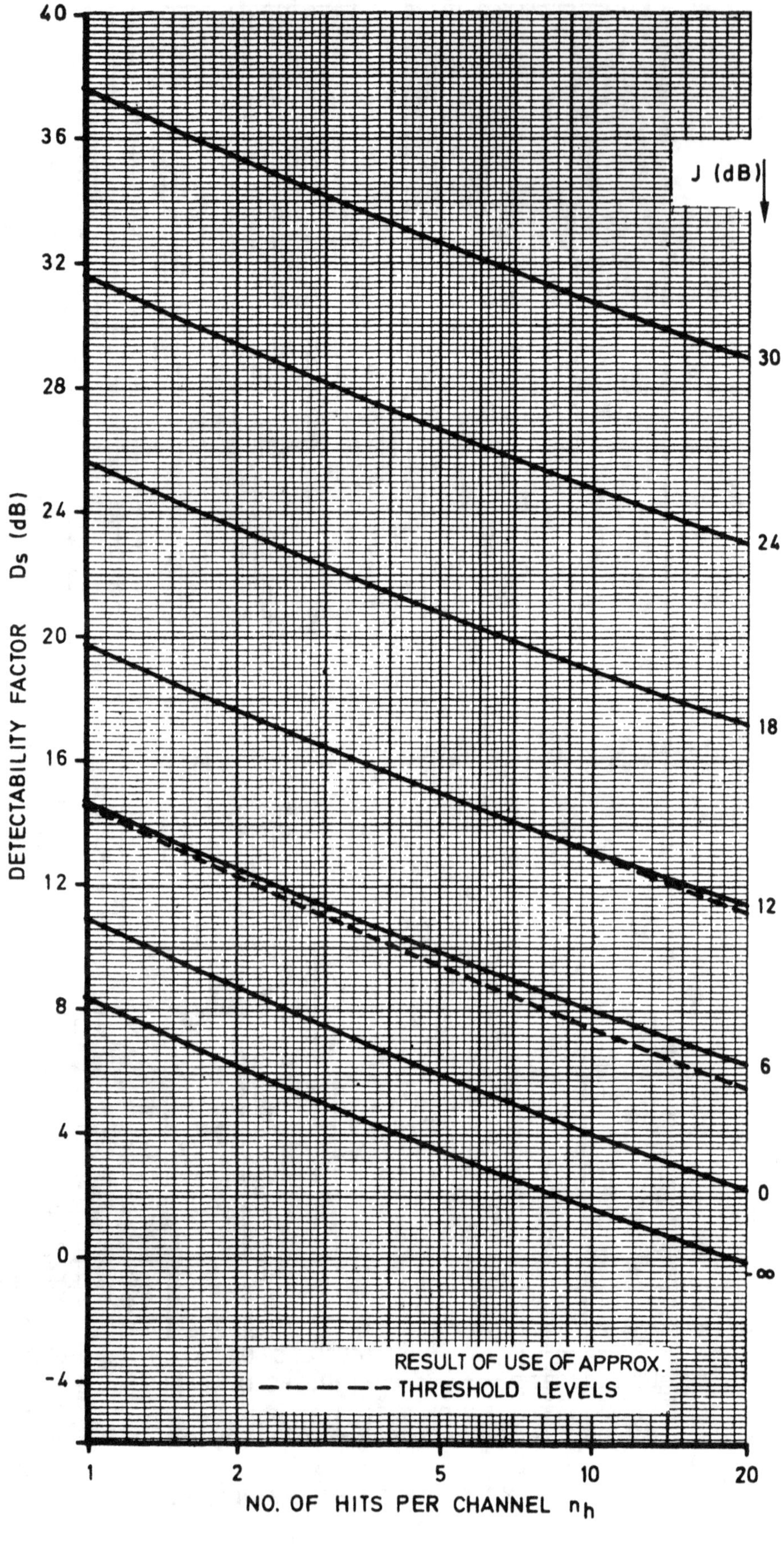

DETECTABILITY FACTOR Ds (dB)
40
36
32
28
24
20
16
12
8
4
0
-4
J (dB)
30
24
18
12
6
0
∞
RESULT OF USE OF APPROX.
THRESHOLD LEVELS
1
2
5
10
20
NO. OF HITS PER CHANNEL n_h

Fig. 20

210

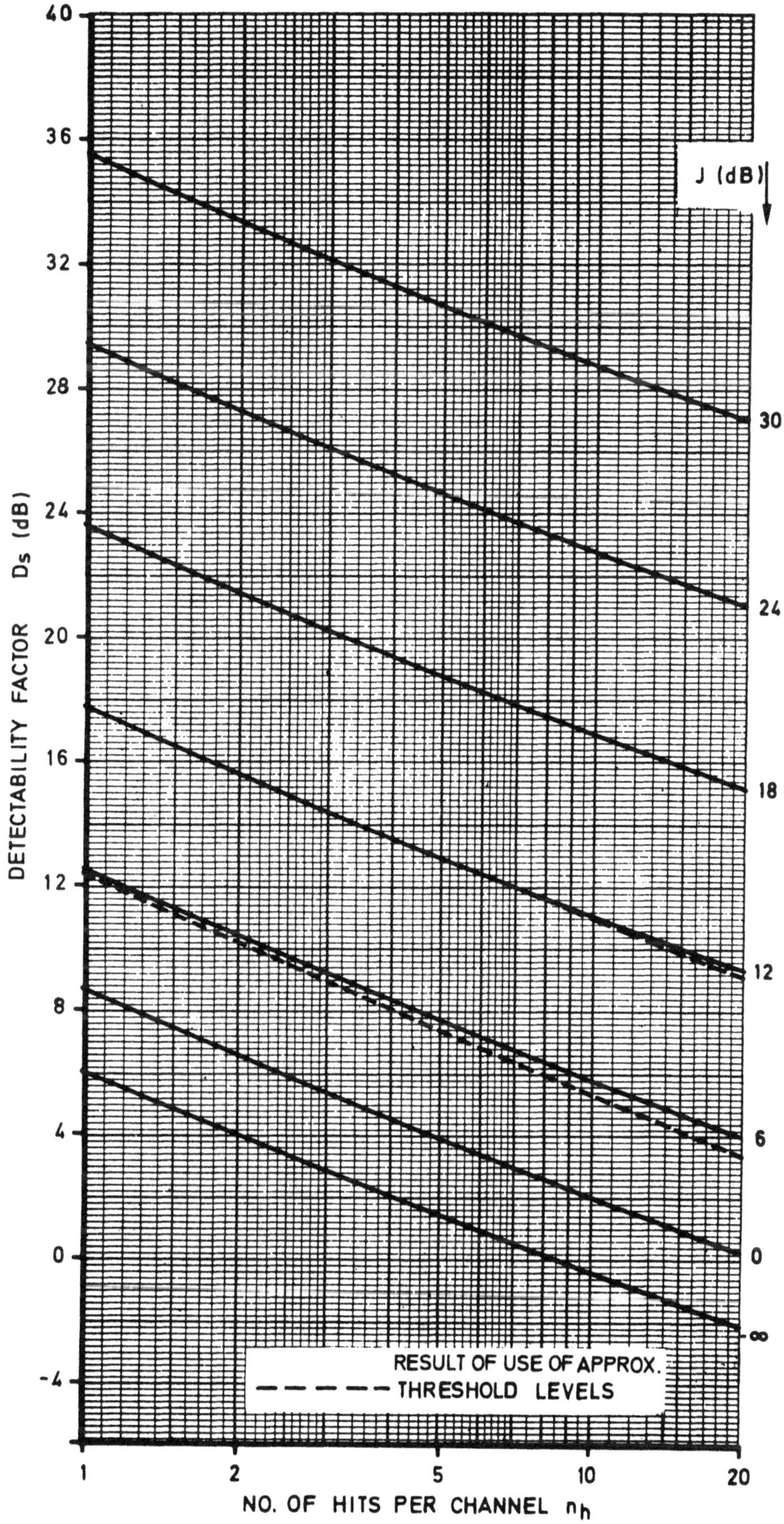
40
36
32
28
24
20
16
12
8
4
0
-4
DETECTABILITY FACTOR Ds (dB)
J (dB)
30
24
18
12
6
0
∞
RESULT OF USE OF APPROX.
THRESHOLD LEVELS
1
2
5
10
20
NO. OF HITS PER CHANNEL n_h

Fig. 21

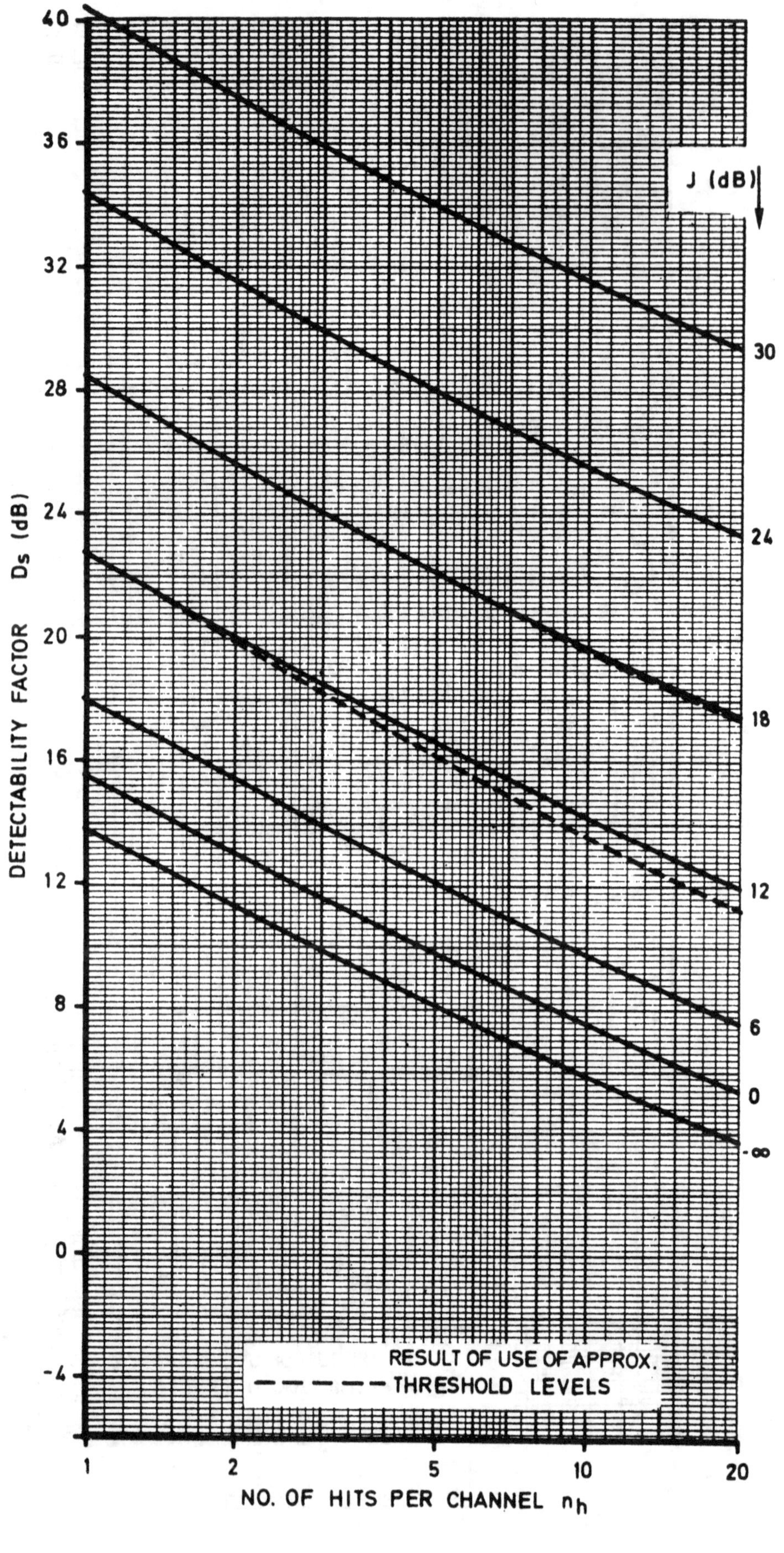
J (dB)
30
24
18
12
6
0
-∞
DETECTABILITY FACTOR Ds (dB)
40
36
32
28
24
20
16
12
8
4
0
-4
1
2
5
10
20
NO. OF HITS PER CHANNEL n_h
RESULT OF USE OF APPROX.
THRESHOLD LEVELS

Fig. 22

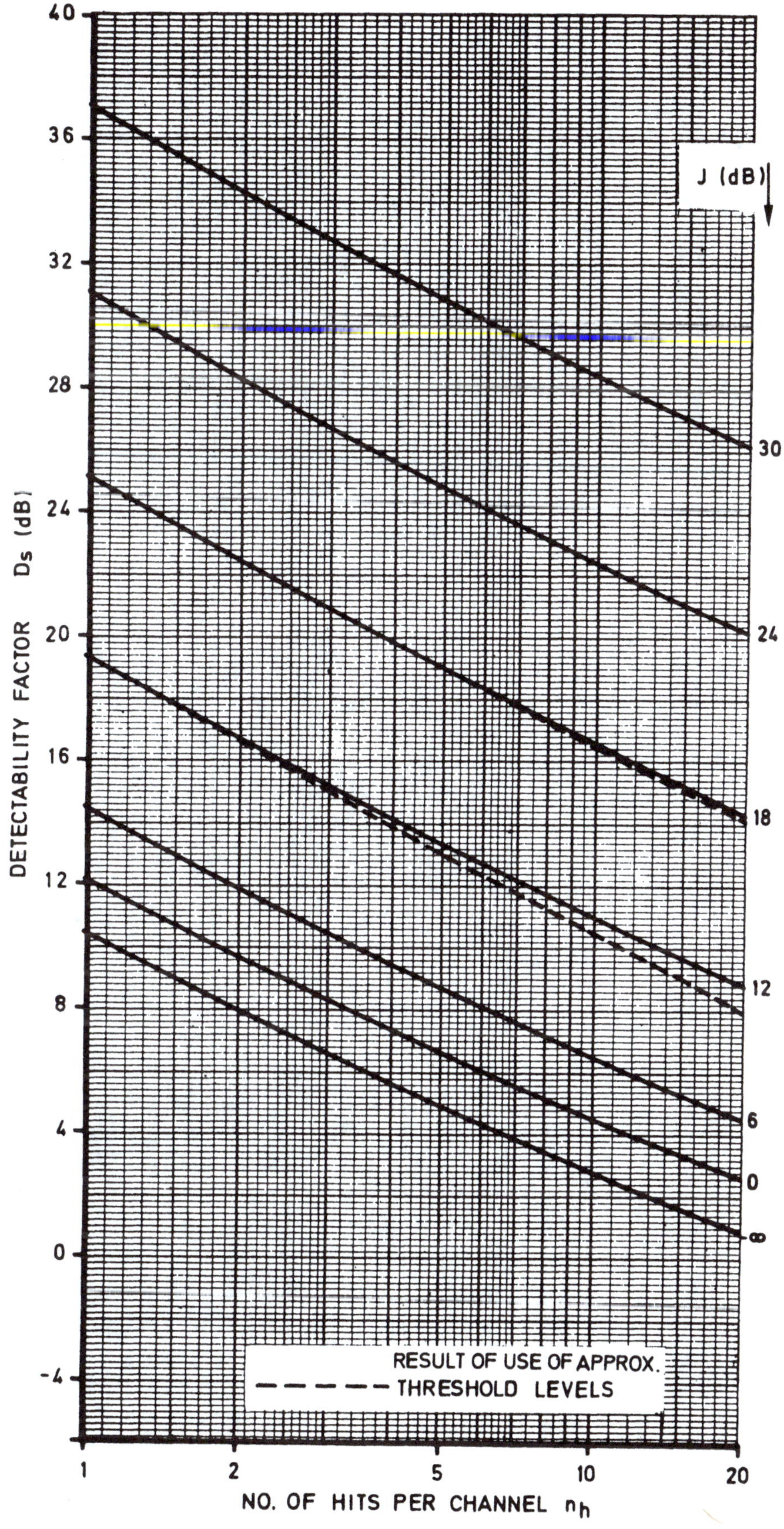

DETECTABILITY FACTOR Ds (dB)
NO. OF HITS PER CHANNEL n_h
J (dB)
30
24
18
12
6
0
∞
RESULT OF USE OF APPROX.
THRESHOLD LEVELS

Fig. 23

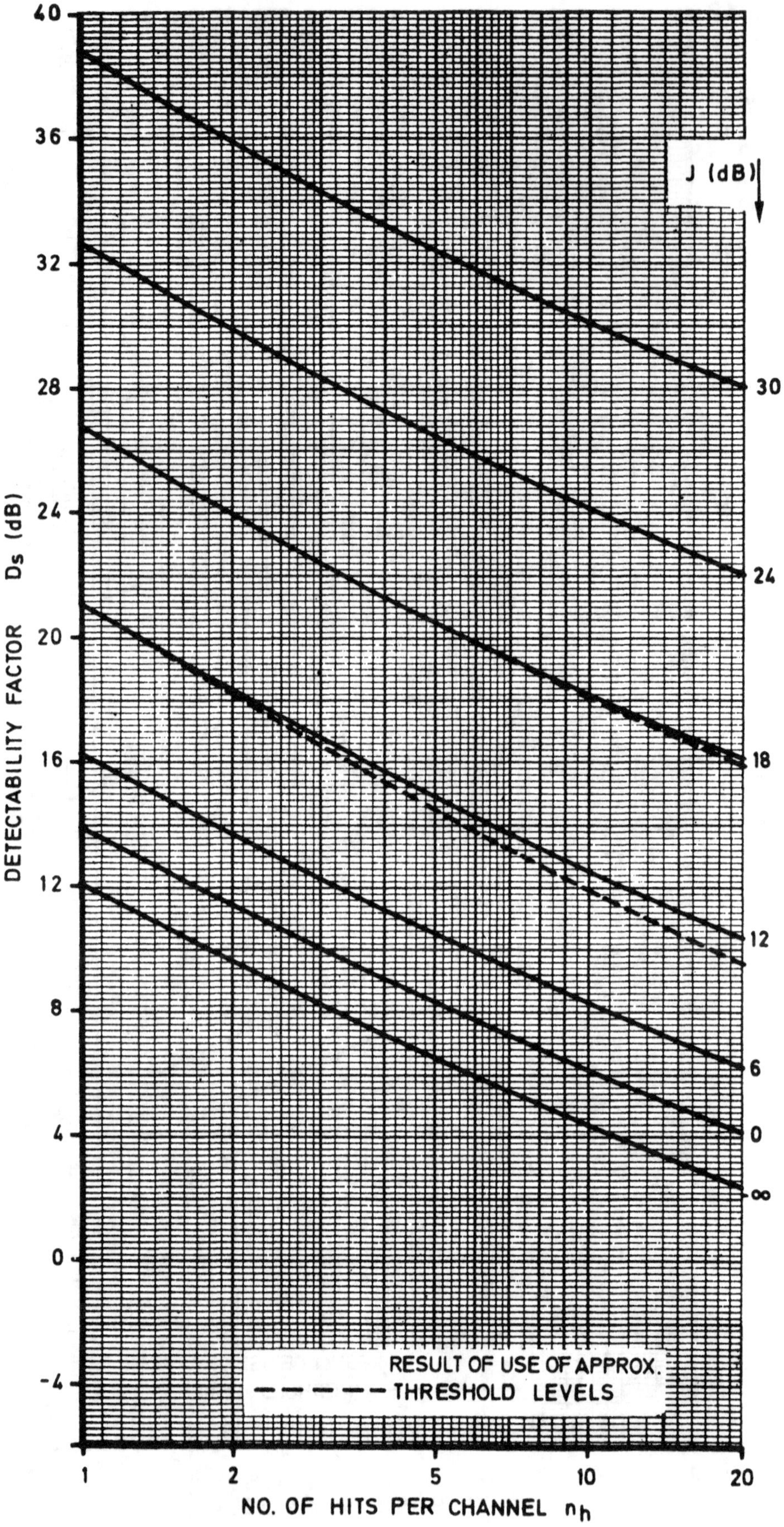

J (dB)
30
24
18
12
6
0
∞
DETECTABILITY FACTOR Ds (dB)
NO. OF HITS PER CHANNEL nh
RESULT OF USE OF APPROX.
THRESHOLD LEVELS

Fig. 24

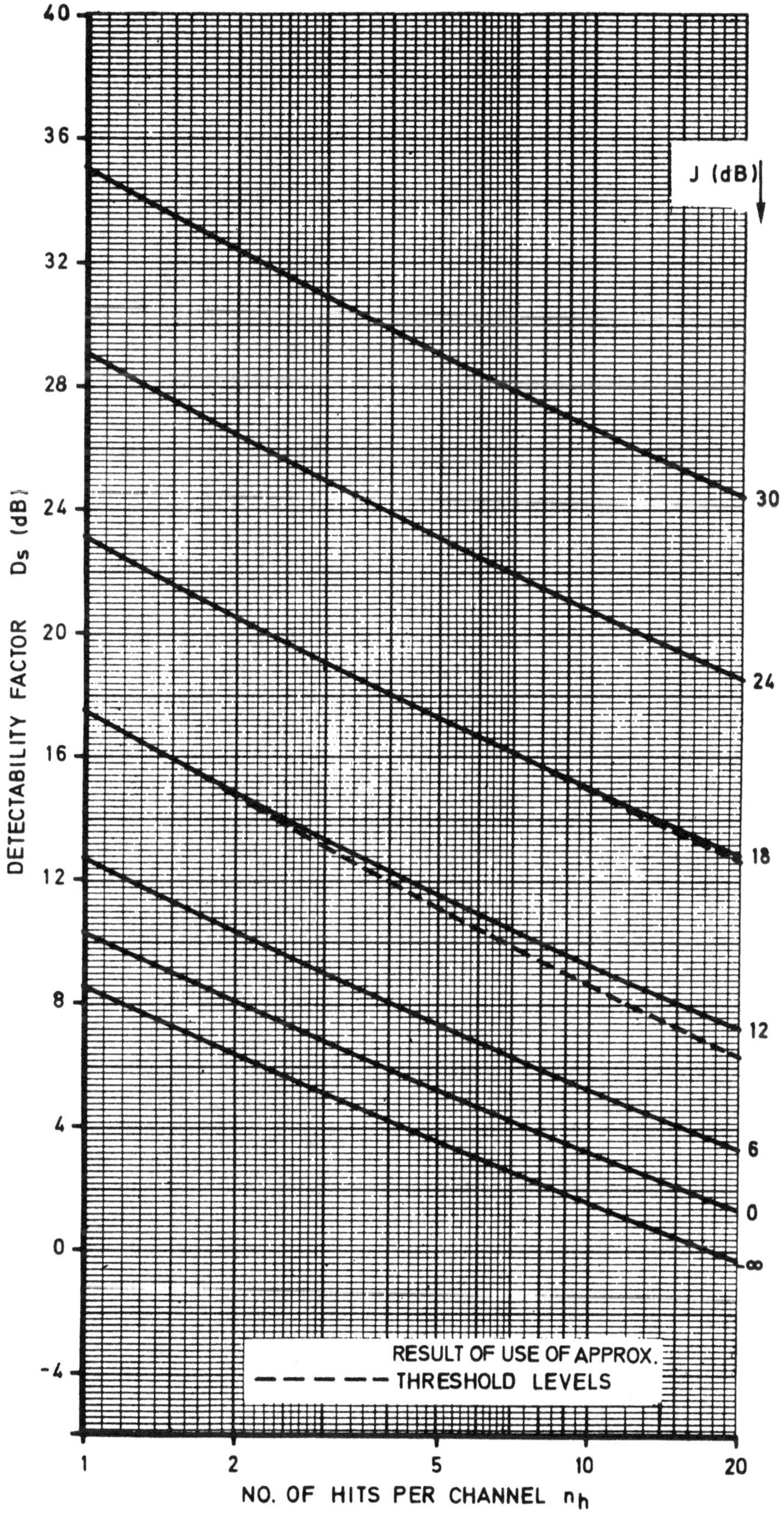
J (dB)
30
24
18
12
6
0
∞
RESULT OF USE OF APPROX.
THRESHOLD LEVELS
DETECTABILITY FACTOR D_s (dB)
NO. OF HITS PER CHANNEL n_h
40
36
32
28
24
20
16
12
8
4
0
-4
1
2
5
10
20

Fig. 25

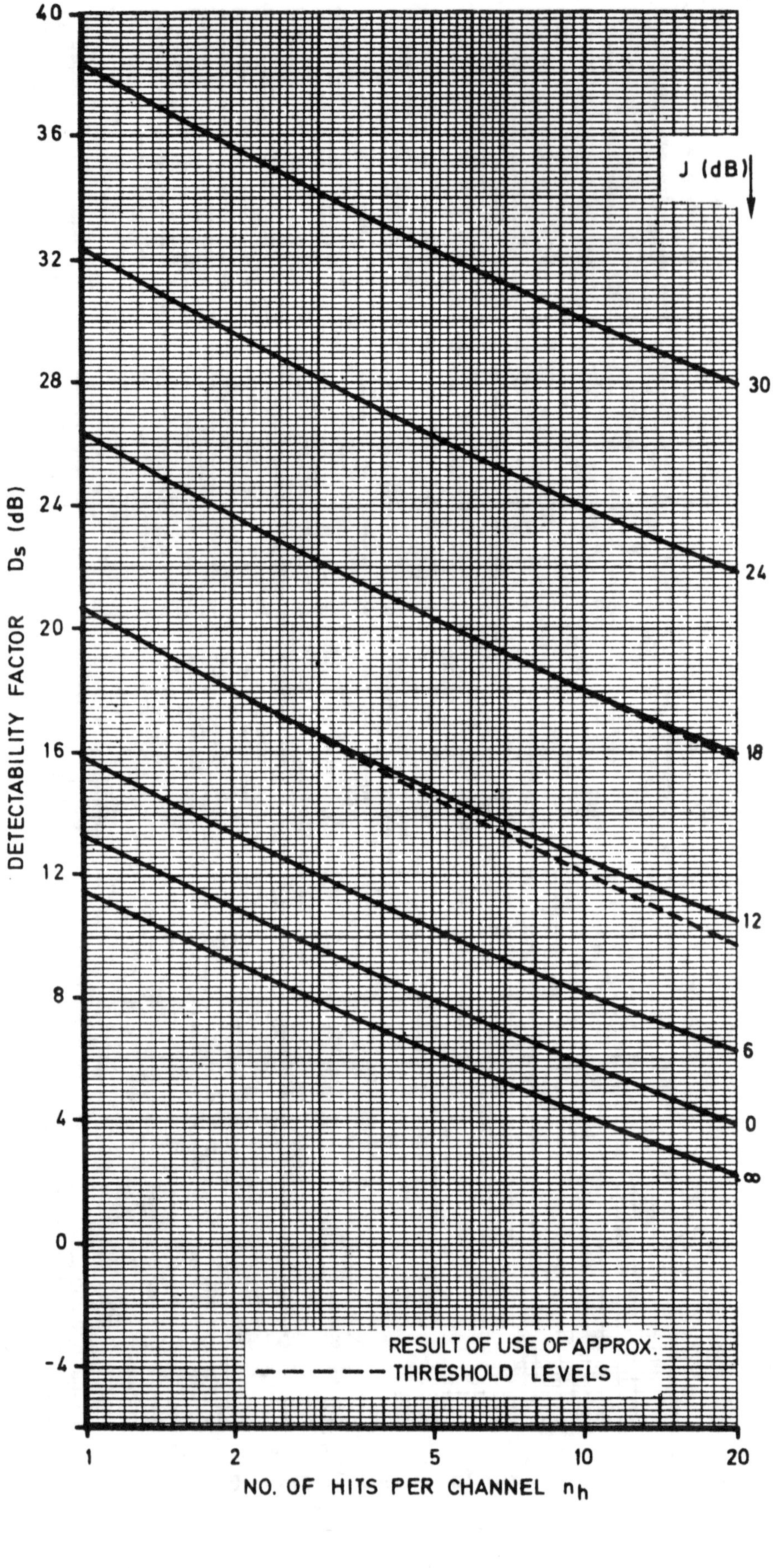

J (dB)
30
24
18
12
6
0
∞
DETECTABILITY FACTOR Ds (dB)
NO. OF HITS PER CHANNEL nh
RESULT OF USE OF APPROX.
THRESHOLD LEVELS

Fig. 26

216

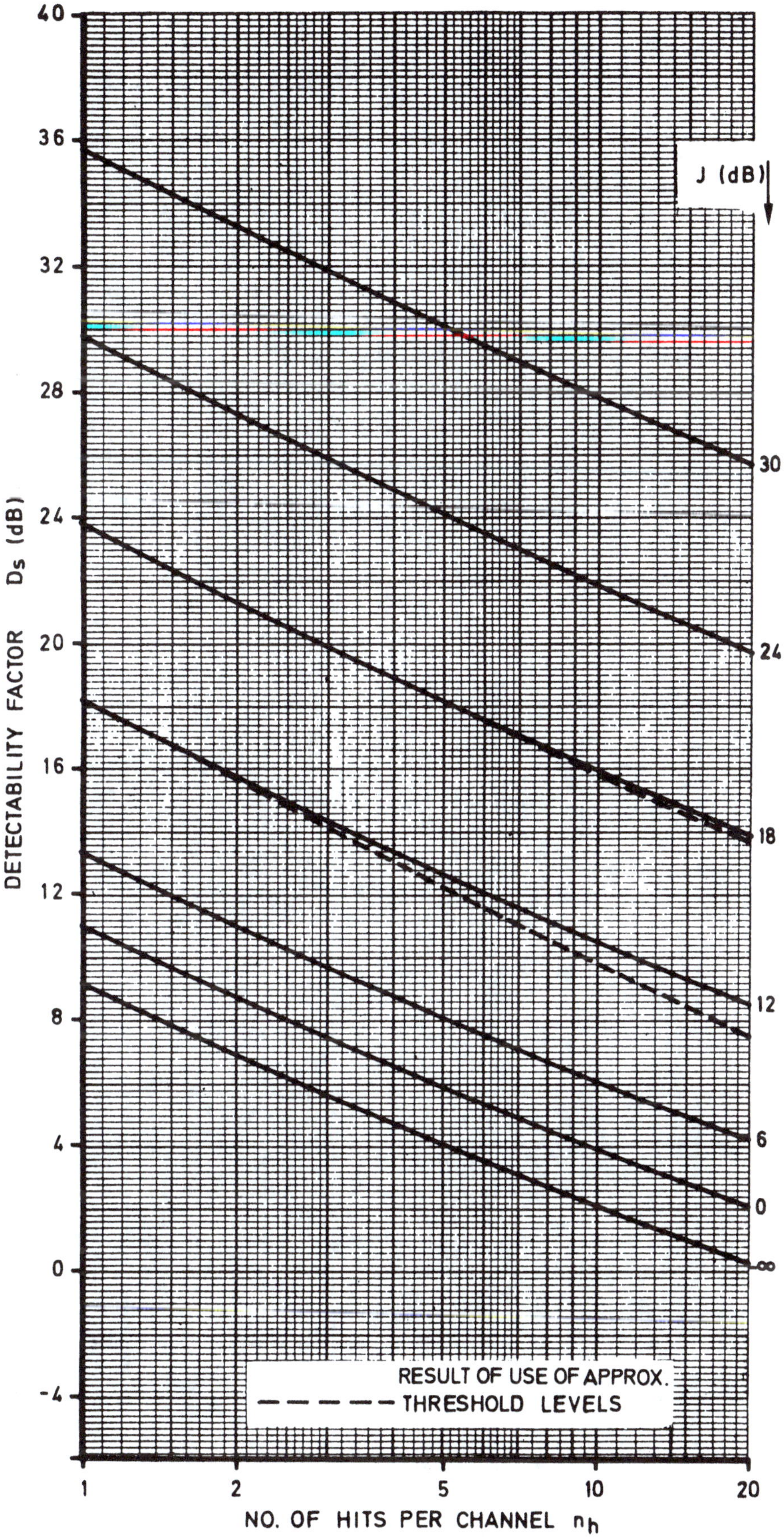
J (dB)
30
24
18
12
6
0
∞
RESULT OF USE OF APPROX.
THRESHOLD LEVELS
DETECTABILITY FACTOR Ds (dB)
NO. OF HITS PER CHANNEL n_h
40
36
32
28
24
20
16
12
8
4
0
-4
1
2
5
10
20

Fig. 27

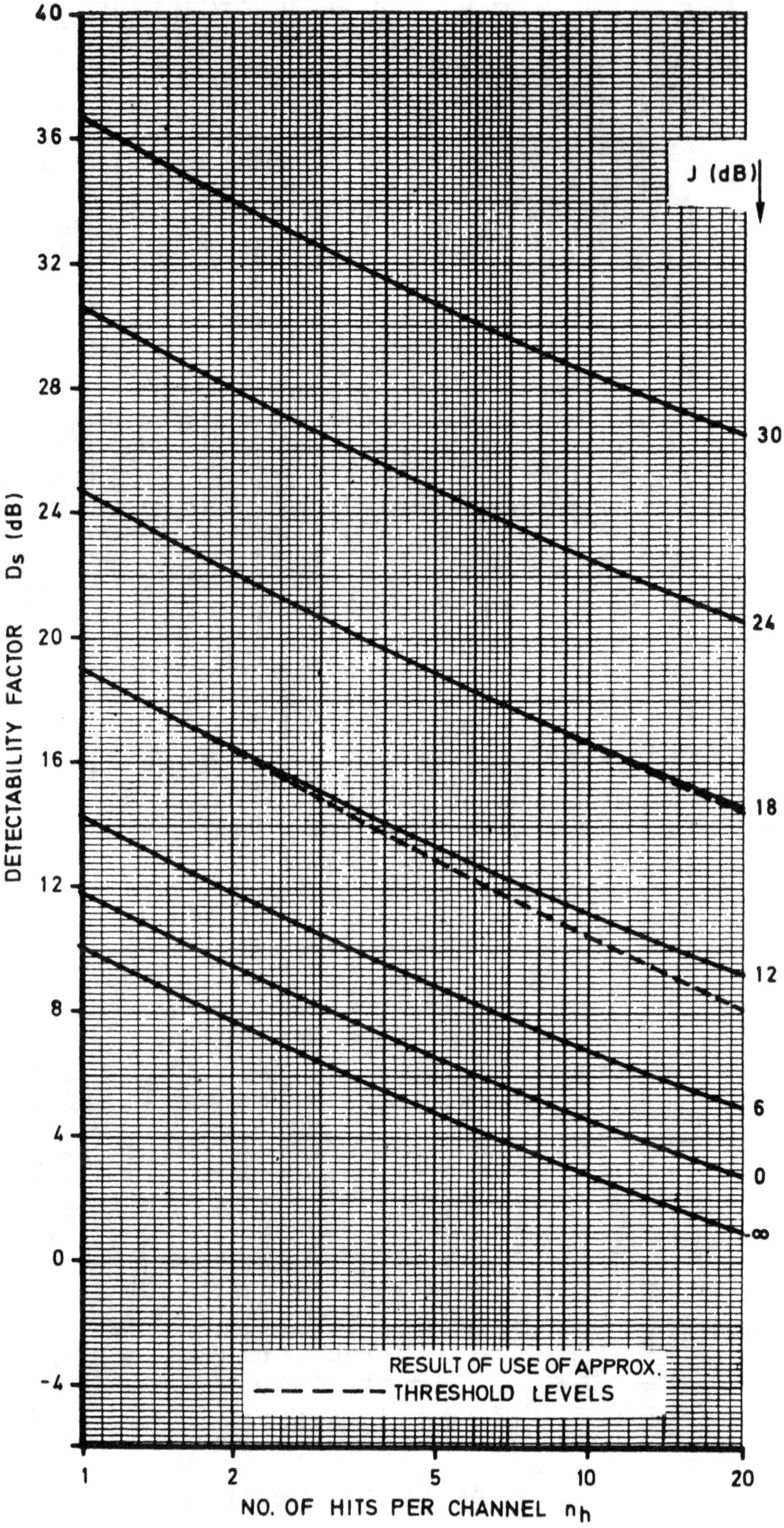
J (dB)
30
24
18
12
6
0
∞
DETECTABILITY FACTOR Ds (dB)
NO. OF HITS PER CHANNEL n_h
RESULT OF USE OF APPROX.
THRESHOLD LEVELS
40
36
32
28
24
20
16
12
8
4
0
-4
1
2
5
10
20

Fig. 28

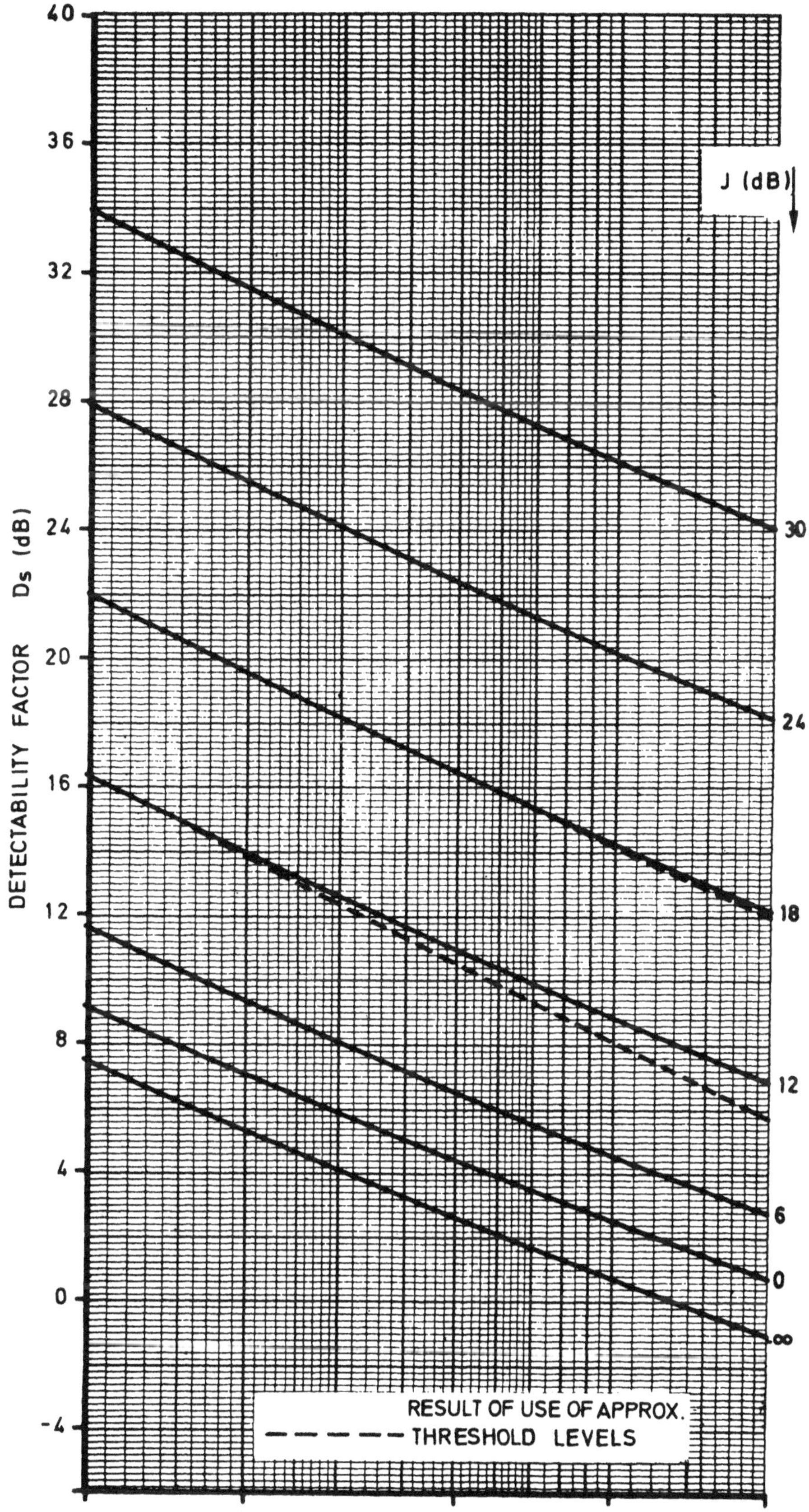
DETECTABILITY FACTOR Ds (dB)
40
36
32
28
24
20
16
12
8
4
0
-4
J (dB)
30
24
18
12
6
0
∞
RESULT OF USE OF APPROX.
THRESHOLD LEVELS
1
2
5
10
20
NO. OF HITS PER CHANNEL nh

Fig. 29

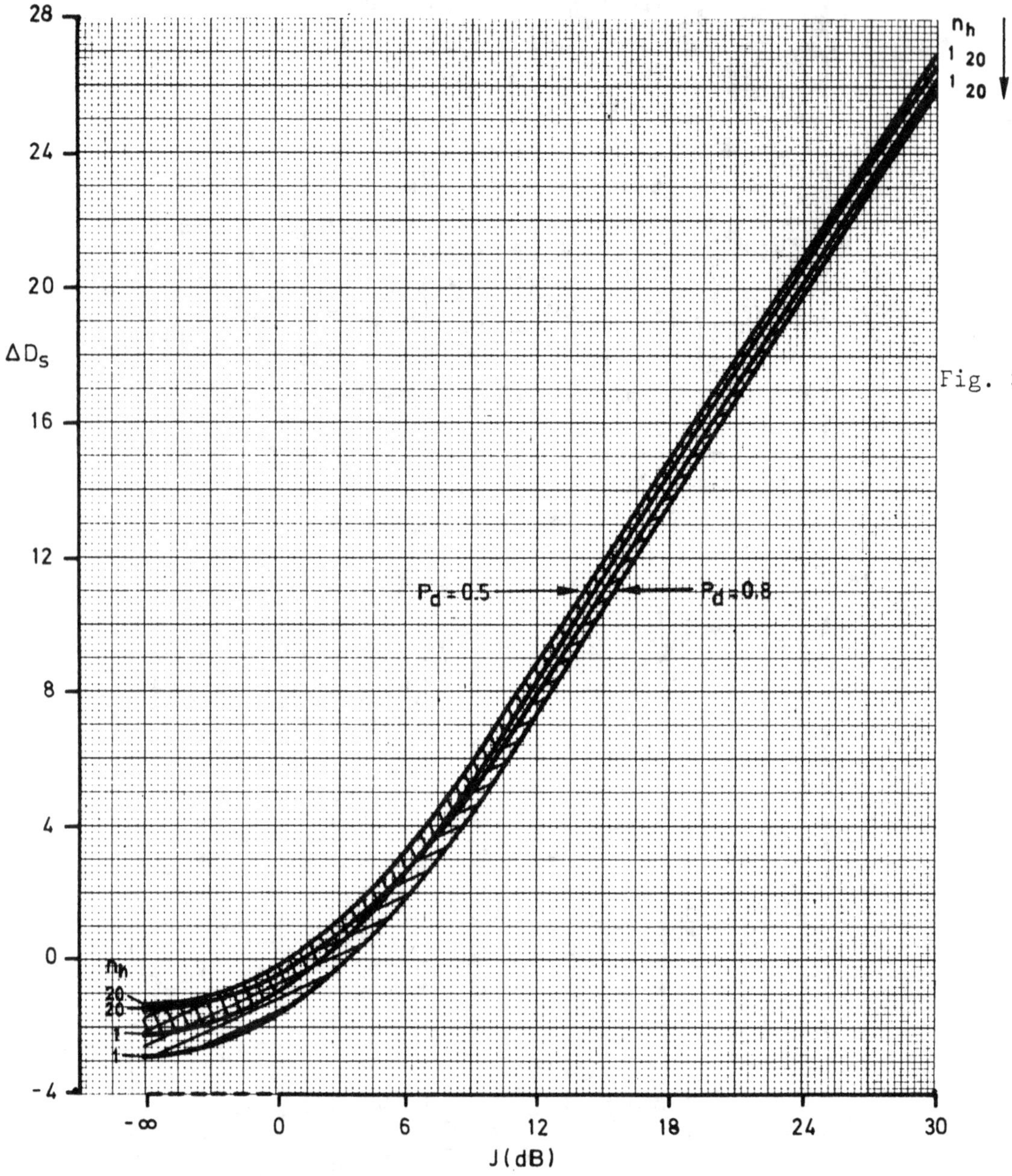
28
24
20
ΔDs
16
12
8
4
0
-4
nh
1 20
1 20
20
20
1
1
Pd = 0.5
Pd = 0.8
-∞
0
6
12
18
24
30
J(dB)

Fig. 30

220

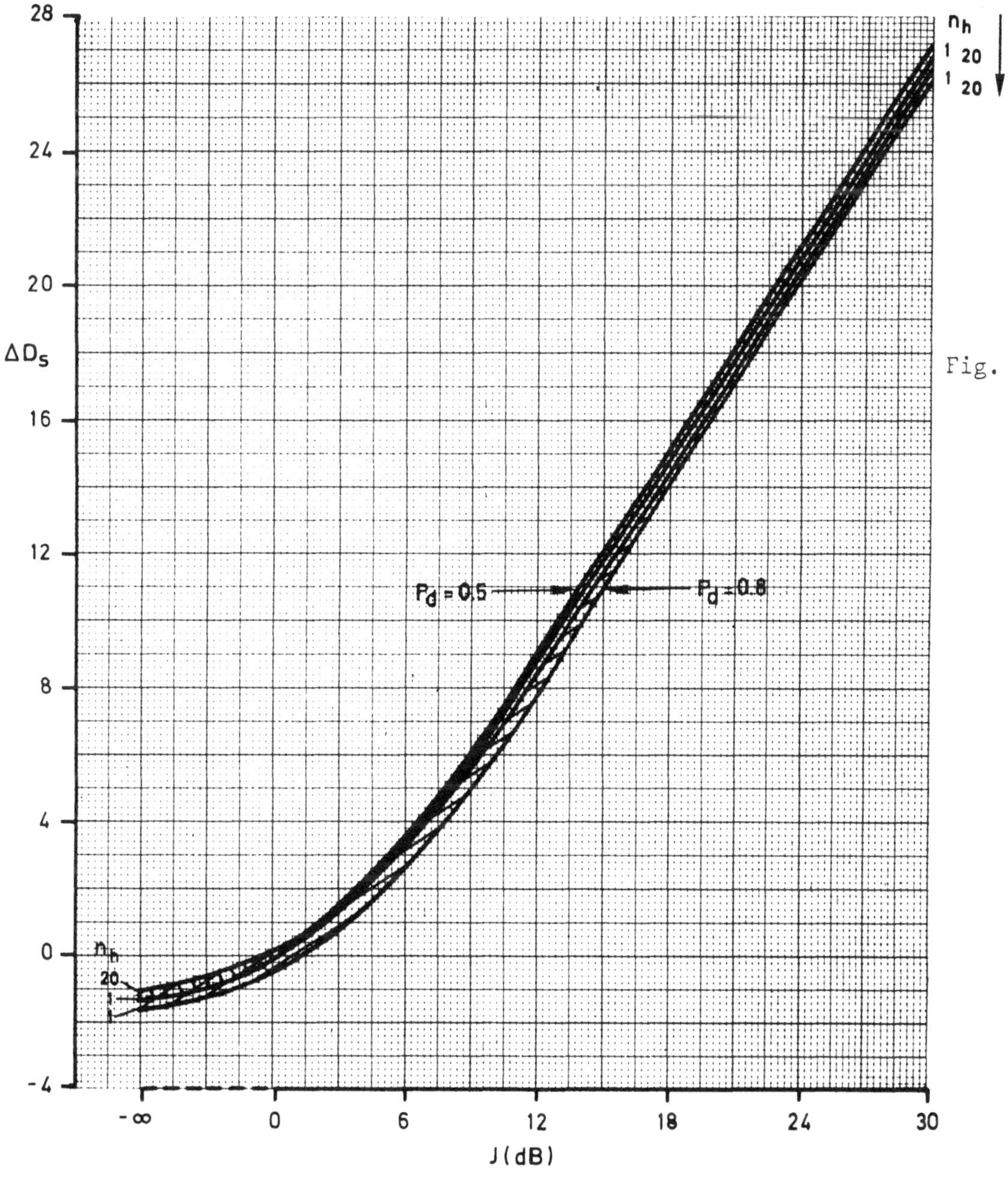

28
24
20
ΔDs
16
12
8
4
0
-4
nh
1 20
1 20
Pd = 0.5
Pd = 0.8
nh
20
1
1
-∞
0
6
12
18
24
30
J(dB)

Fig. 31

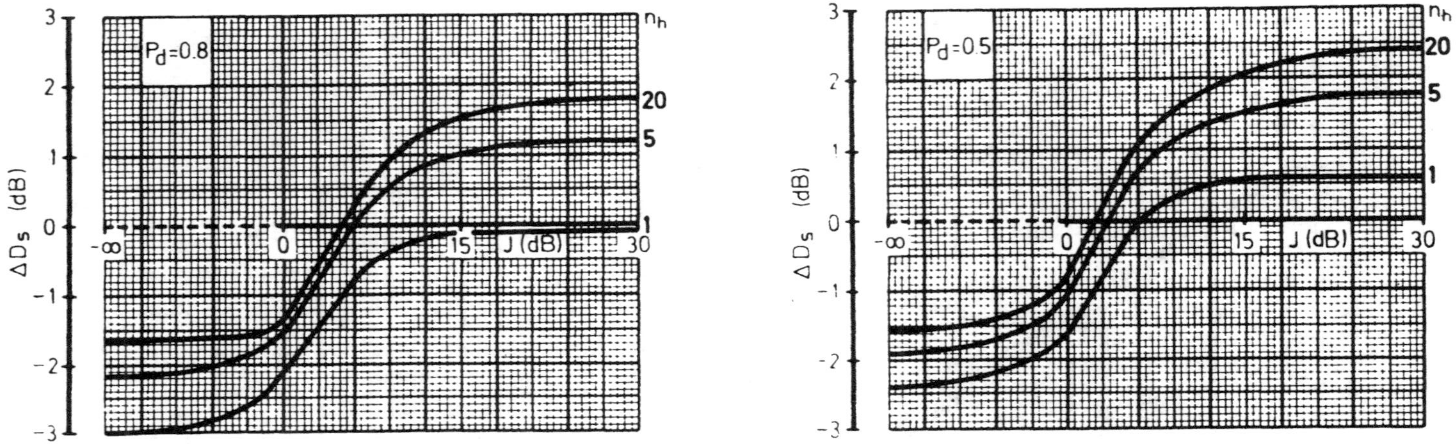
3
2
1
0
-1
-2
-3
ΔD_s (dB)
P_d=0.8
P_d=0.5
n_h
20
5
1
-∞
0
15
J (dB)
30

(a)

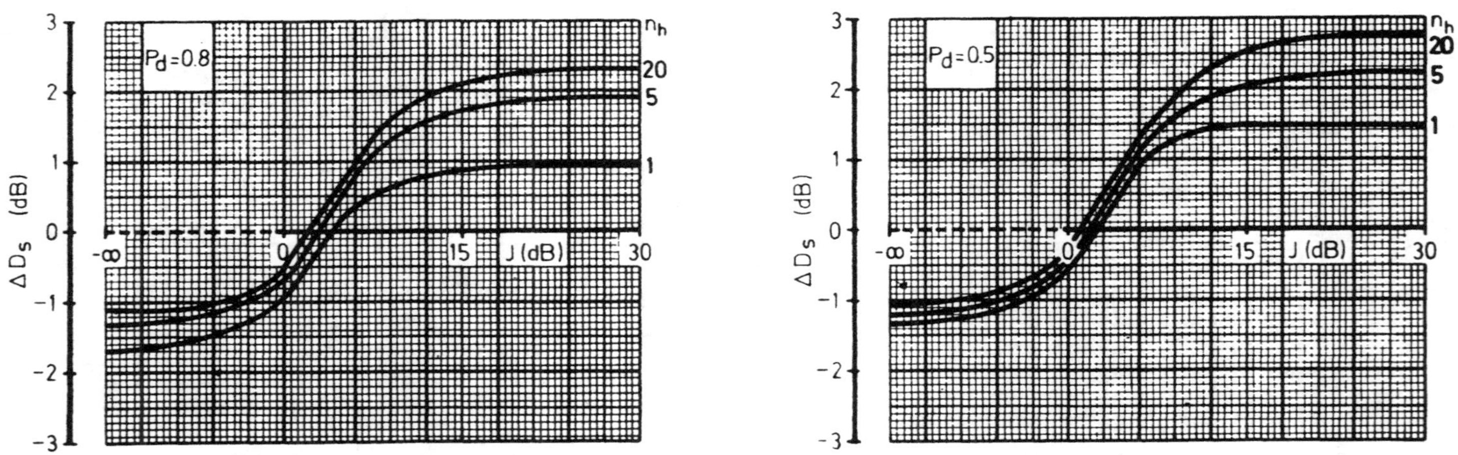
3
2
1
0
-1
-2
-3
ΔD_s (dB)
P_d=0.8
P_d=0.5
n_h
20
5
1
-∞
0
15
J (dB)
30

(b)

Fig. 32

K E Y   T O   F I G U R E S

Variation of D_s with J and n_h:

Fig. No.	Case No.	P_d	P_f	Swerling Case No.
2	IIE-0	0.8	10^{-6}	I/II
3	"	0.5	"	"
4	"	0.8	10^{-4}	"
5	"	0.5	"	"
6	IIE-0	0.8	10^{-6}	III/IV
7	"	0.5	"	"
8	"	0.8	10^{-4}	"
9	"	0.5	"	"
10(a)	IIE-1	0.8	10^{-6}	I/II
(b)	"	0.5	"	"
11(a)	"	0.8	10^{-4}	"
(b)	"	0.5	"	"
12(a)	IIE-1	0.8	10^{-6}	III/IV
(b)	"	0.5	"	"
13(a)	"	0.8	10^{-4}	"
(b)	"	0.5	"	"
14	IEE-0	0.8	10^{-6}	I/II
15	"	0.5	"	"
16	"	0.8	10^{-4}	"
17	"	0.5	"	"
18	IEE-0	0.8	10^{-6}	III/IV
19	"	0.5	"	"
20	"	0.8	10^{-4}	"
21	"	0.5	"	"
22	IEE-1	0.8	10^{-6}	I/II
23	"	0.5	"	"
24	"	0.8	10^{-4}	"
25	"	0.5	"	"
26	IEE-1	0.8	10^{-6}	III/IV
27	"	0.5	"	"
28	"	0.8	10^{-4}	"
29	"	0.5	"	"

Gain in D_s of Mode 1 relative to Mode 0, $10^{-6} \leq P_f \leq 10^{-4}$:

Fig. No.	Environmental condition	Swerling Case No.
30	IIE	I/II
31	"	III/IV
32(a)	IEE	I/II
(b)	"	III/IV

REFERENCES

1. J.A. Bijvoet, "Standard Methods for Predicting and Specifying Performance of Air Surveillance Radar Systems", SHAPE Technical Centre, Technical Report TR-50-U, April 1969

   <u>Supplement 1</u>

   J. Eendebak: "Detectability Curves for Swerling Cases III and IV Target Models", December, 1969

   <u>Supplement 2</u>

   A.J. Poelman: "Performance Evaluation of Frequency Diversity Radars", April 1970

2. S. Herscovici et A. Detape, "Performances d'un Radar Diversité", Annales de Radioelectricité, Tome 12, No. 50, October 1957, pp 285-298

3. J.I. Marcum, "A Statistical Theory of Target Detection by Pulsed Radar", Mathematical Appendix, IRE Trans. on Information Theory, April 1960, pp 149-169. Special Monograph Issue

4. P. Swerling, "Probability of Detection for Fluctuating Targets", IRE Trans. on Information Theory, April 1960, pp. 269-308. Special Monograph Issue.

5. A.J. Poelman and Mrs. E.J. Bennée, "The performance of Triple-Channel Frequency Diversity Radar Systems - An Investigation of the Efficiency of Different Types of Video Recombination", SHAPE Technical Centre, Technical Memorandum TM-377 (to be published mid-1973)

6. M.I. Skolnik, "Introduction to Radar Systems", McGraw-Hill Book Company, Inc., 1962, Section 2.5

The Performance of Triple-Channel Frequency Diversity Radar Systems — An Investigation of the Efficiency of Different Types of Video Recombination in an Interference Environment

A.J. Poelman and E.J. Bennée

Technical Memorandum TM-377, SHAPE Technical Centre, The Hague, Netherlands (November 1973)

SUMMARY

This memorandum deals with deriving and comparing the performance of receiver modes of operation which are or might be used in triple-channel frequency diversity radar systems. These receiver modes are characterized by the following video recombinations, u_i:

Mode 0 : $u_o = y_s + y_i + y_\ell$
Mode 1 : $u_1 = 2y_s + y_i$
Mode 2 : $u_2 = y_s + y_i$
Mode 3 : $u_3 = y_s + 2y_i$
Mode 4 : $u_4 = y_s$,

where y_s, y_i, and y_ℓ are the smallest, the intermediate, and the largest video of the three channel videos y_1, y_2, and y_3 respectively. The applicability of these receiver modes (except Mode 3, which is probably of only academic interest) was first mentioned by Herscovici and Detape in Ref. 1.

Mode 0, the linear-addition mode of video recombination is fundamentally the most efficient one, at least in a quiet environment. Modes 1 to 4 have electronic counter-countermeasure (ECCM) properties. Since Modes 1, 2, and 3 are combinations of only the smallest and the intermediate videos and thus cancel the channel with the largest video, these modes have an ECCM feature when only one channel is jammed. Mode 4 on the other hand processes only the channel with the smallest video, and thus has an ECCM feature when two channels are simultaneously jammed.

Nowadays frequency diversity radars, which usually employ three channels, mostly use Mode 1 in addition to Mode 0 in order to have an ECCM capability. To date there are two STC papers dealing with the performance of these modes of frequency diversity operation. Supplement 2 to TR-50-U (Ref. 2) derives the performance of multi-channel frequency diversity radars operating in receiver Mode 0 in a quiet and clear environment. Supplement 3 to TR-50-U (Ref. 2), recently published, presents curves of the detectability factor for triple-channel frequency diversity radars, operating in Modes 0 and 1 in a jamming environment (one channel jammed, and two channels simultaneously jammed). For reasons of efficiency the derivation of the formulae for the probability of false alarm and the probability of detection are not included in TR-50-U Supplement 3, but are presented in this memorandum.

The reasons for undertaking the work were that, in STC knowledge:

- no performance evaluations exist for triple-channel frequency diversity radar systems operating in the afore-mentioned receiver modes, in a jamming environment,

- although Modes 2 and 4 are not at present used in triple-channel frequency diversity systems, these modes of operation are of practical interest. This is because it is reasonable to suppose that since Mode 2 is a 'linear' combination of two channel videos, it will be more efficient than Mode 1, at least when one channel is heavily jammed; and of course Mode 4 has the ECCM capability when two channels are jammed.

This memorandum contains for receiver Modes 0 to 4:

- complete derivations of the formulae for the probability of false alarm, distinguishing between the environmental conditions "only one channel jammed" and "two channels simultaneously jammed",

- tables of threshold levels corresponding with two practical values for the probability of false alarm ($P_f = 10^{-6}$ and 10^{-4}),

- derivation of the formulae for the probability of detection, where both Swerling target fluctuation models (Swerling Case I/II, and Swerling Case III/IV) are considered (due to mathematical complexities, formulae could not be obtained for all cases),

- graphs of the detectability factor for both Swerling target fluctuation models, and environmental conditions. The majority of the results have been obtained by Monte Carlo simulation.

- a discussion on the computation problems which arose in the determination of threshold levels.

Basically, the results on threshold levels and detectability factors have been obtained by extending the procedures given by Marcum (Ref. 3) and Swerling (Ref. 4).

From the results it is concluded that for triple-channel frequency diversity radar systems the use of receiver Modes 0, 2, and 4 in an adaptive way gives full benefit of the flexibility of such systems and produces an optimum performance in the jamming environment considered. But, as the performances of Modes 1, 2, and 3 do not differ much when only one channel is jammed, the use of either of these modes must be determined by weighing the costs against the efficiency, and taking into consideration the extra insertion losses.

This work was performed under Task V.d of the STC Programme of Work for 1971/1972.

1. INTRODUCTION

In frequency diversity radar systems pulses from different transmitters are transmitted sequentially. The target returns received in the different channels are rephased in time and then suitably recombined at video level before transmission to automatic or manual extractors.

There are several ways in which the video signals in the different channels can be recombined. In the original work of Herscovici and Detape (see Ref. 1) the following recombinations for a triple-channel system are mentioned:

$$\begin{aligned} A_3 &= (a + b + c) \\ D_1 &= (a.b + b.c + c.a) \\ D_2 &= (a + b).\ (b + c).\ (c + a) \\ \text{and} \quad M_3 &= (a\ .\ b\ .\ c), \end{aligned} \tag{1}$$

where a, b, and c are the three channels.

In a quiet environment the linear-addition mode of recombination, A_3 in (1), is fundamentally the most efficient. Herscovici and Detape showed, although not explicitly, that the performances of the mixed combinations D_1 and D_2 are not much worse, and are of importance, since they are more efficient against certain types of jamming.

When logical products are considered in equations 1, however,the electronic counter-countermeasure (ECCM) feature of the mixed combinations D_1 and D_2, and also M_3, becomes quite obvious. In the D_1 and D_2 modes the channel with the largest video will be cancelled, and in the M_3 mode the channels with the largest and intermediate video will be cancelled.

This means that when one channel is jammed, e.g. in the case of a spot jammer or repeater jammer, in general the jammed channel will be cancelled when the system operates in either the D_1, D_2, or M_3 mode. The M_3 mode, moreover, has an ECCM feature when two channels are jammed, since this mode processes only the channel with the smallest video.

The early frequency diversity radar systems, which employed two or three channels, used only the A_3 mode of video recombination. This is the main reason why Supplement 2 of Ref. 2 deals only with the performance evaluation of the linear-addition mode of multi-channel frequency diversity radar systems in a clear and quiet environment. Nowadays frequency diversity radar systems, which usually employ three channels, mostly use in addition to this A_3 mode the D_1 mode of logic-product video recombination. Supplement 3 of Ref. 2 accordingly deals with the performance evaluation of triple-channel frequency diversity radar systems operating in the A_3 mode and D_1 mode, in a jamming environment. This supplement contains threshold levels and graphs of the detectability factor, but does not include the derivation of the formulae for the probability of false alarm and the probability of detection.

The present memorandum is concerned with deriving the performance of triple-channel frequency diversity radar systems operating in the following modes:

- Mode 0 : $u_o = y_1 + y_2 + y_3 = y_s + y_i + y_\ell$

  (Mode A_3 of equations 1)

- Mode 1 : $u_1 = y_1 \cdot y_2 + y_2 \cdot y_3 + y_3 \cdot y_1 = 2y_s + y_i$

  (Mode D_1 of equations 1)

- Mode 2 : $u_2 = (y_1 + y_2).(y_2 + y_3).(y_3 + y_1) = y_s + y_i$

(Mode D_2 of equations 1) (2)

- Mode 3 : $u_3 = y_s + 2y_i = 2u_2 - u_4 = 2u_1 - 3u_4$

- Mode 4 : $u_4 = y_1.y_2.y_3 = y_s$

(Mode M_3 of equations 1),

where y_s, y_i, and y_ℓ are the smallest, the intermediate, and the largest videos respectively of the channel videos y_1, y_2, and y_3 (see Fig. 1), when stationary additive narrow-band white Gaussian noise is present in the channels and where the following environmental conditions are distinguished:

- none of the channels jammed : here called Condition III
- one channel jammed : here called Condition IIE
- two channels jammed simultaneously: here called Condition IEE.

In this context a parameter J is defined for each video channel by:

$$J = \frac{\text{average jamming noise power}}{\text{average internal noise power}} .$$

We can now combine the above notations for operating mode and environmental condition to designate the fifteen cases for performance evaluation:

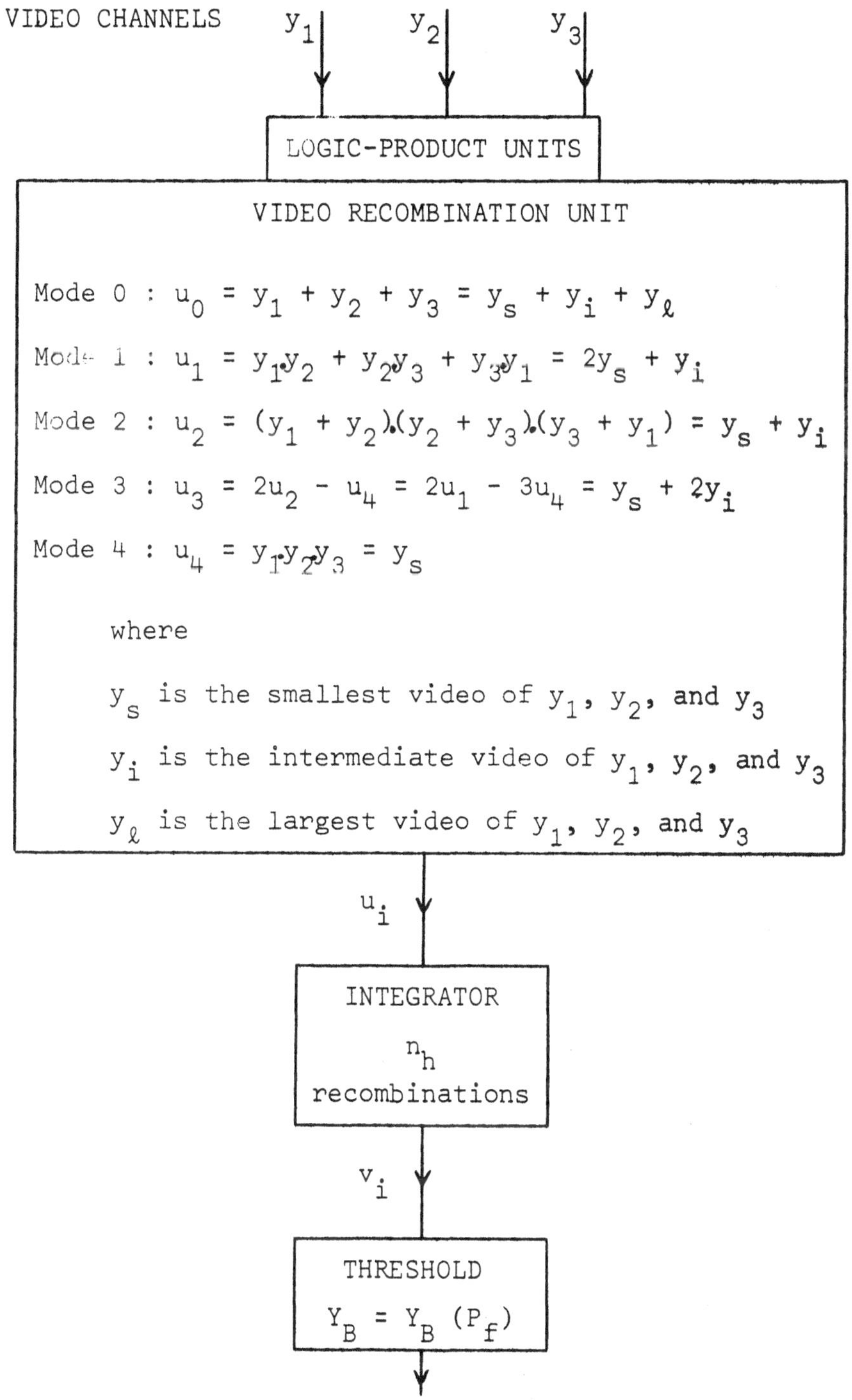

Fig. 1 Triple-channel frequency diversity operation: video recombination techniques considered

Environmental condition	Mode of operation	Case (i = 0, 1, 2, 3, 4)
III	i	III-i
IIE	i	IIE-i
IEE	i	IEE-i

Here capitals I and E denote respectively a channel with only internal noise and a channel with external noise also, due to interference or jamming.

It should be noted that when all three channels are simultaneously jammed with equal power the Cases EEE-i can be considered as special cases of Case III-i, at least when the jamming noise has the same (i.e. Gaussian) distribution as the internal noise.

Particular reasons for including some of the above cases are as follows:

- It is to be expected that in the case of one heavily jammed channel only, Mode 2 will be somewhat more efficient than Mode 1 as it is a "linear addition" combination of two videos.
- Although the performance of Mode 3 may be of academic interest only, its derivation runs completely parallel with that of Mode 1, and so is included in this memorandum.
- As Mode 4 has an ECCM feature even when two channels are jammed, it might be of great practical interest.

The required information on threshold levels and detectability factors is obtained by extending some procedures given by Marcum (Ref. 3) and Swerling (Ref. 4), and applying them to both Swerling target cross-section fluctuation models (Swerling Cases I/II and III/IV).

For simplicity, the following assumptions are made in the analysis:

- the transmitters all have the same peak power and pulse width
- there is complete correlation between the target echoes on each of the individual channels
- there is adequate spacing between the frequencies of the transmitters, so that there is no correlation between the groups of target echoes on the three channels
- the target average cross-sections for the three transmitter frequencies are identical
- the average internal noise powers on the channels are equal
- the external noise or jamming has the same (i.e. Gaussian) distribution as the internal noise
- when two channels are jammed the average jamming noise powers on these channels are equal.

It should be noted that the present study is aimed primarily at deriving threshold levels and detectability factors. The procedures for complete evaluation of the radar coverage provided by these receiving modes of operation of frequency diversity systems reference can be found in Ref. 2.

2. EVALUATION PROCEDURE

2.1 GENERAL

A triple-channel system is considered, with transmitters of equal peak power and pulse width, n_h hits per channel, square-law detection in each of the channels, Mode i (i = 0, 1, 2, 3, 4) video recombination, and a threshold after integration. It is further assumed that:

- stationary additive internal narrow-band white Gaussian noise with zero mean and of equal average power is present in the channels, and the channel noises are independent
- the noises from external origin (interference or jamming), if present in the channels, have the same statistical distribution as the internal noise (channel parameter J = average jamming noise-to-average internal noise power ratio); if two channels are simultaneously jammed, channel noises are independent but of equal average power
- there is complete correlation between the n_h target echoes on the same channel (slow scintillation) and no correlation between the echoes on different channels (fast scintillation)
- target average cross-sections for the three transmitter frequencies are identical
- the receivers operate linearly over the signal-strength range considered.

Square-law detection is considered because the linear envelope detector is more difficult to deal with, and it may be expected (see Ref. 3) that the results for both types of detector do not differ much.

When J >> 1, approximate expressions for the video combinations u_i (i = 0, 1, 2, 3, 4) can easily be obtained from equations (2) on pages 2 and 3.

The expressions are given in the following table, and their accuracy increases with increasing J:

Mode	Condition IIE	Condition IEE
0	$u_o \approx y$	$u_o \approx y_s + y_\ell$
1	$u_1 \approx 2y_s + y_\ell$	$u_1 \approx y_s$
2	$u_2 \approx y_s + y_\ell$	$u_2 \approx y_s$
3	$u_3 \approx y_s + 2y_\ell$	$u_3 \approx 2y_s$
4	$u_4 \approx y_s$	$u_4 \approx y$
Condition J >> 1	y = video in jammed channel; y_s and y_ℓ are the smallest and largest video in the two unjammed channels	y = video in unjammed channel; y_s and y_ℓ are the smallest and largest video in the two jammed channels

From the table we see that for J >> 1 the video recombinations

$$w_1 = 2y_s + y_\ell$$

$$w_2 = y_s + y_\ell \qquad (3)$$

$$w_3 = y_s + 2y_\ell$$

and $$w_4 = y_s,$$

where y_s and y_ℓ are the smaller and larger of y_a and y_b, and y_a and y_b are the videos of an unjammed or jammed dual-channel system, occur as approximate expressions for u_i. They are important because their statistical properties can more easily be determined.

It is noted that here only the case $f_{y_a} = f_{y_b}$ is of interest, where $f_{y_a}(y_a)$ and $f_{y_b}(y_b)$ are the probability density functions of y_a and y_b.

The <u>first step</u> in the derivation of the formulae for the probability of false alarm and the probability of detection is the determination of the following probability density functions (pdf):

(a) $f_{y_s,y_i,y_\ell}(y_s,y_i,y_\ell)$

$$f_{y_s,y_i}(y_s,y_i) \tag{4}$$

and $f_{y_s}(y_s)$,

where y_s, y_i, and y_ℓ are the smallest, the intermediate, and the largest of the channel videos y_1, y_2, and y_3, and where the case (Conditions IIE and IEE)

$f_{y_1} = f_{y_0}$ and $f_{y_2} = f_{y_3} = f_y$ (or $f_{y_2} = f_{y_0}$, or $f_{y_3} = f_{y_0}$)

is of particular interest in this study.

The pdf listed at equ.(4) are determined in Section A.1 of Appendix A.

(b) $f_{y_s,y_\ell}(y_s,y_\ell)$ (5)

and $f_{y_s}(y_s)$,

where y_s and y_ℓ are the smaller and larger of the channel videos y_a and y_b (see equs.(3)) and where the case $f_{y_a} = f_{y_b}$ is of particular interest.

The pdf at equ.(5) are determined in Section A.2 of Appendix A.

The second step in the derivation of the required formulae is the determination of general expressions for the pdf of the video recombinations u_i ($i = 0, 1, 2, 3, 4$) (see equs. 2) and the approximations w_j ($j = 1, 2, 3, 4$) (see equs. 3). These pdf are obtained from the pdf given in equs. (4) and (5) by application of basic rules for the transformation of random variables. The general expressions for the pdf of u_i and w_j are determined in Sections B.1 and B.2 of Appendix B respectively.

2.2 FORMULAE FOR THE PROBABILITY OF FALSE ALARM

The probability density functions and distribution functions of the normalized signals on the individual channels after square-law detection in the no-target-signal case can be written as follows (see Ref. 3, page 157, equ. 9):

If only internal noise is present

$$f_{y'}(y') = \exp(-y'), \quad \text{and} \quad F_{y'}(y') = 1 - \exp(-y'), \quad y' \geq 0 \tag{6}$$

If jamming noise is also present

$$\begin{aligned} f_{y''}(y'') &= \frac{1}{J+1} \exp\left(-y'' \frac{1}{J+1}\right), \quad \text{and} \\ F_{y''}(y'') &= 1 - \exp\left(-y'' \frac{1}{J+1}\right), \quad y'' \geq 0 \end{aligned} \tag{7}$$

In Sections C.2 to C.6 of Appendix C the pdf and the characteristic functions of u_i ($i = 0, 1, 2, 3, 4$) and their approximations w_j ($j = 1, 2, 3, 4$) are determined by substitution of equs. (6) and (7) into the general expressions for the pdf of u_i and w_j derived in Appendix B.

Finally, the formulae for the probability of false alarm P_f for n_h integrated variables u_i (w_j), v_i $(i = 0, 1, 2, 3, 4)$ are obtained from

$$P_f = \int_{Y_B}^{\infty} f_{v_i}(v_i)\, dv_i = 1 - \int_{0}^{Y_B} f_{v_i}(v_i)\, dv_i,$$

and are derived in Appendix D. These formulae are used to calculate the normalized threshold levels Y_B corresponding with desired values for P_f.

2.3 FORMULAE FOR THE PROBABILITY OF DETECTION

The formulae for the probability of detection P_d which are obtained from

$$P_d = \int_{Y_B}^{\infty} f_{v_i}(v_i)\, dvi = 1 - \int_{0}^{Y_B} f_{v_i}(v_i)\, dv_i,$$

where now v_i includes target returns, are derived in Appendix E. They are used to calculate the detectability curves which in turn provide the relevant detectability factors.

In Section E.2 of Appendix E formulae for P_d are derived for Mode 0, and both Swerling target cross-section fluctuation models (Swerling Cases I/II and III/IV) are considered. Here we recall that these fluctuation models result in the following pdf (see Ref. 4):

<u>Swerling Case I/II</u>

$$f_x(x) = \frac{1}{\bar{x}} \exp\left(- \frac{x}{\bar{x}}\right), \quad x \geqslant 0 \qquad (8)$$

Swerling Case III/IV

$$f_x(x) = \frac{4x}{\bar{x}^2} \exp\left(-\frac{2x}{\bar{x}}\right), \quad x \geqslant 0 \tag{9}$$

where x = input signal power-to-average internal noise power ratio in each channel, and

$\bar{x}$ = average of x over all target fluctuations.

The basic relations used in the determination of the pdf of v_0, the signal after linear-addition recombination (u_0) and integration, are the characteristic functions of the normalized signals (u' and u'') after square-law detection and integration in each individual channel. They are as follows:

Swerling Case I

No noise from external origin present

$$\Phi_{u'}(p) = \frac{1}{1 + n_h\bar{x}} \cdot \frac{1}{(p+1)^{n_h - 1}} \cdot \frac{1}{p + \frac{1}{1+n_h\bar{x}}}$$

Noise from external origin present

$$\Phi_{u''}(p) = \frac{1}{J + 1 + n_h\bar{x}} \cdot \frac{1}{(J+1)^{n_h - 1}} \cdot \frac{1}{\left(p + \frac{1}{J+1}\right)^{n_h - 1}} \cdot \frac{1}{p + \frac{1}{J + 1 + n_h\bar{x}}}$$

which follow directly from (III.3) on page 283 of Ref. 4.

Swerling Case III

No noise from external origin present

$$\Phi_{u'}(p) = \frac{1}{(1 + n_h \frac{\overline{x}}{2})^2} \cdot \frac{1}{(p + 1)^{n_h - 2}} \cdot \frac{1}{\left(p + \frac{1}{1 + n_h \frac{\overline{x}}{2}}\right)^2}$$

Noise from external origin present

$$\Phi_{u''}(p) = \frac{1}{(J + 1 + n_h \frac{\overline{x}}{2})^2} \cdot \frac{1}{(J + 1)^{n_h - 2}} \cdot$$

$$\cdot \frac{1}{(p + \frac{1}{J + 1})^{n_h - 2}} \cdot \frac{1}{\left(p + \frac{1}{J + 1 + n_h \frac{\overline{x}}{2}}\right)^2}$$

which follow directly from (III.14) on page 285 of Ref. 4.

Due to mathematical difficulties for Modes 1, 2, 3, and 4 formulae for the probability of detection could be obtained only for $n_h = 1$ and a Swerling Case I/II target: they are derived in Section E3 of Appendix E.

The basic relations used in the determination of the pdf of v_i for Swerling Case I/II (i=1,2,3,4) are the pdf and distribution functions of the normalized signals on the individual channels after square-law detection, which follow directly from (III.4) on page 283 of Ref. 4. They are as follows:

Swerling Case I

No noise from external origin present

$$f_{y'}(y') = \frac{1}{1 + \overline{x}} \exp(-y' \frac{1}{1 + \overline{x}})$$

$$F_{y'}(y') = 1 - \exp(-y' \frac{1}{1 + \bar{x}})$$

Noise from external origin present

$$f_{y''}(y'') = \frac{1}{J + 1 + \bar{x}} \exp(-y'' \frac{1}{J + 1 + \bar{x}})$$

$$F_{y''}(y'') = 1 - \exp(-y'' \frac{1}{J + 1 + \bar{x}})$$

Since detectability curves can easily be obtained by simulation, there is no point in deriving approximate formulae for P_d for $J >> 1$, using the approximations w_j for v_i (see Section 2.1).

3. PRESENTATION OF RESULTS

3.1 GENERAL

Extremely difficult computation problems arise in the accurate determination of the normalized threshold levels Y_B. The degree of difficulty of these computations increases with increasing number of hits per channel n_h, with increasing values for the average external noise-to-average internal noise power ratio J, and with decreasing values for the probability of false alarm P_f.

In order to fill the gap in threshold level information, some approximate relations between P_f and Y_B for $J >> 1$ have been determined. It is believed that in both single-channel radar systems (see Ref. 5) and frequency diversity radar systems the average signal power-to-average internal noise power ratio required for a certain detection quality is not a very sensitive function of P_f and consequently of Y_B.

Since the only Y_B values of practical interest are those which correspond with very low values for P_f ($P_f < 10^{-3}$), in the present context there is no point in trying to obtain results by Monte Carlo simulation. The determination of accurate threshold levels would require millions of samples, which is manifestly impracticable.

The method used for the calculation of the Y_B values and the computation problems are discussed in Sections F2 and F3 of Appendix F.

The detectability curves are obtained by Monte Carlo simulation for all receiver modes of operation except for the cases:

Mode 0	:	$J = 0$, and $n_h = 1$ and $J \neq 0$
Modes 1, 2, 3, and 4	:	$n_h = 1$ and a Swerling Case I/II target

In these cases the derived equations for the probability of detection P_d (Appendix E) are used. It may be noted that for Mode 0 the derived equations for P_d are rather cumbersome to deal with, except for the above-mentioned cases.

Except for the determination of Y_B values, Monte Carlo simulation is a very suitable technique for obtaining the detectability curves and related detectability factors, since the only detectability factors of practical interest are those which correspond to P_d values $\geqslant 0.5$. The simulation program is discussed in Section F.4 of Appendix F.

The number of hits per channel n_h or number of video recombinations considered in this study is limited to a maximum of 20. This is because of the difficulties in determining threshold levels for larger values of n_h (even for low values of J), and the computer time required for the determination of the detectability curves by Monte Carlo simulation, which increases with increasing value of n_h. Moreover, it is considered that information for $n_h > 20$ is of limited practical interest.

3.2 THRESHOLD LEVELS

The normalized threshold levels Y_B corresponding to two practical values for the probability of false alarm, $P_f = 10^{-6}$ ($\pm 10^{-9}$) and 10^{-4} ($\pm 10^{-7}$), for various values of the parameters n_h and J, and for the various cases considered, are presented in Tables 1 to 10. Due to computation problems, threshold levels can not be given for all parameter combinations, except for Mode 4.

The coverage is as follows:

Condition IIE

Curves of ΔD_s for Modes 2 and 4 relative to Mode 0, and for Modes 1 and 3 relative to Mode 2 are presented in Figs. 10 to 13. Since the ΔD_s values for $P_f = 10^{-6}$ and 10^{-4} differ only slightly ($\leqslant$0.3 dB), this information is given for $10^{-6} \leqslant P_f \leqslant 10^{-4}$. The reason why the performances of Modes 1 and 3 are presented relative to Mode 2 is because it was expected that Mode 2 is the most efficient video recombination for Condition IIE, at least when $J \gg 1$.

Condition IEE

Curves of ΔD_s for Mode 4 relative to Mode 0 and for Modes 1, 2, and 3 relative to Mode 0 are presented in Figs. 22 to 25 and Figs. 26 and 27 respectively. These last mentioned curves are given for the range $10^{-6} \leqslant P_f \leqslant 10^{-4}$, since the ΔD_s values for $P_f = 10^{-6}$ and 10^{-4} differ very little ($\leqslant$0.3 dB).

The following additional information is noted about particular cases:

Cases IIE-0 and IEE-i (i = 0, 1, 2, 3)

It seems that for $J \geqslant 12$ dB and n_h up to 20 (Mode 0) and $J \geqslant 18$ dB and n_h up to 20 (Modes 1, 2, and 3), the use of the approximate normalized threshold levels gives reasonably accurate information on D_s. This conclusion can be drawn since the linear relationship between D_s and J, which one expects for large values for J, becomes clearly visible.

For the Cases IIE-i (i = 1, 2, 3, 4) and IEE-4, Y_B approaches an upper limit when J increases to infinity. These upper limits are determined by the internal noise signals in the unjammed channels, and of course by n_h.

The approximate normalized threshold levels for Modes 0, 1, 2, and 3 are presented in Tables 11 to 18. Where possible, corresponding values for the accurate P_f are included: these are obtained by inserting the approximate Y_B values in the accurate equations for P_f. From the results given in these tables it is concluded that for $J \geqslant 18$ dB for all cases the approximate Y_B values correspond reasonably well to the P_f values considered.

3.3 DETECTABILITY FACTORS

The detectability factor D_s is defined as the average signal power-to-average internal noise power ratio that is required in the video channels for a given P_d and P_f. Expressed analytically

$$D_s = \frac{\text{minimum detectable signal energy per pulse}}{\text{internal noise per cycle bandwidth}} \text{ (dB)}$$

in each channel (see Ref. 2, TR-50-U, Fig. 13).

For receiver Mode 0, curves (in the format used in Ref. 2), showing the variation of D_s with the number of hits per channel n_h for both Swerling cases, and with parameters J, P_f, and P_d (0.8 and 0.5), are presented in Figs. 2 to 9 (Condition IIE) and Figs. 14 to 21 (Condition IEE).

For receiver Modes 1, 2, 3, and 4 the information on D_s is presented in curves showing ΔD_s, the relative gain in D_s.

It means that for J > 30 dB information on D_s can be obtained by simple linear extrapolation in the relevant Figures.

In the cases where the information on D_s obtained by using the approximate Y_B values was apparently not accurate enough, the presented curves were obtained by interpolation in the J field.

For general information, curves of D_s for Mode 0 and J = 6 dB and 12 dB, resulting from the use of the approximate Y_B values without correction by interpolation in the J field, are also included in the Figures (dashed curves).

Cases IIE-i (i = 1, 2, 3)

For these cases no approximate relationsbips between P_f and Y_B for J >> 1 could be derived, but then the Y_B values obtained for J = ∞ might be used. However, due to computation problems, even for J = ∞ it appeared only possible to calculate Y_B values up to n_h = 5 (except for Mode 2). Therefore the information on D_s versus J for n_h = 5, 10, and 20 was obtained by extrapolation in the n_h field.

TABLES AND FIGURES

The Key to Figures 2 to 27 is Given on Page 294.

Case	Noise power ratio $J = \frac{\langle (P_n)_E \rangle}{\langle (P_n)_I \rangle}$ in dB	Number of hits per channel, n_h, or number of video recombinations integrated						
		1	2	3	4	5	10	20
III-0	$-\infty$	19.1292	25.4126	30.9571	36.1144	41.0221	63.5482	104.252
	0	30.4036	38.7689	46.1870	53.1102	59.7116	90.0821	144.100
IIE-0	6	70.8983	87.8183	102.247	115.505	128.101	185.221	-
external noise	12	-	-	-	-	-	-	-
present in one	18	-	-	-	-	-	-	-
channel	24	-	-	-	-	-	-	-
	30	-	-	-	-	-	-	-
	0	34.7260	45.3090	54.6510	63.3419	71.6116	109.550	178.009
IEE-0	6	84.5258	109.006	130.315	150.153	169.036	255.025	-
external noise	12	-	-	-	-	-	-	-
present in two	18	-	-	-	-	-	-	-
channels	24	-	-	-	-	-	-	-
	30	-	-	-	-	-	-	-

Table 1

Normalized threshold levels for Mode 0 and $P_f = 10^{-6}$

Case	Noise power ratio $J = \frac{\langle (P_n)_E \rangle}{\langle (P_n)_I \rangle}$ in dB	Number of hits per channel, n_h, or number of video recombinations integrated						
		1	2	3	4	5	10	20
III-0	$-\infty$	13.9282	19.5672	24.5947	29.3065	33.8163	54.7515	93.1630
	0	21.1930	28.8363	35.6395	42.0137	48.1132	76.4045	128.216
IIE-0	6	48.2584	63.1789	76.1981	88.2637	99.6827	152.340	257.553
external noise	12	-	-	-	-	-	-	-
present in one	18	-	-	-	-	-	-	-
channel	24	-	-	-	-	-	-	-
	30	-	-	-	-	-	-	-
	0	24.8461	34.3833	42.8716	50.8196	58.4218	93.6646	158.203
IEE-0	6	59.7851	81.6656	101.095	119.122	136.269	216.046	-
external noise	12	-	-	-	-	-	-	-
present in two	18	-	-	-	-	-	-	-
channels	24	-	-	-	-	-	-	-
	30	-	-	-	-	-	-	-

Table 2

Normalized threshold levels for Mode 0 and $P_f = 10^{-4}$

Case	Noise power ratio $J = \frac{\langle (P_n)_E \rangle}{\langle (P_n)_I \rangle}$ in dB	Number of hits per channel, n_h, or number of video recombinations integrated						
		1	2	3	4	5	10	20
III-1	$-\infty$	14.5087	18.0372	21.1135	23.9567	26.6495	38.8927	60.7026
IIE-1 external noise present in one channel	0	17.4932	21.7959	25.5494	29.0193	-	-	-
	6	20.0785	25.1324	29.5472	33.6303	-	-	-
	12	21.6335	27.1772	32.0242	36.5076	-	-	-
	18	22.1720	27.8920	32.8946	37.5230	-	-	-
	24	22.3200	28.0890	33.1349	37.8034	-	-	-
	30	22.3579	28.1395	33.1966	37.8762	-	-	-
	∞	22.3714	28.1563	33.2174	37.9003	42.3336	-	-
IEE-1 external noise present in two channels	0	22.0361	27.5323	32.3247	36.7524	-	-	-
	6	38.4104	48.4633	57.1927	64.4218	-	-	-
	12	119.712	146.350	169.278	190.316	-	-	-
	18	451.068	546.490	627.859	-	-	-	-
	24	-	-	-	-	-	-	-
	30	-	-	-	-	-	-	-

Table 3

Normalized threshold levels Y_B for Mode 1 and $P_f = 10^{-6}$

Case	Noise power ratio $J = \frac{\langle (P_n)_E \rangle}{\langle (P_n)_I \rangle}$ in dB	Number of hits per channel, n_h, or number of video recombinations integrated						
		1	2	3	4	5	10	20
III-1	$-\infty$	9.9035	13.0889	15.8773	18.4665	20.9279	32.2182	52.6114
IIE-1	0	11.9668	15.8515	19.2542	22.4133	-	-	-
external noise	6	13.7975	18.3611	22.3632	26.0805	-	-	-
present in one	12	14.9198	19.9250	24.3184	28.4009	-	-	-
channel	18	15.3123	20.4761	25.0103	29.2243	-	-	-
	24	15.4206	20.6284	25.2017	29.4523	-	-	-
	30	15.4483	20.6675	25.2508	29.5109	-	-	-
	∞	15.4584	20.6803	25.2676	29.5302	33.5877	-	-
IEE-1	0	15.1247	20.0781	24.4159	28.4420	-	-	-
external noise	6	26.7049	35.6662	43.4990	50.7475	-	-	-
present in two	12	80.5681	104.401	125.011	144.012	-	-	-
channels	18	301.400	386.193	458.815	525.380	-	-	-
	24	-	-	-	-	-	-	-
	30	-	-	-	-	-	-	-

Table 4

Normalized threshold levels Y_B for Receiver Mode 1 and $P_f = 10^{-4}$

Case	Noise power ratio $J = \frac{\langle (P_n)_E \rangle}{\langle (P_n)_I \rangle}$ in dB	Number of hits per channel, nh, or number of video recombinations integrated						
		1	2	3	4	5	10	20
III-2	$-\infty$	10.1314	12.8312	15.1918	17.3750	19.4434	28.8525	46.6324
	0	12.3638	15.6998	18.6113	21.3003	-	-	-
IIE-2	6	14.5338	18.5362	22.1030	-	-	-	-
external noise	12	15.9686	20.3502	-	-	-	-	-
present in one	18	16.4918	-	-	-	-	-	-
channel	24	16.6384	-	-	-	-	-	-
	30	16.6388	-	-	-	-	-	-
	∞	16.6884	21.3505	25.4126	29.1622	32.7103	48.8265	77.5403
	0	-	-	-	-	-	-	-
IEE-2	6	35.8158	44.2076	51.4847	58.1905	-	-	-
external noise	12	118.496	143.970	165.765	185.691	-	-	-
present in two	18	450.020	544.403	624.740	-	-	-	-
channels	24	-	-	-	-	-	-	-
	30	-	-	-	-	-	-	-

Table 5

Normalized threshold levels Y_B for Mode 2 and $P_f = 10^{-6}$

Case	Noise power ratio $J = \frac{<(P_n)_E>}{<(P_n)_I>}$ in dB	Number of bits per channel, n_h, or number of video recombinations integrated						
		1	2	3	4	5	10	20
III-2	$-\infty$	7.0495	9.4793	11.6142	13.5986	15.4870	24.1618	39.8633
IIE-2 external noise present in one channel	0	8.6349	11.6255	14.2522	16.6930	-	-	-
	6	10.2056	13.7785	16.9171	-	-	-	-
	12	11.2408	15.2031	-	-	-	-	-
	18	11.6158	15.7237	-	-	-	-	-
	24	11.7204	-	-	-	-	-	-
	30	11.7470	-	-	-	-	-	-
	∞	11.7564	15.9138	19.5672	22.9624	26.1930	41.0311	67.8913
IEE-2 external noise present in two channels	0	-	-	-	-	-	-	-
	6	24.3026	31.8484	38.4220	44.5075	-	-	-
	12	79.3518	102.044	121.548	139.465	-	-	-
	18	300.352	384.111	455.706	521.249	593.403	-	-
	24	-	-	-	-	-	-	-
	30	-	-	-	-	-	-	-

Table 6

Normalized threshold levels Y_b for Mode 2 and $P_f = 10^{-4}$

Case	Noise power ratio $J = \frac{\langle (P_n)_E \rangle}{\langle (P_n)_I \rangle}$ in dB	Number of hits per channel, n_h, or number of video recombinations integrated						
		1	2	3	4	5	10	20
III-3	$-\infty$	16.6884	21.3505	25.4126	29.1622	32.7103	48.8265	77.5043
	0	21.0062	26.7465	31.5248	-	-	-	-
IIE-3	6	25.6669	32.5759	38.6045	44.4524	-	-	-
external noise	12	28.8135	36.5159	43.2470	49.4105	-	-	-
present in one	18	29.9626	37.9539	44.9321	51.4211	-	-	-
channel	24	30.2837	38.3556	45.4124	51.9539	-	-	-
	30	30.3664	38.4590	45.5343	52.0829	-	-	-
	∞	30.3951	38.4934	45.5758	52.1258	58.2889	-	-
	0	29.0381	36.3091	42.6929	47.4371	-	-	-
IEE-3	6	70.1933	85.7172	99.0972	111.385	-	-	-
external noise	12	235.894	295.775	328.316	367.131	-	-	-
present in two	18	899.016	1,086.77	1,246.43	-	-	-	-
channels	24	-	-	-	-	-	-	-
		-	-	-	-	-	-	-

Table 7

Normalized threshold levels for Mode 3 and $P_f = 10^{-6}$

Case	Noise power ratio $J = \frac{\langle (P_n)_E \rangle}{\langle (P_n)_I \rangle}$ in dB	Number of hits per channel, n_h, or number of video recombinations integrated						
		1	2	3	4	5	10	20
III-3	$-\infty$	11.7564	15.9138	19.5672	22.9624	26.1930	41.0311	67.8913
IIE-3 external noise present in one channel	0	14.7163	19.8528	24.3631	29.7294	-	-	-
	6	17.9141	24.1111	29.5541	34.6148	-	-	-
	12	20.0685	26.9924	33.0745	38.7266	-	-	-
	18	20.8539	28.0435	34.3600	40.2314	-	-	-
	24	21.0732	28.3371	34.7191	40.6515	-	-	-
	30	21.1296	28.4126	34.8116	40.7596	-	-	-
	∞	21.1496	28.4374	34.8436	40.7938	46.4671	-	-
IEE-3 external noise present in two channels	0	19.8867	26.4969	32.3016	37.6824	-	-	-
	6	47.1674	61.0751	73.1255	84.2515	-	-	-
	12	157.606	201.935	239.910	274.724	-	-	-
	18	599.680	766.184	908.365	1,038.45	1,138.28	-	-
	24	-	-	-	-	-	-	-
	30	-	-	-	-	-	-	-

Table 8

Normalized threshold levels for Mode 3 and $P_f = 10^{-4}$

Case	Noise power ratio $J = \frac{\langle (P_n)_E \rangle}{\langle (P_n)_I \rangle}$ in dB	Number of hits per channel, n_h, or number of video recombinations integrated						
		1	2	3	4	5	10	20
III-4	$-\infty$	4.6053	5.5628	6.3764	7.1168	7.8105	10.9034	16.2755
	0	5.5264	6.6754	7.6517	8.5402	9.3726	13.0841	19.5306
IIE-4	6	6.2800	7.5857	8.6951	9.7048	10.6507	14.3683	22.1939
external noise	12	6.7106	8.1058	9.2913	10.3702	11.3810	15.3879	23.7157
present in one	18	6.8553	8.2805	9.4916	10.5937	11.6263	16.2303	24.2269
channel	24	6.8946	8.3280	9.5460	10.6545	11.6930	16.3234	24.3658
	30	6.9046	8.3402	9.5599	10.6700	11.7100	16.3472	24.4013
		6.9080	8.3442	9.5646	10.6752	11.7158	16.3552	24.4132
	0	6.9080	8.3442	9.5646	10.6752	11.7158	16.3552	24.4132
IEE-4	6	9.8686	11.9203	13.6637	15.2503	16.7368	23.3645	34.8760
external noise	12	12.3617	14.9318	17.1156	19.1030	20.9650	29.2671	43.6868
present in two	18	13.4036	16.1903	18.5581	20.7131	22.7321	31.7339	47.3690
channels	24	13.7093	16.5596	18.9815	21.1856	23.2506	32.4577	48.4494
	30	13.7891	16.6559	19.0919	21.3089	23.3859	32.6466	48.7314
	∞	13.8160	16.6884	19.1292	21.3505	23.4315	32.7103	48.8265

Table 9

Normalized threshold levels Y_B for Mode 4 and $P_f = 10^{-6}$

Case	Noise power ratio $J = \frac{\langle (P_n)_E \rangle}{\langle (P_n)_I \rangle}$ in dB	Number of hits per channel, n_h, or number of video recombinations integrated						
		1	2	3	4	5	10	20
III-4	$-\infty$	3.0703	3.9188	4.6427	5.3046	5.9273	8.7310	13.6770
IIE-4 external noise present in one channel	0	3.6843	4.7026	5.5713	6.3655	7.1128	10.4772	16.4125
	6	4.1867	5.3438	6.3310	7.2336	8.0827	11.9059	18.6505
	12	4.4738	5.7102	6.7651	7.7296	8.6370	12.7223	19.9294
	18	4.5703	5.8333	6.9109	7.8962	8.8231	12.9965	20.3590
	24	4.5965	5.8668	6.9506	7.9415	8.8737	13.0711	20.4757
	30	4.6032	5.8753	6.9607	7.9530	8.8867	13.0901	20.5056
	∞	4.6054	5.8782	6.9641	7.9569	8.8910	13.0965	20.5156
IEE-4 external noise present in two channels	0	4.6054	5.8782	6.9641	7.9569	8.8910	13.0965	20.5156
	6	6.5792	8.3974	9.9487	11.3670	12.7014	18.7093	29.3080
	12	8.2413	10.5189	12.4620	14.2387	15.9102	23.4358	36.7121
	18	8.9359	11.4055	13.5124	15.4388	17.2512	25.4111	39.8063
	24	9.1397	11.6656	13.8206	15.7909	17.6447	25.9907	40.7143
	30	9.1929	11.7335	13.9010	15.8828	17.7474	26.1420	40.9512
	∞	9.2108	11.7564	13.9138	15.9138	17.7820	26.1930	41.0311

Table 10

Normalized threshold levels Y_B for Mode 4 and $P_f = 10^{-4}$

Case	Noise power ratio $J = \frac{\langle (P_n)_E \rangle}{\langle (P_n)_I \rangle}$ in dB	Number of hits per channel, n_h, or number of video recombinations integrated						
		1	2	3	4	5	10	20
	6	69.0801	83.4422	95.6459	106.752	117.158	163.552	244.132
IIE-O	$P_f \rightarrow$	1.56×10^{-6}	2.29×10^{-6}	3.25×10^{-6}	4.48×10^{-6}	6.11×10^{-6}	21.6×10^{-6}	145×10^{-6}
external noise	12	234.872	283.703	325.196	362.958	398.336	556.076	830.050
present in one	18	898.041	1,084.75	1,243.40	1,387.78	1,523.05	2,126.17	3,173.72
channel	24	3,550.71	4,288.93	4,916.20	5,487.07	6,021.90	8,406.56	12,548.4
	30	14,161.4	17,105.6	19,607.4	21,884.2	24,017.3	33,528.1	50,047.1
	6	83.4422	106.752	127.063	145.811	163.552	244.132	387.701
IEE-O	$P_f \rightarrow$	1.23×10^{-6}	1.45×10^{-6}	1.71×10^{-6}	1.92×10^{-6}	2.18×10^{-6}	3.72×10^{-6}	-
external noise	12	283.703	362.958	432.015	495.757	556.076	830.050	1,318.18
present in two	18	1,084.75	1,387.78	1,651.82	1,895.54	2,126.17	3,173.72	5,040.12
channels	24	4,288.93	5,487.07	6,531.04	7,494.68	8,406.56	12,548.4	19,927.8
	30	17,105.6	21,884.2	26,047.9	29,891.2	33,528.1	50,047.1	79,478.8

Table 11

Approximate normalized threshold levels for Mode O and $P_f = 10^{-6}$, and the corresponding accurate P_f values

Case	Noise power ratio $J = \frac{<(P_n)_E>}{<(P_n)_I>}$ in dB	Number of hits per channel, n_h, or number of video recombinations integrated						
		1	2	3	4	5	10	20
	6	46.0542	58.7820	69.6409	79.5691	88.9100	130.965	205.156
IIE-0	$P_f \rightarrow$	1.56×10^{-4}	2.21×10^{-4}	3.05×10^{-4}	4.07×10^{-4}	4.59×10^{-4}	19.7×10^{-4}	68.4×10^{-4}
external noise	12	156.584	199.859	236.779	270.535	302.294	445.281	697.530
present in one	18	598.705	764.166	905.331	1,034.40	1,155.83	1,702.54	2,667.02
channel	24	2,311.92	3,021.39	3,579.54	4,089.85	4,569.98	6,731.60	10,545.0
	30	9,441.11	12,050.3	14,276.4	16,311.7	18,226.5	26,847.8	42,056.9
	6	58.7820	79.5691	97.8360	114.812	130.965	205.156	339.456
IEE-0	$P_f \rightarrow$	1.20×10^{-4}	1.41×10^{-4}	1.58×10^{-4}	1.78×10^{-4}	2.45×10^{-4}	3.15×10^{-4}	-
external noise	12	199.859	270.535	332.642	390.362	445.281	697.530	1,154.15
present in two	18	764.165	1,034.40	1,271.87	1,492.56	1,702.54	2,667.02	4,412.93
channels	24	3,021.39	4,089.85	5,028.77	5,901.35	6,731.60	10,545.0	17,448.0
	30	12,050.3	16,311.7	20,056.4	23,536.5	26,847.8	42,056.9	69,588.5

Table 12

Approximate normalized threshold levels for Mode 0 and $P_f = 10^{-4}$, and the corresponding accurate P_f values

Case	Noise power ratio $J = \frac{\langle (P_n)_E \rangle}{\langle (P_n)_I \rangle}$ in dB	Number of hits per channel, n_h, or number of video recombinations integrated						
		1	2	3	4	5	10	20
IEE-1 external noise present in two channels	6	35.5400	41.7211	47.8229	53.3761	58.5788	81.7758	122.066
	$P_f \rightarrow$	4.62×10^{-6}	11.6×10^{-6}	24.1×10^{-6}	44.3×10^{-6}	-	-	-
	12	117.436	141.852	162.598	181.479	199.168	278.038	415.025
	$P_f \rightarrow$	1.31×10^{-6}	1.65×10^{-6}	2.03×10^{-6}	2.45×10^{-6}	-	-	-
	18	449.020	542.374	621.698	693.890	761.524	1,063.09	1,586.86
	$P_f \rightarrow$	1.07×10^{-6}	1.13×10^{-6}	1.19×10^{-6}	-	-	-	-
	24	1,775.36	2,144.46	2,458.10	2,743.53	3,010.95	4,203.28	6,274.20
	30	7,080.71	8,552.83	9,803.70	10,942.1	12,008.6	16,764.0	25,023.6

Table 13

Approximate normalized threshold levels Y_B for Mode 1 and $P_f = 10^{-6}$ and the corresponding accurate P_f values

Case	Noise power ratio $J = \frac{\langle (P_n)_E \rangle}{\langle (P_n)_I \rangle}$ in dB	Number of hits per channel, n_h, or number of video recombinations integrated						
		1	2	3	4	5	10	20
IEE-1 external noise present in two channels	6	23.0271	29.3910	34.8204	39.7845	44.4550	65.4825	102.578
	$P_f \rightarrow$	4.18×10^{-4}	8.76×10^{-4}	15.6×10^{-4}	25.1×10^{-4}	-	-	-
	12	78.2921	99.9293	118.389	135.267	151.147	222.640	348.765
	$P_f \rightarrow$	1.31×10^{-4}	1.62×10^{-4}	1.96×10^{-4}	2.32×10^{-4}	-	-	-
	18	299.352	382.083	452.665	517.199	577.915	851.272	1,333.51
	$P_f \rightarrow$	1.07×10^{-4}	1.13×10^{-4}	1.18×10^{-4}	1.23×10^{-4}	1.92×10^{-4}	-	-
	24	1,183.59	1,510.70	1,789.77	2,044.92	2,284.99	3,365.80	5,272.50
	30	4,720.56	6,025.15	7,138.19	8,155.83	9,113.28	13,423.9	21,028.5

Table 14

Approximate normalized threshold levels Y_B for Mode 1 and $P_f = 10^{-4}$, and the corresponding accurate P_f values

Case	Noise power ratio $J = \frac{\langle (P_n)_E \rangle}{\langle (P_n)_I \rangle}$ in dB	Number of hits per channel, n_h, or number of video recombinations integrated						
		1	2	3	4	5	10	20
IEE-2 external noise present in two channels	6	34.5400	41.7211	47.8229	53.3761	58.5788	81.7758	122.066
	$P_f \rightarrow$	1.67×10^{-6}	2.55×10^{-6}	3.70×10^{-6}	5.19×10^{-6}	-	-	-
	12	117.436	141.852	162.598	181.479	199.168	278.038	415.025
	$P_f \rightarrow$	1.13×10^{-6}	1.26×10^{-6}	1.40×10^{-6}	1.53×10^{-6}	-	-	-
	18	449.020	542.374	621.698	693.890	761.524	1,063.09	1,586.86
	$P_f \rightarrow$	1.03×10^{-6}	1.06×10^{-6}	1.09×10^{-6}	-	-	-	-
	24	1,775.36	2,144.46	2,458.10	2,743.53	3,010.95	4,203.28	6,274.20
	30	7,080.71	8,552.83	9,803.70	10,942.1	12,008.6	16,764.0	25,023.6

Table 15

Approximate normalized threshold levels Y_B for Mode 2 and $P_f = 10^{-6}$, and the corresponding accurate P_f values

Case	Noise power ratio $J = \frac{\langle (P_n)_E \rangle}{\langle (P_n)_I \rangle}$ in dB	Number of hits per channel, n_h, or number of video recombinations integrated						
		1	2	3	4	5	10	20
IEE-2 external noise present in two channels	6	23.0271	29.3910	34.8204	39.7845	44.4550	65.4825	102.578
	$P_f \rightarrow$	1.67×10^{-4}	2.46×10^{-4}	3.42×10^{-4}	4.58×10^{-4}	-	-	-
	12	78.2921	99.9293	118.389	135.267	151.147	222.640	348.765
	$P_f \rightarrow$	1.13×10^{-4}	1.26×10^{-4}	1.38×10^{-4}	1.50×10^{-4}	-	-	-
	18	299.352	382.083	452.665	517.199	577.915	851.272	1,333.51
	$P_f \rightarrow$	1.03×10^{-4}	1.06×10^{-4}	1.08×10^{-4}	1.11×10^{-4}	1.35×10^{-4}	-	-
	24	1,183.59	1,510.70	1,789.77	2,044.92	2,284.99	3,365.80	5,272.50
	30	4,720.56	6,025.15	7,138.19	8,155.83	9,113.28	13,423.9	21,028.5

Table 16

Approximate normalized threshold levels Y_B for Mode 2 and $P_f = 10^{-4}$, and the corresponding accurate P_f values

Case	Noise power ratio $J = \frac{<(P_n)_E>}{<(P_n)_I>}$ in dB	Number of hits per channel, n_h, or number of video recombinations integrated						
		1	2	3	4	5	10	20
	6	69.0801	83.4422	95.6459	106.752	117.158	163.552	244.132
	$P_f \rightarrow$	1.25×10^{-6}	1.54×10^{-6}	1.86×10^{-6}	2.23×10^{-6}	-	-	-
IEE-3	12	234.872	283.703	325.196	362.958	398.336	556.076	830.050
external noise	$P_f \rightarrow$	1.06×10^{-6}	1.12×10^{-6}	1.18×10^{-6}	1.24×10^{-6}	-	-	-
present in two	18	898.041	1,084.75	1,243.40	1,387.78	1,523.05	2,126.17	3,173.72
channels	$P_f \rightarrow$	1.02×10^{-6}	1.03×10^{-6}	1.04×10^{-6}	-	-	-	-
	24	3,550.71	4,288.93	4,916.20	5,487.07	6,021.90	8,406.56	12,548.4
	30	14,161.4	17,105.6	19,607.4	21,884.2	24,017.3	33,528.1	50,047.1

Table 17

Approximate normalized threshold levels Y_B for Mode 3 and $P_f = 10^{-6}$, and the corresponding accurate P_f value

Case	Noise power ratio $J = \frac{\langle (P_n)_E \rangle}{\langle (P_n)_I \rangle}$ in dB	Number of hits per channel, n_h, or number of video recombinations integrated: 1	2	3	4	5	10	20
IEE-3 external noise present in two channels	6	46.0542	58.7820	69.6409	79.5691	88.9100	130.965	205.156
	$P_f \rightarrow$	1.25×10^{-4}	1.53×10^{-4}	1.83×10^{-4}	2.16×10^{-4}	-	-	-
	12	151.584	199.859	236.779	270.535	302.294	445.281	697.530
	$P_f \rightarrow$	1.06×10^{-4}	1.12×10^{-4}	1.17×10^{-4}	1.22×10^{-4}	-	-	-
	18	598.705	764.166	905.331	1,034.40	1,155.83	1,702.54	2,667.02
	$P_f \rightarrow$	1.02×10^{-4}	1.03×10^{-4}	1.04×10^{-4}	1.05×10^{-4}	-	-	-
	24	2,311.92	3,021.39	3,579.54	4,089.85	4,569.98	6,731.60	10,545.0
	30	9,441.11	12,050.3	14,276.4	16,311.7	18,226.5	26,847.8	42,056.9

Table 18

Approximate normalized threshold levels Y_B for Mode 3 and $P_f = 10^{-4}$, and the corresponding accurate P_f values

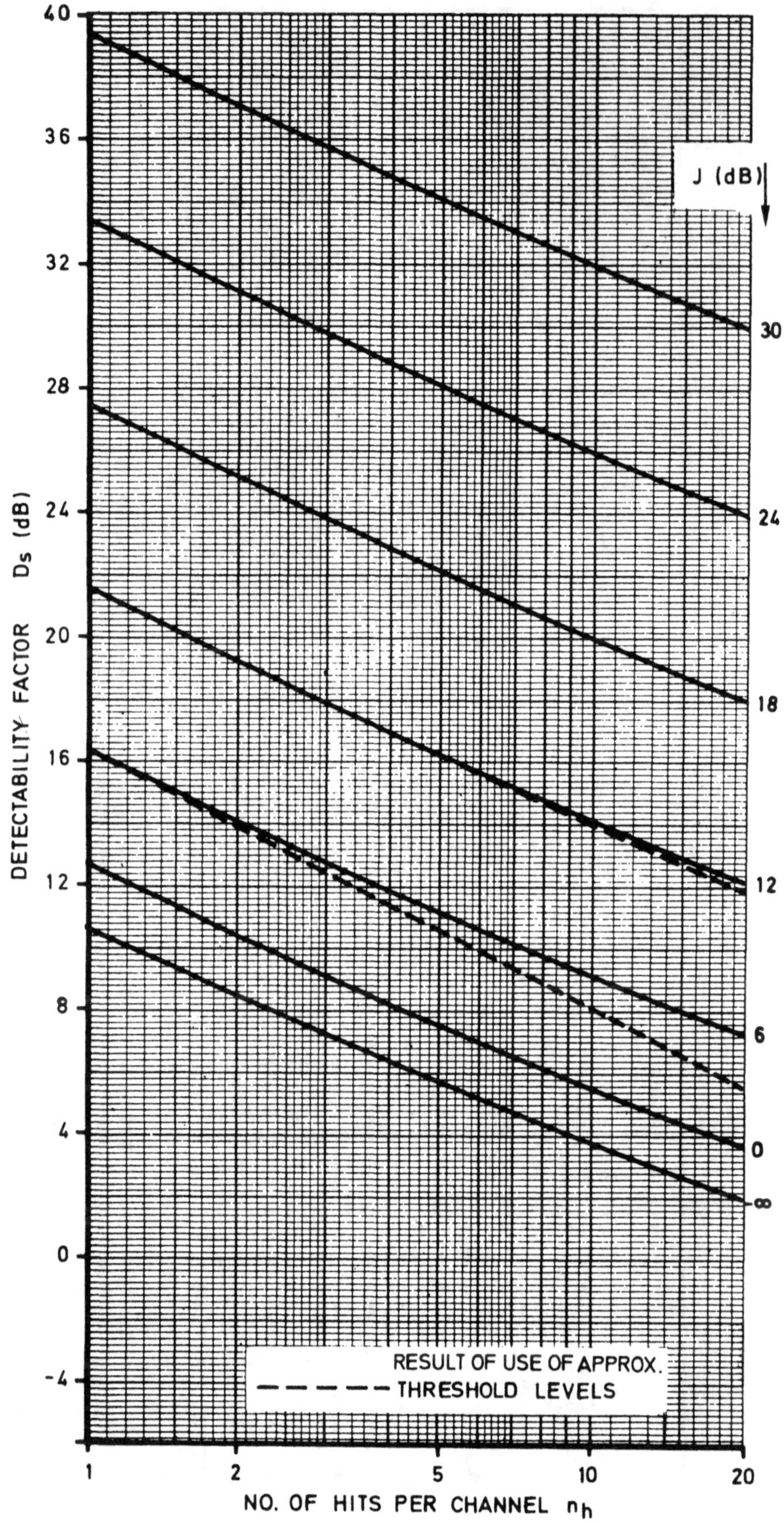
DETECTABILITY FACTOR Ds (dB)
NO. OF HITS PER CHANNEL n_h
J (dB)
30
24
18
12
6
0
∞
RESULT OF USE OF APPROX.
THRESHOLD LEVELS
40
36
32
28
24
20
16
12
8
4
0
-4
1
2
5
10
20

Fig. 2

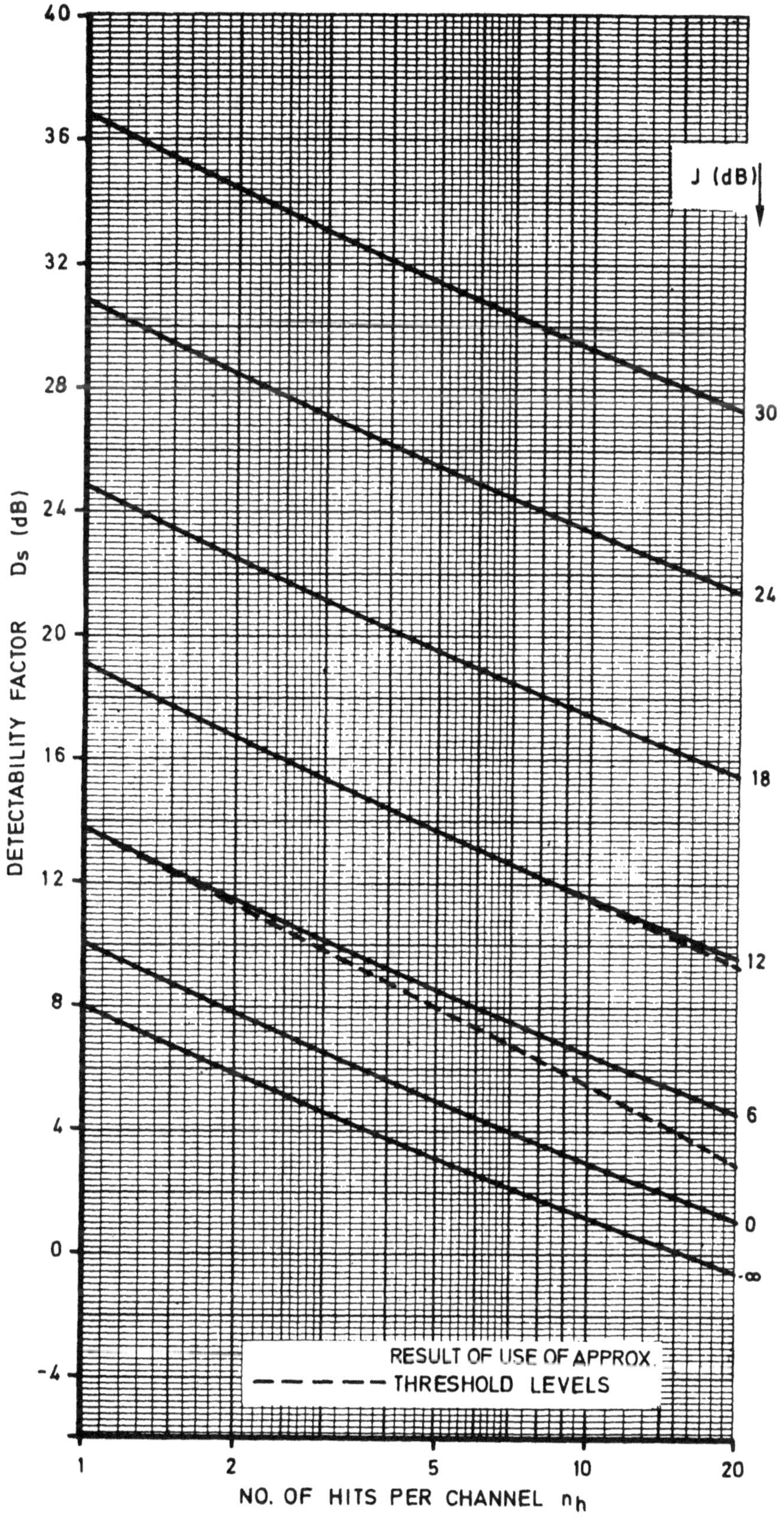

DETECTABILITY FACTOR Ds (dB)
40
36
32
28
24
20
16
12
8
4
0
-4
J (dB)
30
24
18
12
6
0
-∞
RESULT OF USE OF APPROX.
THRESHOLD LEVELS
1
2
5
10
20
NO. OF HITS PER CHANNEL n_h

Fig. 3

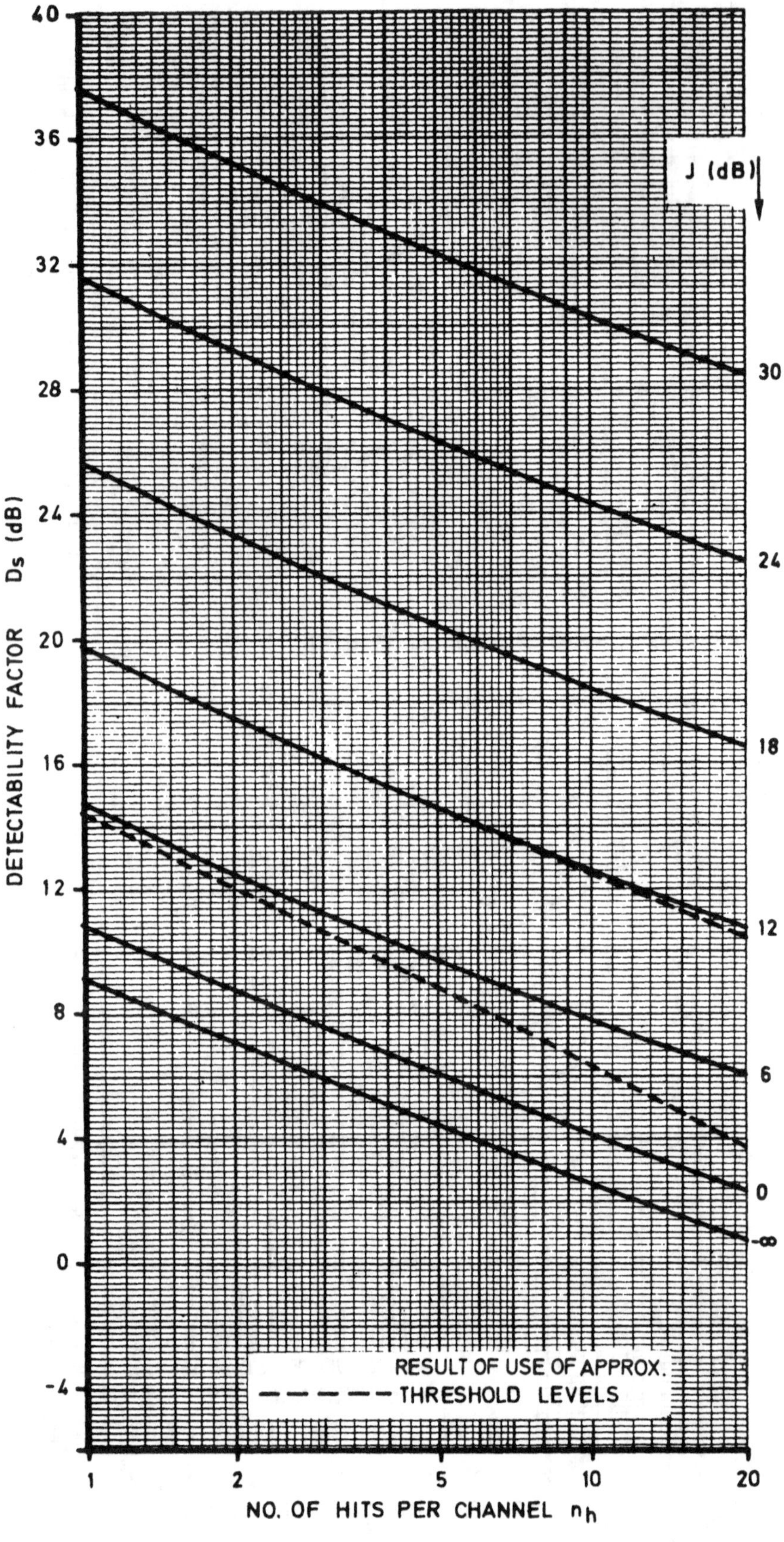
DETECTABILITY FACTOR Ds (dB)
NO. OF HITS PER CHANNEL n_h
J (dB)
30
24
18
12
6
0
-∞
RESULT OF USE OF APPROX.
THRESHOLD LEVELS

Fig. 4

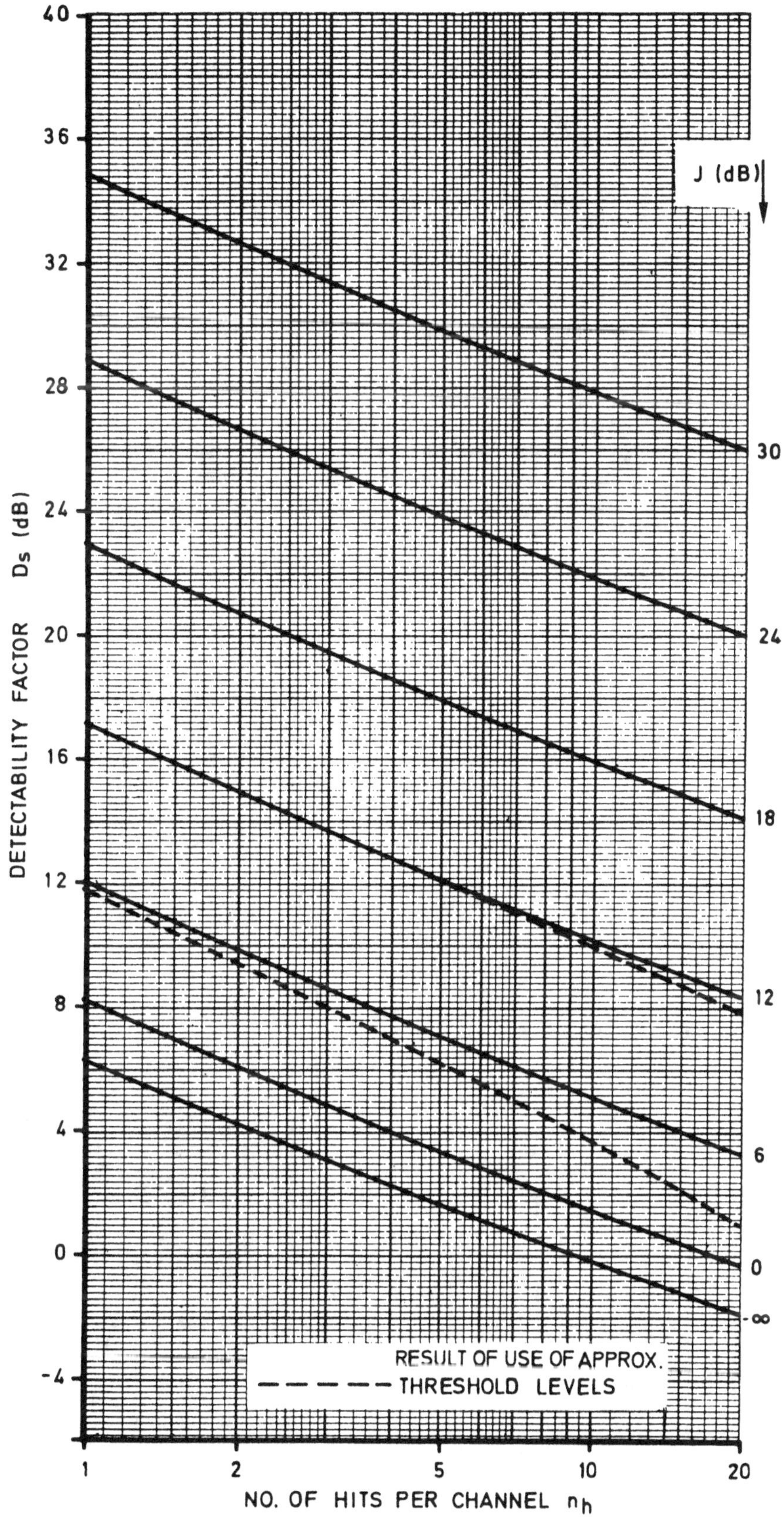

J (dB)
DETECTABILITY FACTOR Ds (dB)
NO. OF HITS PER CHANNEL n_h
40
36
32
28
24
20
16
12
8
4
0
-4
1
2
5
10
20
30
24
18
12
6
0
-∞
RESULT OF USE OF APPROX.
THRESHOLD LEVELS

Fig. 5

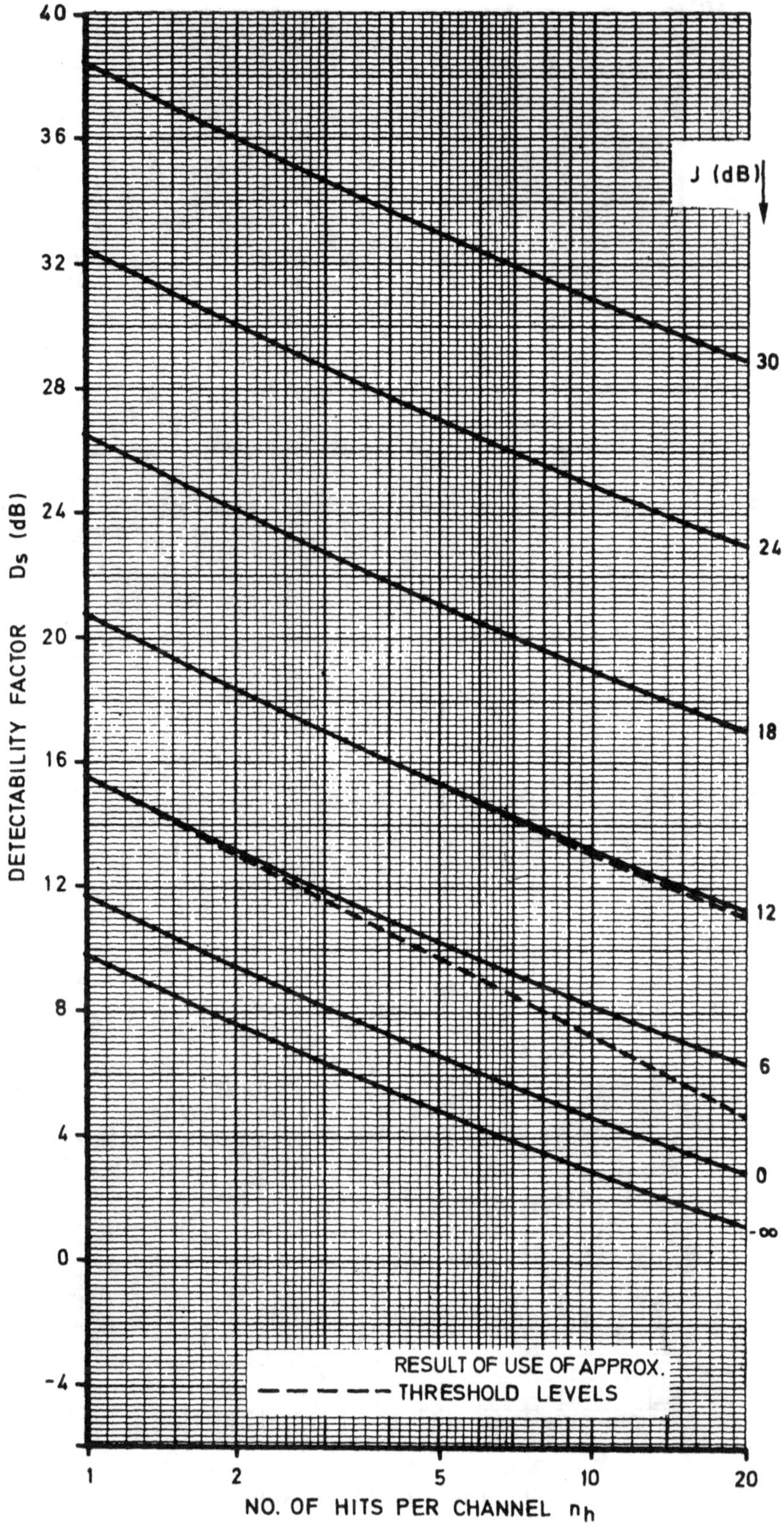

J (dB)
30
24
18
12
6
0
-∞
DETECTABILITY FACTOR Ds (dB)
NO. OF HITS PER CHANNEL n_h
RESULT OF USE OF APPROX.
THRESHOLD LEVELS

Fig. 6

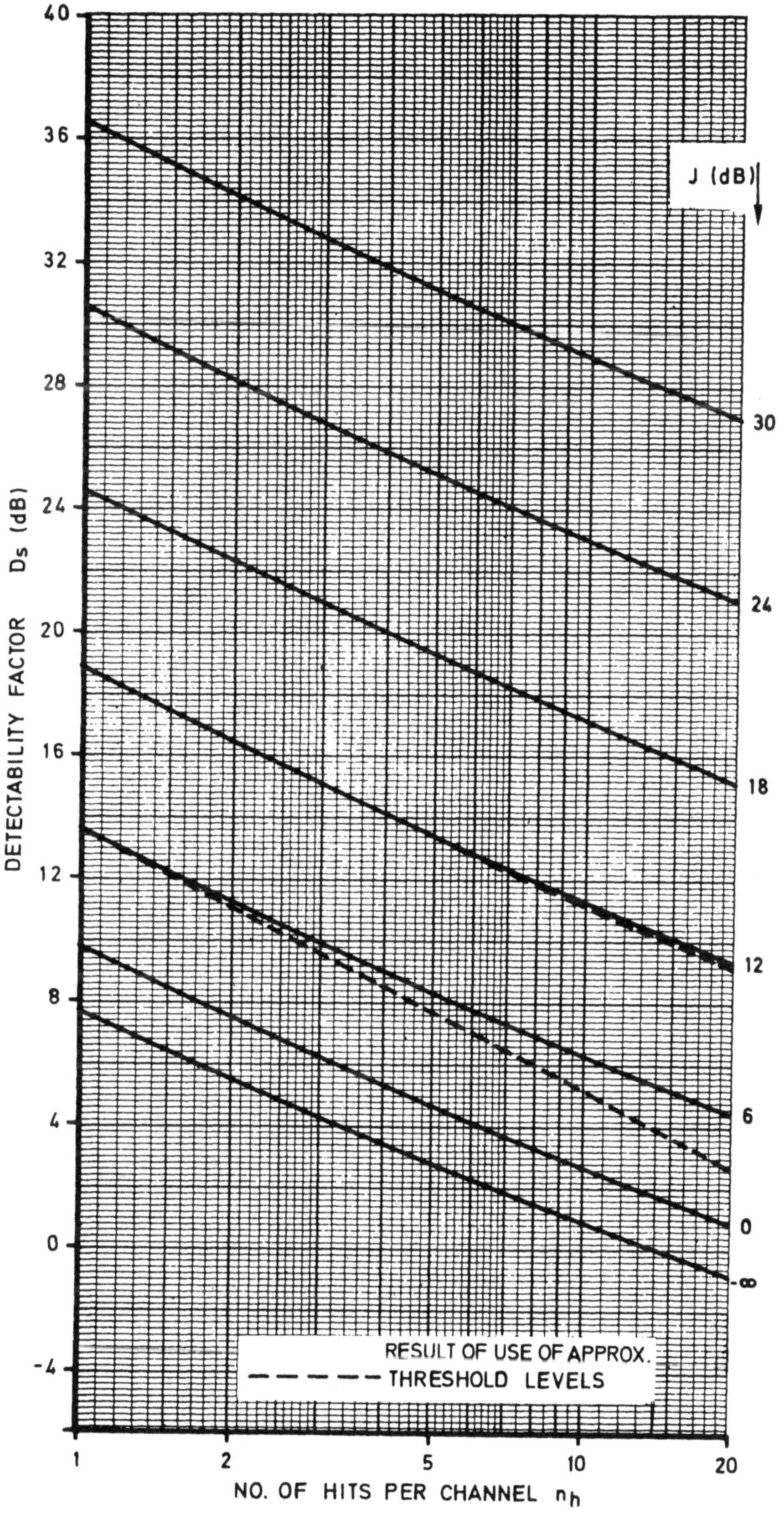
J (dB)
30
24
18
12
6
0
-∞
RESULT OF USE OF APPROX.
THRESHOLD LEVELS
DETECTABILITY FACTOR Ds (dB)
NO. OF HITS PER CHANNEL n_h
40
36
32
28
24
20
16
12
8
4
0
-4
1
2
5
10
20

Fig. 7

273

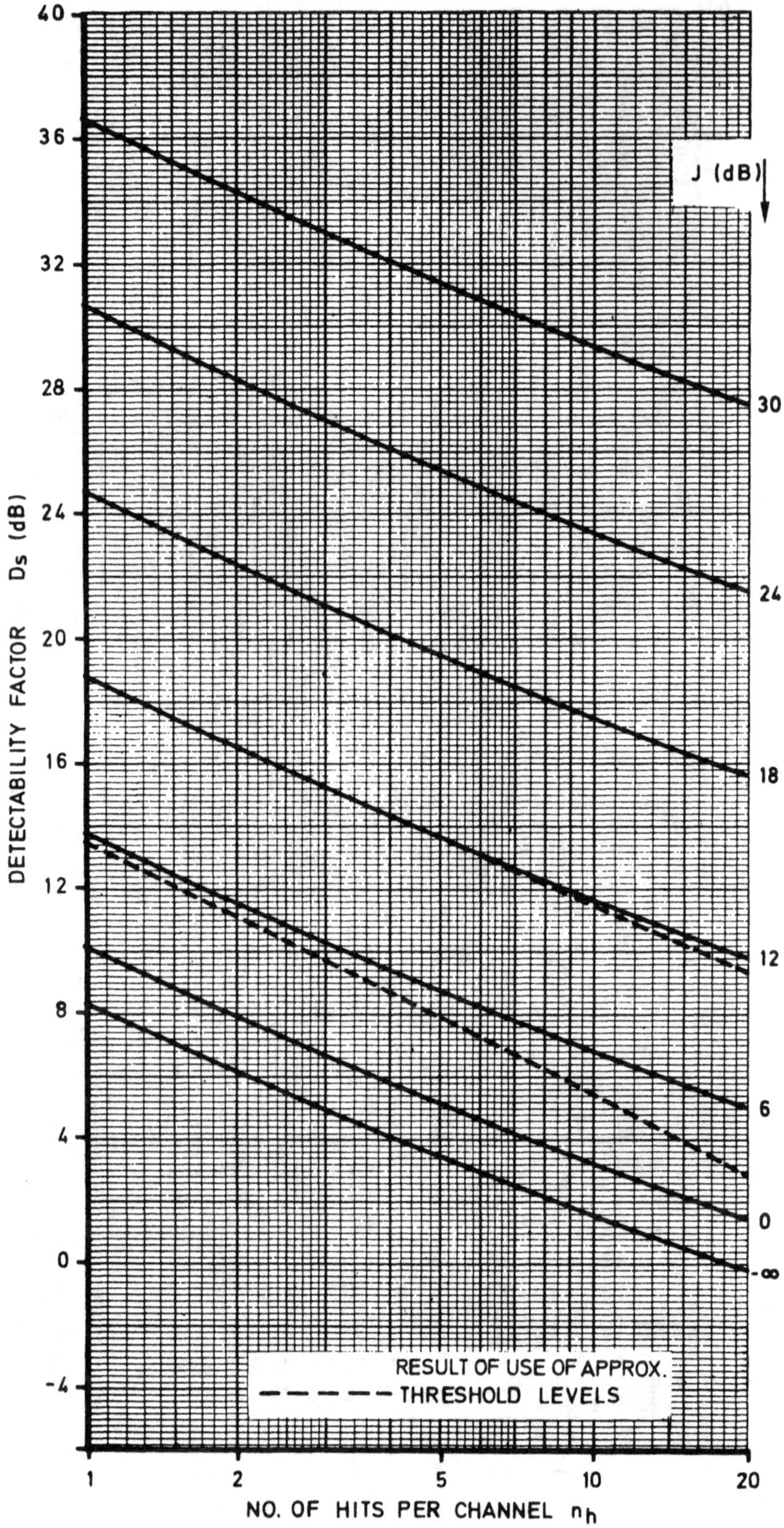
J (dB)
30
24
18
12
6
0
-∞
RESULT OF USE OF APPROX.
THRESHOLD LEVELS
DETECTABILITY FACTOR Ds (dB)
NO. OF HITS PER CHANNEL n_h
40
36
32
28
24
20
16
12
8
4
0
-4
1
2
5
10
20

Fig. 8

274

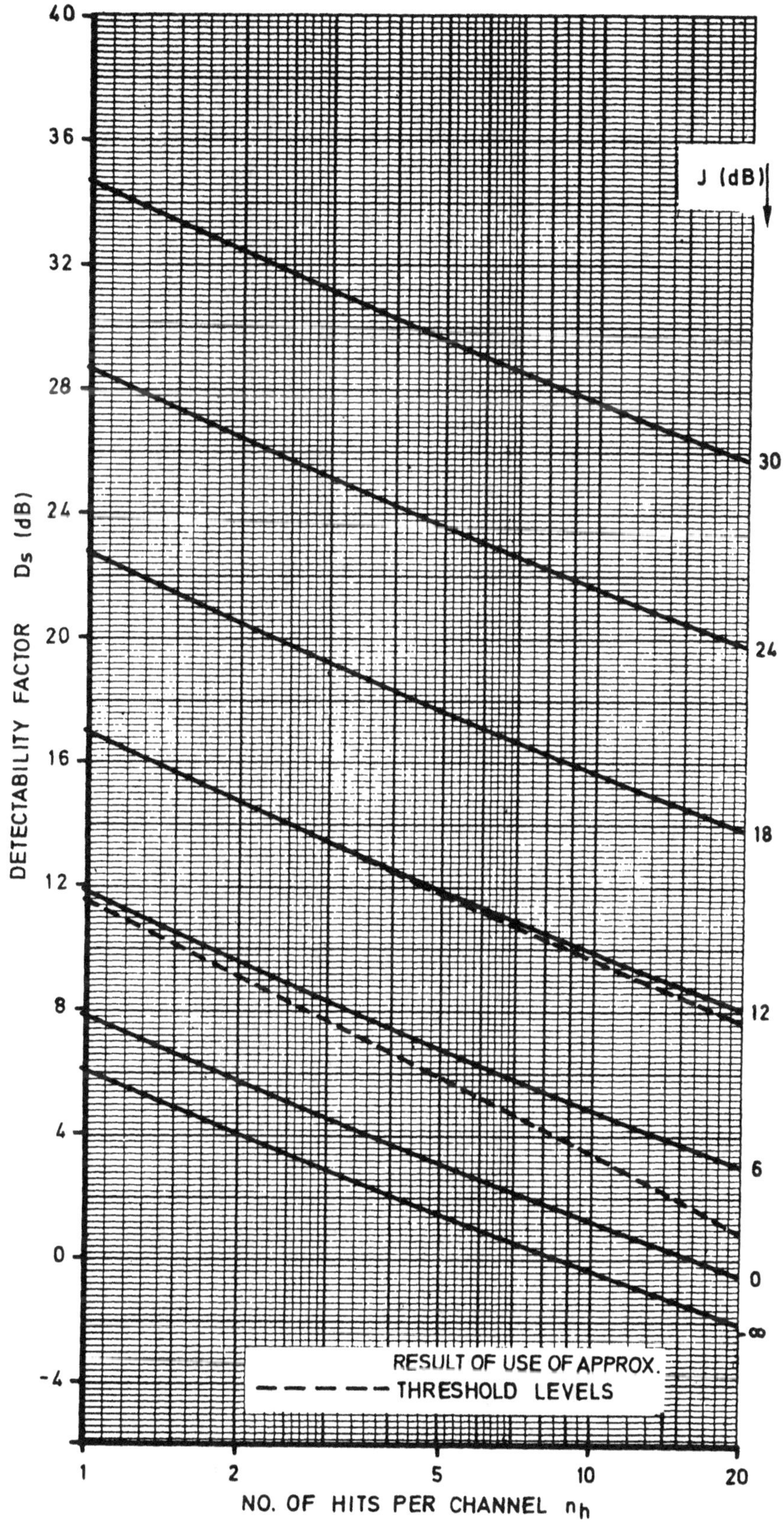

J (dB)
30
24
18
12
6
0
∞
DETECTABILITY FACTOR Ds (dB)
NO. OF HITS PER CHANNEL n_h
RESULT OF USE OF APPROX.
THRESHOLD LEVELS

Fig. 9

276

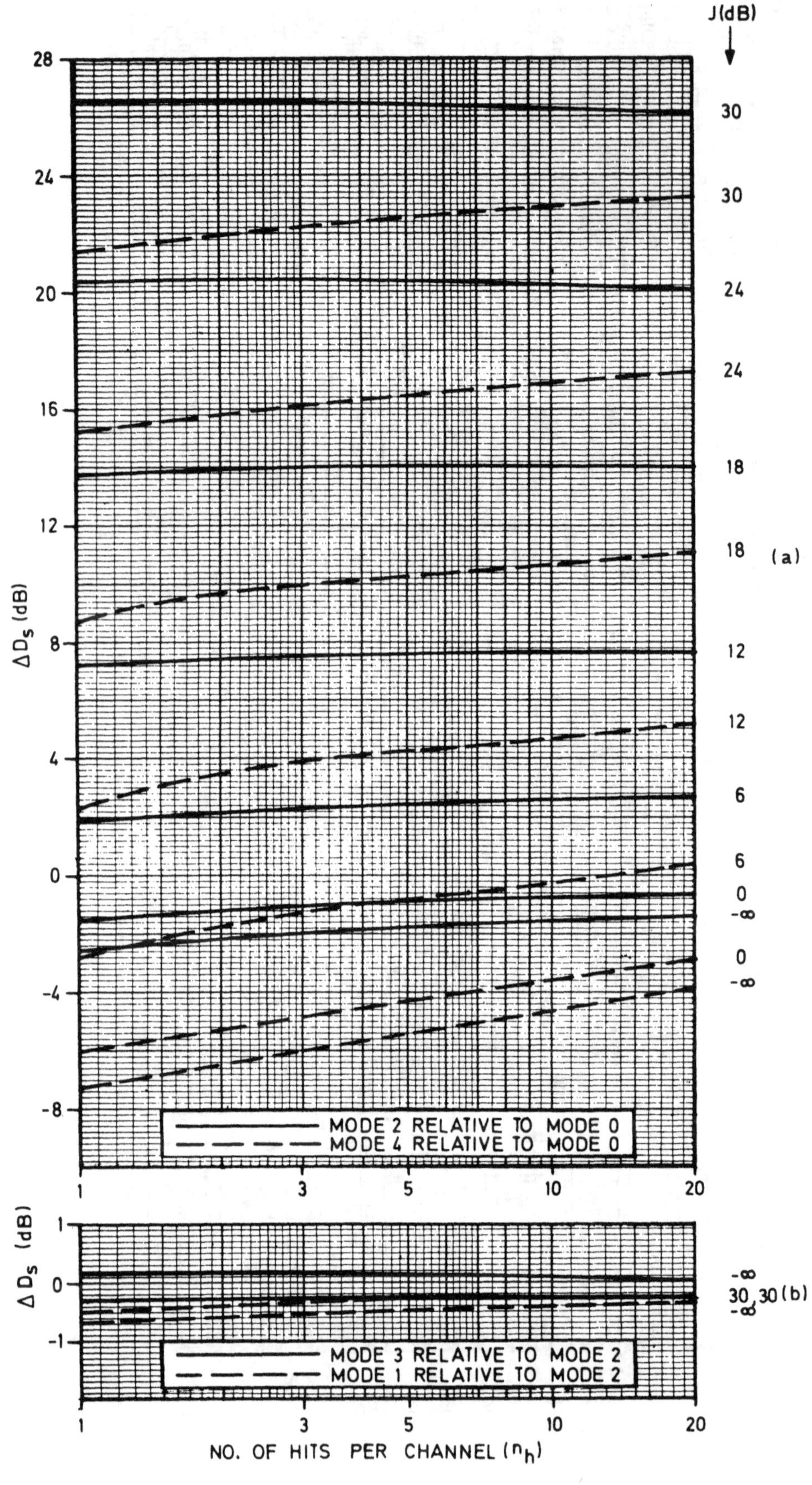

J(dB)
28
24
20
16
12
8
4
0
-4
-8
ΔDs (dB)
30
30
24
24
18
18
(a)
12
12
6
6
0
-∞
0
-∞
MODE 2 RELATIVE TO MODE 0
MODE 4 RELATIVE TO MODE 0
1
3
5
10
20
1
0
-1
-∞
30,30 (b)
-∞
MODE 3 RELATIVE TO MODE 2
MODE 1 RELATIVE TO MODE 2
NO. OF HITS PER CHANNEL (nh)

Fig. 10

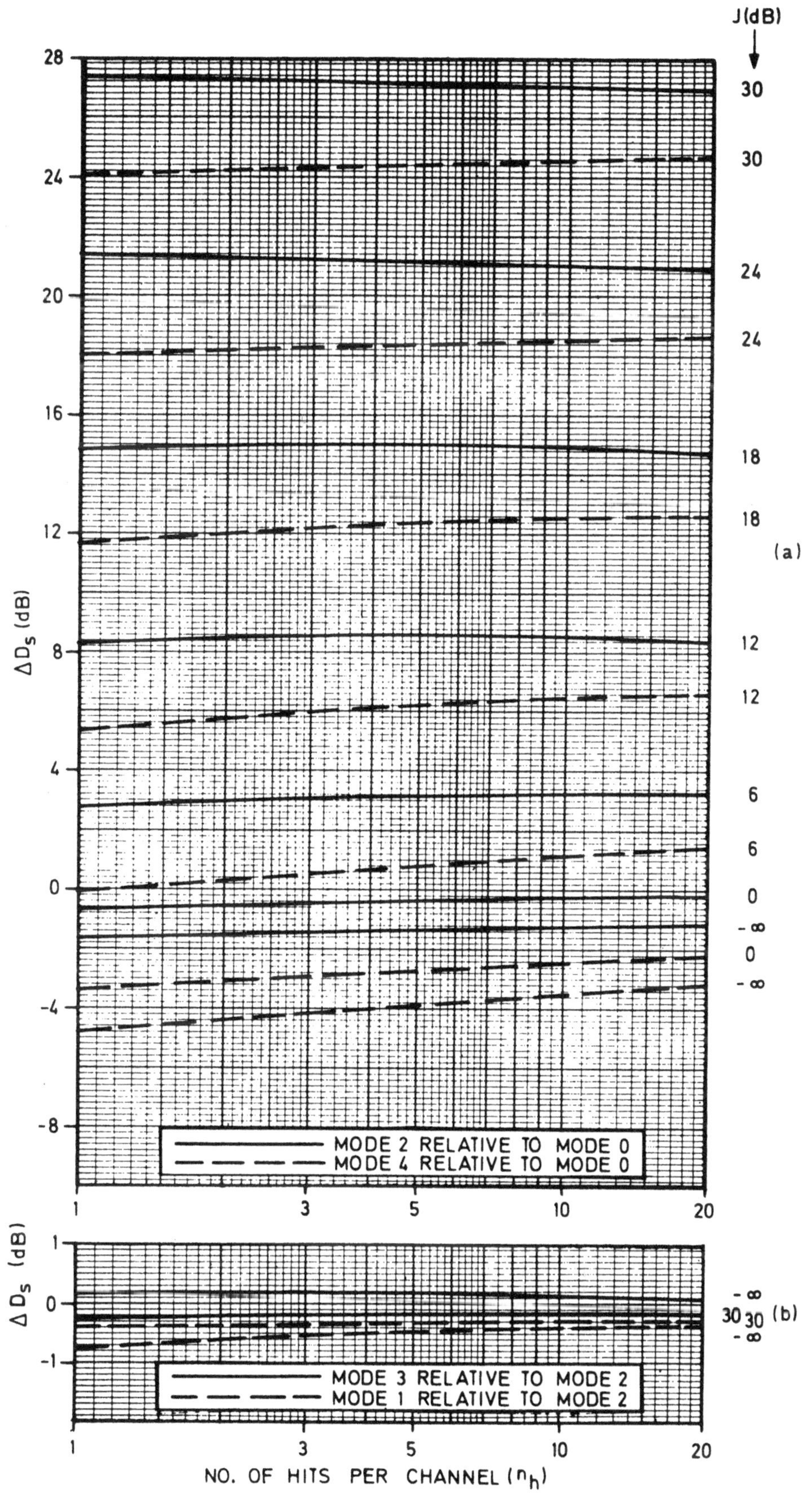
J(dB)
30
30
24
24
18
18
(a)
12
12
6
6
0
- ∞
0
- ∞
ΔDs (dB)
MODE 2 RELATIVE TO MODE 0
MODE 4 RELATIVE TO MODE 0
MODE 3 RELATIVE TO MODE 2
MODE 1 RELATIVE TO MODE 2
(b)
NO. OF HITS PER CHANNEL (nh)

277

Fig. 11

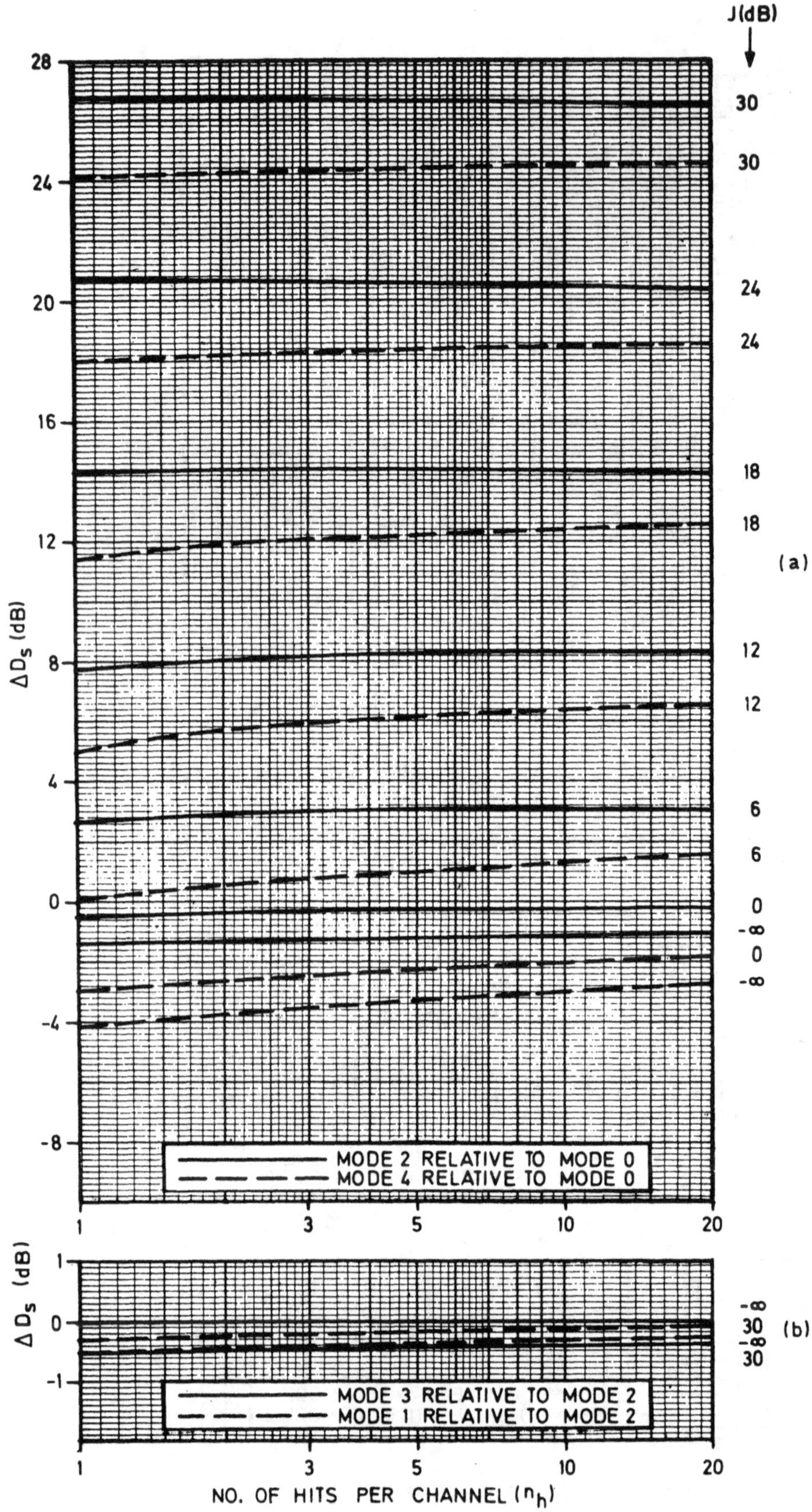

J(dB)
28
24
20
16
12
8
4
0
-4
-8
ΔDs (dB)
30
30
24
24
18
18
(a)
12
12
6
6
0
-∞
0
-∞
MODE 2 RELATIVE TO MODE 0
MODE 4 RELATIVE TO MODE 0
1
3
5
10
20
1
0
-1
-∞
30
-∞
30
(b)
MODE 3 RELATIVE TO MODE 2
MODE 1 RELATIVE TO MODE 2
NO. OF HITS PER CHANNEL (nh)

Fig. 12

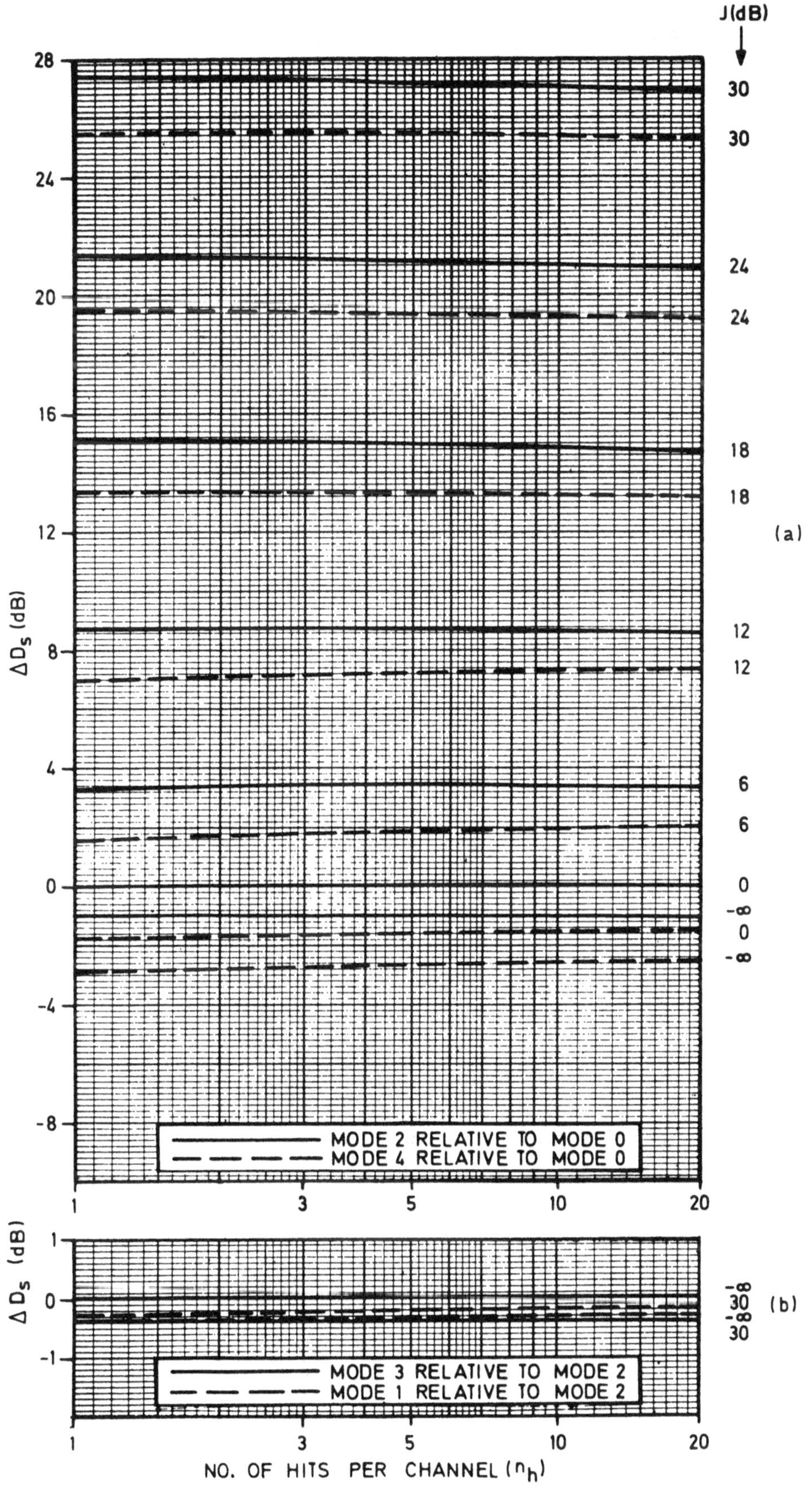
J(dB)
28
24
20
16
12
8
4
0
-4
-8
ΔDs (dB)
30
30
24
24
18
18
12
12
6
6
0
-∞
0
-∞
(a)
MODE 2 RELATIVE TO MODE 0
MODE 4 RELATIVE TO MODE 0
1
3
5
10
20
1
0
-1
ΔDs (dB)
-∞
30
-∞
30
(b)
MODE 3 RELATIVE TO MODE 2
MODE 1 RELATIVE TO MODE 2
1
3
5
10
20
NO. OF HITS PER CHANNEL (nh)

Fig. 13

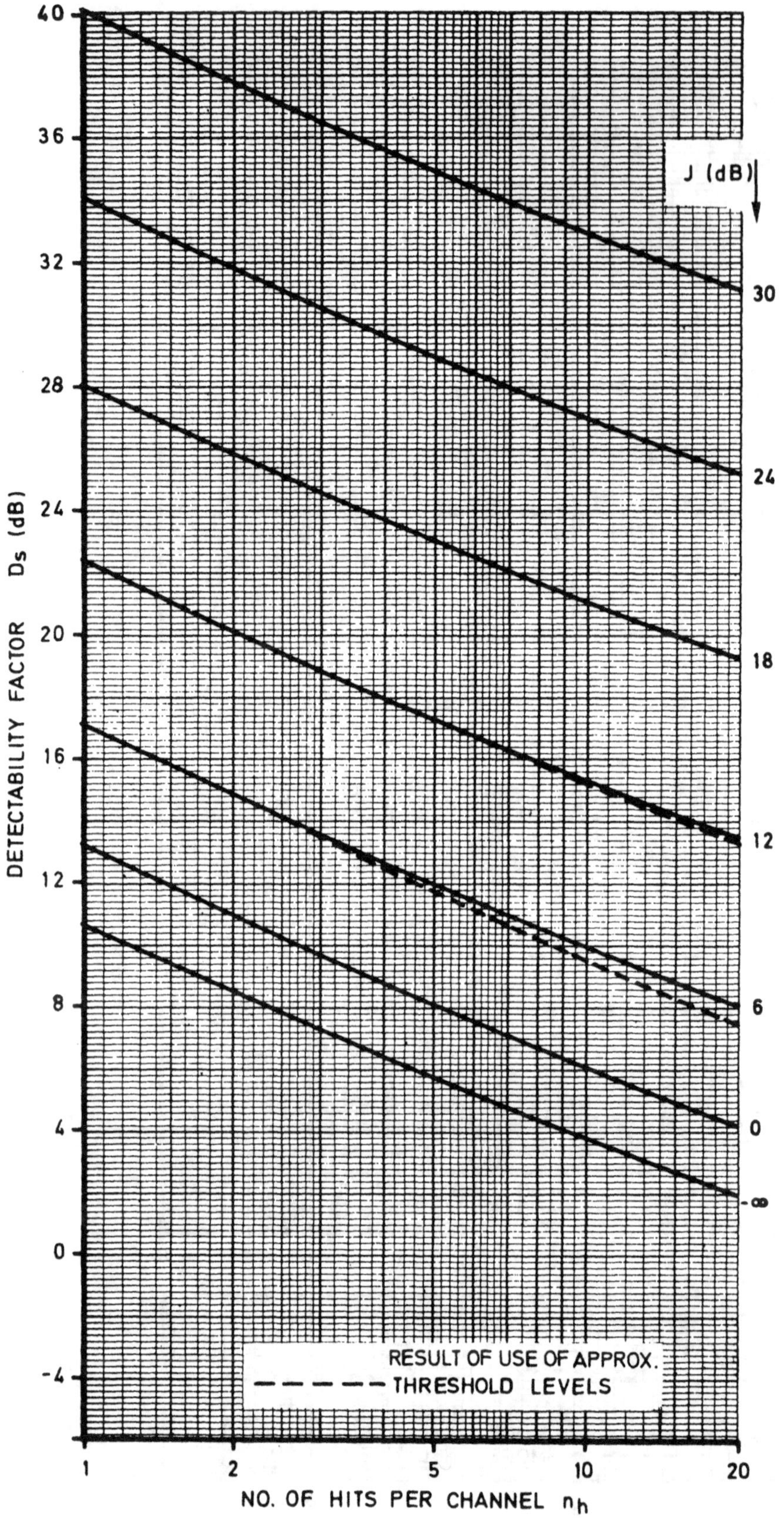

DETECTABILITY FACTOR Ds (dB)
40
36
32
28
24
20
16
12
8
4
0
-4
J (dB)
30
24
18
12
6
0
-∞
RESULT OF USE OF APPROX.
THRESHOLD LEVELS
1
2
5
10
20
NO. OF HITS PER CHANNEL n_h

Fig. 14

280

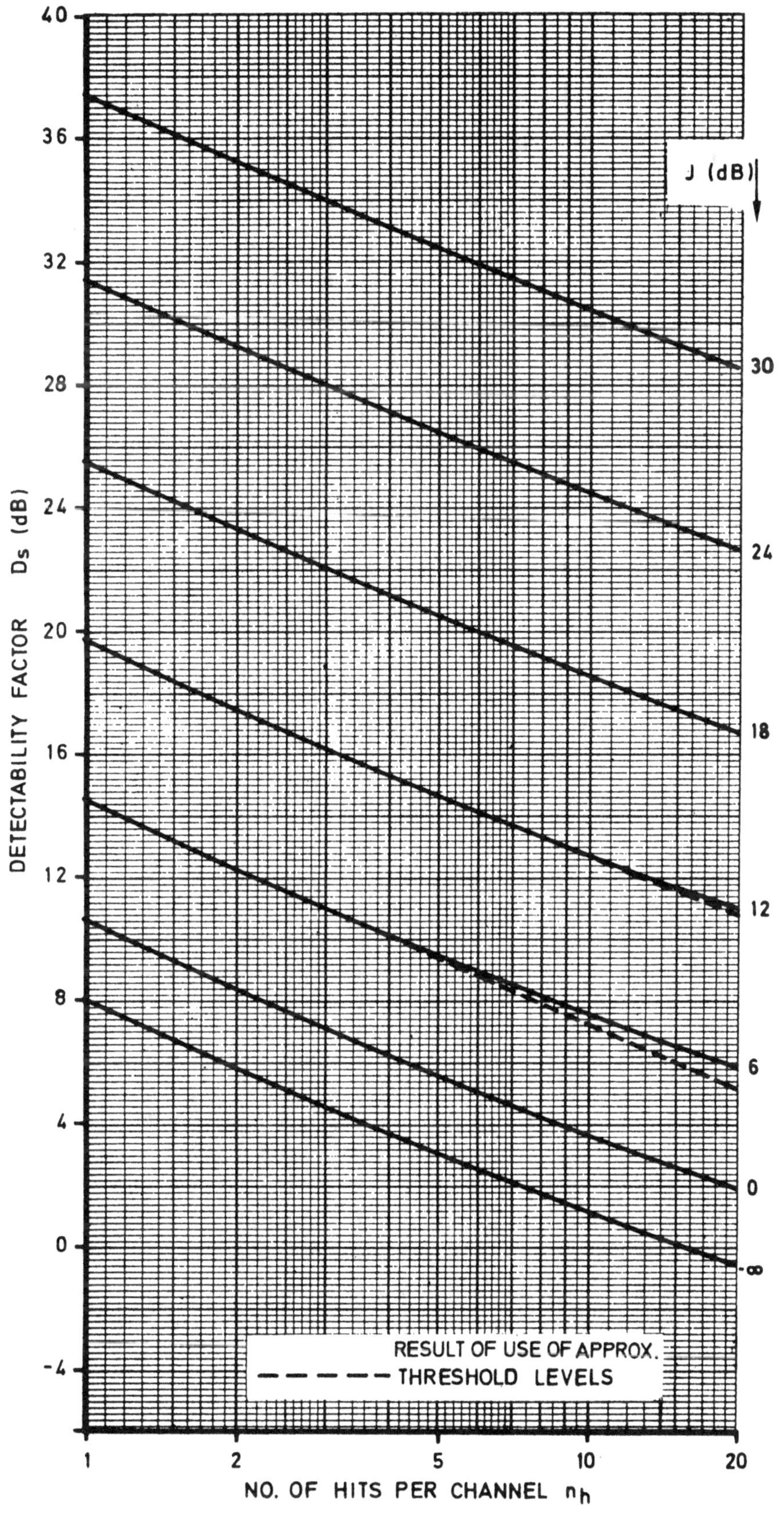

DETECTABILITY FACTOR Ds (dB)
40
36
32
28
24
20
16
12
8
4
0
-4
J (dB)
30
24
18
12
6
0
-∞
RESULT OF USE OF APPROX.
THRESHOLD LEVELS
1
2
5
10
20
NO. OF HITS PER CHANNEL n_h

Fig. 15

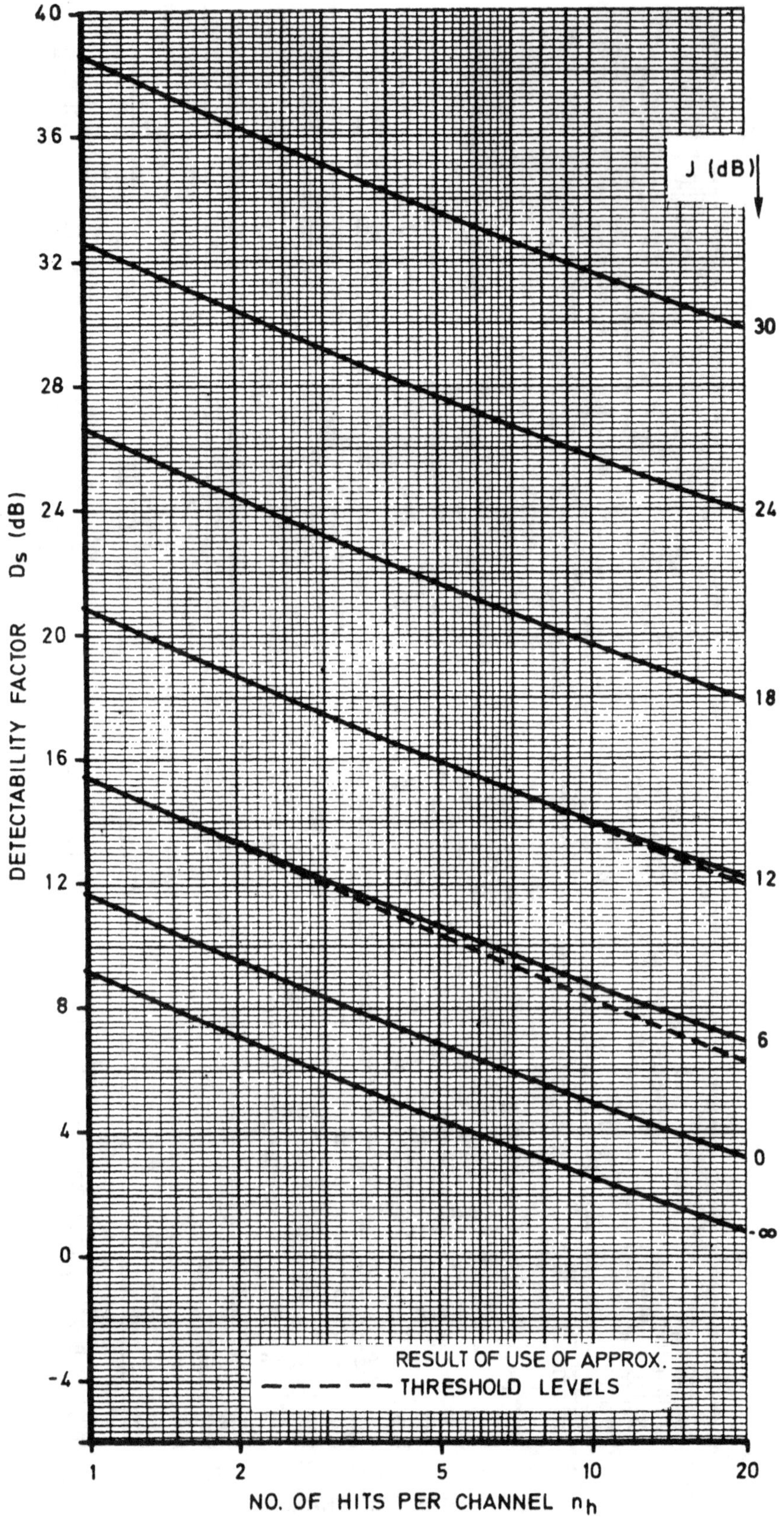

40
36
32
28
24
20
16
12
8
4
0
-4
DETECTABILITY FACTOR Ds (dB)
J (dB)
30
24
18
12
6
0
-∞
RESULT OF USE OF APPROX.
THRESHOLD LEVELS
1
2
5
10
20
NO. OF HITS PER CHANNEL nh

Fig. 16

282

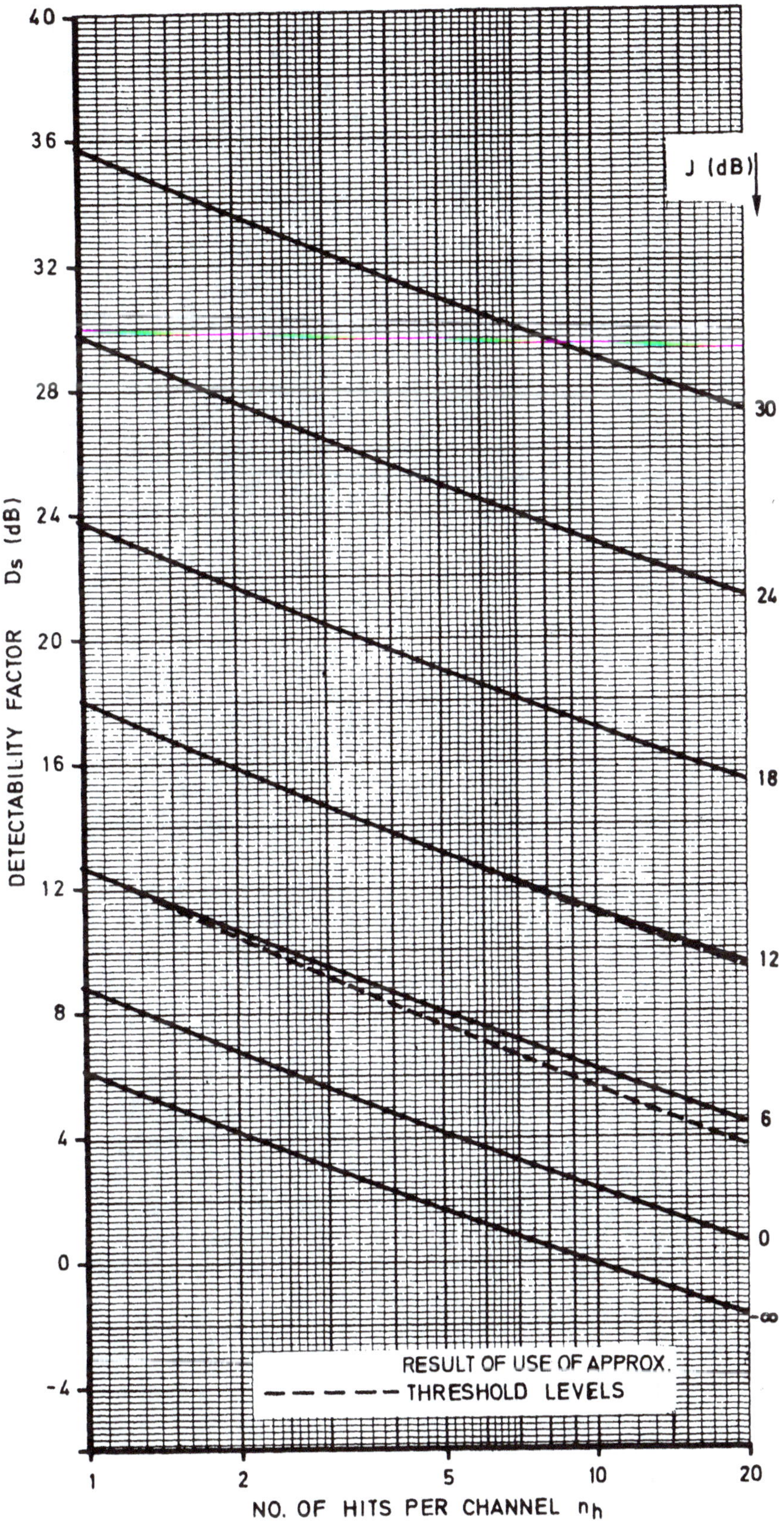
DETECTABILITY FACTOR Ds (dB)
40
36
32
28
24
20
16
12
8
4
0
-4
J (dB)
30
24
18
12
6
0
-∞
RESULT OF USE OF APPROX.
THRESHOLD LEVELS
1
2
5
10
20
NO. OF HITS PER CHANNEL n_h

Fig. 17

283

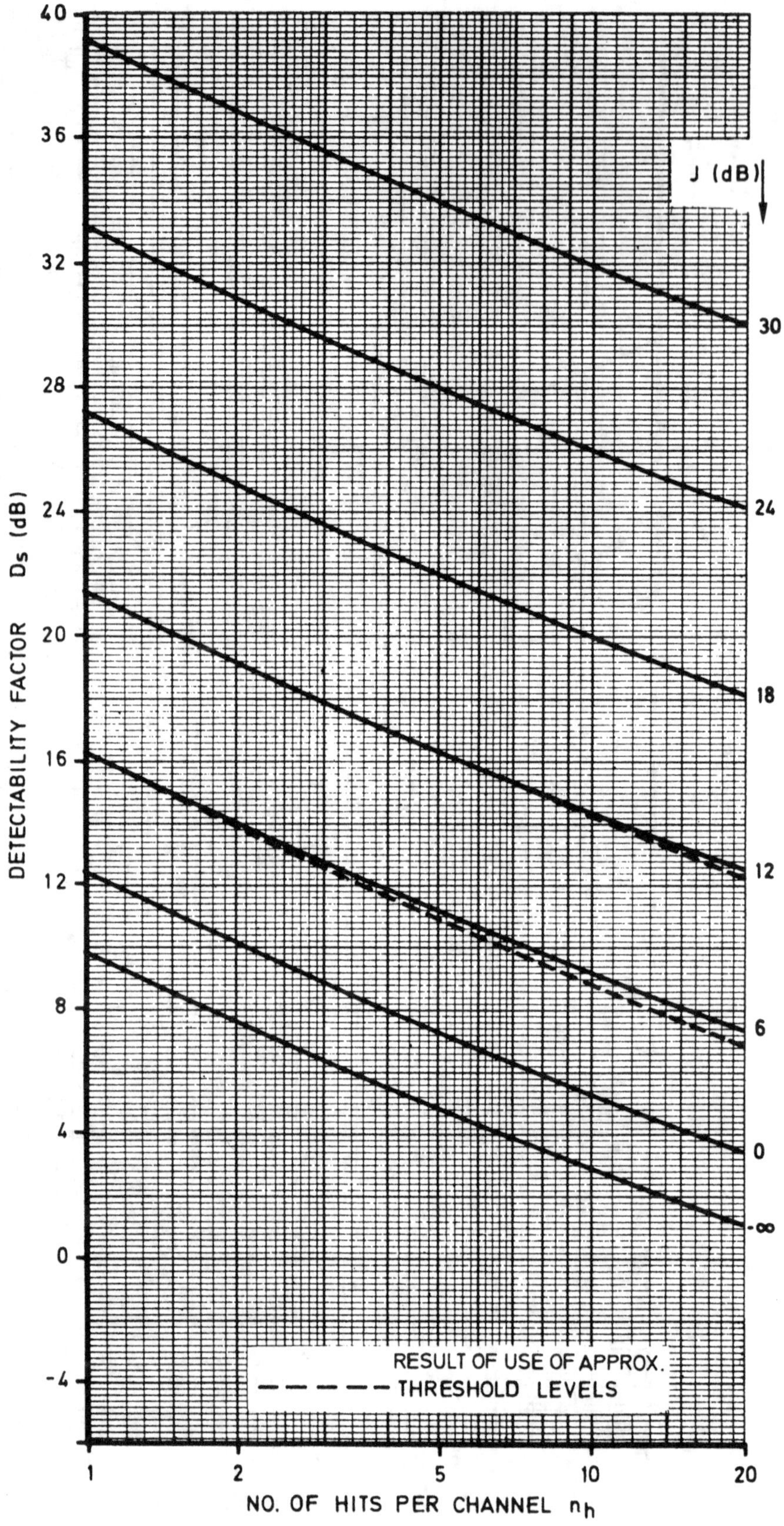

40
36
32
28
24
20
16
12
8
4
0
-4
DETECTABILITY FACTOR Ds (dB)
J (dB)
30
24
18
12
6
0
-∞
RESULT OF USE OF APPROX.
THRESHOLD LEVELS
1
2
5
10
20
NO. OF HITS PER CHANNEL nh

Fig. 18

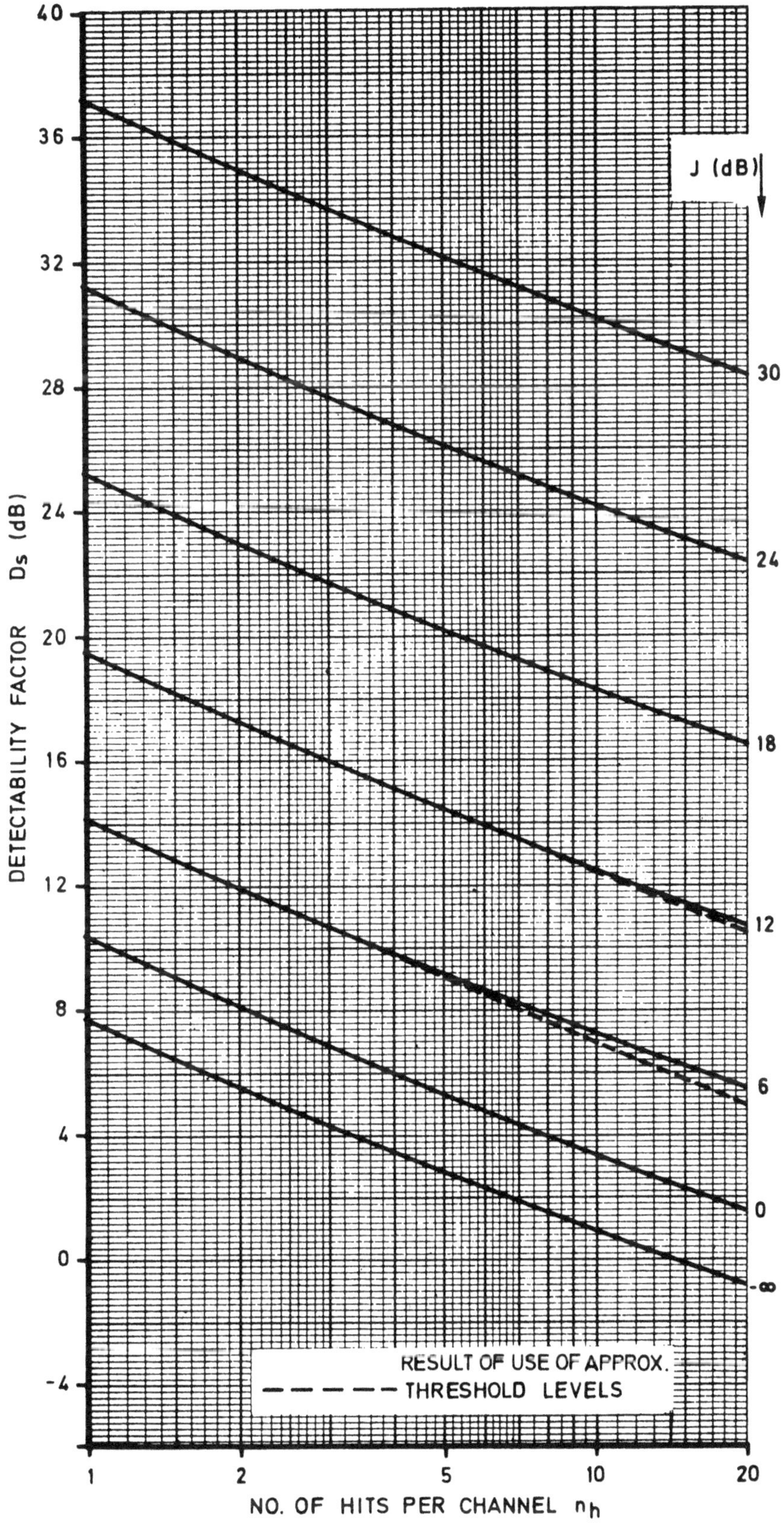

DETECTABILITY FACTOR Ds (dB)
40
36
32
28
24
20
16
12
8
4
0
-4
J (dB)
30
24
18
12
6
0
-∞
RESULT OF USE OF APPROX.
THRESHOLD LEVELS
1
2
5
10
20
NO. OF HITS PER CHANNEL nh

Fig. 19

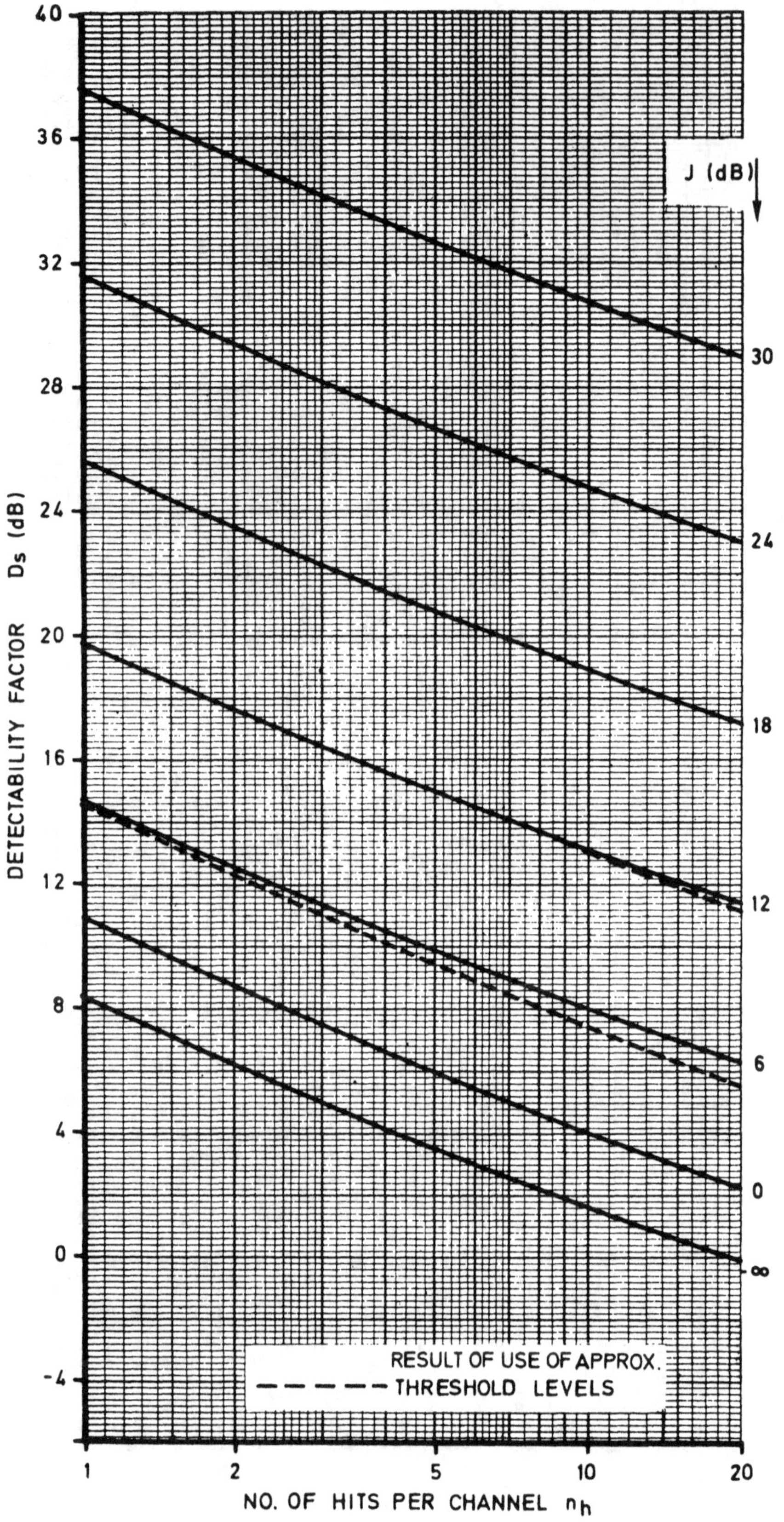
40
36
32
28
24
20
16
12
8
4
0
-4
DETECTABILITY FACTOR Ds (dB)
J (dB)
30
24
18
12
6
0
∞
RESULT OF USE OF APPROX.
THRESHOLD LEVELS
1
2
5
10
20
NO. OF HITS PER CHANNEL n_h

Fig. 20

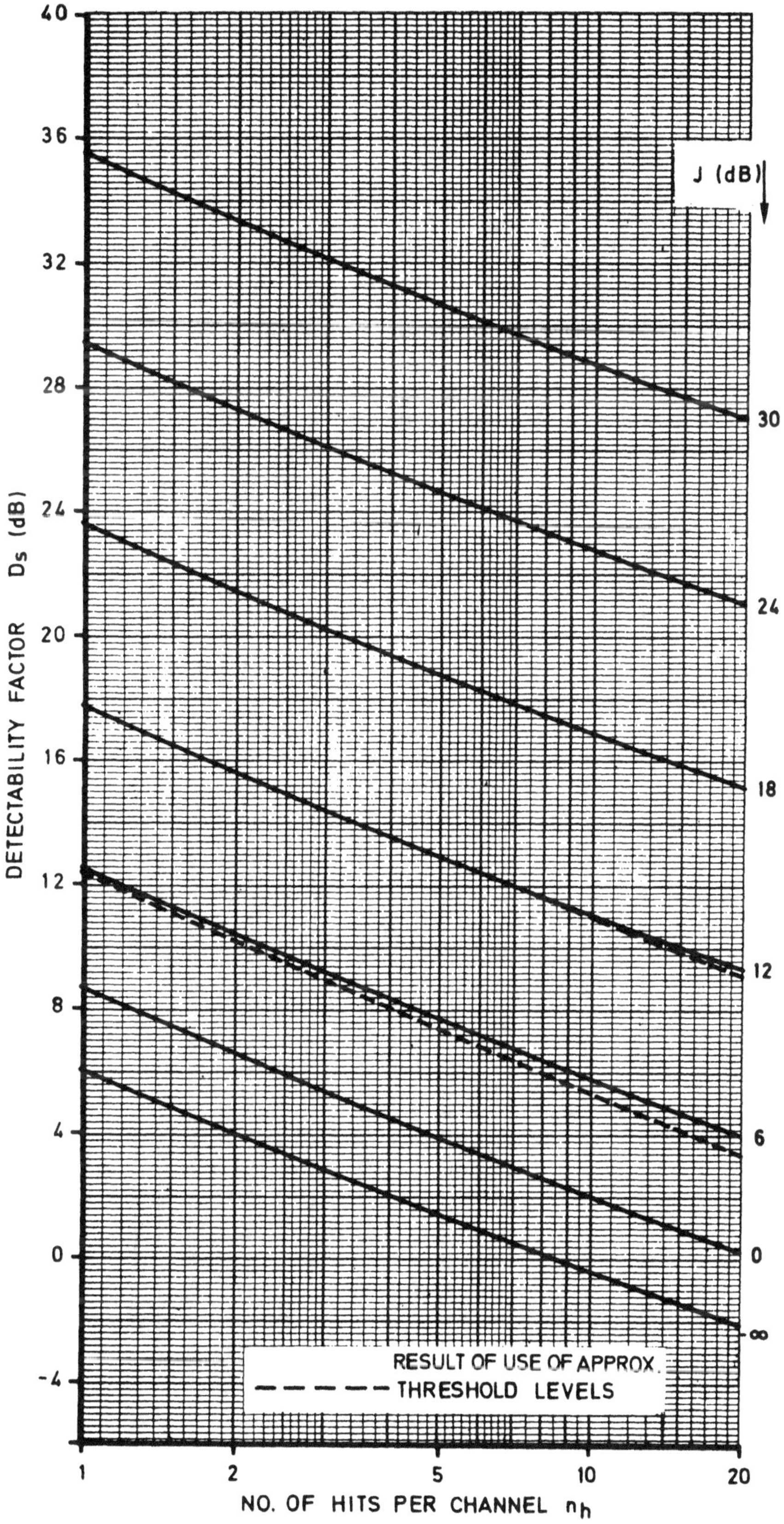
J (dB)
30
24
18
12
6
0
-∞
DETECTABILITY FACTOR Ds (dB)
40
36
32
28
24
20
16
12
8
4
0
-4
RESULT OF USE OF APPROX.
THRESHOLD LEVELS
1
2
5
10
20
NO. OF HITS PER CHANNEL n_h

Fig. 21

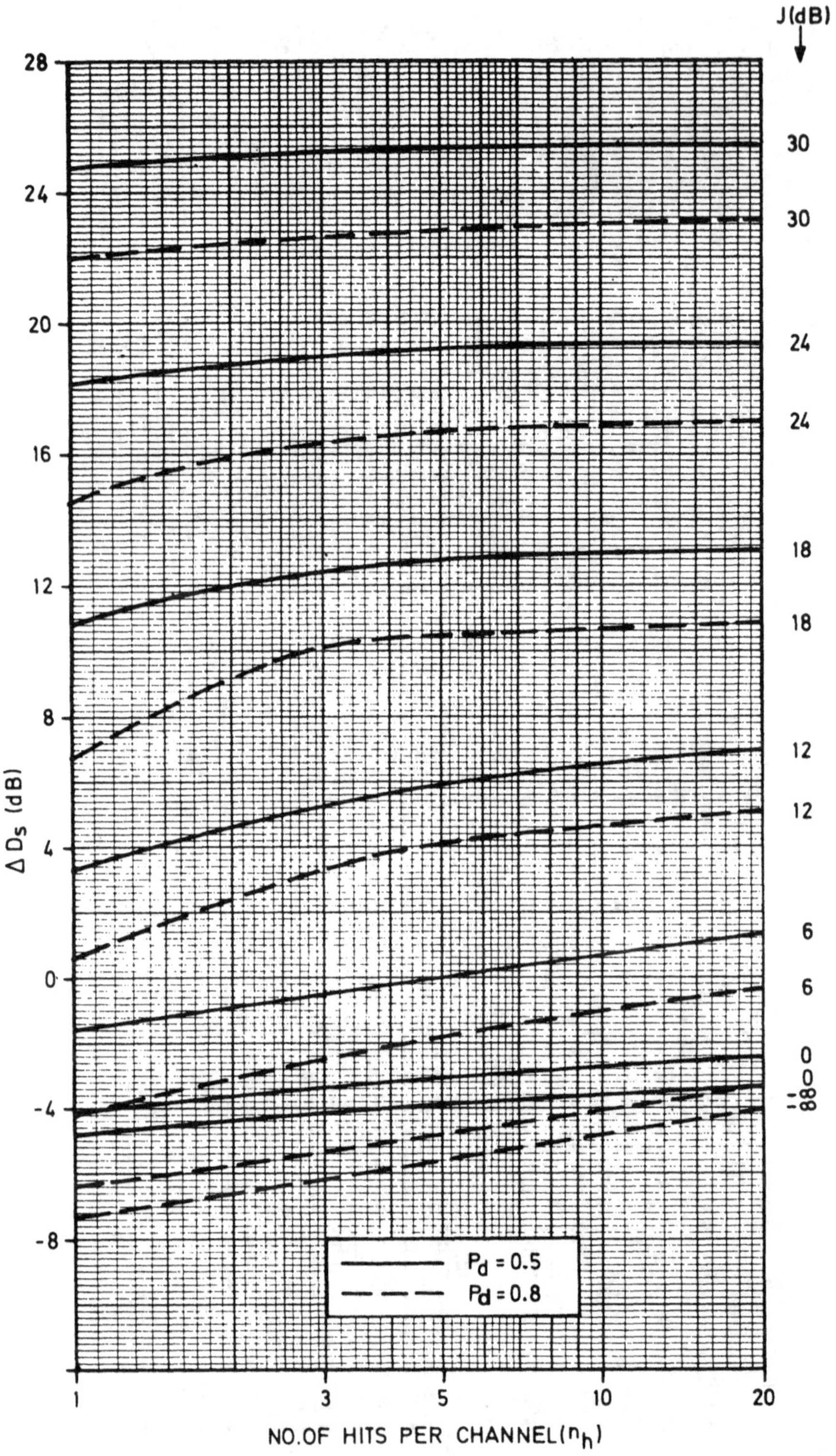
J(dB)
30
30
24
24
18
18
12
12
6
6
0
0
28
24
20
16
12
8
4
0
-4
-8
ΔDs (dB)
1
3
5
10
20
NO.OF HITS PER CHANNEL(nh)
Pd = 0.5
Pd = 0.8

Fig. 22

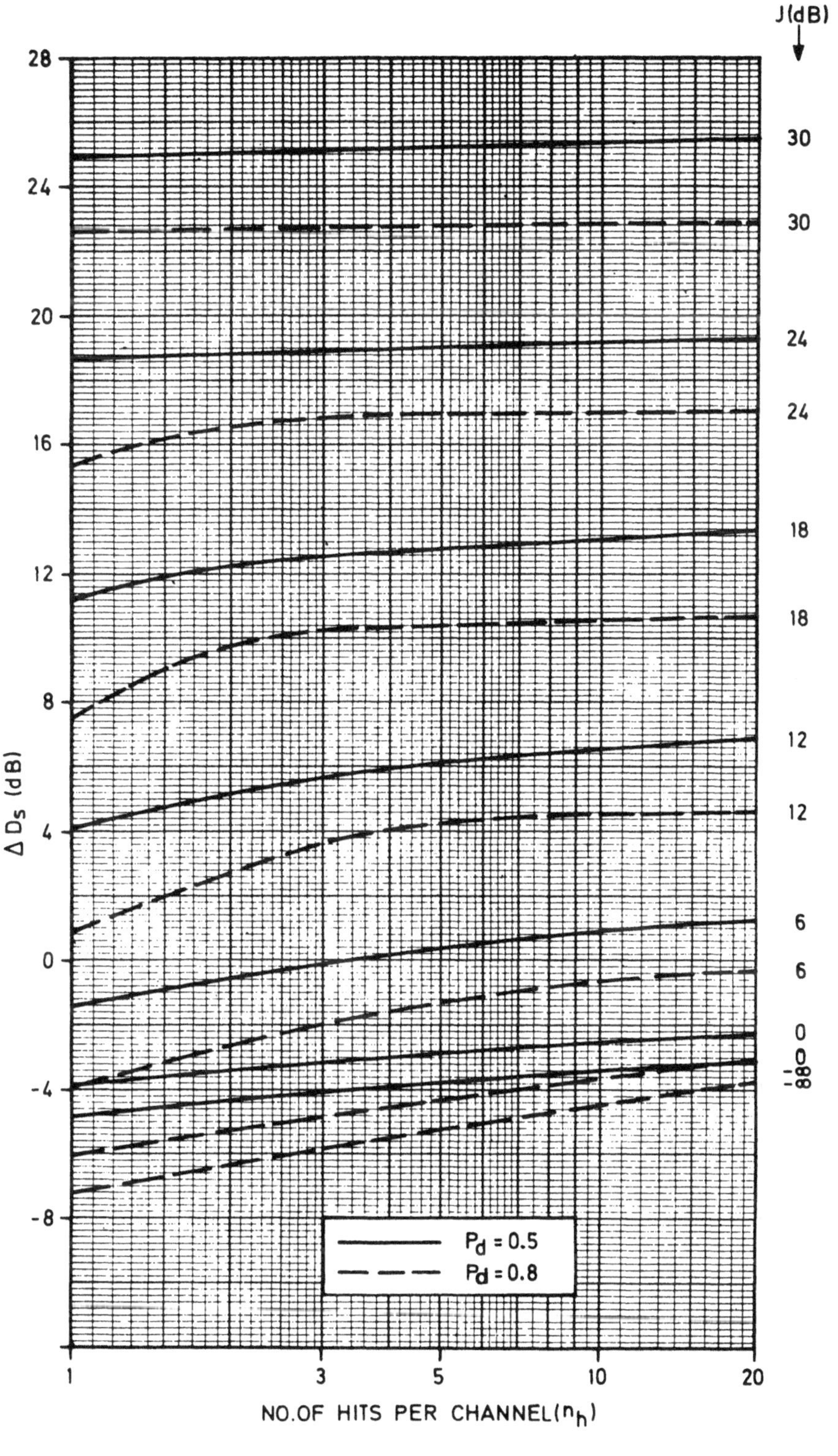
J(dB)
30
30
24
24
18
18
12
12
6
6
0
0
-88
ΔDs (dB)
28
24
20
16
12
8
4
0
-4
-8
$P_d = 0.5$
$P_d = 0.8$
1
3
5
10
20
NO.OF HITS PER CHANNEL(n_h)

Fig. 23

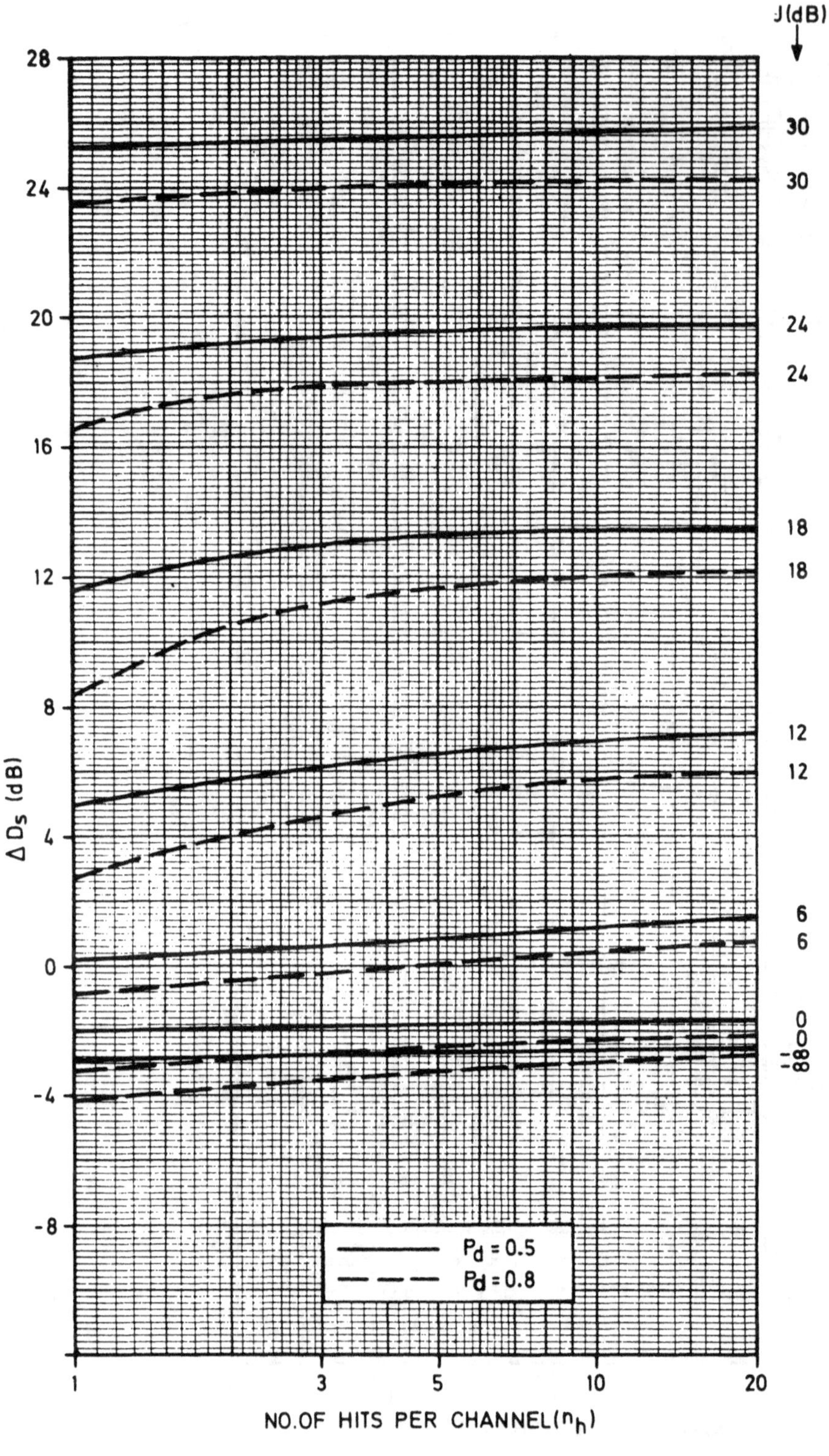
J(dB)
28
24
20
16
12
8
4
0
-4
-8
ΔDs (dB)
30
30
24
24
18
18
12
12
6
6
0
0
-88
Pd = 0.5
Pd = 0.8
1
3
5
10
20
NO.OF HITS PER CHANNEL(nh)

Fig. 24

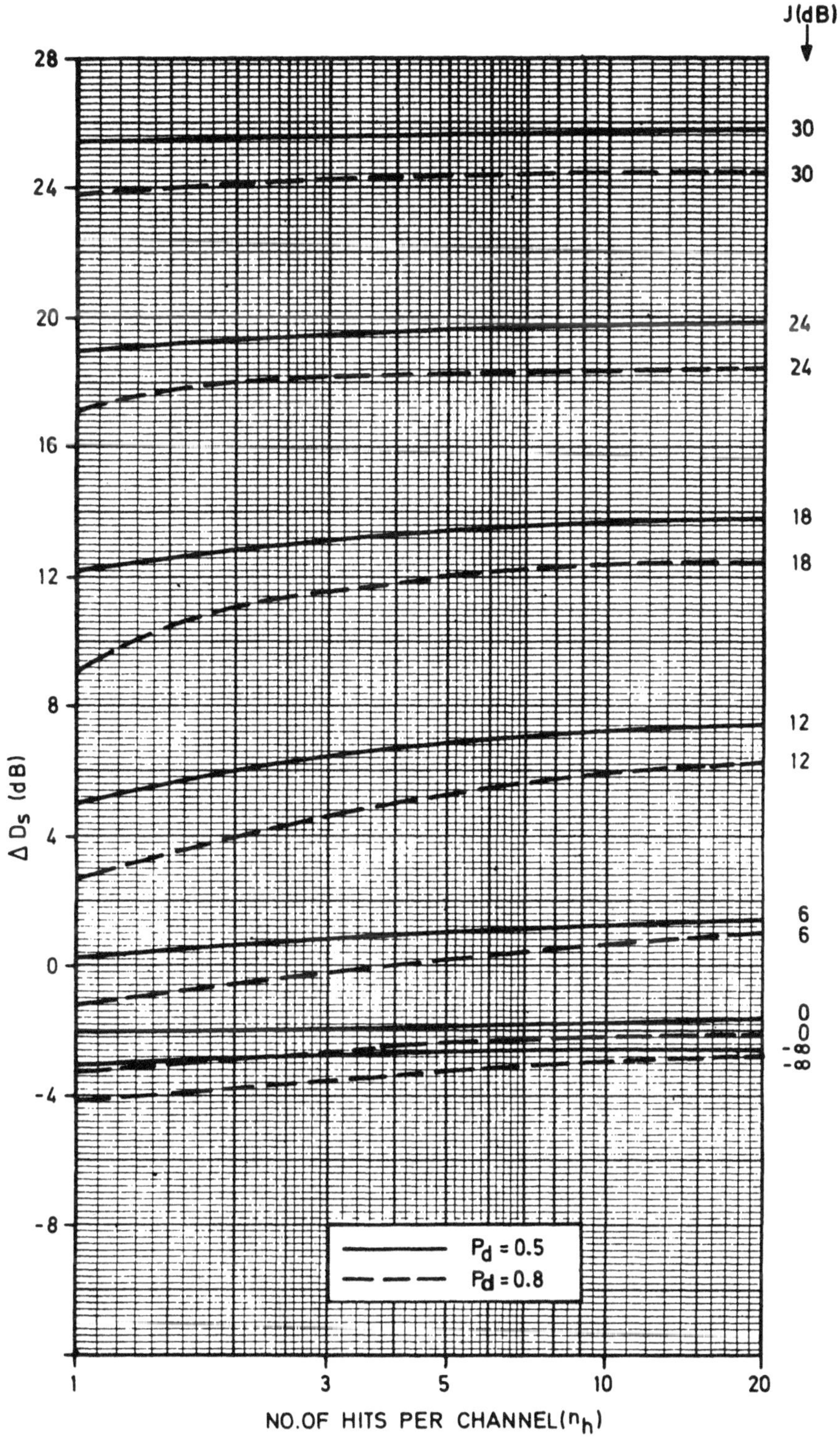
J(dB)
28
24
20
16
12
8
4
0
-4
-8
ΔDs (dB)
30
30
24
24
18
18
12
12
6
6
0
0
-8
-8
Pd = 0.5
Pd = 0.8
1
3
5
10
20
NO.OF HITS PER CHANNEL(nh)

Fig. 2[illegible]

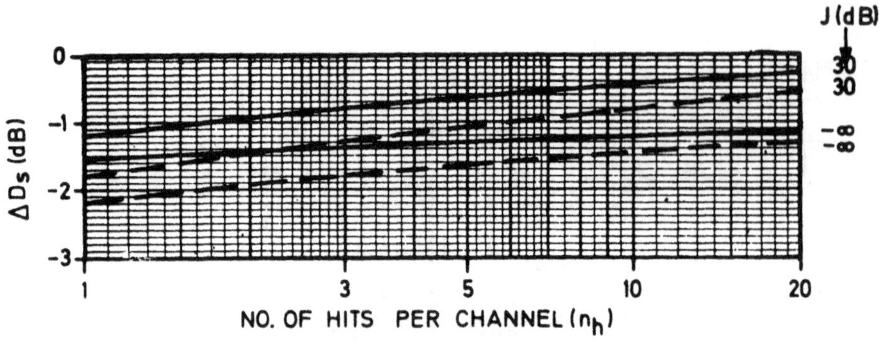

(A) MODE 3 RELATIVE TO MODE 0

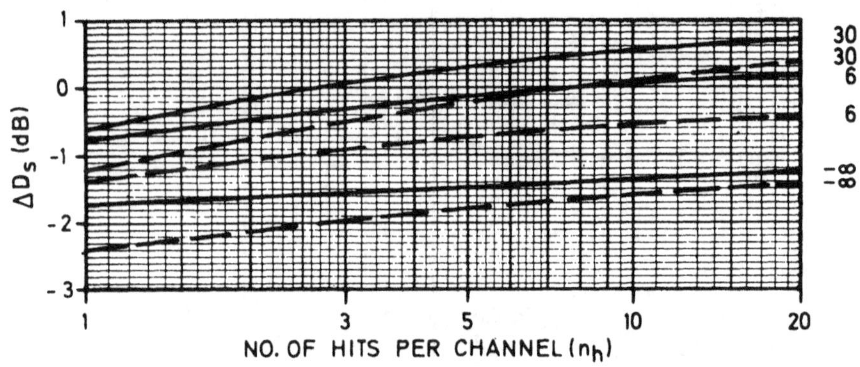

(B) MODE 2 RELATIVE TO MODE 0

Fig. 26

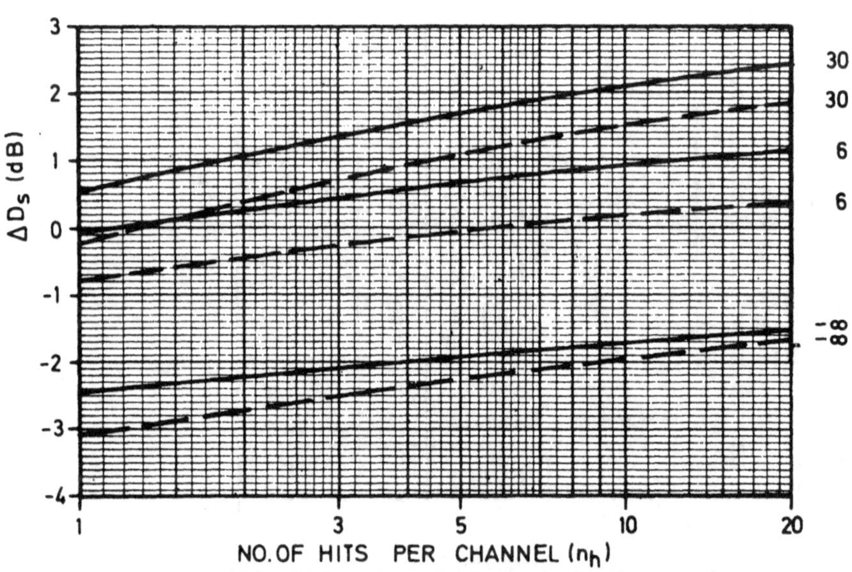

(C) MODE 1 RELATIVE TO MODE 0

——— $P_d = 0.5$, — — — $P_d = 0.8$

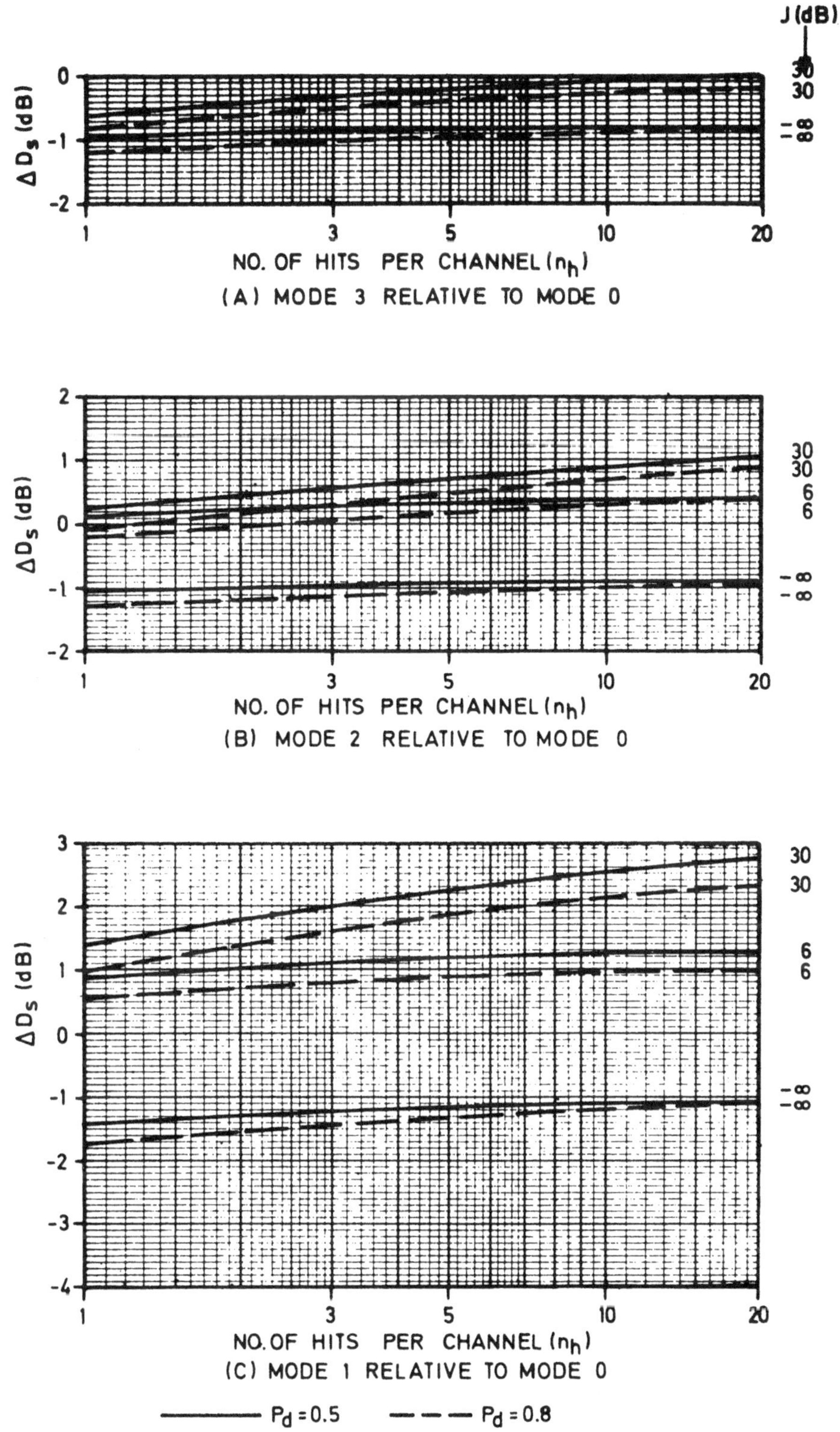
J(dB)
ΔDs (dB)
NO. OF HITS PER CHANNEL (nh)
(A) MODE 3 RELATIVE TO MODE 0
(B) MODE 2 RELATIVE TO MODE 0
(C) MODE 1 RELATIVE TO MODE 0
Pd = 0.5
Pd = 0.8

Fig. 27

KEY TO FIGURES 2 TO 27

Fig. No.	Case No.	P_d	P_f	Swerling Case No.
Variation of D_s with J and n_h				
2	IIE-0	0.8	10^{-6}	I/II
3	IIE-0	0.5	10^{-6}	I/II
4	IIE-0	0.8	10^{-4}	I/II
5	IIE-0	0.5	10^{-4}	I/II
6	IIE-0	0.8	10^{-6}	III/IV
7	IIE-0	0.5	10^{-6}	III/IV
8	IIE-0	0.8	10^{-4}	III/IV
9	IIE-0	0.5	10^{-4}	III/IV
Relative gain of the various modes				
10(a)	IIE-2,4	0.8	10^{-6}, 10^{-4}	I/II
(b)	IIE-1,3	0.8	10^{-6}, 10^{-4}	I/II
11(a)	IIE-2,4	0.5	10^{-6}, 10^{-4}	I/II
(b)	IIE-1,3	0.5	10^{-6}, 10^{-4}	I/II
12(a)	IIE-2,4	0.8	10^{-6}, 10^{-4}	III/IV
(b)	IIE-1,3	0.8	10^{-6}, 10^{-4}	III/IV
13(a)	IIE-2,4	0.5	10^{-6}, 10^{-4}	III/IV
(b)	IIE-1,3	0.5	10^{-6}, 10^{-4}	III/IV
Variation of D_s with J and n_h				
14	IEE-0	0.8	10^{-6}	I/II
15	IEE-0	0.5	10^{-6}	I/II
16	IEE-0	0.8	10^{-4}	I/II
17	IEE-0	0.5	10^{-4}	I/II
18	IEE-0	0.8	10^{-6}	III/IV
19	IEE-0	0.5	10^{-6}	III/IV
20	IEE-0	0.8	10^{-4}	III/IV
21	IEE-0	0.5	10^{-4}	III/IV
Gain ΔD_s in D_s of Mode 4 relative to Mode 0				
22	IEE-4	0.8, 0.5	10^{-6}	I/II
23	IEE-4	0.8, 0.5	10^{-4}	I/II
24	IEE-4	0.8, 0.5	10^{-6}	III/IV
25	IEE-4	0.8, 0.5	10^{-4}	III/IV
Gain ΔD_s in D_s of Modes 1,2 and 3 relative to Mode 0				
26	IEE-1,2,3	0.8, 0.5	$\leq 10^{-4}$, $\geq 10^{-6}$	I/II
27	IEE-1,2,3	0.8, 0.5	$\leq 10^{-4}$, $\geq 10^{-6}$	III/IV

Appendices A-F have been omitted from this paper.

REFERENCES

1. S. Herscovici et A. Detape, "Performances d'un Radar Diversité", Annales de Radioélectricité, Tome 12, No. 50, October 1957, pp. 285 - 298

2. J.A. Bijvoet, "Standard Methods for Predicting and Specifying Performance of Air Surveillance Radar Systems", SHAPE Technical Centre, Technical Report TR-50-U, April 1969 (NATO Unclassified)

   <u>Supplement 1</u>

   J. Eendebak, "Detectability Curves for Swerling Cases III and IV Target Models", December 1969 (NATO Unclassified)

   <u>Supplement 2</u>

   A.J. Poelman, "Performance Evaluation of Frequency Diversity Radars", April 1970 (NATO Unclassified)

   <u>Supplement 3</u>

   A.J. Poelman and Mrs. E.J. Bennée, "Further Notes on Performance Evaluation of Frequency Diversity Radars", May 1973 (NATO Unclassified)

3. J.I. Marcum, "A Statistical Theory of Target Detection by Pulsed Radar", Mathematical Appendix, IRE Trans. on Information Theory, April 1960, pp. 145 - 169, Special Monograph Issue

4. P. Swerling, "Probability of Detection for Fluctuating Targets", IRE Trans. on Information Theory, April 1960, pp. 269 - 308, Special Monograph Issue

5. M.I. Skolnik, "Introduction to Radar Systems", McGraw-Hill Book Company, Inc., 1962, Section 2.5

6. A. Papoulis, "Probability, Random Variables and Stochastic Processes", McGraw-Hill Book Company, Inc., 1965

7. G.A. Campbell and R.M Foster, "Fourier Integrals for Practical Applications", D. van Nostrand Company, Inc., 1957

8. A. Erdélyi et al, "Tables of Integral Transforms, Volume I", McGraw-Hill Book Company, Inc., 1954

Section IV Effect of Diversity on Clutter

The theoretical advantages of frequency diversity in detection of targets against a clutter background are dependent partly upon decorrelation of the target signals and partly up on decorrelation of clutter samples at the different frequencies. With respect to a target, the improvement for any fixed threshold level can be obtained from results of the previous sections. The effect of diversity on the clutter statistics, which control either the false-alarm probability or, for CFAR systems, the variable threshold level, is much more difficult to assess. The papers reprinted in this section, and others listed in the bibliography, are addressed to this subject.

In rain clutter (and, by analogy, in chaff) the effect of diversity is relatively clear. Distribution of the scattering sources tends to be uniform over several adjacent range cells. The correlation time is controlled by the wind shear and turbulence factors, and the frequency correlation function for a rectangular pulse follows the theoretical (sin x)/x form (see Nathanson papers listed in the bibliography). Thus, in cases in which the correlation time is greater than the pulse repetition interval, frequency shifts in increments of $1/\tau$ provide decorrelated samples.

The case of sea clutter is more difficult because, for low grazing angles, the scatterers are not distributed uniformly, but rather are concentrated near the tops of the higher waves. Radars using long pulses or wide beams, such that the resolution cell extends over many individual waves, can achieve decorrelation with frequency shifts on the order of $1/\tau$. This method is the basis of the analysis by Beasley and Ward (Reprint 4-1). Airborne radars operating with grazing angles of more than a few degrees also can depend on essentially uniform scatterer distributions; as the range increases, thereby producing lower grazing angle, the wider illuminated area tends to cover more waves and retain the homogeneous Rayleigh distribution for which diversity is most advantageous.

As the grazing angle and size of the cell decrease, the scattering regions become isolated into individual resolution cells, often separated by cells containing little or no clutter. The two papers by Lind (Reprints 4-2 and 4-3, the latter coauthored by Bergkvist) describe both experimental data and the mathematical model of the phenomenon. Further supporting data and analysis is given in the papers by Trunk and by Bishop (see bibliography). The recent Trunk analysis shows that frequency diversity over a broad bandwidth can produce decorrelation of that portion of the variation which corresponds to the change in phase between capillary waves on a given major wave top, but the underlying modulation by the wave structure remains. The use of the time-varying Rayleigh distribution (and its space-varying counterpart) provides insight into the physical nature of the phenomenon beyond what can be obtained from the several non-Rayleigh amplitude distributions which lack information on correlation properties. More work on interrelationships between correlation intervals, cell sizes, and detection properties is needed before radar performance can be estimated with confidence.

If the correlation properties of sea clutter are not entirely clear, those of land clutter are almost entirely opaque. In his early paper (Reprint 1-2), Ray refers to improvement in mapping radar performance resulting from frequency agility. Whitlock (Reprint 1-3) presents some frequency correlation data on agricultural land; Rigden (1973; see bibliography) presents data on the spatial extent of clutter patches of different magnitude when viewed with 15-ns resolution; and Nathanson (1967; see bibliography) presents some typical A-scope traces of land clutter. However, the statistics of land clutter correlation intervals, from which advantages of frequency diversity could be computed, remain unexplored, at least in available literature.

ANNOTATED BIBLIOGRAPHY

Voles, R.: "Frequency Correlation of Clutter," *Proceedings of the IEEE, Vol. 54, No. 6,* June 1966, pp. 881-882c. Derives frequency correlation functions for uniform and uniform-plus-sinusoidal range distributions of scatterers.

Nathanson, F. E.; and Reilly, J. P.: "Clutter Statistics Which Affect Radar Performance," *IEEE Transactions of Aerospace and Electronic Systems, Vol. AES-3, No. 6 (Supplement),* November 1967, pp. 386-398. Broad survey of clutter parameters including frequency correlation of rain clutter. Reprinted in *Radars* (Volume 5 - *Radar Clutter*), Artech House, 1975.

Beasley, E. W.; and Ward, H. R.: "A Quantitative Analysis of Sea Clutter Decorrelation with Frequency Agility," *IEEE Transactions of Aerospace and Electronic Systems, Vol. AES-4, No. 3,* May 1968, pp. 468-473. Considers frequency and time-correlation functions of sea clutter and number of independent samples obtained with sinusoidal frequency modulations. Reprint 4-1.

Nathanson, F. E.; and Reilly, J. P.: "Radar Precipitation Echoes," *IEEE Transactions of Aerospace and Electronic Systems, Vol. AES-4, No. 4,* July 1968, pp. 505-514. General description of rain clutter, including correlation intervals in space and in frequency. Reprinted in *Radars* (Volume 5 - *Radar Clutter*), Artech House, 1975.

Pidgeon, V. W.: "Time, Frequency, and Spatial Correlation of Radar Sea Return," *Use of Space Systems for Planetary Geology and Geophysics,* Volume 17 of AAS Science and Technology Series, 1968, pp. 455-458. Presents experimental data on correlation functions of sea clutter at C-band, confirming for 10° grazing angle the (sin x)/x form of correlation predicted for uniform scatterer distribution over the resolution cell.

Nathanson, F. E.: *Radar Design Principles: Signal Processing and the Environment*, McGraw-Hill, 1969. Basic text on radar equation and signal processing considerations, including frequency correlation of rain and sea clutter and spatial correlation of sea clutter.

Lind, G.: "Measurement of Sea Clutter Correlation with Frequency Agility and Fixed Frequency Radar," *Philips Telecommunications Review, Vol. 29, No. 1,* April 1970, pp. 32-38. Experimental data are analyzed to determine the number of independent samples available from sea clutter with and without frequency agility. Reprint 4-2.

Trunk, G. V.; and George, S. F.: "Detection of Targets in Non-Gaussian Sea Clutter," *IEEE Transactions on Aerospace and Electronic Systems, Vol. AES-6, No. 5,* September 1970, pp. 620-628. Discusses the statistics of sea clutter echoes in high resolution radar and illustrates the need for many independent samples. Reprinted in *Radars* (Volume 5 - *Radar Clutter*), Artech House, 1975.

Skolnik, M. I.: "Sea Echo," Chapter 26 in *Radar Handbook*, McGraw-Hill 1970. Use of frequency agility is compared qualitatively with high range resolution as a means of improving detection of targets in sea clutter, with range resolution as the preferred technique. In Chapter 31, J. Croney presents display data showing effects of scan-to-scan integration, but notes that commercial maritime radar has not yet exploited frequency agility or fast scan for this purpose.

Trunk, G. V.: "Further Results on the Detection of Targets in Non-Gaussian Sea Clutter," *IEEE Transactions on Aerospace and Electronic Systems, Vol. AES-7, No. 3,* May 1971, pp. 553-556. Extends previous work (1970) to fluctuating targets and to trimmed-mean detector, increasing emphasis on need for many independent samples.

Trunk, G. V.: "Detection of Targets in Non-Rayleigh Sea Clutter," *1971 IEEE EASCON Record*, pp. 239-245. Includes discussion of time and frequency decorrelation of sea clutter echoes in high-resolution radar data, indicating presence of long-period modulation of reflectivity that is not smoothed by frequency diversity. Reprinted in *Radars* (Volume 5 - *Radar Clutter*), Artech House, 1975.

Trunk, G. V.: "Radar Properties of Non-Rayleigh Sea Clutter," *IEEE Transactions on Aerospace and Electronic Systems, Vol. AES-8, No. 2,* March 1972, pp. 196-204. Continues discussion of time-varying Rayleigh model for high-resolution sea clutter. Test data are presented showing limited decorrelation of clutter with frequency diversity. Reprinted in *Radars* (Volume 5 - *Radar Clutter*), Artech House, 1975.

Bishop, G.: "Radar Sea Clutter," *AGARD Lecture Series No. 59*, September 1973, AD769277. Describes results of high-resolution X-band measurements of sea clutter, including correlation functions in time.

Rigden, C. J.: "High Resolution Land Clutter Characteristics," *IEE Conference Publication No. 105*, October 1973. Reprinted in *Radars*, (Volume 5 - *Radar Clutter*), Artech House, 1975.

Lind, G.; and Bergkvist, B.: "A Method to Calculate the Density of Sea Clutter Echoes at Low Grazing Angles," *Proceedings of the 1975 IEEE International Radar Conference*, pp. 230-234. Provides a basis for computing the effectiveness of frequency agility and of range resolution in improving detectability of targets in sea clutter. Reprint 4-3.

Trunk, G. V.: "Non-Rayleigh Sea Clutter: Properties and Detection of Targets," *Naval Research Laboratory Report 7986*, 25 June, 1976, ADA028875. Reviews and expands previous references on detection in sea clutter.

A Quantitative Analysis of Sea Clutter Decorrelation with Frequency Agility

E.W. Beasley and H.R. Ward

IEEE Transactions on Aerospace and Electronic Systems, Vol. AES-4, No. 3 (May 1968)
Pages 468-473

Abstract

A brief statement of the sea clutter problem in surface-search radar operation illustrates the need for some form of signal-to-clutter enhancement. Post-detection integration used in the simpler radars is limited by the pulse-to-pulse correlation of the clutter. Analysis of the effect of changing frequency from pulse to pulse leads to an expression for the correlation between pulses in the sequence. Knowing this correlation, the reduction in the fluctuating clutter component produced by integration can be determined. This is described by an equivalent number of independent pulses, N_e. For the particular case of sinusoidal modulation of the transmitted frequency, N_e is computed. The critical dependence of N_e upon the modulating frequency f_m is illustrated by spectrum photographs. Choice of an optimum f_m is discussed. The results of computations of N_e for optimum f_m are presented as a family of normalized curves. These data permit the tradeoff of the radar parameters against their quantitative effect on radar performance.

Introduction

The ability of a surface-search radar to detect small targets near the sea surface is often limited by the pulse-to-pulse correlation of the sea clutter. This can be reduced by changing the radar frequency between pulses. When a series of pulses is returned from a target, the question arises as to how much frequency deviation is needed and at what rate. This paper answers the question, more quantitatively than earlier analyses,[1] by establishing as a performance criterion N_e, the effective number of independent hits. N_e is closely related to target detectability and provides a convenient means for evaluating alternative techniques or parameter settings. Sinusoidal frequency modulation is studied in detail by computing N_e for a wide range of parameters.

Manuscript received August 7, 1967.

E. W. Beasley is now with the MITRE Corporation, Bedford, Mass.

Sea Clutter (Background)

In surface-search radars, the wanted target return is accompanied by energy reflected from the sea. The effective surface area illuminated is determined by the size of the radar resolution cell, which in the case of long range and low elevation angles is given by Barton[2] as

$$A_c \approx \frac{\theta_a R}{\sqrt{2}} \frac{\tau c}{2} \tag{1}$$

where θ_a is the azimuth beamwidth (radians), R is the range, τ is the pulse length, c is the velocity of propagation.

The clutter energy is returned from a number of scattering elements located within A_c; its level is affected by many factors, as summarized by Barton.[2] We think of the clutter within the resolution cell as having an effective radar cross section σ_c given by

$$\sigma_c = \sigma^0 A_c \tag{2}$$

where σ^0 is a dimensionless factor known as the clutter reflectivity. Its value has been investigated by many authorities.[2] From such data it can be seen that values of σ^0 as high as 0.001 could be encountered quite often. Thus for a radar with azimuth beamwidth $\theta_a = 2°$, and pulse length 0.25 μs, the value of A at range 1 mile would be about 1700 square meters. With $\sigma^0 = 0.001$ the effective clutter cross section would thus be 1.7 square meter. It would clearly be difficult to detect a target of 1.0-square-meter cross section with a single pulse having the above parameters.

To improve target detectability, advantage is taken of the number of pulses N hitting the target as the beam sweeps past. The integrating effect of a PPI display, or other means, provides some enhancement when the masking noise is not correlated, or partly correlated, from pulse-to-pulse. Blake[3] shows, for example, that when $N = 30$, a typical value, the improvement over thermal noise is about 12 dB for a PPI display.

Unfortunately, sea clutter may remain correlated over several radar periods. Data presented by Kerr[4] suggests that the sea clutter autocorrelation function is roughly Gaussian in shape with a standard deviation of about 3 ms for a radar wavelength of 9.2 cm. Other sources have reported values up to 5 ms. Since short-range surface-search radars may have interpulse periods of about 250 to 500 μs, it is seen that the clutter may remain substantially correlated for as many as 10 pulses. With total hits on target equal to 30, the effective number of independent samples would be only 3.

In the case of a target masked by sea clutter, improvements in detectability can be made if the successive clutter returns can be made uncorrelated.

Clutter Decorrelation by Interpulse Frequency Change

The correlation between two successive sea clutter returns can be reduced by changing the transmitted frequency from pulse to pulse. The relation between correlation and frequency shift is derived in the following analysis.

Let the normalized correlation between the ith and jth clutter returns be $R_x(i, j)$, and let x_i, x_j be the ith and jth received voltages. It is assumed that the transmitted waveform is a rectangular pulse τ seconds long. Then

$$R_x(i, j) = \frac{\overline{x_i x_j}}{\overline{x^2}}. \tag{3}$$

The received voltage is obtained by convolving the transmitted waveform, $\mu(t)$, with the clutter impulse response, $c(t)$:

$$x_i = c(t) * \mu_i(t) \tag{4}$$

and using Woodward's notation:[5]

$$\mu_i(t) = \text{rect}\left(\frac{t}{\tau}\right) \cos(2\pi f_i t). \tag{5}$$

The autocorrelation function of x is

$$\overline{x_i x_j} = \overline{[c(t) * \mu_i(t)][c(t) * \mu_j(t)]} \tag{6}$$

$$\overline{x_i x_j} = \mathfrak{F}\langle[c(t) * \mu_i(t)][c(t) * \mu_j(t)]\rangle_{f=0}$$

$$\overline{x_i x_j} = \int_{-\infty}^{\infty} c^2(f)\mu_i(f)\mu_j(f)df. \tag{7}$$

Here, $c(f)$ and $\mu(f)$ are the Fourier transforms of $c(t)$ and $\mu(t)$.

If the clutter return from a resolution cell is assumed to be the sum of many random point reflections, then $c(t)$ is the sum of closely spaced delta functions in time. According to Rice,[6] this is a correct representation for white noise. The spectrum $c(f)$ will therefore be uniform of power density c_0. We can thus replace $c^2(f)$ by c_0.

For the ith transmitted pulse,

$$\mu_i(f) = \frac{\tau}{2} \text{sinc}\{\tau(f + f_i)\} + \frac{\tau}{2} \text{sinc}\{\tau(f - f_i)\} \tag{8}$$

where

$$\text{sinc}(x) = \frac{\sin(\pi x)}{(\pi x)}.$$

Therefore

$$\overline{x_i x_j} = \frac{c_0 \tau^2}{4} \int_{-\infty}^{\infty} \{[\text{sinc}\,\tau(f + f_i) + \text{sinc}\,\tau(f - f_i)]$$
$$\cdot [\text{sinc}\,\tau(f + f_j) + \text{sinc}\,\tau(f - f_j)]\} df. \tag{9}$$

Noting that integrals of products of sinc $\langle\tau(f+f_{i,j})\rangle$ and sinc $\langle\tau(f-f_{i,j})\rangle$ are equal to zero, we obtain, after some manipulation,

$$\overline{x_i x_j} = \frac{c_0 \tau^2}{4} \text{sinc}[\tau(f_i - f_j)]. \tag{10}$$

Also

$$\overline{x^2} = \frac{c_0 \tau^2}{4}. \tag{11}$$

We can therefore write

$$R_x(i, j) = \text{sinc}[\tau(f_i - f_j)]. \tag{12}$$

In addition to the effect of frequency change, the clutter return may decorrelate with time due to relative motion of the point scatterers. If this effect can be described as $R_x(t_i - t_j)$ at one frequency, then the total correlation function, since these processes are independent, is given by $R_x(f_i - f_j)\,R_x(t_i - t_j)$.

Under some sea conditions the clutter model used above is inaccurate. For example, there may be returns from one or two larger wavefronts within the illuminated area on the sea surface. Equation (12) would then not hold exactly.

Decorrelation of Multiple Returns

Assume a simplified model of the radar detector–integrator shown in Fig. 1. The received signal x is square-law detected, and N successive pulses received (as the beam scans a target) are integrated with equal weighting.

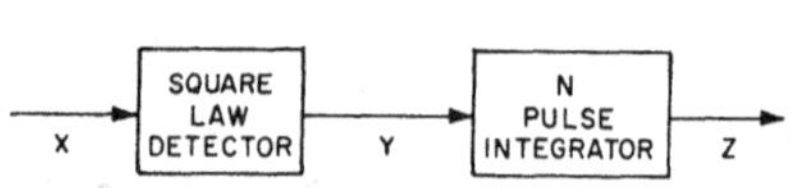

Fig. 1. Functional diagram.

We have shown that the introduction of a frequency shift between two transmitted pulses reduces the received signal correlation. In a sequence of N pulses, for any given form of frequency modulation we can calculate the correlation between each pair of pulses; we can also make allowance for the time decorrelation of the clutter. But how can we express the overall effect of this decorrelation on radar performance?

We have chosen to measure performance in terms of an equivalent number of independent samples N_e:

$$N_e \equiv \frac{\left[\frac{\text{Fluctuation}}{\text{Mean}}\right]^2 \text{ at integrator input}}{\left[\frac{\text{Fluctuation}}{\text{Mean}}\right]^2 \text{ at integrator output}}. \tag{13}$$

If all samples are independent, $N_e=N$, then the effect of the integrator is to reduce the ratio of the fluctuating component to the steady component by a factor N. If all samples are completely correlated, $N_e=1$ and there is no reduction of this ratio. N_e is thus a useful measure of the effectiveness of the integrator in reducing the fluctuating component.

If the received signal voltage is x, the square-law detector output y, and the integrator output z, we can write

$$N_e = \frac{\sigma_y^2/\bar{y}^2}{\sigma_z^2/\bar{z}^2} = \frac{\sigma_y^2\bar{z}^2}{\sigma_z^2\bar{y}^2}. \tag{14}$$

Here σ_y and σ_z are the standard deviations of y and z. Since

$$z = \sum_{i=1}^{N} y_i \tag{15}$$

and

$$\sigma_y^2 = \overline{y^2} - (\bar{y})^2 \tag{16}$$

we can express N_e in terms of the correlation function $R_y(i, j)$ as follows:

$$N_e = \frac{\overline{y^2}/\bar{y}^2 - 1}{[\overline{y^2}/\bar{y}^2]\frac{1}{N^2}\cdot\sum_{i=1}^{N}\sum_{j=1}^{N} R_y(i,j) - 1}. \tag{17}$$

For a square-law detector, according to Davenport and Root,[7]

$$\overline{y^2}/\bar{y}^2 = 3. \tag{18}$$

Therefore

$$R_y(i,j) = \frac{\overline{y_iy_j}}{\bar{y}^2} = \frac{1}{3} + \frac{2}{3}R_x^2(i,j). \tag{19}$$

Substituting for y and R_y in (17), we obtain, after some manipulation,

$$N_e = \frac{N^2}{\sum_{i=1}^{N}\sum_{j=1}^{N} R_x^2(i,j)} \tag{20}$$

and using the diagonal symmetry to simplify the double sum,

$$N_e = \frac{N^2}{N + 2\sum_{k=1}^{N-1}(N-k)R_x^2(k)}. \tag{21}$$

Example of N_e Calculation

To illustrate how N_e may be computed in a particular case, consider a radar in which the transmitted frequency is modulated sinusoidally. Assume also that the clutter decorrelates in time (at a fixed frequency) according to a Gaussian law:

$$R_x(t_i - t_j) = \exp\left[-\left(\frac{t_i - t_j}{t_c}\right)^2\right]. \tag{22}$$

Here, $t_i - t_j$ is the time between the ith and jth pulses, and t_c is $\sqrt{2}$ times the standard deviation of the Gaussian curve. The correlation between ith and jth pulses as a function of both frequency and time is

$$R_x(i,j) = \text{sinc}\,[\tau(f_i - f_j)]\exp\left[-\left(\frac{t_i - t_j}{t_c}\right)^2\right]. \tag{23}$$

The transmitted frequency is determined by sampling a sine wave at regular intervals Δt (the radar repetition period). The sine wave has peak amplitude A, in hertz, angular frequency ω_m, and starting phase θ:

$$f = A\sin(\omega_m t + \theta). \tag{24}$$

The frequency separation of the ith and jth pulses is thus

$$f_i - f_j = A\{\sin(i\omega_m\Delta t + \theta) - \sin(j\omega_m\Delta t + \theta)\}. \tag{25}$$

Therefore

$$R_x^2(i,j) = \text{sinc}^2\{\tau A[\sin(i\omega_m\Delta t+\theta) - \sin(j\omega_m\Delta t+\theta)]\}\cdot\exp\left[-2\left(\frac{i\Delta t - j\Delta t}{t_c}\right)\right]. \tag{26}$$

Given the pulse length τ, the amplitude and frequency of modulation A and ω_m, the radar period Δt, the clutter correlation time t_c, and an assumed value of initial phase θ, (25) enables us to compute $R_x^2(i, j)$ for any i and j. Then for a sequence of N pulses, (20) yields the value of N_e as a measure of the effectiveness of the decorrelation process, expressed as (13).

Results of Computations

In order to plot results in a usable form and reduce the number of variables involved, some of the parameters must be normalized. From inspection of (25), it can be seen that $R_x^2(i, j)$ depends only upon τA, $\omega_m\Delta t$, $\Delta t/t_c$, i, j, and θ.

We could proceed to compute $R_x^2(i, j)$ from these quantities and then compute N_e as a function of these quantities and of N^2. To simplify the presentation of results, we choose further to use the parameters

$$\tau A/N,\ \Delta t/t_c,\ N_e/N,$$

and to average results over a set of starting phases θ. The dependence on $\omega_m\Delta t$ was evaluated as a separate issue.

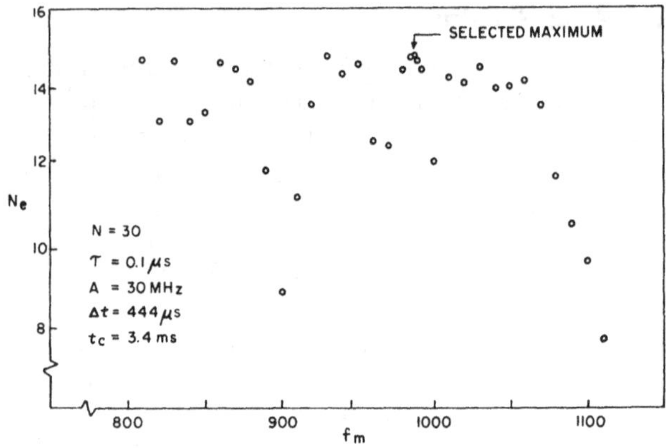

Fig. 2. Optimum frequency selection.

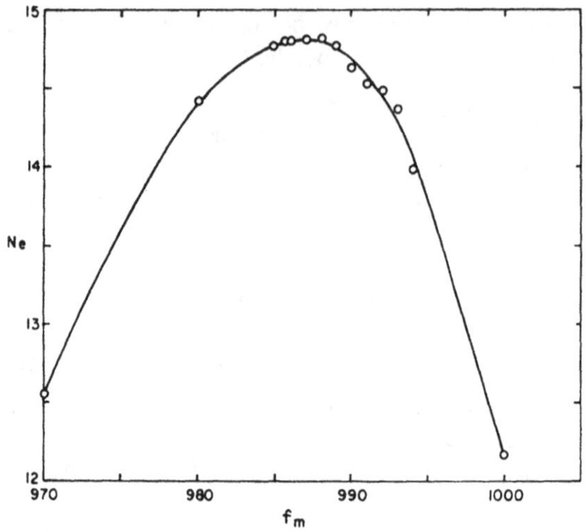

Fig. 4. N_e in optimum region.

Fig. 3. Output spectra $f_m = \frac{1}{2} f_R$ (one division = 5 MHz).

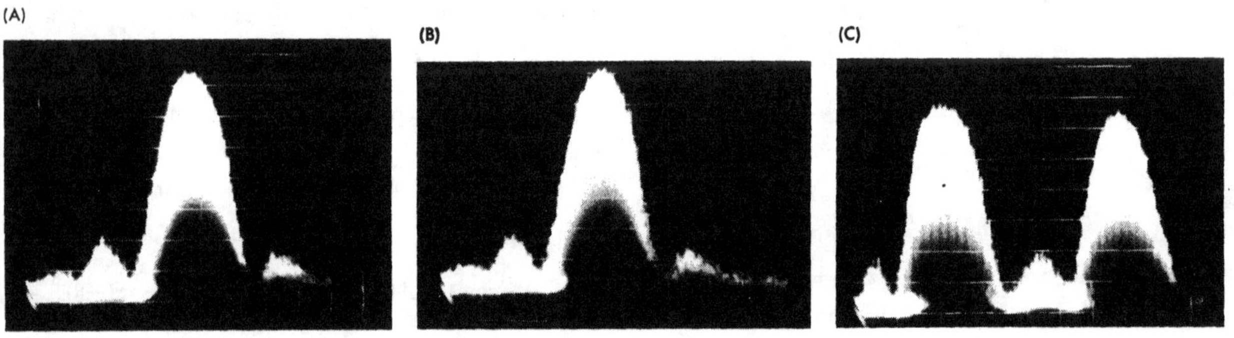

A FORTRAN program was written to compute N_e for a given set of variables N, ω_m, Δt, A, τ, and t_c, for values of θ ranging from 0 to 90° at intervals of 18°. This was used initially for a specific case, with $N=30$, $A=70$ MHz, $\tau=0.1$ μs, $\Delta t=440$ μs, and $t_c=2.4$ ms, to determine the effect of varying f_m. (For each value of t_m, N_e was averaged over the range of starting phase θ.) The results of these computations, made at intervals of 10 Hz with f_m going from 210 to 1150 Hz is shown in Fig. 2. It is seen that N_e depends critically upon f_m. The values of f_m for which N_e is low corresponds to cases where there is a relation between f_m and $f_R = 1/\Delta t$ of the form

$$f_m = \frac{j}{i} f_R, \tag{27}$$

j and i being integers.

For example, since in this case $f_R = 2250$ Hz, $f_m = 2250/2 = 1125$ Hz corresponds to the case where the transmitted frequency is the same for every other pulse. This case is illustrated by Fig. 3, which gives photographs of the transmitter output spectrum. In 3(A) there is no dither applied; in Fig. 3(B) and (C) the dither frequency f_m is equal to one-half the PRF, f_R, but with two different relative phases.

There are many values of f_m for which N_e is as high as possible, so that an optimum value can be chosen on the basis of that giving the broadest tolerance in f_m for a given degradation of N_e. Fig. 4 is an expanded view of the region around $f_m = 988$ Hz, which is an optimum for these parameters. If we require N_e to stay within ± 10 percent of the peak value of 14.81 at 978.5 Hz, f_m must be held to a tolerance of about ± 1.2 percent.

Fig. 5(A) shows the spectrum at the optimum dither frequency; Fig. 5(B) and (C) was observed for dither frequencies at 1 percent above and below this, respectively.

The spectra of Fig. 6 were observed at $f_m = 3/5\, f_R$ for various phase relations between the dither waveform and transmitter firing.

To produce a set of data of general use, f_m was held at the optimum value

$$f_m = 0.4388\, f_R. \tag{28}$$

The computations were then performed in terms of the parameters $\Delta t / t_c$ and $A\tau / N$, which both lie in the range zero to one for most cases. N_e / N was computed for a wide range of values, and the results are given in Fig. 7.

Strictly speaking, this set of curves holds only for $N=30$. They hold with good accuracy, however, for a range of values $20 < N < 40$.

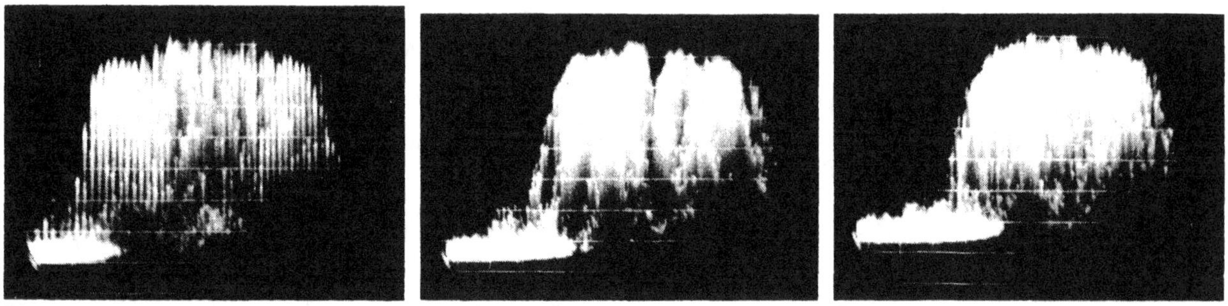

(A) (B) (C)

Fig. 5. Spectra for f_m close to optimum (one division = 5 MHz).

Fig. 6. Spectra for $f_m = 3/5 f_R$ (one division = 5 MHz).

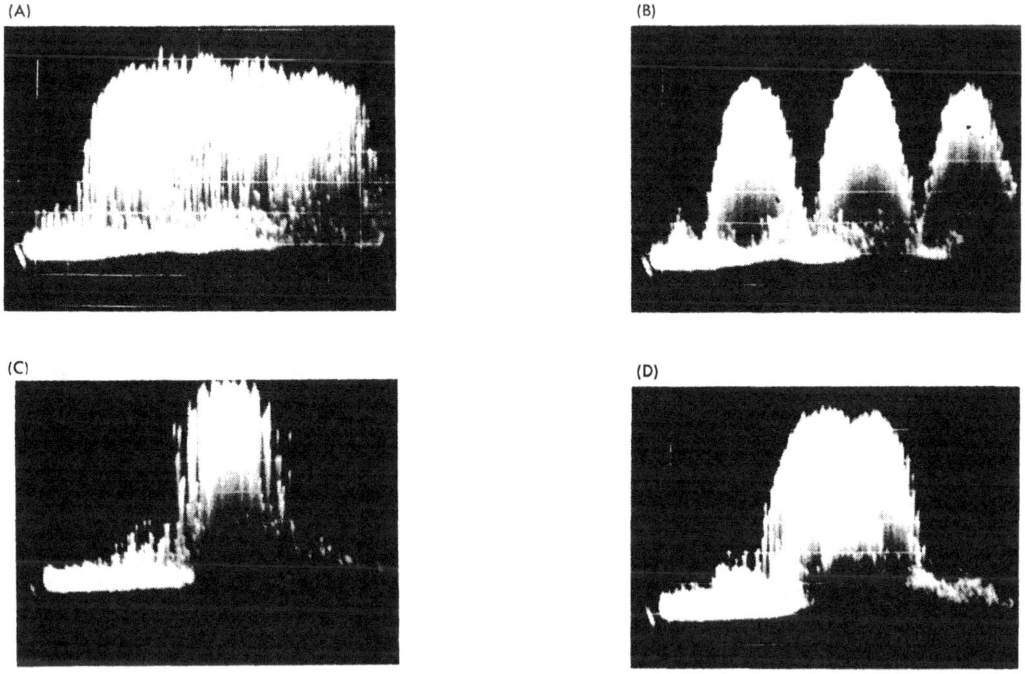

(A) (B) (C) (D)

Inspection of Fig. 7 shows that for $\Delta t \to 0$, i.e., when the clutter does not decorrelate significantly in time at a single frequency, $A\tau/N$ must approach 1.0 to achieve good overall decorrelation ($N_e/N > 0.8$). Thus $A \approx N/\tau$, which implies the frequency must shift by $1/\tau$ from each pulse to the next.

But if $\Delta t/t_c \approx 1$, i.e. the clutter decorrelates well in a radar period, the value of $A\tau/N$ does not have much effect.

A set of typical values might be

$$\Delta t = 500\ \mu s$$

$$t_c = 3.40\ ms$$

$$\tau = 0.1\ \mu s,$$

so that $\Delta t/t_c = 0.147$.

Then at the optimum f_m (877.6 Hz) if we require $N_e/N > 0.8$, we must make $A\tau/N > 0.320$, or $A > 96$ MHz. For $N_e/N > 0.5$, $A > 27$ MHz.

In any practical system we have to balance the detection improvement resulting from increased N_e/N against the cost of increasing A.

Conclusion

The decorrelation of sea clutter return resulting from interpulse frequency modulation has been analyzed for an idealized clutter model. As a criterion in judging the effective decorrelation of a sequence of pulses, the concept of an effective number of independent hits N_e was introduced. A set of curves relating N_e/N to the radar and clutter parameters has been computed for the sinusoidal modulation case.

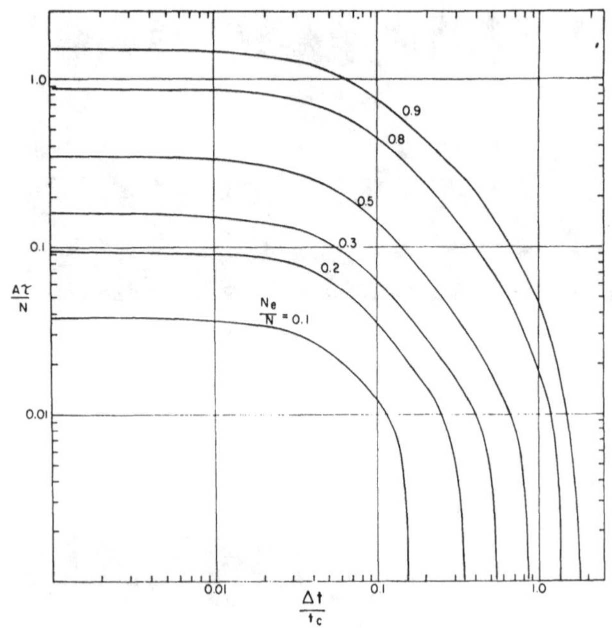

Fig. 7. Parameter trade-off characteristics.

REFERENCES

[1] H. Ray, "Improving radar range and angle detection with frequency agility," *Microwave J.*, p. 63, May 1966.

[2] D. Barton, *Radar System Analysis.* New York: Prentice-Hall, 1964.

[3] L. V. Blake, "A guide to basic pulse radar maximum range calculation," NRL Rept. 5868, December 28, 1962, p. 8.

[4] D. E. Kerr, *Propagation of Short Radiowaves.* Boston: Boston Technical Publishers, 1964, pp. 579–581.

[5] P. M. Woodward, *Probability and Information Theory.* New York: Pergamon, 1953.

[6] S. O. Rice, "Mathematical analysis of random noise," in *Selected Papers on Noise and Stochastic Processes*, N. Wax, Ed. New York: Dover, 1954.

[7] W. B. Davenport and W. L. Root, *Random Signals and Noise.* New York: McGraw-Hill, 1958.

ADDITIONAL REFERENCES

[8] H. M. Deutch, "Intrapulse frequency diversity for detection improvement with fluctuating targets or targets in clutter," *12th Tri-Service Radar Symp. Rec.*, August 1966.

[9] R. D. Hayes and J. L. Edwards, "Some statistical properties of radar sea returns," *12th Tri-Service Radar Symp. Rec.*, August 1966.

[10] F. E. Nathanson, "Report of radar clutter signal processing committee—Part I, Radar clutter effects," Applied Physics Lab., Tech. Memo. TG 842-1, September 1966.

[11] B. G. Gustafson and B-O. Ås, "System properties of frequency jumping radars," *Philips Telecommun. Rev.*, vol. 25, no. 1. July 1964.

Eric W. Beasley was born in Manchester, England, on February 28, 1925. He studied electrical engineering at the University of Manchester and received the B.Sc.Tech. and A.M.C.T. degrees in 1945.

From 1945 to 1952 he was a development engineer at Marconi's Wireless Telegraph Co., Chelmsford, England, engaged mostly in broadcast transmitter and radar receiver design. Emigrating to Canada in 1952, he joined the Canadian Marconi Company, Montreal, as engineering supervisor of the Aviation Department, where he participated in the development of a new line of avionics equipment. In 1957 he joined the Raytheon Company, Wayland, Mass., first as a group head of radar receiver design, and then as a staff engineer engaged in radar system analysis. This activity included the AN/FPS-28 and AN/SPS-55 radars, and 3D radar studies. He is presently a Member of the Technical Staff of the MITRE Corporation, Bedford, Mass., and is engaged in airborne radar system studies in the Radar Technology Department.

Harold R. Ward (S'53–A'54–M'60) was born in Lancaster, N. Y., on July 31, 1931. He received the B.E.E. degree from Clarkson College, Potsdam, N. Y., in 1953, and the M.S. in electrical engineering from the University of Southern California, Los Angeles, in 1957.

Since graduation, he has been employed by Hughes Aircraft Company, the U. S. Army, and Sylvania Electric Products where he has been engaged in electronic development and system design, analysis, test and evaluation. Since 1964 he has been a Principal Engineer at Raytheon's Wayland Laboratories, Wayland, Mass., and has served as a consultant on a number of radar systems problems.

Mr. Ward is a member of Tau Beta Pi and Eta Kappa Nu.

Measurement of Sea Clutter Correlation with Frequency Agility and Fixed Frequency Radar

G. Lind

Philips Telecommunication Review, Vol. 29, No. 1 (April 1970)
Pages 32-38

Summary

After a general presentation of frequency agility as a means of improving radar performance the paper discusses the advantage of frequency agility in a sea clutter limited detection case. Measurements are presented showing a strong reduction of clutter correlation with frequency agility compared to fixed frequency radar. The decorrelation measured will, in a typical case, let the frequency-agile radar tolerate 6 dB more sea clutter than the fixed frequency radar for the same detection performance.

1 Frequency agility radar

Radar as an instrument of measurement or vision is often compared with optical vision. There are a number of similarities but also a number of differences. One important difference is that radar operates in a small relative bandwidth, normally less than 0.1 %, whereas optical vision using back-ground light from the sun uses a relative band of about one octave.

The limited bandwidth of radar compared to the size of radar targets gives rise to a number of phenomena arising from interference between the partial reflectors of a complex target. The same type of phenomena occur in optics with light of extreme monochromaticity, i.e. with laser light, and are used for holograms.

In radar the interference phenomenon normally causes lower performance. Examples of this are lower range (fluctuation loss) and lower accuracy (glint error).

Frequency agility radar [1, 2] presents a solution to the above problem. By pulse-to-pulse random frequency modulation over an adequate band, all interference pattern effects such as fluctuation and glint error become very rapid. Their detrimental effects on performance are eliminated by the averaging that takes place on the PPI by detection and in the antenna servo by tracking. Increased jamming immunity is obtained due to the fact that the jammer must spread its power over the whole frequency agility band.

Second-time-around targets are automatically eliminated since the receiver is only tuned to the frequency of a transmitted pulse for one listening period.

Another advantage of frequency agility which is similar to the elimination of fluctuation loss is decorrelation of clutter echoes. This will be discussed in more detail in the next part.

2
The value of clutter decorrelation

In a clutter limited radar detection case, range is determined by the signal/clutter ratio. The signal/clutter ratio is, however, not improved by integration as much as it is in the noise limited case. With noise, the one pulse (IF) signal/noise ratio is improved by adding a number of pulses. The improvement is proportional to the number of pulses added, with a correction for the so-called integration loss.

If the clutter and the signal are steady during integration, no signal/clutter ratio improvement occurs. If the clutter is steady for a certain period longer than the pulse interval, this gives a lower number of independent clutter samples. The integration improvement will then be decided by this lower number instead of by the number of pulses.

With frequency agility, rain and sea clutter can be decorrelated more rapidly, thereby increasing the number of independent clutter samples and securing more integration gain. A numerical example shows this.

Data for a typical case:

Pulse repetition frequency	2000 pps
Time on target	10 ms
Hits per scan	20
Radar band	X
Probability of detection	90 %
Probability of false alarm	10^{-5}

The following values are measured data on sea clutter taken from chapter 4.

Number of independent clutter pulses for 20 transmitted pulses:

Fixed frequency	2.3 (measured values 2.2; 2.4)
Frequency agility	13.4 (measured values 11.6; 15.1)

With the formulas derived by MARCUM and SWERLING [3] a total improvement of 14 dB is calculated. This is to say that a frequency agility radar can tolerate 14 dB more clutter at the 90 % detection probability range.

Phrasing the same situation in a different manner, a frequency agile radar can see in the clutter at the 90 % detection probability range a 14 dB smaller target than a fixed frequency radar. Of this 14 dB, however, not all stems from the decorrelation of clutter. About 8 dB of gain can be referred to the decorrelation of the target [4], which leaves 6 dB for the clutter decorrelation.

This 6 dB is quite valuable. Sea clutter echo strength normally varies as R^{-3}, with R for range. Target echo strength varies as R^{-4}, so that signal/clutter ratio varies as R^{-1}. 6dB gain in signal/clutter ratio therefore means four times more clutter limited range.

3
Sea clutter measurements

A sea clutter measurement with fixed frequency and frequency agility took place on the Swedish west coast in 1964 [5]. The transmitter-receiver used was designed around Philips 200 kW spintuned magnetron YJ1180. The main data of this transmitter-receiver are:

Pulse length	0.5 μs
Pulse interval	525 μs
Pulse interval random jitter	±40 μs
IF amplifier	linear characteristic, bandwidth 3 MHz

The magnetron tuning band was 8.9-9.4 GHz. In frequency agility operation the frequency varied randomly from pulse to pulse over the band. In fixed frequency operation the same set was used and the transmitter could operate at any frequency within the tuning band. The antenna was a quarter cheese type designed for shipborne use. The beam widths were horizontally 1.4° and vertically 30°.

The radar antenna was situated approximately 80 metres above the sea level. The wind velocity was about 10 m/s. The wave height was 3-5 ft., sea state No. 3, moderate. During the measurement the antenna was kept non-rotating in such a direction that the radar beam was directed towards the wave front of the sea waves.

The video signal was connected to a range-gating unit. The gate was 1 μs. The video signal from the gate was taken through a pulse stretcher (or boxcar generator) to the y-plates of an oscilloscope. In the x-direction the oscilloscope beam was swept over the screen in 50 ms. The pulse stretcher served to facilitate observation of the amplitude of the echo-pulse on the oscilloscope screen. The screen of the oscilloscope was photographed with the exposure time adjusted to the time of one sweep, 50 ms. Two sets of measurements were taken, with range in one case 3 km and in the other 6 km.

A total of 50 photos were taken, distributed in the way indicated in Table I. Fig. 1 shows a selection of the photographs obtained.

Table I

Range km	Frequency	No. of photos	No. of pulses
3	Fixed	12	1080
3	Agility	10	900
6	Fixed	15	1350
6	Agility	13	1170

4 Data evaluation and results

The purpose of the evaluation was to compute the correlation function of sea clutter video for fixed frequency and frequency agility operation. To start with, each photo was manually transferred to a list of numbers.

The correlation function for an infinite sample is defined by LAWSON and UHLENBECK [6] as:

$$\varrho(\tau) = \frac{\overline{P(t)\,P(t+\tau)} - (\bar{P})^2}{\overline{P^2(t)} - (\bar{P})^2}, \tag{1}$$

where P(t) is the signal and a bar indicates averaging. For a finite sample an observed correlation function is defined as:

$$\varrho_{\text{obs}}(n) = \frac{\frac{1}{N}\sum_{j=1}^{N} P_j\,P_{j+n} - \left(\frac{1}{N}\sum_{j=1}^{N} P_j\right)^2}{\frac{1}{N}\sum_{j=1}^{N} P_j^2 - \left(\frac{1}{N}\sum_{j=1}^{N} P_j\right)^2}, \tag{2}$$

where the correlation time has been replaced by the more convenient experimental parameter n defined as equal to τ/T where T is the time between pulses.

Now equation (2) cannot be used for the whole observation material at one range with or without frequency agility. This total material comprises typically 12 photos and 1100 pulses. But the photos were taken with interruptions for film feed so that there are 12 independent series of 90 readings. The method used was to compute $\varrho_{\text{obs}}(n)$ for each photo according to equation (2), and then take the average of $\varrho_{\text{obs}}(n)$ for all photos of the same group.

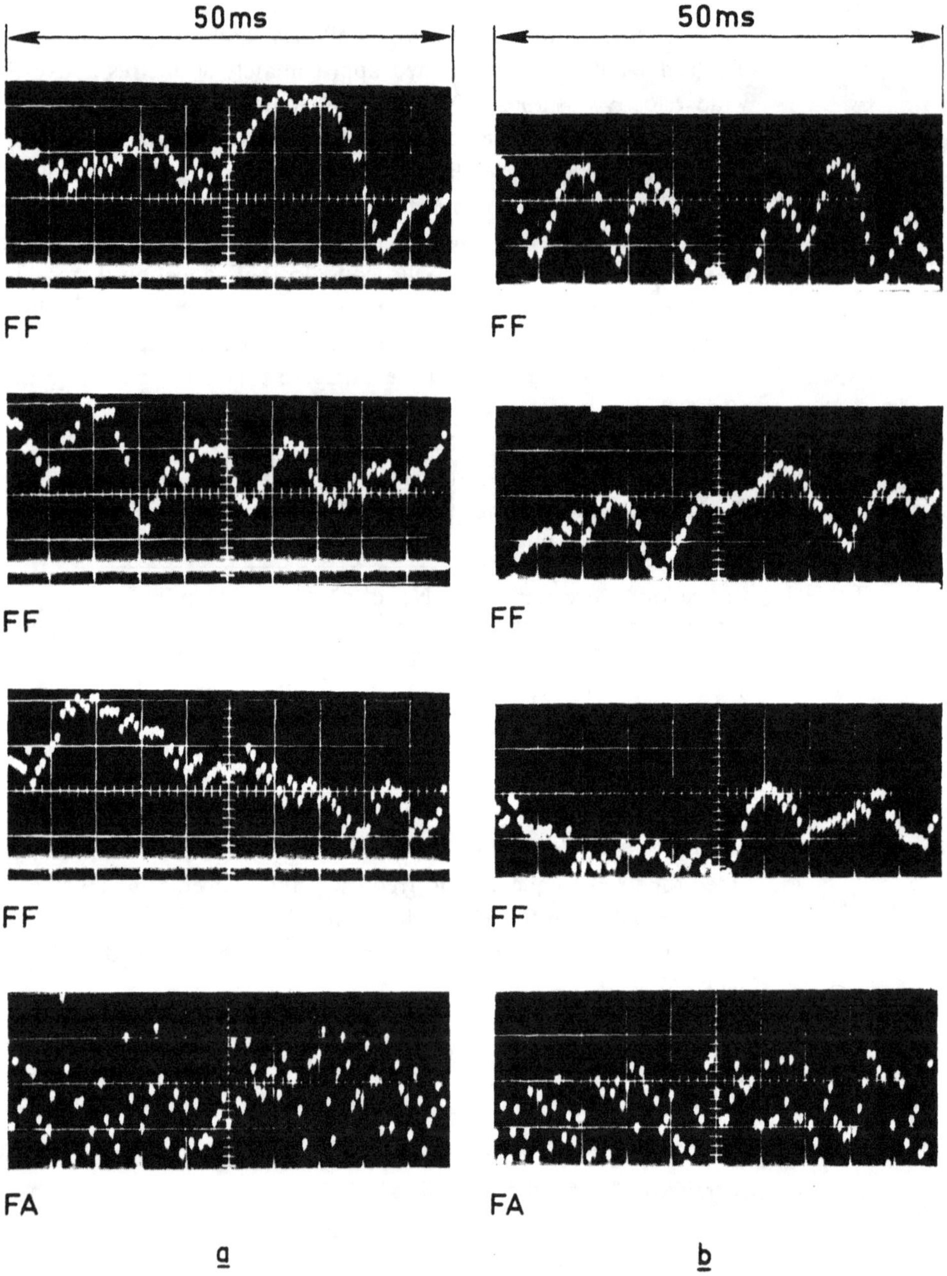

Fig. 1. Clutter echoes from range gate as a function of time; (a): range 3 km; (b) range 6 km; FF = fixed frequency; FA = frequency agility

$\varrho_{obs}(n)$ from one photo has a rather irregular behaviour due to the short sample, but the average seems to be fairly stable.

The computation was performed on a Bull/GE time-sharing terminal. Figs. 2-5 show the calculated average of $\varrho_{obs}(\tau)$ for $0 < \tau < 20$ ms. Correction has been made for receiver noise and numerical quantum noise using the method given by LAWSON and UHLENBECK [6].

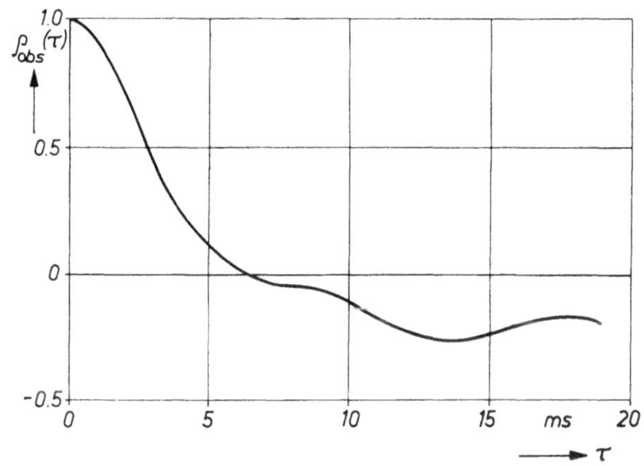

Fig. 2. $\varrho_{\text{obs}}(\tau)$ for fixed frequency, range 3 km

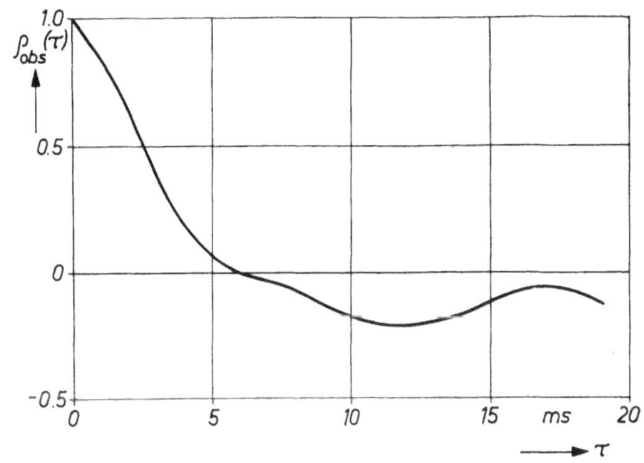

Fig. 4. $\varrho_{\text{obs}}(\tau)$ for fixed frequency, range 6 km

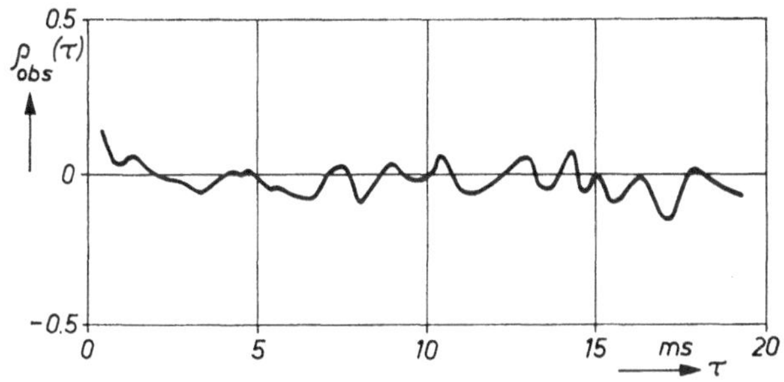

Fig. 3. $\varrho_{\text{obs}}(\tau)$ for frequency agility, range 3 km

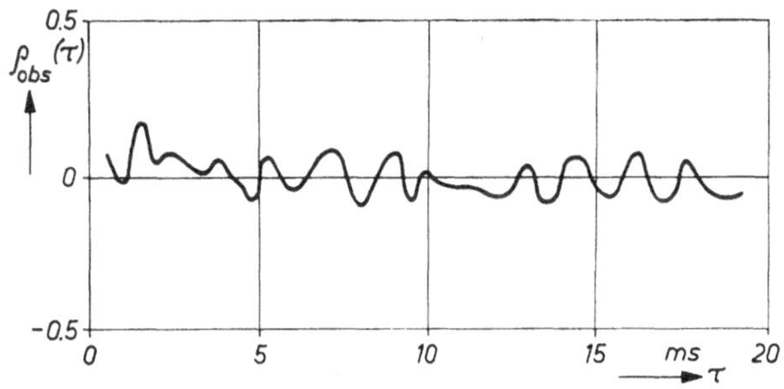

Fig. 5. $\varrho_{\text{obs}}(\tau)$ for frequency agility, range 6 km

The statistical significance of the results in Figs. 2-5 is fairly good. The standard deviation of the average is found to be approximately ± 0.07 for Figs. 2 and 4 (fixed frequency) and ± 0.03 for Figs. 3 and 5 (frequency agility).

From the curves a number of entities can be measured or calculated. For fixed frequency the root mean square correlation time σ_τ is estimated by approximating the correlation function by a Gaussian function.

The standard deviation of the spectral density function σ_f is related to the correlation time by the equation:

$$\sigma_f = \frac{1}{2\pi\,\sigma_\tau}.$$

The corresponding velocity standard deviation is related by the Doppler equation:

$$\sigma_v = \frac{\lambda\,\sigma_f}{2},$$

where λ is the radar wave length.

The results are shown in Table II.

The above results are in good accordance with earlier published results. Barton [7] cites four different measurements on sea echoes giving values of σ_v between 0.5 and 1.2 m/s.

The number of degrees of freedom for a certain number of pulses can also be calculated. If this number is defined as the number of independent pulses that gives the

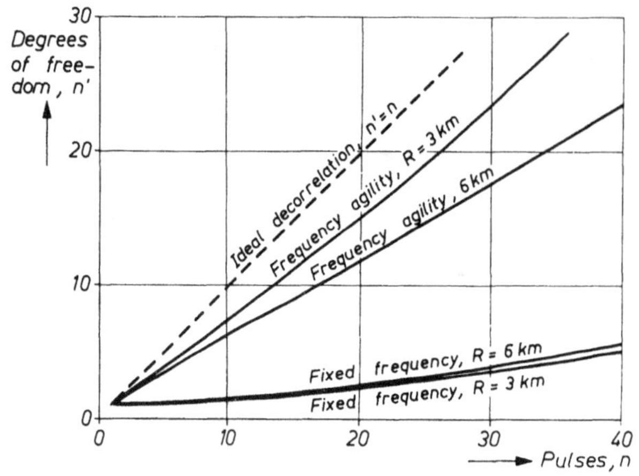

Fig. 6. Degrees of freedom (independant samples) as a function of the number of pulses, prf 1900

same relative variance after summing as the actual partially correlated case, it can be shown that:

$$n' = \frac{n}{1 + \frac{2}{n} \sum_{v=1}^{n-1} (n-v)\,\varrho_v}, \tag{3}$$

where n' is the number of independent pulses, n is the actual number of pulses and ϱ_v is defined by: $\varrho_v = \varrho\,(vT)$, where T is the time interval between pulses.
As an example the number of independent pulses has been calculated from the material measured with a prf of 1900 pulses per second. The result is shown in Fig. 6.
For the specific example in section 2, with 20 pulses, the result is shown in Table III. As seen in the figure, the number of degrees of freedom for frequency agility is lower at 6 km. This might be explained in the following way. It is supposed that nonideal decorrelation of clutter with frequency agility is caused by the fact that there are not always many clutter returns from the illuminated area. If there is only one clutter reflector in the illuminated area, no decorrelation occurs, and many reflectors give ideal decorrelation (complete decorrelation for a frequency shift of one inverse pulselength) [8].

Table II

Fig. no.	Range km	σ_τ ms	σ_f Hz	σ_v m/s
2	3	2.5	64	1.05
5	6	2.4	66	1.10

Table III

Fig. no.	Range km	Frequency	n'
2	3	Fixed	2.2
3	3	Agility	15.1
4	6	Fixed	2.4
5	6	Agility	11.6

At 6 km the illuminated area has two times the size of the area at 3 km. However, the total clutter response (and the density of clutter reflectors) is lower due to the lower grazing angle, which is 0.75° at 6 km and 1.5° at 3 km. As the variation in clutter return between 0.75° and 1.5° can be as much as 10 dB, this effect is greater than the increase in illuminated area, which would explain the lower decorrelation at 6 km.

5 Conclusions

The measurement and the evaluation show that frequency agility gives a substantial decorrelation of sea clutter under conditions similar to the experimental situation described in section 3. While fixed frequency radars at X-band have strong clutter correlation for about 3ms, a frequency agility radar gives efficient decorrelation between the pulses with correlation coefficients in the order of 0.05-0.10. As an example, a search radar with 20 hits/target will experience an improvement of 6 dB for 90 % target detection probability due to the clutter decorrelation. To this improvement another 8 dB will be added from the target decorrelation in comparing fixed frequency and frequency agility performance.

6 Literature

1 B. G. GUSTAFSON and B-O. ÅS: System properties of jumping frequency radar, *Philips Telecommunication Review*, Vol. 25, No. 1, pp. 70-76, 1964.

2 G. LIND: Reduction of radar tracking errors with frequency agility, *IEEE Trans. on Aerospace and Electronic Systems*, Vol. AES-4, No. 3, pp. 410-416, May 1968.

3 J. I. MARCUM and P. SWERLING: Studies of target detection by pulsed radar, *IRE Trans. on Information Theory*, Vol. IT-6, April 1960.

4 P. F. GUARGUAGLINI: A unified analysis of diversity radar systems, Correspondence, *IEEE Trans. on Aerospace and Electronic Systems*, March 1968, pp. 318-320.

5 J.-O. THULIN and N. WÄSTERLID: Studies of Clutter and Comparison of Contrast between Targets and Clutter Background with Fixed Frequency and Frequency Agility Operation. Philips Teleindustri AB, Lab. Report 2700-R36, March 17, 1967.

6 J. L. LAWSON and G. E. UHLENBECK: Threshold signals, MIT Radiation Lab. Series No. 24, McGraw-Hill, New York 1950, pp. 132-136.

7 D. K. BARTON: Radar System Analysis, Prentice-Hall, N.J., 1964, page 100, Table 3.3.

8 E. W. BEASLEY and H. R. WARD: A Quantitative Analysis of Sea Clutter Decorrelation with Frequency Agility, *IEEE Trans. on Aerospace and Electronic Systems*, Vol. AES-4, No. 3, May 1968, pp. 468-473.

A Method to Calculate the Density of Sea Clutter Echoes at Low Grazing Angles

G. Lind and B. Bergkvist

IEEE International Radar Conference (1975)
Pages 230-234

ABSTRACT

At low grazing angles, shadowing will obscure some of the waves of the sea, while others are seen by the radar and cause sea-clutter reflections. A method of calculating the density of echoes is suggested, based on a computer simulation of the sea surface.

The results can be used for a radar with a certain resolution cell to assess whether the sea clutter is best described as point clutter or as distributed clutter. This influences the degree of lognormality (or spikiness) and is important in evaluating designs such as increased resolution, multiple bin or non-parametric CFAR-systems and frequency agility.

INTRODUCTION

Sea clutter returns set the performance limits in a great number of naval radar applications. Therefore, ever since the beginning of radar technology great efforts have been made to give descriptions of the sea environment suitable for design and performance prediction.

The present paper deals with sea clutter at very low grazing angles, which occur at interesting ranges for radars located 5-30 m above the sea surface. This case includes a great portion of existing radars, e.g. navigation radars for merchant ships, radars onboard naval ships and radar seekers in sea skimming missiles.

One aspect of sea clutter which has received great attention in the last years is the questions of deviations from a Rayleigh distribution of the sea clutter return. Measured probability distributions have been observed to have higher density of large returns than commensurable with a Rayleigh description. The phenomenon is often referred to as lognormality due to the fact that measured data in these cases often can be described by a lognormal distribution. Lognormality has been reported to occur especially in systems using very short pulse lengths. Nathanson (Ref. 13) discusses indications from a number of measurements and concludes that a pulse length of 1 μs will give a Rayleigh distribution but a much shorter pulse length, say 0.1 μs will lead to a lognormal distribution.

It is proposed that at low grazing angles the phenomenon called lognormality of sea clutter is mainly caused by shadowing in combination with high resolution. Shadowing causes the sea surface to be broken up into subsurfaces. Some of these surfaces will be visible from the radar and thus can give rise to sea clutter return. Other surfaces will be masked by the sea between the observation point and the surface, and these will not give any contribution to the total sea clutter return. This gives a point-reflector structure instead of a distributed reflector structure.

With low resolution, i.e. a wide azimuth beam and a long pulse there will be many reflecting areas within the resolution cell. The reflector distribution will then be Rayleigh.

With high resolution, i.e. a narrow beam and a short pulse, there will be few reflecting areas within the resolution cell and deviation from a Rayleigh distribution can be expected.

With a given resolution of the radar, the number of reflecting areas within the resolution cell will depend on the grazing angle and the sea state. The lower the grazing angle is, the higher is the probability that a wave crest will be in the shadow of other wave crests, and the lower will the average density of reflecting areas be. The higher the sea-state, the longer is the average distance between wave crests, which also leads to a lower density of reflectors. From the above, low grazing angle and high sea state will increase the tendency of deviation from a Rayleigh distribution of the sea clutter.

The paper will discuss this mechanicm. A simulation model of the sea surface will be used to obtain approximate values of the average reflector density as a function of grazing angle and sea state.

The average reflector density (reflectors/km^2) can be used with a given resolution to calculate the average number of reflectors in the resolution cell. If this number is small (<< 1) the clutter is point clutter, if it is large (>> 1) the clutter is distributed.

Consequences if the clutter is point clutter or distributed are discussed. Among others, the following factors are influenced

- the value of increased resolution
- the function of CFAR techniques
- the value of broadband techniques or frequency agility

A more general treatment of sea clutter at low grazing angle would have to take not only shadowing, but also multipath, into account. Both effects are certainly important. Only shadowing has been treated here, because only a first order approximation of reflector density has been sought. This is still an improvement to the zeroeth order approximation that can be found in several places in the literature, which only compares the pulse length to the wavelength of the sea, but does not take the grazing angle into consideration.

SEA SURFACE AT VERY LOW GRAZING ANGLES

General

In order to get a general feeling for how a sea surface may appear at very low grazing angles, a simple model as illustrated by figure 1 has been evaluated. An observer is placed at a point A. At short ranges he will be able to see most of the corrugated "sea surface", but at longer distances the greater waves may mask an increasing part of it. Shadowing effects will occur if waves with steeper slopes than the observation angle exist (ref. 5). Cox and Munk (ref. 12) have measured the rms slope of waves and their observations show that this is normally much greater than the actual grazing angles for radar sites discussed in this paper.

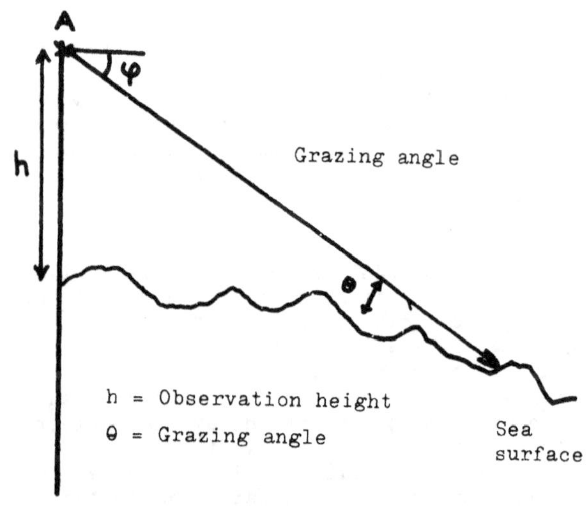

Fig. 1. Simulation model geometry

A wind velocity of 10 m/s should for instance generate an rms slope of 7° for the long wave-length part of the wave spectrum (gravity waves) which determines the large scale shadowing properties of the surface. For radars, observing the sea surface at lower grazing angles, the surface will be broken up in a number of separated subsurfaces. The total size of these surfaces as well as their number within a given range interval will decrease with the grazing angle and thus with the distance. It is thus likely that shadowing effects will create a similar point reflector environment for ship borne radars as the one encountered by extreme resolution radars at higher angles.

In order to get a rough idea of how the reflector density will vary with range and sea state a model based on the simple geometry of Fig. 1 was implemented in a digital computer.

Simulation model

For simplicity, the study was made using a one-dimensional model of the sea in the upwind-downwind direction. The extension to two dimensions is discussed below.

The spectrum of a sea surface is still a controversial subject among the oceanographers. Several forms have been proposed (reference 10, 11) but the number of actual measurements available are few (reference 9). Since they all have a resonance peak in the spectrum, the simplest nontrivial model would be a single resonance circuit with some Q-value, fed with white noise. The same model has been used by some oceanographers (reference 3) in the study of ship´s rolling.

The "sea surface" was thus generated by filtering white noise by a second order filter. The filter output (the waves) was defined by the resonance frequency (wave number), Q value and its rms output signal.

The Q value was chosen to be 1.25, which gave a fairly good fit to available experimental data (reference 9) and simultaneously a rather wavelike appearance of the output. Runs with different Q values ranging from 0.7 to 3 showed, that the exact value of the Q was of minor importance for the reflector density which probably partly was due to the fact that only large scale variations of the surface were evaluated.

The resonant frequency of the filter was chosen to give a correct wave length for the actual simulated sea state. The associated wave height was obtained by setting an appropriate filter gain. The data used are illustrated in table 1.

Table 1

Sea state	Wave length m	Wave height σ-value m	Q
3	25	0.3	1.25
5	65	0.75	1.25

From an observation point A a vector was pointed at the sea surface (Fig. 1), which was curved in order to take earth curvature and radar refraction into account.

Every time the vector angle (φ) was increased above an earlier maximum value, the range as well as the size of the unshadowed surface was registered. The sampling rate of the digital filter was matched to obtain a good simulation of the large scale sea ondulations, which were assumed to be of importance for shadowing. The high frequency ripple of the process was eliminated by joining all observable subsurfaces with "gaps" less than 5 m (20 noise inputs to the filter).

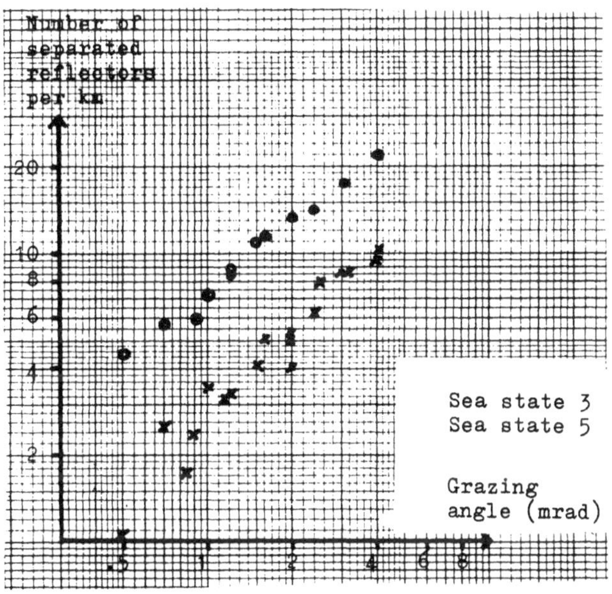

Fig. 2. Number of reflectors as a function of grazing angle.

The surface was observed in the range interval 0 to 10 km from an observation height of 5, 10 and 15 m. Observations were made with data simulating sea state 3 and 5. 10 runs were made for each parameter and the average reflector density was plotted as a function of actual grazing angle in figure 2.

Extension to two dimensions

The number of observable surfaces is, as can be seen from Fig. 2, directly proportional to the grazing angle. The average linear distance between reflectors at sea state 3 is 15 m for a grazing angle of 10 mrad and 150 m for a grazing angle of 1 mrad. For sea state 5 the figures are 35 and 350 m, respectively.

The extension to two dimensions requires knowledge of the coherency length of sea waves. This is determined by the variation of the direction of travel of actual sea waves. There exist few reliable measurements of this parameter in the oceanographic literature. We have used the estimate that the coherency length is about two wavelengths (Ref. 1, 9). The number of reflections in a resolution cell is then calculated by assuming stripes according to the linear model above, at a crosswind distance of two seawave wavelengths, each giving an independent set of reflections.

According to this, the surface required for 1 reflector should be 370 and 3700 m^2 for sea state 3 and 10 resp 1 mrad observation angle and 2250 resp 22500 m^2 for sea state 5 at the same angles.

For the case of the radar looking crosswind, no model has been developed. We assume, however, that properties such as reflector density are not dramatically changed.

Examples

Two examples are shown, both with data which could be used by a navigation radar.

Data for example 1

Antenna height:	15 m
Distance:	3 nautical miles
Angular resolution:	2°
Range resolution:	0.5 μs

The radar looks upwind.

The grazing angle is calculated to be (assuming 4/3 equivalent earth radius) 2.4 mrad.

The density of reflectors at this grazing angle is calculated, with results from above, to be one reflector per

$1.6 \cdot 10^3$ m^2 at sea state 3

$9.5 \cdot 10^3$ m^2 at sea state 5

The size of the resolution is $1.5 \cdot 10^4$ m^2.

The average number of reflectors is then

at sea state 3 9.3

at sea state 5 1.5

Data for example 2

are the same as for example 1 except for resolution, which is

Angular resolution 0.8°

Range resolution 0.1 μs

The average number of reflectors is then

at sea state 3 0.8

at sea state 5 0.1

We see that in example 1 the average number of reflectors in both sea-states is higher than one. We can expect the sea-clutter to behave like distributed clutter,

32

at least in sea state 3, where there are almost ten reflectors in the resolution cell.

By increasing the resolution of the radar as in example 2, we get an average number of reflectors less than one in both sea-states. We can then expect the sea-clutter to be point-clutter, with all the consequences of that, especially in sea-state 5 where there is an average density of 0.1 reflectors per cell. This also means that on the average there is sea clutter only in 10 % of the resolution cells and no-clutter in 90 % of the cells (with the exception of the weak clutter echoes resulting from diffraction and sidelobes).

DISCUSSION AND CONCLUSIONS

The paper has presented a method, whereby the average number of sea clutter reflectors in the resolution cell can be estimated, if radar resolution data, antenna height, range and sea state are given. Of special interest are two cases, distributed clutter with many reflectors in the cell and point clutter with on the average less than one reflector in the cell. Two examples with realistic data have been given, one resulting in distributed clutter, the other in point clutter.

What are the consequences? First it is obvious that with many reflectors in the cell, the clutter is bound to be close to Rayleigh distributed, since a Rayleigh distribution always results from many reflectors of approximately the same size. Point clutter, with on the average less than one reflector in the cell, can on the other hand be expected to give deviation from a Rayleigh distribution, lognormal or other.

Secondly, the performance of a number of proposed designs is influenced. We will here shortly discuss three, namely increased resolution, multiple bin or nonparametric CFAR receivers and frequency agility.

Increased resolution is generally considered to be a method of improving the signal to clutter ratio and thereby make it possible for the radar to detect smaller targets. This is, however, only certain as long as the clutter is distributed. At some limiting resolution the clutter becomes point clutter, and then further increases in resolution will only decrease the proportion of the resolution cells where clutter exists. Detection and tracking of small targets through the clutter is then in some cases possible (interclutter visibility), especially when the targets are moving much faster than the clutter and leave visible trails on an indicator. In other cases, with fixed or slowly moving targets (buoys or fishing boats to be seen with a navigation radar) it is no longer possible to detect smaller targets. The limit beyond which further resolution improvement will not always pay off can be estimated by the method given in this paper.

To resolve the situation when the clutter distribution is no longer Rayleigh, and to counter the increase in false alarms due to the longer tails of distribution such as the log-normal, special types of robust CFAR receivers have been proposed, e.g. the multiple bin or nonparametric CFAR recievers (Ref. 4). Receivers of this type rely on using a number of adjacent resolution cells to set the threshold or in other keep a constant false-alarm ratio. As we have seen above, log-normal or other non-Rayleigh types of sea clutter echoes have (in the low-grazing-angle case) low density. Accurate statistical estimates of the tail of the distribution, which is necessary in order to control the false-alarm ratio, requires a great number of bins. This is not always observed in the analysis of this type of receiver.

Finally, some CFAR receivers of the type described above need decorrelation of the clutter between pulses for their proper function. One way to get decorrelation is to use frequency agility. However, in just that case when decorrelation from frequency agility is needed by the above mentioned receivers, it becomes more difficult to obtain. When the clutter is distributed with many reflectors spread over the whole resolution cell, a frequency shift of one inverse pulse length is enough to give complete decorrelation. With point clutter, there is often one reflector in the resolution cell. If this reflector is much smaller than the cell itself, which is quite common, e.g. in the simulated cases discussed in this paper, a much wider frequency shift than one inverse pulse length is required. Thus, wide-band frequency agility becomes desirable for the above described CFAR receivers. In extreme cases, the clutter reflectors might be so small in depth that only partial decorrelation can be obtained at all.

The method of estimating the boundary between the point-clutter case is approximate and based on simulations only. It is proposed that improved estimates are obtained by measurements.

REFERENCES

1. Robert S. Arthur: "Variability in direction of wave travel," Annals New York Academy of Sciences, V. 51, 1949, pp 511-522.

2. H.C. Freisleben: "Untersuchung von Seegangreflexen (radar) im Nordatlantik," DGON Heft II 1968.

3. Voznessensky and Firsoff: "Statistics of data concerning rolling of ships," Proceedings Symp on the Behaviour of Ships in a Seaway. Vol. 1. Wageningen 1957, pp 878-881.

4. G. M. Dillard, C. E. Antoniak: "A Practical Distribution free Procedure for Multiple Range-Bin Radars," IEEE Trans. AES-6, No. 5, Sept. 1970, pp 629-635.

5. Peter Beckman: "Shadowing of Random Rough Surfaces," IEEE Trans. AP-13, May 1965, pp 384-388.

6. G. V. Trunk: "Radar Properties of Non Rayleigh Sea," IEEE Trans. AES-8, No. 2, March 1972, pp 196-204.

7. Norman W. Guinard, John C. Daley: "An Experimental Study of a Sea Clutter Model," Proc. IEEE, Vol. 58, No. 4, April 1970, pp 543-550.

8. J. Croney: "Clutter and its reduction on shipborne radars," Proc. Conf. on Modern Radar, London Oct. 1973, pp 213-220.

9. B. Kinsman: "Wind waves," Prentice Hall, Englewood Cliffs NJ, 1965.

10. Phillips: "The dynamic of the upper ocean," Cambridge University Press, 1966.

11. Gerhard Neuman, Willard J. Pierson, Jr: "Principle of Physical Oceanography," Prentice Hall Inc., 1966.

12. Cox Munk: "Measurements of the roughness of the sea from photographs of the sun´s glitter," J Opt. Soc. Amer., Vol. 44, Nov 54, pp 838-850.

13. F. Nathanson: "Radar Design Principles," Mc Graw Hill Book Company, 1969.

Section V
Glint and Error Reduction by Diversity

Use of frequency agility or diversity to improve the angular accuracy of tracking radar has been a subject of considerable work in the United States, starting with that reported in the 1963 papers of Birkemeier and Wallace (Reprint 5-1) and of Kosowsky et al. (Reprint 5-2). Early European work was reported by Gustafson and Ås (Reprint 1-1), and a quantitative theory was given by Lind (Reprint 5-3). In addition to this published work, Kosowsky refers to a 1959 report by Shapiro on angular resolution improvement, and early reference to diversity techniques was made in a report by Weimer and Peters. An excellent summary of the historical developments in the area of glint modelling and reduction is given by Sims and Graf (Reprint 5-5), which cites the Weimer and Peters report as a reference. That summary, along with the Lind paper and subsequent Loomis and Graf work (Reprint 5-6) is sufficiently complete that no detailed survey is needed here.

Two subjects related to glint have received somewhat less thorough discussion. The effect of amplitude scintillation on measurements by scanning radars was described by Ray (Reprint 1-2), who showed an advantage for agile systems. Birkemeier and Wallace note the increased scintillation component introduced by agility in conical-scan systems. This subject is carried further by Guarguaglini (Reprint 5-4), with typical examples comparing monopulse and conical scan trackers. The conical-scan errors may be larger with agility than at fixed frequency for longer ranges where glint is not the dominant error component.

The other related subject concerns multipath errors, which are especially important in low-angle tracking. This area was covered by papers in Volume 4 of *Radars*, particularly those of Symonds and Smith (1973) and of Barton (1974), cited in the bibliography below.

An unresolved problem is the prediction of error reduction through use of agility. The 1972 correspondence by Lind, cited in the bibliography, provides a quantitative method of calculating improvement, based on the amount of spectral spreading of an assumed constant glint variance. Nicholls, in a 1975 correspondence, deduces by analogy to phase characteristics of narrowband noise that only about 10 dB of error reduction is available for the widest agile bandwidth. His result depends upon the assumption that agile operation generates an additional, wideband component rather than simply redistributing the constant variance of the target that would be seen by a fixed-frequency system. At this writing, the limit of improvement is not clearly established, and the effect of possible variation in the power-averaged "center of gravity" of the multiple scatterers is also uncertain. Practical improvements on the order of 10 dB (a reduction factor of 3.2 in angular error) appear attainable, but more work is needed to determine whether greater factors are available.

ANNOTATED BIBLIOGRAPHY

Berkemeier, W. P.; and Wallace, N. D.: "Radar Tracking Accuracy Improvement by Means of Pulse-to-Pulse Frequency Modulation," *AIEE Transactions on Communications Electronics, Vol. 81, January 1963*, pp. 571-575. Considers effects of linear frequency shift (giving N independent target samples before repeating) on glint error of trackers and scintillation error in conical scan, assuming uniform distribution of scatterers over target. Reprint 5-1.

Kosowsky, L. H.; et al.: "The Reduction of Angle-of-Arrival Scintillation by a Frequency Shifting Techniques," *Proceedings of the NEC, Vol. 19*, 1963, pp. 52-71. Presents theory and experimental data on apparent angle-of-arrival variations from terrain, with and without frequency shift. Reprint 5-2.

Lind, G.: "Reduction of Radar Tracking Errors with Frequency Agility," *IEEE Transactions on Aerospace and Electronic Systems, Vol. AES-4, No. 3*, May 1968, pp. 410-416. Derives spectra of glint error and of amplitude scintillation when frequency agility is used and gives typical results on error reduction. Reprint 5-3.

Barton, D. K.; and Ward, H. R.: *Handbook of Radar Measurement*, Prentice-Hall, 1969. Gives correlation functions in frequency and time for glint and scintillation, and shows error reduction for search radar with agility.

Guarguaglini, P. F.: "Tracking Errors in a Frequency Agility System," *AGARD Conference Proceedings No. 66 - Advanced Radar Systems*, May 1970. The correlation function and spectrum of target scintillation are derived for random frequency selection with uniform distribution over the tuning band. Typical errors for monopulse and conical scan trackers are presented. Reprint 5-4.

Sims, R. J.; and Graf, E. R.: "The Reduction of Radar Glint by Diversity Techniques," *IEEE Transactions on Antennas and Propagation, Vol. AP-19, No. 4*, July 1971, pp. 462-468. This paper includes an extensive bibliography and historical review of glint modelling, along with description and results of a computer simulation for a multiple-source scattering model. Reprint 5-5.

Lind, G.: "A Simple Approximate Formula for Glint Improvement with Frequency Agility," *IEEE Transactions on Aerospace and Electronic Systems, Vol. AES-8, No. 6*, November 1972, pp. 854-855c. Based on the author's work in a 1968 paper, a simple formula is given for the factor by which the low-frequency value of glint spectral density is reduced by agility over a given bandwidth.

Symonds, M. D.; and Smith, J. M.: "Multi-Frequency Complex-Angle Tracking of Low-Level Targets," *IEE Conference Publication No. 105*, October 1973, pp. 166-171. Gives results of simulation in which frequency diversity is used to resolve ambiguity in complex angle processing of targets in a multipath environment. Reprinted in *Radars* (Volume 4 - *Radar Resolution and Multipath Effects*) Artech House, 1975.

Barton, D. K.: "Low-Angle Radar Tracking," *Proceedings of the IEEE, Vol. 62, No. 6*, June 1974, pp. 687-704. Gives calculated time-delay distribution of typical multipath reflections, indicating limited advantage of frequency diversity in reducing multipath error in tracking. Reprinted in *Radars* (Volume 4 - *Radar Resolution and Multipath Effects*) Artech House, 1975.

Loomis, J. M.; and Graf, E. R.: "Frequency-Agility Processing to Reduce Radar Glint Pointing Error," *IEEE Transactions on Aerospace and Electronic Systems, Vol. AES-10, No. 6*, November 1974, pp. 811-820. Discusses implementation and effectiveness of several procedures for selecting frequencies and combining the resulting measurement samples, suggests selection of the sample with largest amplitude as the preferred approach. Reprint 5-6.

Nicholls, L. A.: "Reduction of Radar Glint for Complex Targets by Use of Frequency Agility," *IEEE Transactions on Aerospace and Electronic Systems, Vol. AES-11, No. 4*, July 1975, pp. 647-650c. By analogy to phase characteristics of narrowband and wideband noise, a maximum reduction of 9.8 dB in zero-frequency spectral density of glint is predicted for frequency-agile radar, as compared to fixed-frequency radar.

Radar Tracking Accuracy Improvement by Means of Pulse-to-Pulse Frequency Modulation

W.P. Birkemeier and N.D. Wallace

AIEE Transactions on Communications Electronics, Vol. 81 (January 1963)
Pages 571-575

Summary: Angular scintillation tracking noise for simplified radar target models has been analyzed by Delano, Muchmore, Freeman, and others for monofrequency pulse radars. This paper extends their work to the case of pulse-to-pulse frequency-modulated radars and demonstrates that such radars offer a significant reduction in angle tracking error. Tracking improvement is derived as a function of tracking servo bandwidth, scintillation spectrum, and FM (frequency modulation) regimen. A quantity, Δf_c, is defined as the minimum pulse-to-pulse frequency shift required to achieve maximum tracking improvement, and is derived as a function of target model and radar parameters.

Paper 61-1114, recommended by the AIEE Radio Communication Systems Committee and approved by the AIEE Technical Operations Department for presentation at the AIEE Fall General Meeting, Detroit, Mich., October 15–20, 1961. Manuscript submitted July 18, 1961; made available for printing January 16, 1962.

W. P. Birkemeier is with the University of Wisconsin, Madison, Wis., and N. D. Wallace is with the Emerson Electric Company, St. Louis, Mo.

The authors would like to express their appreciation for the many helpful discussions with co-workers at the Emerson Electric Company, and especially to Allen Furfine, whose suggestions and comments contributed greatly to the paper.

NOISE components in radar tracking error arising from phase interference of the reflected signals from various points on the target have been investigated by several authors: Muchmore,[1] Freeman,[2] Delano,[3] Dunn,[4] and Brockner.[5] In order to standardize the notation and nomenclature, the results of some of these authors will be reviewed.

There are two causes of target generated noise. The amplitude of the returned wave is known to fluctuate, and the apparent center of reflectivity of the target wanders randomly. These two sources of noise have been known collectively as scintillation. The wander of the apparent center of reflectivity of the target is known as glint and the perturbations in the received signal due to glint have been given the notation $U(t)$ by Delano and Muchmore. In these studies glint has been considered to be due to phase interference between elements of the target with gross cross-sectional changes occurring slowly enough to allow the glint to be considered stationary. The elevation and azimuth axes are considered independent of each other with a target model as shown in Fig. 1 for the azimuth axis. The elements of the target are considered to be uniformly distributed along the strip, each with an independent random phase angle uniformly distributed from zero to 2π. For the model of Fig. 1 located at a fixed range and undergoing constant rotational motion through small angles at the rate Ω_o, the variance of $U(t)$ can be shown to be independent of Ω_o while the width of the spectrum of $U(t)$ is proportional to Ω_o. Increased target turning rate thus results in decreased spectral density at low frequencies.

Specifically, Muchmore[1] and Freeman[2] have shown the spectrum $W_U(f)$ for the case of constant turning motion to be

$$W_U(f)=\frac{\overline{U^2}}{2f_m}\left(2-3\frac{f}{f_m}+\frac{3f^2}{f_m^2}-\frac{f^3}{f_m^3}\right),\quad f\leq 2f_m \tag{1}$$

$$=0 \quad \text{elsewhere}$$

where $\overline{U^2}$ is the mean-square value of $U(t)$

$$f_m=\frac{f_0 L\Omega_0}{c} \tag{2}$$

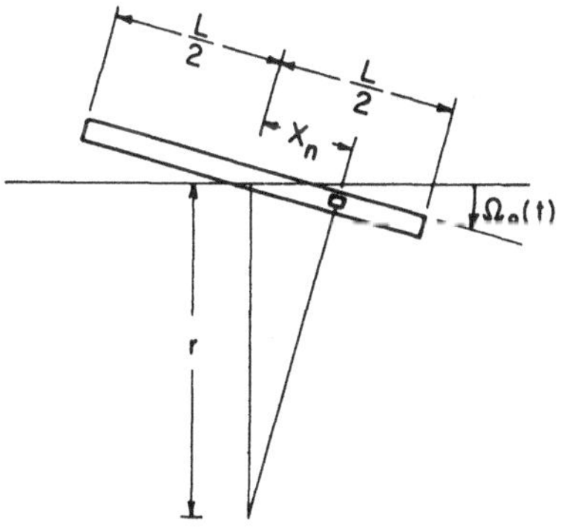

Fig. 1. Line target model

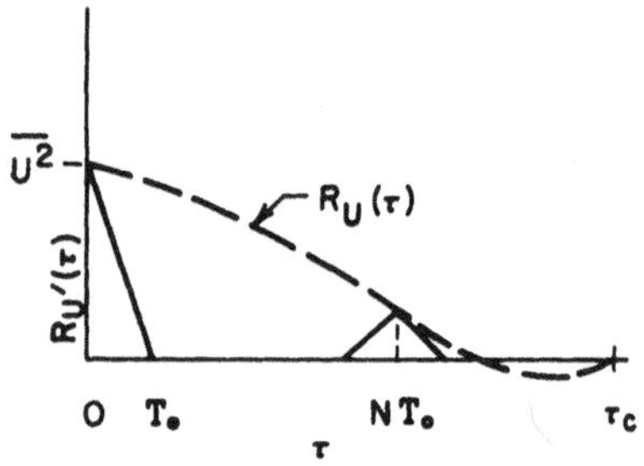

Fig. 2. Glint autocorrelation function with FM

c is the velocity of light, and f_o is the radar operating frequency. For a target of length $L=50$ feet rotating at 0.1 radian/second and an operating frequency $f_o=10^{10}$ cps (cycles per second), this yields $f_m=50$ cps.

Delano[3] points out that, for a system with an AGC (automatic gain control) that is slow compared with the fluctuations in U, the servo differential equation has constant coefficients, with $U(t)$ as the forcing function. Under these conditions, the variance in the tracking error due to glint for a servo whose transfer function is $Y(f)$ is given by

$$\sigma_U^2=\int_0^\infty W_U(f)|Y(f)|^2df \tag{3}$$

For a narrow band servo where the noise can be considered flat across the servo passband, equation 3 simplifies to

$$\sigma_U^2=W_U(0)\int_0^\infty |Y(f)|^2df=W_U(0)B \tag{4}$$

where B is the servo equivalent noise bandwidth.

The phase interference fluctuations in the rectified amplitude of the returned wave $E(t)$, known as fading, also perturb the tracking function. In both monopulse and conical scan tracking systems, this low-frequency fading noise is troublesome; however, in the conical scan case, noise components of $E(t)$ near the conical scan frequency f_s cannot be distinguished from the actual tracking modulation and so must also be considered. Muchmore[1] has shown the spectrum of $E(t)$ for constant turning motion to be

$$W_E(f)=\frac{(\overline{E^2}-\bar{E}^2)}{f_m}\left(1-\frac{f}{2f_m}\right),\ f\leq 2f_m \tag{5}$$
$$=0 \text{ elsewhere}$$

where f_m is given by equation 2. Delano[3] has shown that $E(t)$ and $U(t)$ are linearly independent random variables for a target symmetrical about the tracking axis. Consequently, the total variance in the tracking error due to target scintillation is the sum of the variances due to glint and fading. Assuming a narrow servo passband, the variance due to $E(t)$ for a conical scan system has been shown to be[6]

$$\sigma_E^2=\epsilon^2W_E(0)B+2W_E(f_s)B \tag{6}$$

where ϵ is the tracking error. The factor of 2 in the right-hand term in equation 6 is obtained because frequencies above and below f_s contribute noise. It can be seen from equation 6 that for a close tracking system with additional AGC smoothing in the low-frequency region these components will tend to be small. Subsequently, therefore, the specific analysis concerning low-frequency noise will be applied to glint; however, all remarks concerning glint can be applied equally well to low-frequency fading noise.

Since the tracking error variance in a monopulse system is due to low-frequency noise arising from slow, target-induced, phase-interference phenomena, it seems likely that such phase interference can be made to occur more rapidly by progressively varying the radar carrier frequency. The same noise variance would therefore be expected, but with a spectral density spread over a wider band. This would result in less spectral density at low frequencies and improved tracking.

For a cw (continuous wave) radar, the faster the carrier frequency f_o is varied, the more rapidly the fluctuations in $U(t)$ and $E(t)$ should occur. However, with a pulsed radar which is analyzed here, f_o need only be changed sufficiently between pulses to insure linear independence of the video pulse amplitudes comprising the error signal. This change in f_o will be called the critical frequency change, Δf_c.

In this paper, the implications of frequency changing regimens on the unimproved tracking for both monopulse and conical scan tracking systems are discussed, and the critical frequency change for a simplified target model is derived.

Tracking Improvement With Frequency Modulation

Monopulse System

Next, the improvement in tracking for a monopulse system is calculated, for an FM regimen where the carrier frequency is changed progressively in one direction for N-pulses and then repeated. One proceeds by finding the autocorrelation function and thus the low-frequency spectral density for the glinting signal with FM. If the frequency change between pulses is Δf_c, then, by definition, all video pulses in the N-pulse sequence will have independent amplitudes. However, pulses having identical frequencies in adjacent sequences will be correlated according to their time separation as determined by the autocorrelation function $R_U(\tau)$ for the carrier frequency in question. Since Δf_c will be shown to be small compared with the mean carrier frequency f_o the autocorrelation function evaluated for f_o is assumed to hold for all N-frequencies. The video pulse version of the glinting signal with FM thus has the appearance of a time-multiplexed set of samples of the glinting signals which would be obtained by N monofrequency radars using the individual frequencies of the FM regimen. With a pulse repetition period T_o, the correlated pulses are separated in time by multiples of NT_o, and the autocorrelation function for the glint with FM can be shown to be[7]

$$R_{U'}(\tau)=\frac{t_o}{T_o}R_U(\tau)\sum_{n=-\infty}^{\infty}P(\tau-nNT_o) \tag{7}$$

where t_o is the pulse width and $P(\tau)$ represents a triangular function of unit height and base $2T_o$. $R_U(\tau)$ is the autocorrelation function for $U(t)$ and is obtained by the cosine transform of equation 1. This yields[1]

$$R_U(\tau)=\overline{U^2}\left[\frac{3\sin^2 y}{y^2}+\frac{3\sin^2 y}{y^3}-\frac{6\sin^2 y}{y^4}\right] \tag{8}$$

where $y=2\pi f_m\tau$. Inspection of equation 8 reveals that samples of U can be considered effectively uncorrelated beyond the first zero of $\sin^2 y$. Calling this the correlation time, τ_c, yields

$$\tau_c=\frac{1}{2f_m} \tag{9}$$

$R_{U'}(\tau)$ is sketched in Fig. 2 for the case where the pulse width t_o has been stretched to T_o.

The improvement in tracking-error variance due to FM is now defined as the ratio of the tracking-error variance for the

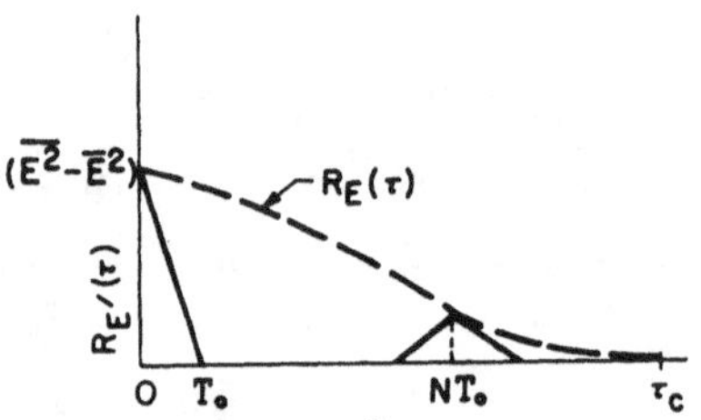

Fig. 3. Fading autocorrelation function with FM

monofrequency case to that for the FM case. Calling this ratio q, one has from equation 4 for a narrow band servo

$$q=\frac{W_U(0)}{W_{U'}(0)} \tag{10}$$

Since the zero-frequency spectral density for a one-sided spectrum is twice the Fourier transform of $R(\tau)$ evaluated at $f=0$, it follows from equation 10 that

$$q=\frac{W_U(0)}{2\int_{-\infty}^{\infty} R_{U'}(\tau)d\tau} \tag{11}$$

It is apparent in Fig. 2 that increasing N results in fewer correlation pulses occurring under the major lobe of the envelope $R_U(\tau)$, which in turn results in decreased zero-frequency spectral density for $U'(t)$. It can be seen further, that maximum improvement occurs when NT_o approaches or exceeds τ_c, since negligible area under $R_U(\tau)$ exists beyond this point. For this optimum FM regimen, all video pulses of $U'(t)$ are essentially uncorrelated and $R_{U'}(\tau)$ reduces to just the correlation pulse at the origin shown in Fig. 2.

In this case, $W_{U'}(0)$ becomes, with the aid of Fig. 2 and the preceding discussion,

$$W_{U'}(0)=2\int_{-\infty}^{\infty} R_{U'}(\tau)d\tau=2\overline{U^2}T_o \tag{12}$$

from which it is clear that $W_{U'}(0)$, and thus the tracking-error variance with optimum FM, is proportional to the radar pulse-repetition period T_o, and is independent of target motion. Then, from equations 1, 9, and 10, the maximum improvement in tracking variance is found:

$$q_{max}=\frac{1}{2f_mT_o}=\frac{\tau_c}{T_o} \text{ for } N\geq\frac{\tau_c}{T_o} \tag{13}$$

For NT_o just equal to τ_c, it follows from equation 13 that

$$q_{max}=N \tag{14}$$

the number of individual frequencies used.

Using the target conditions of the previous example where $f_m=50$ cps with $T_o=0.5\times10^{-3}$, the minimum required number of frequencies N for q_{max} is 20.

Under certain conditions the bandwidth of $W_U(f)$ may be narrow compared with the system bandwidth, for which the preceding analysis is not applicable. For this case, the noise power in the servo passband for a monofrequency radar is simply $\overline{U^2}$. With optimum FM, however, the noise spectrum extends to $f=1/T_o$, and can still be assumed wide compared with the servo bandwidth B. As such, the noise power in the servo passband can again be approximated by $2BW_{U'}(0)$. Since the zero-frequency spectral density with FM is reduced in proportion to the spectrum widening, the improvement in the servo noise power for $B>2f_m$ becomes approximately

$$q_{max}=\frac{\overline{U^2}}{2BW_{U'}(0)}=\frac{1}{2BT_o} \tag{15}$$

Conical Scan System

In a conical scan system not only low-frequency fading noise and glint must be considered, but also the components of E that fall around the conical scan frequency. It is apparent that a spreading of the fading spectrum by FM could introduce more noise at the conical scan frequency than perhaps was already there. However, it will be shown that this depends upon the shape of the original monofrequency spectrum and the frequency changing regimen.

The autocorrelation function for $E(t)$ with FM has the same form as in equation 7, which is

$$R_E'(\tau)=\frac{t_o}{T_o}R_E(\tau)\sum_{n=-\infty}^{\infty}P(\tau-nNT_o) \tag{16}$$

$R_E(\tau)$ can be obtained by the inverse Fourier transform of equation 5, and is given by

$$R_E(\tau)=(\overline{E^2}-\bar{E}^2)\frac{\sin^2 y}{y^2} \tag{17}$$

where $y=2\pi f_m\tau$ and the bar indicates the average. Equation 16 is plotted in Fig. 3, with the video pulses stretched to width T_o.

The spectrum corresponding to $E'(t)$ is given by the Fourier transform of equation 16 which yields for $t_o=T_o$

$$W_E'(f)=\frac{1}{NT_o}\sum_{n=-\infty}^{\infty}W_P\left(\frac{n}{NT_o}\right)\times\left[W_E\left(f+\frac{n}{NT_o}\right)+W_E\left(f-\frac{n}{NT_o}\right)\right] \tag{18}$$

where $W_P(n/NT_o)$ is the spectrum of a unit triangular pulse of width $2T_o$ centered at n/NT_o. Equation 18 is shown in Fig. 4 for the case of $NT_o<\tau_c$ corresponding to N less than optimum. Although practical design considerations may limit the choice of N, it is of interest to note the consequences of various values in this case.

From an examination of Fig. 4, it is apparent that a choice of $1/NT_o$ approximately equal to, or a submultiple of the conical scan frequency, would clearly introduce additional noise at the scan frequency f_s. With N large enough to insure optimum FM, the fading spectrum becomes continuous as shown by the dotted envelope in Fig. 4, and the noise at f_s becomes approximately $W_E(0)/N$. Whether the combined glint and fading noise for this case is more or less than the original noise depends upon the width of the monofrequency fading spectra and the value of N. The final degradation in tracking improvement due to this FM induced fading noise depends also upon the angular sensitivity of the radar. Under appropriate conditions of angular sensitivity and tactical mission, the glint is a much more important source of error than fading noise; therefore, a small increase in fading noise at the conical scan frequency would tend to be unimportant compared with the decrease in glint.

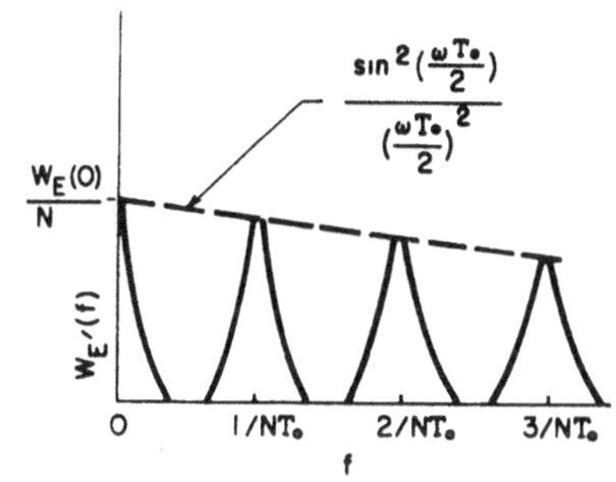

Fig. 4. Fading spectrum with FM

In other cases, where it is necessary to avoid introducing noise into the scan frequency circuits, certain signal processing techniques can be used. Such techniques are being analyzed by the authors, but are beyond the scope of this paper.

Derivation of Critical Frequency Change

As has been the case with previous discussions of target generated noise, it is assumed that the most important cause of signal fluctuations is the phase interference between reflecting elements of the target. Any variation of the scattering cross section of the reflecting elements due to frequency changes is neglected since the frequency changes will be small compared with the center operating frequency. The analysis that follows is made for the time-varying case, but here it is assumed that the carrier frequency is varied so rapidly that fluctuations due to target motion may be neglected, and only fluctuations due to frequency change noted. In this case, the roles of frequency and time are simply interchanged.

The target model in Fig. 5 is the same as in Fig. 1 but with time "frozen" when the target is offset at an angle θ_o. This is done in order to show a more representative frequency sensitivity of the target. Thus, following Muchmore,[1] the voltage return from the target as a function of frequency is given by

$$V(f)=\sum_n a_n e^{i\left(\frac{4\pi f r_n}{c}+\phi_n\right)} \tag{19}$$

where the ϕ_n's are independent random phase angles uniformly distributed from zero to 2π, r_n is the distance in the range direction from the nth reflecting element to a plane perpendicular to a line connecting the radar with the mean center of reflectivity of the target, and a_n is the amplitude of the return from the nth reflecting element. Since only one tracking axis being considered in this analysis, the voltages of interests are the sum and difference antenna voltages

$$V_s(f)=\sum_n a_n \cos\left(\frac{4\pi r_n f}{c}+\phi_n\right) \tag{20}$$

and

$$V_d(f)=\sum_n b_o a_n \frac{X_n}{r} \cos\left(\frac{4\pi r_n f}{c}+\phi_n\right) \tag{21}$$

where b_o is the per-cent modulation per radian of error or the equivalent in a monopulse system, and X_n/r is the offset angle of the nth reflecting element with respect to the antenna tracking axis. $V_d(f)$ is the signal due to the instantaneous tracking error of each of the n reflecting elements. The total signal is the sum of $V_s(f)$ and $V_d(f)$. The problem now is to determine autocorrelation functions for fading and glint as functions of frequency.

It has been shown by Muchmore,[1] Freeman,[2] and Delano,[3] that a proper choice of the center of reflectivity and a symmetric reflectivity density insures independence of V_s and V_d. Under these conditions the autocorrelation function for the glint is given by

$$R_U(\Delta f)=\frac{2\overline{V_s V_s' V_d V_d'}}{\sigma_s^2} \tag{22}$$

where in this case V_s' and V_d' are voltages shifted in frequency from f to $f+\Delta f$, and σ_s^2 is the variance in V_s. From equation 20,

$$V_s V_s'=\sum_n\sum_m \overline{a_n a_m \cos\left(\frac{4\pi f r_n}{c}+\phi_w\right) \cos\left(\frac{4\pi f' r_n}{c}+\phi_m\right)} \tag{23}$$

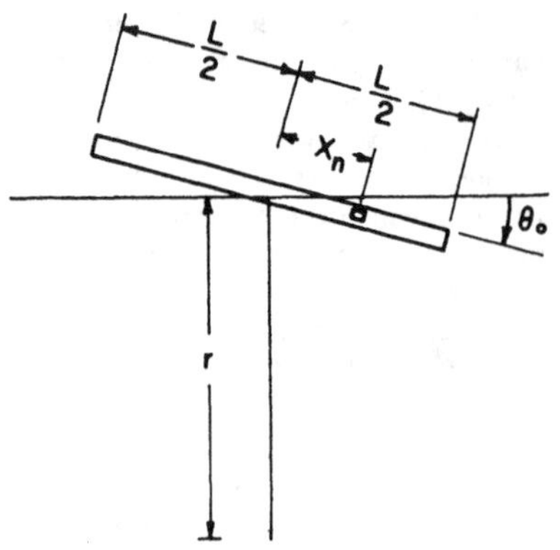

Fig. 5. Offset target model

where $f'=f+\Delta f$. Next, it is helpful to note that with the ϕ_n's independent

$$\overline{\cos\phi_n \cos\phi_m}=\frac{1}{2}\delta_m{}^n$$

$$\text{and } \overline{\sin\phi_n \sin\phi_m}=\frac{1}{2}\delta_m{}^n$$

where $\delta_m{}^n=1$ if $m=n$, $\delta_m{}^n=0$ if $m\neq n$.

Then, using a trigonometric identity, it follows immediately that

$$V_s V_s'=\frac{1}{2}\sum_n \overline{a_n^2 \cos\frac{4\pi r_n}{c}(f'-f)} \tag{24}$$

In a similar manner,

$$V_d V_d'=\frac{1}{2}\sum_n \overline{a_n^2 b_o^2\left(\frac{X_n}{r}\right)^2 \cos\frac{4\pi r_n}{c}(f'-f)} \tag{25}$$

If the probability is 1/2 that Δf is positive and 1/2 that Δf is negative, then

$$\overline{\cos\left(\frac{4\pi r_n \Delta f}{c}\right)}=\frac{1}{2}\cos\left(\frac{4\pi\, {}_n\Delta f}{c}\right)+\frac{1}{2}\left(\frac{-4\pi r_n \Delta f}{c}\right) \tag{26}$$

$$=\cos\left(\frac{4\pi r_n \Delta f}{c}\right)$$

The summations of equations 24 and 25 can be approximated as integrals over the target length L. Noting that $r_n=X_n \sin\theta_o$, and that the integrals are even functions of X,

$$\overline{V_s V_s'}=\frac{\overline{V^2}}{2}\int_0^{L/2} \cos\left(\frac{4\pi \sin\theta_o \Delta f X}{c}\right) dX$$

$$=\frac{\overline{V^2}}{2}\frac{\sin(\alpha\Delta f)}{(\alpha\Delta f)} \tag{27}$$

where

$$\alpha=\frac{2\pi L \sin\theta_o}{c}$$

and $\overline{V^2}$ is the mean square value of the returned voltage. Similarly,

$$\overline{V_d V_d'}=\frac{b_o^2 \overline{V^2}}{Lr^2}\int_0^{L/2} X^2 \cos\left(\frac{4\pi\Delta f \sin\theta_o X}{c}\right) dX$$

$$=\frac{b_o^2\overline{V^2}}{Lr^2}\;\frac{2\Delta f\alpha \cos(\alpha\Delta f)+[\alpha^2(\Delta f)^2-2]\sin(\alpha\Delta f)}{(\alpha\Delta f)^3} \tag{28}$$

Let $E(f)$ be the voltage as a function of frequency after square law envelope detection. Since the spectrum of $E(f)$ is given by the convolution of the spectrum of V_s on itself, the correlation coefficient of $E(f)$ is given by the square of the V_s. This yields

$$\rho_E(\Delta f)=\frac{R_E(\Delta f)}{R_E(0)}=\frac{\sin^2(\alpha\Delta f)}{(\alpha\Delta f)^2} \tag{29}$$

The critical frequency change for E will be taken as the first zero of the correlation coefficient, since by the nature of the $\sin^2(\alpha\Delta f)/(\alpha\Delta f)^2$ function, samples of E at greater frequency changes can be considered essentially uncorrelated. Thus,

$$\Delta f_c=\frac{c}{2L\sin\theta_o} \tag{30}$$

It is interesting to note that this frequency change is just that amount which is necessary to change by 180 degrees the difference in phase of an element at the center of the target and the most distant element from it.

Similarly, from equations 22, 27, and 28, the correlation function of the glint becomes

$$\rho U(\Delta f)=\frac{RU(\Delta f)}{RU(0)}=\frac{3\sin(\alpha\Delta f)}{(\alpha\Delta f)^4}[2\alpha\Delta f \cos(\alpha\Delta f)+(\alpha^2\Delta f^2-2)\sin\alpha\Delta f] \tag{31}$$

Comparing equations 31 and 29, it is apparent again that samples of U at frequency changes of Δf_c or greater are also essentially uncorrelated. For our previous example of a 50-foot target, fixed at an offset angle of 0.1 radian, Δf_c is approximately 100 megacycles. While these results are based on a highly simplified model, they display frequency dependency relationships which should be useful in considering actual target conditions.

Conclusions

It has been shown that under certain conditions a considerable improvement in tracking can be obtained by a pulse-to-pulse frequency changing scheme. In particular, for a monopulse system, maximum improvement is dependent upon the correlation time of the monofrequency fading or glint record, or the integration time of the servo, whichever is smaller. For the conical scan system, no general conclusions predicting tracking improvement can be made. In order to predict the improvement in tracking for a particular conical scan system, consideration must be given to the fading spectra for targets of interest, the relative effect of glint and fading noise on tracking, and the frequency-changing regimen. It may be practical to consider methods of processing the signal to reduce the noise at the conical scan frequency which is introduced by frequency changing.

The value of Δf_c that was found is, admittedly, based on a very simplified target model; however, this analysis was made to determine dependency relationships. Any consideration of Δf_c for a particular target must be based upon the physical characteristics of the target in question and typical target aspects with respect to the radar.

References

1. Aircraft Scintillation Spectra, **R. B. Muchmore.** *Transactions on Antennas and Propagation*, Institute of Radio Engineers, New York, N. Y., vol. AP-8, Mar. 1960, pp. 201–12.

2. Principles of Noise (book), **J. J. Freeman.** John Wiley & Sons, Inc., New York, N. Y., 1958.

3. A Theory of Target Glint or Angular Scintillation in Radar Tracking, **R. H. Delano.** *Proceedings*, Institute of Radio Engineers, New York, N. Y., vol. 41, Dec. 1953, pp. 1778–84.

4. Phenomena of Scintillation Noise in Radar Tracking Systems, **J. H. Dunn.** *Ibid.*, vol. 47, May 1959, pp. 855–64.

5. Angular Jitter in Conventional Conical-Scanning Automatic Tracking Radar Systems, **E. C. Brockner.** *Ibid.*, vol. 39, Jan. 1951, pp. 54–6.

6. Theory of Servomechanisms (book), **H. M. James, N. B. Nichols, R. S. Phillips.** McGraw-Hill Book Company, Inc., New York, N. Y., 1947, p. 293.

7. Introduction to Statistical Communication Theory (book), **David Middleton.** McGraw-Hill Book Company, Inc., 1960, p. 220.

The Reduction of Angle-of Arrival Scintillation by a Frequency Shifting Technique

L.H. Kosowsky, S.S. Brody, L. Chanzit, and S. Saslovsky

Proceedings of the National Electronics Conference, Vol. 19 (1963)
Pages 52-71

ABSTRACT

The radar echo from a group of distributed radar scatterers exhibits a non-spherical phase front or equi-phase surface whose slope at a point in space is a function of the state of the scatterers. The angle between the normal to the wavefront, and the antenna boresight is defined to be the apparent angle-of-arrival.

The measured angle-of-arrival from a group of unconstrained scatterers exhibits a random nature, whose correlation time is a function of the mobility, size, and arrangement of the scatterers, as well as the location and aspect of the observer. From a study of the interference phenomenon which produces angle scintillation, it has been found that the transmitter frequency plays a significant role in determining the instantaneous value of the measured angle-of-arrival. In the classic two reflector problem, the observed angle-of-arrival is shown to be a sharply tuned function of frequency whose maximum value and "bandwidth" are determined by the intensity and orientation of the reflectors. In radar systems in which the available time of observation is much less than the natural correlation time of the scatterers, the sensitivity of the angle-of-arrival to frequency deviation may be exploited to shift the scintillation spectrum to higher frequencies where filtering techniques can remove most of the scintillation energy.

An experimental radar transmitter capable of rapid frequency shifting has been constructed. Angle-of-arrival experiments utilizing a phase interferometer were conducted against natural terrain, and scintillation data has been obtained both with and without frequency shifting. Results are presented to illustrate the redistribution of the scintillation energy which is obtained with this technique. The low frequency spectrum is substantially reduced over that obtained without frequency shifting, and after filtering leads to an improved estimate of the true angle-of-arrival in a finite observation time.

I. INTRODUCTION

The measurement of the angle-of-arrival of radiation from a distributed group of scatterers has been accomplished by a variety of radar techniques, e.g., a phase interferometer (1)*, an amplitude monopulse radar (2), and a phase monopulse radar (3). In each of these systems, fluctuations of the angle-of-arrival have been reported, the intensity and frequency of which appear to be a function of the composition of the scatterers as well as the wavelength of the radiation. It has been shown (4), that the observed scintillations are due to the perturbation in the wavefront resulting from the granularity of the distributed target, and that any system which measures the slope of the wave propagated from the target area will yield an erroneous angle measurement when operating on this perturbed wavefront.

The angle measurement problem is most easily visualized by considering that in the vicinity of the receiver, the radia-

*Numbers in parenthesis refer to references at the end of the paper.

tion scattered from a target area is composed of a large number of plane waves of assorted amplitudes and phases. In the general case, the individual scatterers in the complex move independently, with different velocities, and scatter energy with different doppler shifts. In the vicinity of the receiver, the superposition of all the scattered waves results in a composite wave train whose slope can be determined by a variety of techniques. The normal to the wavefront defines the "apparent target direction" or the direction of the apparent center-of-return. As has been observed in high accuracy angle measuring radars, the normal to the composite wavefront can differ appreciably from that of the separate wavelets whose sum gave rise to the wave. Thus, the apparent direction of the complex may, instantaneously, be very different from the direction of any part of the target.

The scattered energy arriving at the receiver from a given direction is characterized by three parameters: amplitude, initial phase, and doppler shift. A change in the relative phases of the component wavetrains will result in a changed slope of the composite wavefront, giving an apparently different target direction. Since each of the phases depends on the total path length, in wavelengths, between transmission, scattering, and reception, the slope of the wavefront is sensitive to scatterer motion and the location of the receiver, as well as, the transmitted frequency. As the scatterers in a complex rearrange themselves, either because of internal or external influences, the measured angular location of the area can be expected to scintillate as a function of time. When viewed for a period of time, the measured angle-of-arrival from a group of unconstrained scatterers exhibits a random nature, whose correlation time and power spectral density are functions of the mobility, size and arrangement of the scatterers as well as the aspect of the observer.

In radar systems in which the available time of observation of the scatterers is limited by scanning of the antenna or the changing aspect of the observer, the gradual reorientation of the scatterers can lead to large instantaneous errors which may be strongly correlated over the available observation time. For natural terrain, the correlation time of angle scintillation has been observed to be 30 seconds or longer, on days in which the winds were very calm. It is apparent then, that systems which have a limited observation time, cannot make accurate instaneous measurements unless some means are found of obtaining independent samples of the angle-of-arrival within the natural correlation time of the scatterers.

The random nature of the scatterer motion, and the dependence of the wavefront tilt on the path length changes executed by the individual scatterers, makes it apparent that angle scintillation is dependent on the frequency of the transmitted radiation. A study of terrain models consisting of discrete scatterers has demonstrated that angle scintillation is strongly related to the interference phenomenon which produces fading. The classic two reflector problem, which has been used in the past to illustrate angle scintillation, is found to produce an angular error which is a sharply tuned function of frequency whose maximum value and bandwidth, are determined by the intensity ratio of the two radiators. A periodic shifting of the transmitted frequency about its nominal value, produces a series of angle-of-arrival pulses*, whose shape is related to the frequency response function. If the position of the two radiators changes slowly compared to the sweep rate of the transmitted frequency, the measured angle-of-arrival will consist of a series of pulses that are modulated by the slowly changing phase.

Thus, it can be seen that although the orientation of the radiators may be such as to yield a large instantaneous error at the nominal frequency, f_o, deviation of the transmitted frequency yields independent measurements of the true angle of arrival. For an array of scatterers, the frequency shifting technique produces a complex interference pattern which acts as a pulsed carrier for the slowly changing orientation of the scatterers. The spectrum of the pulse train consists of a low frequency portion and bands around the harmonics of the sweep repetition frequency. The low frequency spectrum is substantially reduced over the obtained without modulation, and after filtering leads to an improved estimate of the true angle-of-arrival in a finite observation time.

An experimental radar transmitter capable of rapid frequency shifting has been constructed. Angle-of-arrival experiments utilizing a phase interferometer were conducted against natural terrain, and scintillation data was obtained both with and without frequency

*Since the transmitted signal is itself of a pulsed nature, the term angle-of-arrival pulse is taken as the envelope of the received pulses observed at a particular range.

shifting. Data is presented to illustrate the redistribution of the scintillation energy which is obtained with this technique.

II. THE TWO REFLECTOR PROBLEM

Although it has been treated extensively in the literature, it is instructive to review the results of the classic two reflector problem in which the apparent center of return can be caused to fall far outside the angle subtended by the two radiators. This is, of course, a special case, but it points up the very important point that an observer located at the receiving antenna is sensitive to the tilt of the wavefront passing the receiving system and that the apparent angle-of-arrival is a function of both the location of the reflectors and the relative phase and intensity of the scattered energy from each reflector.

Consider two scatterers, as shown in Figure 1, one located at a depression angle β_1 and the other at an angle β_2. We define the angular separation of the two scatterers to be $\Delta\beta$ and the mean depression angle, $\bar{\beta}$, as $(B_1 + B_2)/2$. The angle measured by the radar is defined to be,

$$\beta = \bar{\beta} + \delta(t)$$

where $\delta(t)$ is the instantaneous angular error. It has been shown (4) that the angle-of-arrival scintillation as measured at a point in space is given as:

$$\delta(t) = \frac{\Delta\beta}{2}\left[\frac{1 - a^2}{1 + a^2 + 2a\cos(\phi(t))}\right] \quad (1)$$

where: δ is the angular error
$\Delta\beta$ is the angle subtended by the two radiators
a is the ratio of the two signal amplitudes as seen at the receiver. $0 < a < 1$
$\phi(t)$ is the relative phase difference of the two radiating particles as seen at the receiving point. (Path length difference to each of the particles are included in $\phi(t)$.)

The intensity of the signal measured at the receiver is given as:

$$E = (1 + a^2 + 2a\cos\phi(t))^{1/2} \quad (2)$$

where E is a function of the intensity of the radiation from each particle, the range, and the transmissivity of the medium between the particles and the receiver. The normalized function, $2\delta/\Delta\beta$, and E are plotted in Figure 2 and 3 as functions of the relative phase, $\phi(t)$, for particular values of "a" to illustrate the large indicated error in the measured angle-of-arrival which can result. Several important conclusions can be drawn from an examination of Figures 2 and 3. For two particles initially in a plane perpendicular to the observation point, and radiating in phase, any motion of the radiators which results in a path length change of an odd number of half wavelengths will yield the maximum angular error. If the particles are illuminated by a transmitted radar pulse, a difference in path length of a quarter wavelength will result in a maximum measured angular error. Since at Ku-band, a quarter wavelength corresponds to a range difference of approximately 0.5 cm, it can be seen that relatively small motions of the radiators can produce significant errors in the measured angle-of-arrival particularly when the radiators are of comparable strength. It can be seen, that large angular errors in the measured angle-of-arrival are accompanied by a sharp drop in the measured intensity of the signal. The intensity variation is called fading and is a common phenomenon in systems designed to measure the intensity of the signal reflected from radar targets. Because of the rapid fluctuations which can occur during interference conditions, the design of AGC systems in amplitude monopulse radars is complicated by the intensity scintillation spectrum.

In order to separate effectively the phenomenon of angle-of-arrival scintillation from that of fading, a phase interferometer was designed and constructed. Such a system determines the angle-of-arrival by measuring the phase difference between the signals in two closely spaced receiving antennas and is insensitive to the intensity of the received radiation over a wide dynamic range of signals. The principles of operation of a phase interferometer are discussed in Section 5.

III. FREQUENCY DEPENDENCE OF ANGULAR SCINTILLATION

Thus far we have been concerned with the angular error in the measured angle-of-arrival which results from the interference of the radiation from two unconstrained particles. We shall now examine the dependence of this error on the frequency of radiation.

Equation 1 may be rewritten in terms of the radiated frequency and the path length difference shown in Figure 1. When this is done equation 3 results:

$$\delta = \frac{\Delta\beta}{2}\left[\frac{1 - a^2}{1 + a^2 + 2a\cos\left(\frac{2\pi f\Delta R}{C} + \gamma\right)}\right] \quad (3)$$

where: ΔR is the instantaneous path length difference
f is the frequency of the radiation in cycles per second
γ is the initial phase relationship between the radiators
C is the speed of light in the medium

Examination of equation 3 reveals that δ is a periodic function of the applied frequency, f, and moreover closely resembles the familiar "comb" structure associated with the transfer characteristics of sampled data systems.

A sketch of δ as a function of frequency is shown in Figure 4. It should be noted that each "tooth" of the comb response is identical in shape to Figure 2. The separation between peaks of the error function is equal to $C/\Delta R$. For ΔR equal to 200 feet, $C/\Delta R$ is approximately 5 megacycles. (An analog simulation of $\delta(f)$ for possible use on analog computers is given in Appendix A.)

An interesting quantity is the average value of the error function over a complete period as shown in Figure 4. Averaging over a cycle of $\delta(f)$, we have:

$$\delta_{av} = \frac{1}{\Delta f}\int_{f_o}^{f_o + \Delta f} \delta(f)\, df$$

where: $\Delta f = C/\Delta R$

Substitution of equation 3 yields,

$$\delta_{av} = \frac{\Delta\beta\ \Delta R}{2C}\int_{f}^{f + C/\Delta R} \frac{(1-a^2)\, df}{1+a^2+2a\cos\left(\frac{2\pi\ \Delta R}{C}f+\gamma\right)} \quad (4)$$

Utilizing the identity:

$$\int \frac{dx}{A + B\cos x} = \quad (5)$$

$$\frac{2}{(A^2 - B^2)^{1/2}}\tan^{-1}\left[\frac{\sqrt{A^2 - B^2}\tan\left(\frac{X}{2}\right)}{A + B}\right]$$

we have:

$$\delta_{av} = \frac{\Delta\beta}{2}, \quad 0 < a < 1 \quad (6)$$

The average error as a function of frequency is independent of the intensity ratio of the two radiators and is equal to one half the angular separation of the radiators. δ_{av} is an extremely useful quantity since it defines the angular extent of a body without scanning and without the necessity of a narrow beamwidth. The acquisition of information on the structure of target areas from an examination of their response to various applied frequencies has been the subject of much research in azimuth resolution where two objects in a beamwidth may be defined by their resultant interference patterns and their angular locations established to a high degree of accuracy. Since in conventional radar systems the ability of a system to resolve two objects is determined by the antenna beamwidth, we have here an example of resolution improvement by a data processing technique (8).

IV. FREQUENCY SHIFTING IN THE TWO RADIATOR PROBLEM

We have observed in the preceding section that during interference the instantaneous angle-of-arrival as measured at the receiver can depart significantly from the true angle and that in fixed frequency systems, only scatterer motion can alter the interference condition. We have also seen that the measured angle-of-arrival is a function of the frequency of the radiation and, indeed, a shift of frequency can negate or enhance an interference condition which occurs at some frequency f_o and path length difference ΔR. In this section, we discuss the concept of frequency shifting and its ability to provide a more reliable measurement of the angle-of-arrival.

For the present study, we shall assume that the frequency is varied in a linear fashion between the limits $(f_o - f_m) < f < (f_o + f_m)$ where f_o is the center frequency and f_m is the peak frequency shift. (Although the present discussion treats linear frequency shifts, non-linear frequency shifting schemes may be considered in more detailed analyses. No attempt is made here to optimize the frequency shift as a function of scintillation characteristics.) As the frequency is varied, we can expect to see the error function, $\delta(t)$, fluctuate in response to the changing frequency. The frequency spectrum of $\delta(t)$ will consist of a d-c term and harmonics of the modulating frequency, the exact nature of the spectrum being determined by the amount of frequency shift and the path length difference as well as the intensity ratio of the two radiators. If we subsequently filter $\delta(t)$ such that only the d-c term is passed, the average angular deviation for $f_m = C/2\Delta R$ is equal to $\Delta\beta/2$ as indicated earlier and the minimum value of the measured angle is equal to the correct angle-of-arrival. However, much more information is contained in the frequency response function. Not only can the angular separation of the radiators be determined, but the intensity ratio and the path length difference may be determined as well. The maximum value of the frequency response function contains information on the intensity ratio, and the bandwidth of the function, or the separation between peaks of the interference function, defines the path length difference.

Thus, it is seen that under the application of frequency shifting, the dependence of angle scintillation on phasing of the radiators may be exploited to yield detailed information on the structure of the two radiators regardless of their initial phasing.

V. INTERFEROMETER PRINCIPLES

The principles of an interferometer have been discussed in considerable detail in the literature and are considered briefly here.

Consider the two receivers shown in Figure 5, which are illuminated by a uniform plane wave of constant frequency. The electrical phase difference between the signal received in each receiver is given as:

$$\text{Electrical Phase Difference} = \text{Time Delay } (2\pi \text{ Frequency}) \tag{7}$$

The time required for a wavefront to pass from one receiving element to the next is given as:

$$\text{Time Delay} = \frac{D \sin \beta}{C} \tag{8}$$

where: D is the spacing between receivers
C is the speed of light in free space
β is the apparent angle-of-arrival

Substitution of equation 8 into equation 7 yields,

$$\text{Electrical Phase Difference} = \frac{2\pi f\, D \sin \beta}{C} \tag{9}$$

Solving for β, we have:

$$\beta = \sin^{-1} \left[\frac{(EPD)C}{2\pi f\, D}\right] \tag{10}$$

Although the above analysis is adequate for the far field of a single radiating element, it becomes grossly incorrect when two or more radiators are arranged such that the interferometer lies in an interference region. The resultant erroneous measurement is angle scintillation or glint. In an interferometer, the measurement of angle-of-arrival under interference conditions approximates the angular error predicted in equation 3. The measurement of the angle-of-arrival by an interferometer is discussed in the following section.

VI. ANGLE-OF-ARRIVAL MEASUREMENTS WITH A PHASE INTERFEROMETER

Throughout the remainder of the paper we shall confine our attention to a receiving system made up of two point receivers lying in a vertical plane, as shown in Figure 6. For simplicity of analysis, the radiators are taken to lie along a straight line perpendicular to the line joining the two receivers. If the signal from the first particle, as received in the lower receiver is taken as the arbitrary reference vector, then we have the signals in each receiver given as:

$$E_1 = \sum_{i=1}^{N} A_i \exp\left[\theta_i + \frac{2\pi}{\lambda}(R_i - R_1)\right] \tag{11}$$

$$E_2 = \sum_{i=1}^{N} A_i \exp\left[\theta_i + \frac{2\pi}{\lambda}(R_{ix} - R_1)\right] \tag{12}$$

where: θ_i is the initial phase of the ith radiator
R_i is the distance from the ith radiator to the lower receiving antenna
R_{ix} is the distance from the ith radiator to the upper receiving antenna
A_i is the intensity of the ith radiator
λ is the wavelength of the radiation

The phase angle of each vector quantity is given as:

$$\text{ANG 1} = \tan^{-1}\left[\frac{\sum_{i=1}^{N} A_i \sin\left(\theta_i + \frac{2\pi}{\lambda}(R_i - R_1)\right)}{\sum_{i=1}^{N} A_i \cos\left(\theta_i + \frac{2\pi}{\lambda}(R_i - R_1)\right)}\right] \tag{13}$$

$$\text{ANG 2} = \tan^{-1}\left[\frac{\sum_{i=1}^{N} A_i \sin\left(\theta_i + \frac{2\pi}{\lambda}(R_{ix} - R_1)\right)}{\sum_{i=1}^{N} A_i \cos\left(\theta_i + \frac{2\pi}{\lambda}(R_{ix} - R_1)\right)}\right] \tag{14}$$

The Electrical Phase Difference of the signals received in each of the antenna is given as:

$$\text{EDP} = \text{ANG 1} - \text{ANG 2} \tag{15}$$

Substitution of equation 15 into equation 10 yields,

$$\beta = \sin^{-1}\left[\frac{(\text{ANG 1} - \text{ANG 2})C}{2\pi f\, D}\right] \tag{16}$$

For a large group of radiators having arbitrary amplitudes and phases, equations 11, 12 and 16 are solvable by a digital computer such as the IBM 1620 which is used at Norden.

Equation 16 has been solved for as many as 100 elemental radiators in a radiating area. The results for N = 2 and N = 11 are shown in Figure 7 and Figure 8.

For N = 2, the results are almost equal* to the angle-of-arrival errors predicted by equation 3. However, for N = 11, it is observed that the nature of the error function changes considerably. Firstly, the error becomes bipolar and secondly the error function becomes asymmetrical with respect to the maximum interference frequency. In addition, the distribution of intensity across the radiating area becomes extremely significant in determining the shape of the frequency response function. A detailed analysis of the intensity and phase in the multiple radiator problem is found in Reference (9).

VII. FREQUENCY SHIFTING IN THE PRESENCE OF SCATTERER MOTION

Thus far we have assumed that the nominal phasing of the scatterers remained unchanged during the frequency shifting process. Let us now consider the effects of scatterer motion on the results thus far obtained. As noted in equation 3, for the two radiator case,

$$\delta = \frac{\Delta\beta}{2}\left(\frac{1 - a^2}{1 + a^2 + 2a\cos\left(\frac{2\pi\Delta R f}{C} + \gamma\right)}\right)$$

If ΔR varies as a function of time, the system becomes a filter with a time varying parameter. Essentially ΔR determines the spacing between the peaks of the frequency response function. If ΔR varies slowly as compared to the sweep rate of the frequency, f, we can expect the "teeth" of $\delta(t)$ to be position modulated by ΔR. Figure 9 indicates the waveform to be expected for a slowly varying ΔR.

It is to be noted that since the frequency deviation is generally fixed, the number of peaks of δ, per second, varies directly with the path length difference ΔR. The angular error, $\delta(t)$, is a complicated frequency and phase modulated function which has only been treated statistically for the case of scatterers distributed along a line (10). The phase angle of the radiation from each scatterer is uniformly distributed from zero to 2π and the ensemble rotates about the center of the line with an angular velocity ω. The utilization of such a model to represent natural or man-made objects has been the subject of some speculation (5, 6). Few definitive statements may be made about the general problem of frequency shifting when large distributed areas consisting of many moving radiators are considered. These references should be consulted for further information.

*The reason for the discrepancy is that equation 3 has been derived on the basis of a point receiver in space.

VIII. ANGLE MEASURING EQUIPMENT

A block diagram of the instrumentation required for the measurement of the angle-of-arrival is shown in Figure 10. Briefly, a pulse of energy is transmitted by the radar and illuminates the area ahead of the transmitter. After a period of time corresponding to the minimum range to a reflecting area, a signal is received at the phase interferometer and the angle-of-arrival is determined as a continuous function of range. Since we wish to examine the scintillation at a particular range, a boxcar circuit (7) selects the data at the desired range and stores the information for an interpulse period. The output of the boxcar circuit is then displayed on an oscillograph and analyzed by the spectrum analyzer.

IX. FREQUENCY SHIFTING RADAR

The experimental frequency shifting transmitter was built around a magnetron which was available in the laboratory. Since the primary consideration of the experiments was the demonstration of the effects of frequency shifting on the measured angle-of-arrival no attempt was made to optimize the performance of the system.

For the particular magnetron used, a frequency pulling scheme was devised as the most efficient and easily constructed method of obtaining the desired frequency shift of several megacycles equivalent to a frequency shift of $C/\Delta R$. Static tests were performed to determine the maximum frequency deviation of the magnetron under 1.5:1 VSWR and variation of the phase of the reflected voltage. A standing wave introducer was inserted into the main waveguide run to produce a 1.5:1 VSWR and the phase was changed by moving the reflection point in the guide. A maximum frequency variation of 5.75 megacycles was obtained.

The experimental set-up is shown in Figure 11. Major elements include the Faraday rotator, ferrite phase shifter, and differential circulator. It is to be noted that as the power to the phase shifter increases, the losses in the phase shifter increase at a rate greater than the input power. When the losses become excessive, it is impossible to maintain the VSWR at 1.5:1. In order to minimize loss in the phase shifter the system of Figure 11 was developed. The amount of power reflected to the magnetron determines the VSWR, while the electrical delay which is introduced by the ferrite phase shifter primarily determines the phase seen by the magnetron. By adjusting the bias current to the differential circulator, a small amount of power, (about 5 percent of transmitted power), is coupled to the port with the Faraday rotator. All of this power is directed toward the short by adjusting the bias on the circulator. Reflection from the short appears at port 3 on the Faraday rotator, flows through the ferrite phase shifter and out through port 4. The power reflected from the phase shifter goes back through port 1 of the differential circulator and is directed towards the magnetron. By adjusting the short, the physical length of the path to the magnetron is altered, and provides a means for setting the initial phase shift. To prevent foldover of the frequency excursion with phase shifter current, initial phase shift should be set to the point on the Rieke diagram corresponding to the maximum frequency shift at the operating VSWR.

X. EXPERIMENTAL RESULTS

Experiments utilizing a phase interferometer in conjunction with an experimental frequency shifting radar were conducted in January, 1963 from a hilly, forested region near the Bear Mountain State Park. Data was obtained from a fixed transmitter located in the trailer laboratory. Figure 12a shows a typical power spectrum of angle scintillation as obtained by boxcarring the output of the phase detector at a range proportional to 19.5 microseconds. Examination of Figure 12a indicates that almost all the scintillation energy lies in the frequency region below 20 cycles per second for the particular terrain under examination. It is to be noted that the shape of the scintillation spectrum is strongly dependent on weather conditions, and the geometry and mobility of the scatterers. A more complete description of the scintillation characteristics of various types of scatterers under different weather conditions forms a significant portion of the research program (1).

Figure 12b shows the power spectrum of the angle-of-arrival from the same terrain with the frequency of the transmitted shifted at a 60 cycle per second rate over a frequency band of 5 megacycles. With frequency shifting, the magnitude of the low frequency portion of the spectrum was considerably attenuated while a new spectrum appeared around the sweep rate of 60 cps. No attempt has been made to optimize the reduction of the low frequency spectrum as a function of sweep rate.

An important conclusion to be drawn from Figure 12a and 12b is that frequency shifting does indeed shift the angle scintillation energy from the low frequency portion of the spectrum to a band of frequencies surrounding the frequency sweep rate where filtering may be utilized to provide a more reliable estimate of the true angle-of-arrival under adverse wind conditions and slowly moving distributed scatterers. With frequency shifting and filtering, RMS angle-of-arrival fluctuations can be expected to decrease as the energy in the low frequency portion of the spectrum is reduced.

XI. CONCLUSION

The two-radiator model has been utilized to describe the interference effects which produce angle scintillation. It has been shown that frequency shifting generates a comb filter response when applied to the two reflectors and can yield data on the intensity and arrangement of the radiators. A phase interferometer was shown to yield a complex frequency response function when used against more than two radiators. An experimental radar transmitter capable of rapid frequency shifting has been constructed. Angle-of-arrival experiments utilizing a phase interferometer were conducted against natural terrain and scintillation data was obtained both with and without frequency shifting. Results are presented to demonstrate the redistribution of the scintillation energy which is obtained with this technique.

APPENDIX A

AN ANALOGUE FOR ANGLE SCINTILLATION

Utilizing the comb filter concept, an interesting analogue for angle scintillation may be developed for use on analogue computers. Consider the transfer function of the comb filter shown in Figure 13.

From feedback theory, the voltage transfer function may be written as:

$$H(s) = \frac{P}{(1 + k\epsilon^{-\tau s})^2} \qquad (17)$$

where s is the complex frequency

In terms of the real frequency, f, we have:

$$H(f) = \frac{P}{(1 + k\epsilon^{-j2\pi f\tau})^2} \qquad (18)$$

The magnitude of H(f) is given as:

$$H(f) = \frac{P}{(1+k\cos 2\pi f\tau)^2+(k\sin 2\pi f\tau)^2} \qquad (19)$$

Simplifying, we have:

$$H(f) = \frac{P}{1 + k^2 + 2k\cos 2\pi f\tau} \qquad (20)$$

Comparison of equation 20 with equation 3 with $\gamma = 0$, shows that the two systems are analagous when the following identities are satisfied:

$$P = \frac{\Delta\beta\,(1-a^2)}{2}, \quad k = a, \quad \tau = \frac{\Delta R}{C}$$

Scaling of the problem can be readily accomplished by reducing the operating frequency and increasing τ to keep the product, $f\tau$, at any desired value.

ACKNOWLEDGEMENT

The authors wish to acknowledge the assistance of Mr. Lester Kaiser, Mr. Phillip Bernstein, Mr. Salvatore Amoroso, and Mr. Joseph Gessaroli of the microwave section for their assistance in developing the frequency shifting transmitter and phase interferometer and to Mr. W. Fenn and Mr. A. David Wunsch for several valuable discussions. The work described in this paper was supported in part by the Grumman Aircraft Engineering Corporation whose assistance is gratefully acknowledged.

REFERENCES

1. L. Kosowsky, L. Chanzit - Angle-of-Arrival Measurements with a Ku-band Phase Interferometer - Norden Internal Report - June 1962.*

2. J. H. Dunn, D. Howard, and A. King - Phenomena of Scintillation Noise in Radar Tracking Systems - Proc. IRE, Vol. 47, May 1959.

3. D. Rhodes - Introduction to Monopulse - McGraw Hill, 1959.

4. D. Howard - Radar Target Angular Scintillation In Tracking and Guidance System Based on Echo Signal Phase Front Distortion - Proc. NEC, October 1959.

5. L. Peters and F. Weimer - Concerning the Assumption of the Random Distribution of Scatterers as a Model of an Aircraft for Tracking Radar - IRE Trans. on Antenna and Propagation - Vol. AP-9, P. 110, January 1961.

6. R. Muchmore - Reply to Comments by L. Peters and F. Weimer - IRE Trans. on Antennas and Propagation - Vol. AP-9, p. 112, January 1961.

7. J. Truxal - Control System Synthesis - McGraw Hill Book Co. - 1955.

8. S. Shapiro - Improvement of Angular Resolution by Means of Interpulse Frequency Modulation - Air Force Cambridge Research Center Report TN-59-754, June 1959.

9. L. Kosowsky - A Study of Amplitude and Phase In a Two Receiver Interferometer - Norden Internal Report, January 1963.*

10. W. P. Birkemeir, N. D. Wallace - Radar Tracking Accuracy Improvement by Means of Pulse-to-Pulse Frequency Modulation - AIEE - CP 61-1114, 1961.

* Norden internal reports are not available for general distribution.

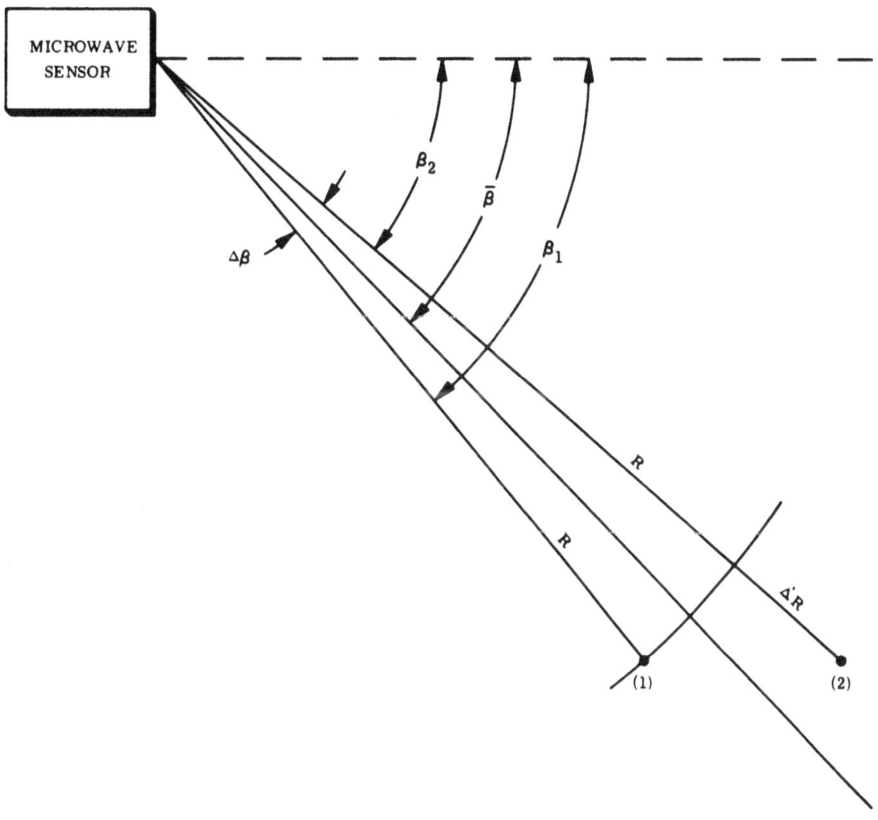

FIGURE 1. GEOMETRY OF THE TWO RADIATOR PROBLEM

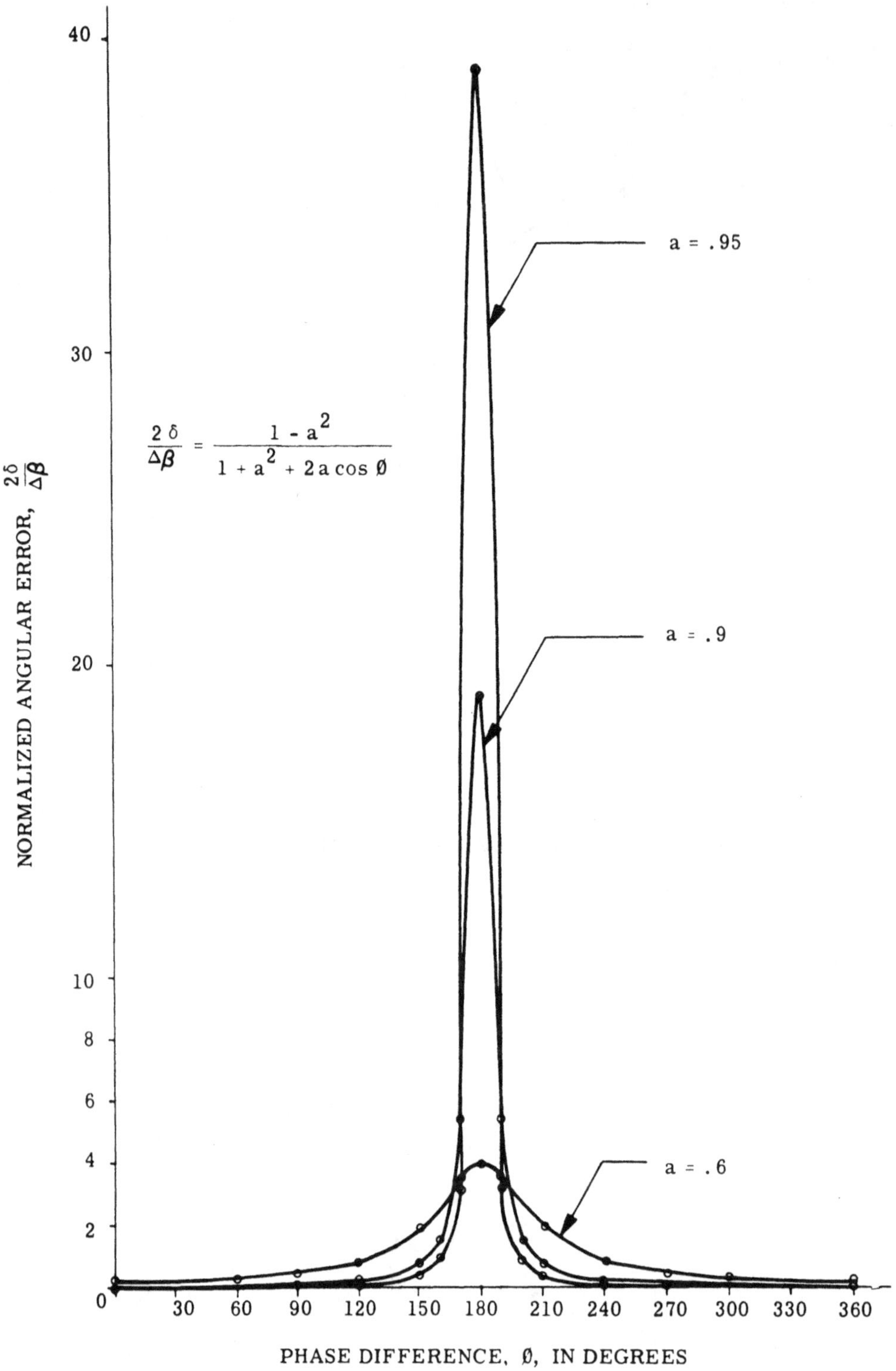

FIGURE 2. NORMALIZED ANGULAR ERROR VERSUS PHASE DIFFERENCE (TWO RADIATOR PROBLEM)

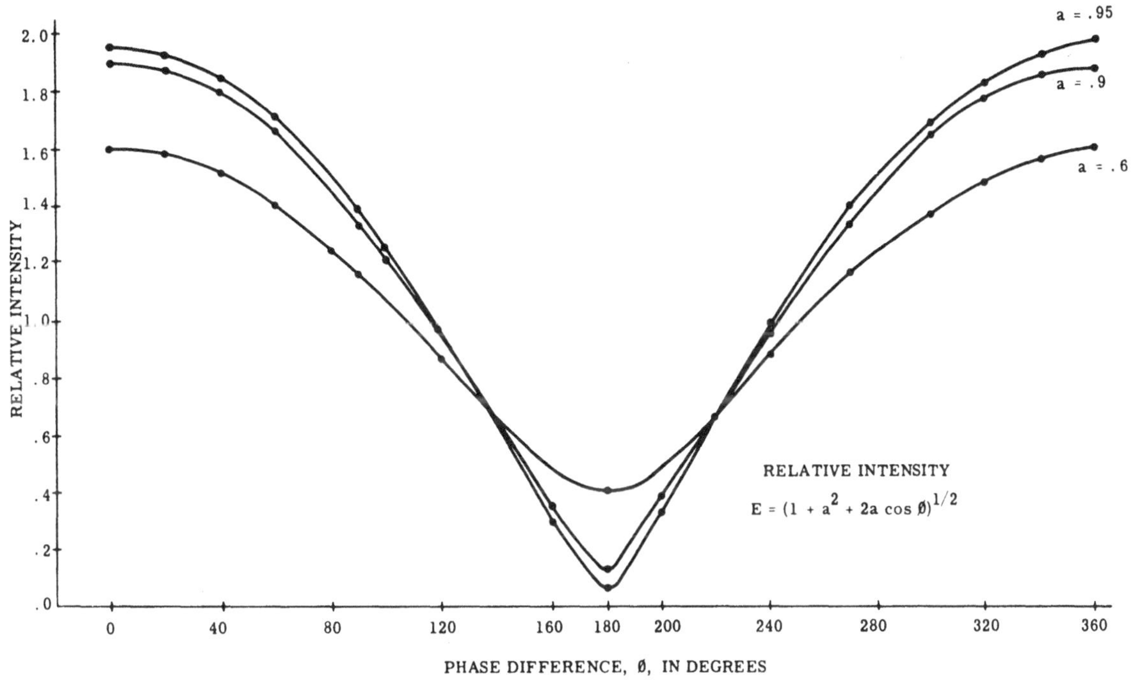

FIGURE 3. RELATIVE INTENSITY VERSUS PHASE DIFFERENCE (TWO RADIATOR PROBLEM)

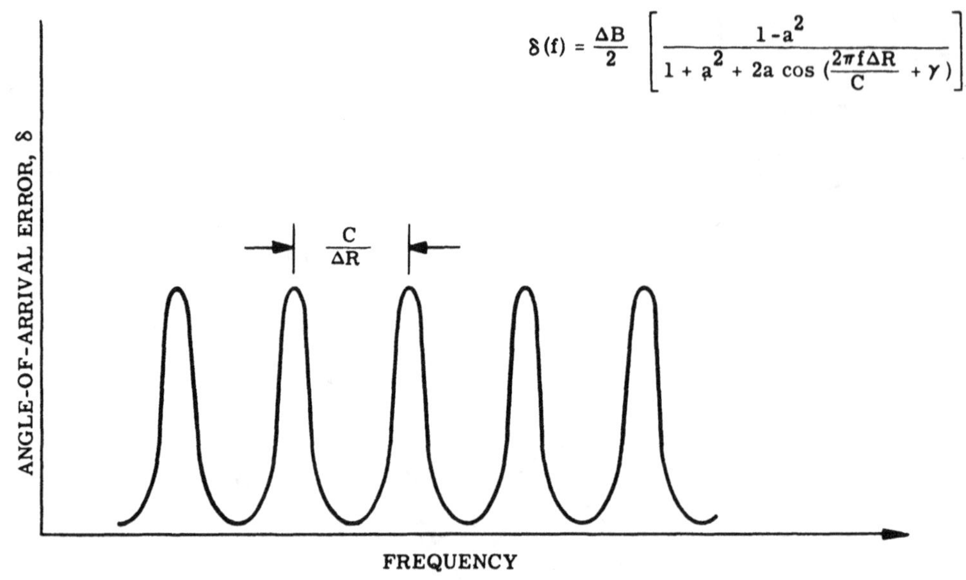

FIGURE 4. ANGLE-OF-ARRIVAL ERROR AS A FUNCTION OF RADIATED FREQUENCY (TWO RADIATOR CASE)

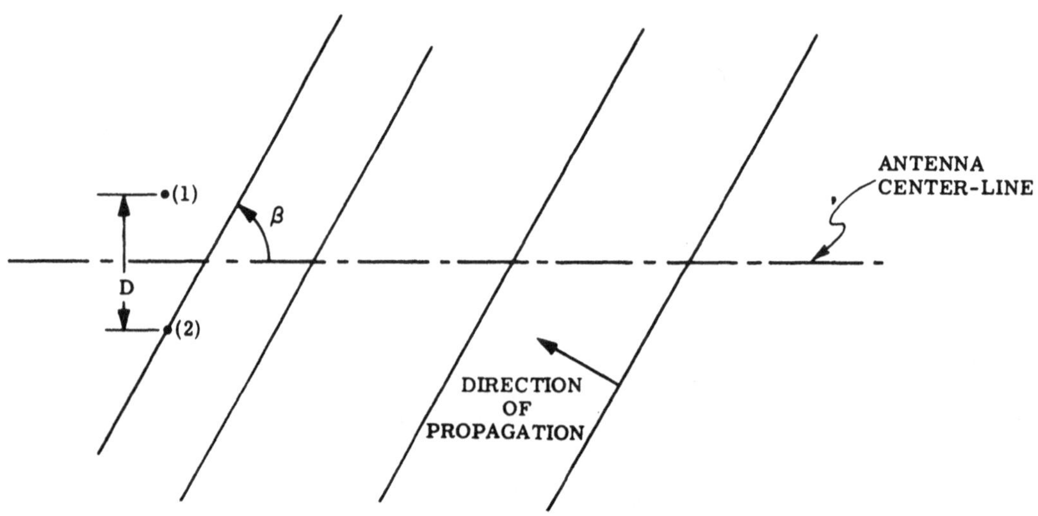

FIGURE 5. BASIC TWO RECEIVER INTERFEROMETER

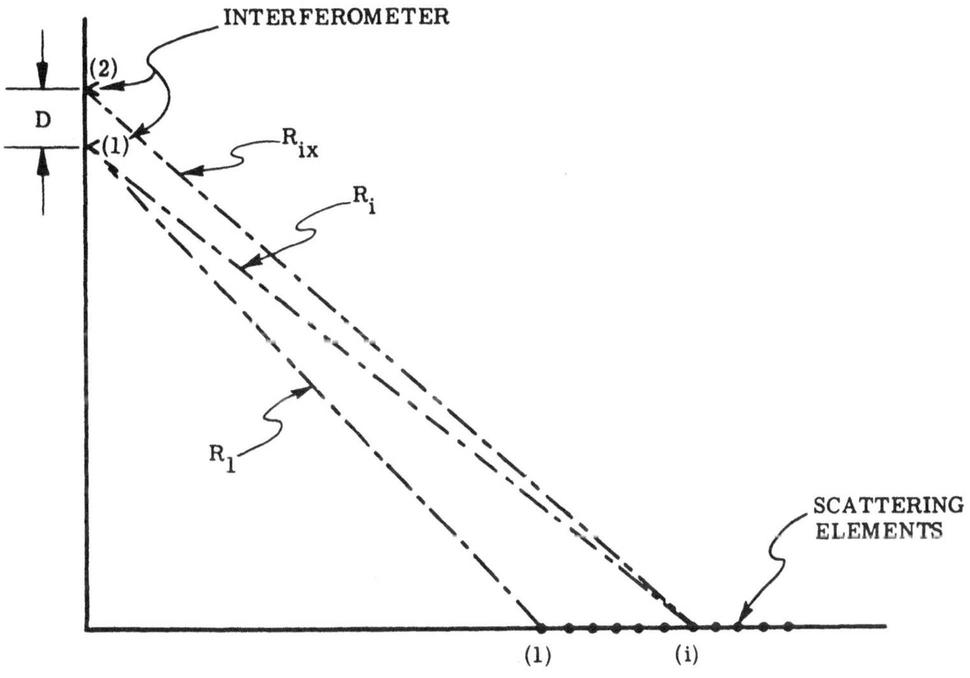

FIGURE 6. GEOMETRY OF THE TWO RECEIVER INTERFEROMETER

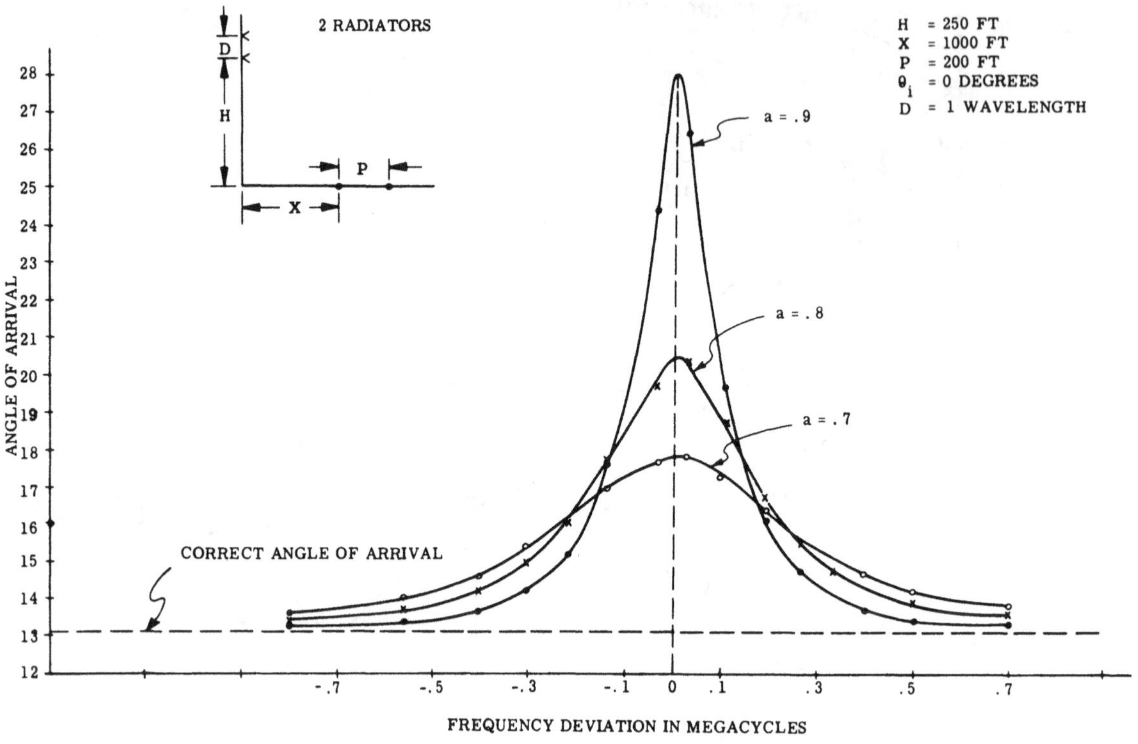

FIGURE 7. ANGLE-OF-ARRIVAL AS A FUNCTION OF FREQUENCY DEVIATION (TWO RADIATORS)

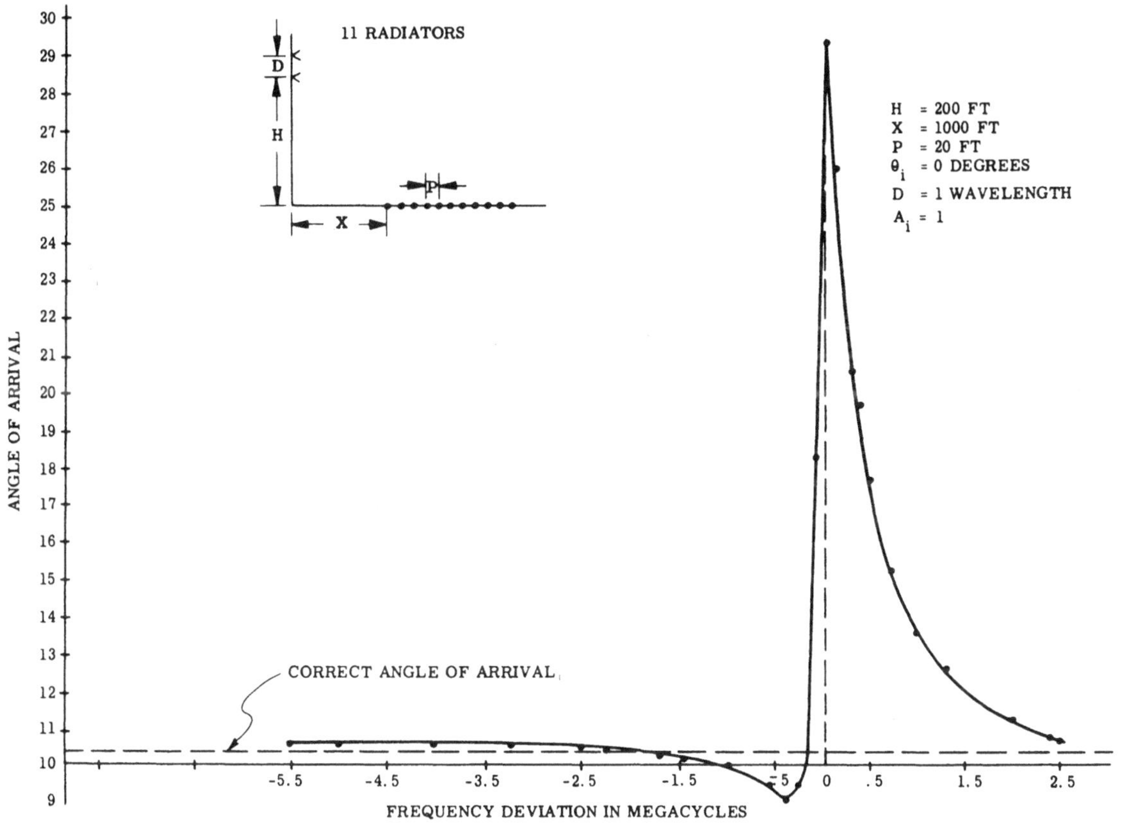

FIGURE 8. ANGLE-OF-ARRIVAL AS A FUNCTION OF FREQUENCY DEVIATION (ELEVEN RADIATORS)

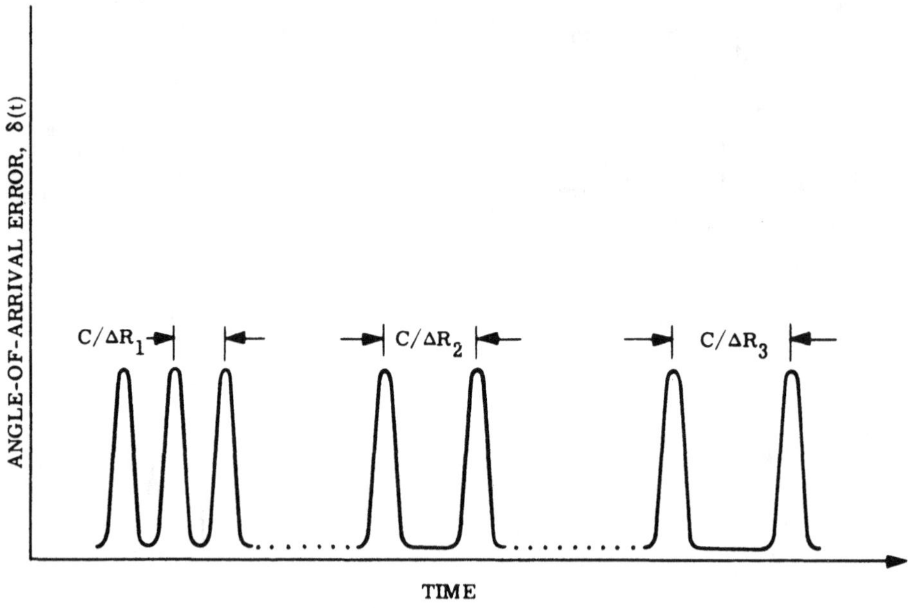

FIGURE 9. MODULATION OF THE ANGULAR ERROR BY ΔR AND FREQUENCY SHIFTING

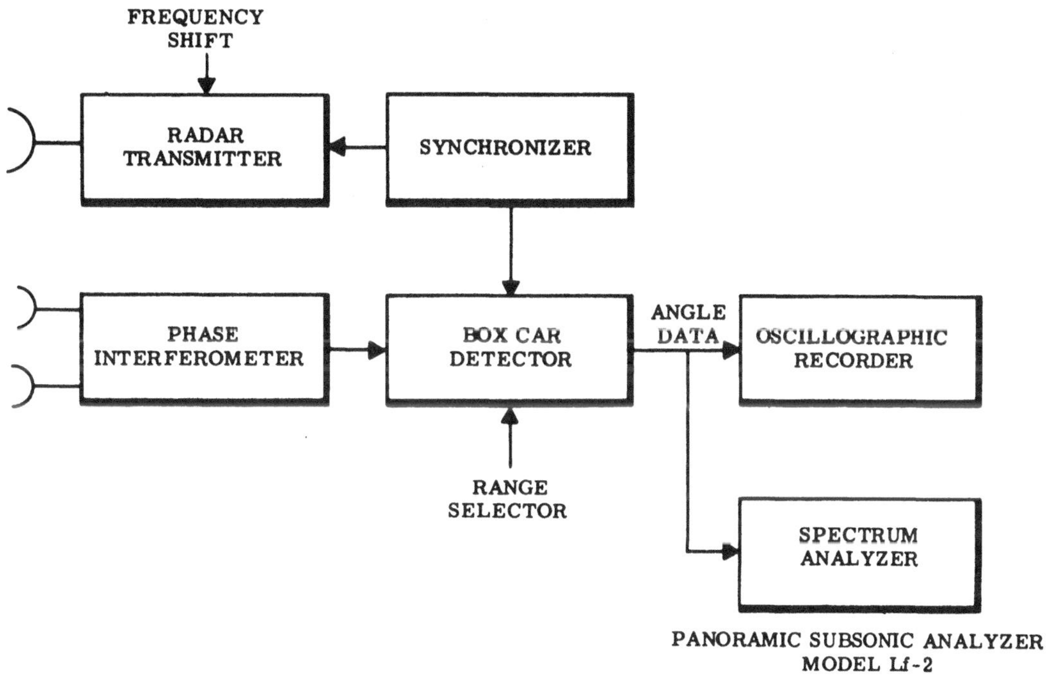

FIGURE 10. BLOCK DIAGRAM OF INSTRUMENTATION FOR ANGLE-OF-ARRIVAL STUDIES

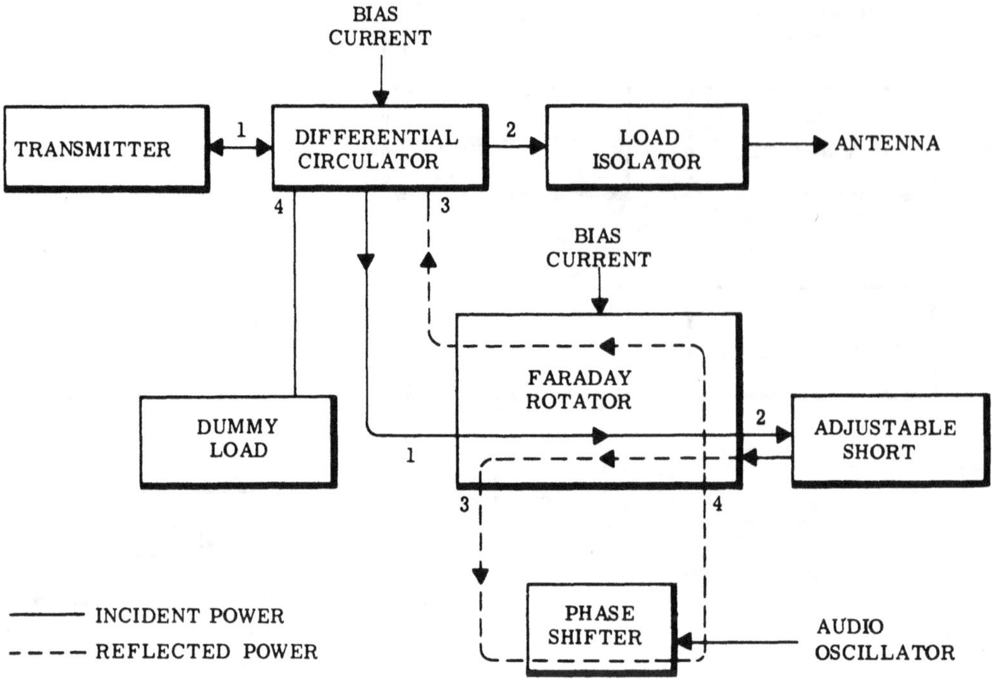

FIGURE 11. EXPERIMENTAL FREQUENCY SHIFTING TRANSMITTER

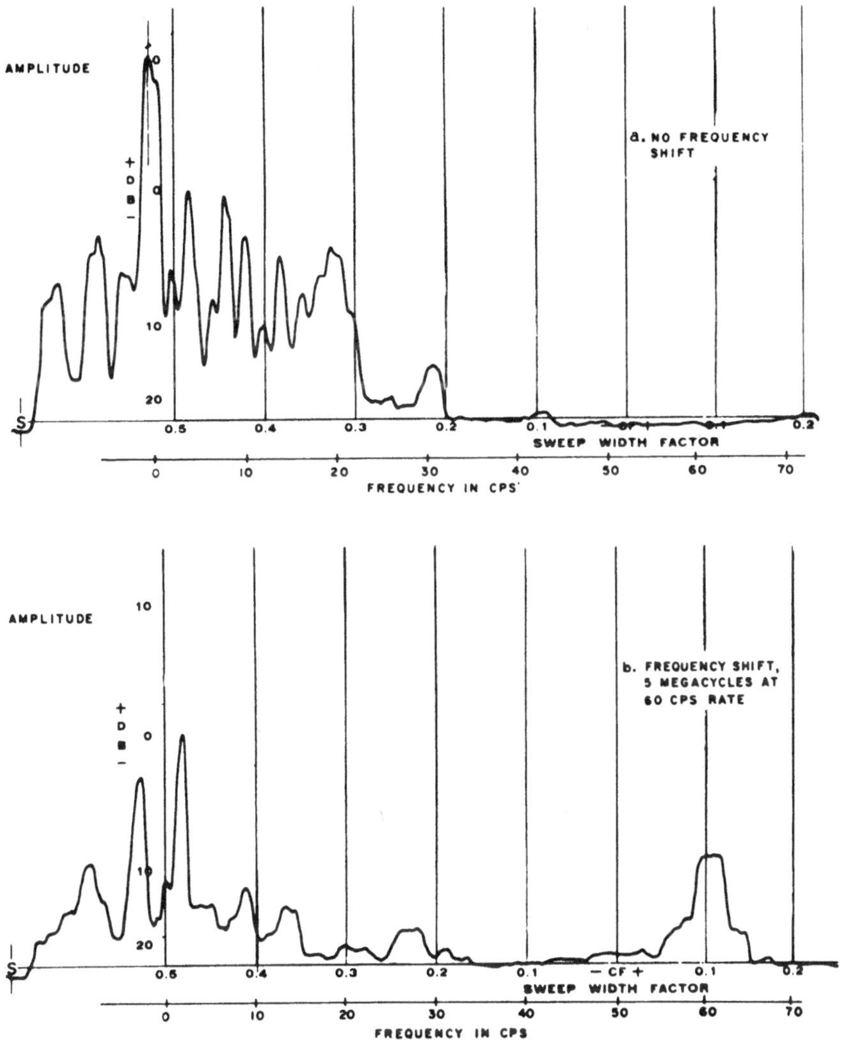

FIGURE 12. SCINTILLATION SPECTRA. (TREE COVERED TERRAIN, SLIGHT WIND)

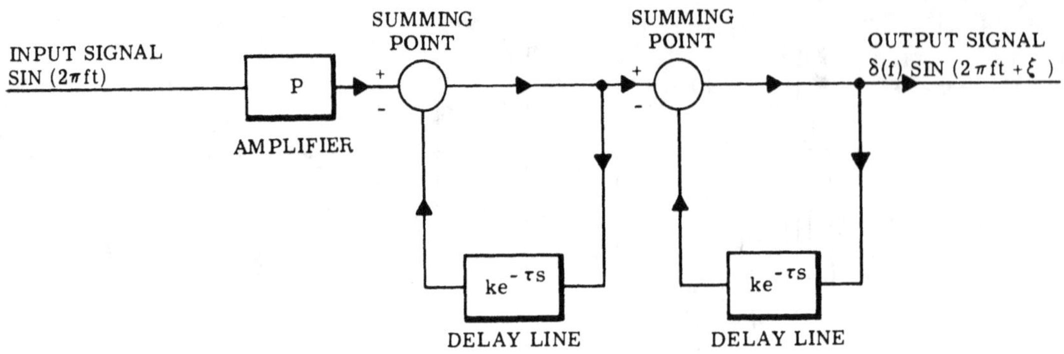

$P = \frac{\Delta\beta(1-a^2)}{2}$, $k = a$, $\tau = \frac{\Delta R}{C}$

FIGURE 13. A SIMULATOR FOR ANGLE-OF-ARRIVAL ERROR (TWO RADIATOR CASE)

Reduction of Radar Tracking Errors with Frequency Agility

G. Lind

IEEE Transactions on Aerospace and Electronic Systems, Vol. AES-4, No. 3 (May 1968)
Pages 410-416

Abstract

Frequency agility with random frequency in each pulse gives an improvement in radar angle tracking with a monopulse radar. With a conical-scan tracking radar, the glint error is reduced but fading error can be increased, and the net result must be studied in each case.

A theory, usable for calculating angle tracking errors with a frequency agile radar, is given, and two examples showing the error reduction are presented. According to the theory, one part of the glint or fading spectra is "smeared out" to half the pulse repetition frequency. Another part, the size of which depends on the degree of correlation between pulses, keeps the form of the original spectrum.

I. Introduction

A tracking radar is a radar that measures the angles to a target by pointing a radar lobe on the target, in most cases with a moving antenna. The radar system includes means for measuring the deviation of the target from the antenna axis. This can be done by comparing signals from different positions of a lobe near the target (lobe scanning or switching) or by comparing signals in different adjacent lobes simultaneously (monopulse). In both cases, the result is a signal representing the target deviation. This signal is used as an error signal in the antenna positioning servo, whereby the antenna is pointed at the target. The target angles are represented by the mechanical position of the antenna.

The accuracy of radar angle tracking as described above is limited by the effects of glint and fading,[1] two results of the interference that occurs between echoes from different parts of a complex target. These effects are serious limitations on the accuracy of many systems. Frequency modulation, regular or random, has been proposed as a method of increasing the accuracy of radar angle tracking.

A study of angle tracking error reduction with frequency modulation has been presented by Birkemeier and Wallace.[2] In their study conditions with a regular pulse-to-pulse frequency modulation are investigated. This case has also been studied by Kosowsky *et al.*[4]

With a so-called jumping frequency or frequency agile radar, a somewhat different situation occurs. A frequency agile radar[3],[10] is characterized by a random pulse-to-pulse frequency modulation. This gives excellent ECCM performance, and also a reduction of radar tracking errors. This paper will deal with the problem of calculating the radar tracking errors with a frequency agile radar.

With random frequency modulation the frequencies of any two pulses in a sequence are independent. The statistical properties of the transmitted frequencies are then completely described by the probability density function (PDF). Two types of PDF for the frequencies will be discussed, each one connected with a special mechanization of the frequency agile radar.

In the system described by Thourel[6] a backward wave oscillator (BWO) is frequency modulated by a random noise voltage, which is sampled and held constant in each pulse interval. The BWO output is then amplified in a power amplifier transmitter tube. In this system, any distribution of the frequencies transmitted can be achieved by manipulating the distribution of the noise voltage. Very often, one is forced to keep the transmitted frequencies within a specified frequency band. To do this it is best to use a rectangular (equal) distribution. Therefore, the rectangular distribution is chosen to represent the first mechanization.

In another type of system, a so-called spin-tuned magnetron is used. The output frequency, or actually the frequency a pulse would have if the magnetron were modulated to give a pulse, is rapidly varied in a nearly sinus-

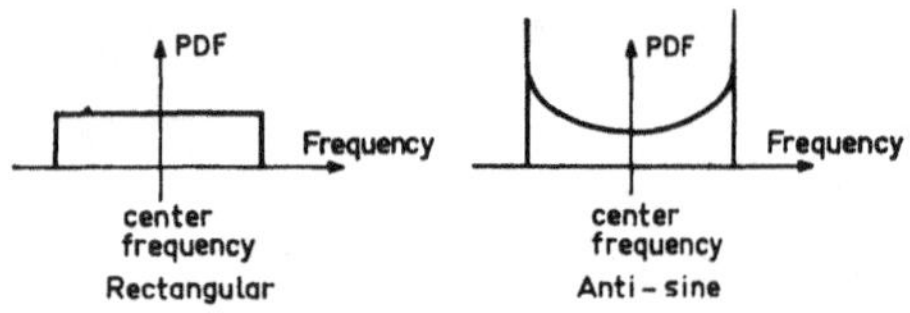

Fig. 1. Probability density function for rectangular and anti-sine distribution of frequency.

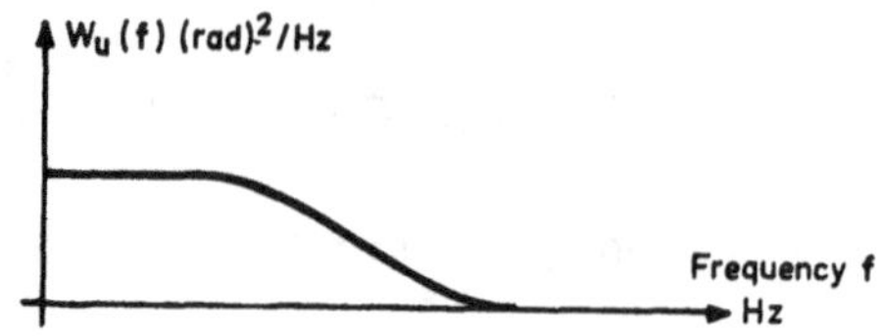

Fig. 2. Example of spectral density of glint.

oidal way. By using a random variation in the pulse repetition frequency, the output frequency of the pulses will be completely random and uncorrelated from pulse to pulse. The distribution of frequencies will in this case be the same as the distribution from sampling a regular sinusoidal function at random intervals. This latter distribution will be referred to as the anti-sine distribution. The two distributions, the rectangular and the anti-sine, are shown in Fig. 1.

II. Glint Spectrum

"Glint" denotes the wandering of the apparent center of reflectivity of a target when measured with a tracking radar. It is caused by interference between the echoes from the different small reflectors of a complex target. For usual targets with a fixed frequency tracking radar (monopulse or conical-scan), glint will have a strong influence on the tracking accuracy.

The standard deviation of the total glint is connected with target size. It has been studied for different reflector distributions by Delano.[7] The glint spectrum is dependent on the relative motions of the small reflectors of a complex target. Such motions are caused by vibration, target maneuvers, range closure, the aspect variation as seen by the radar when the target makes a straight passing course, and course instability caused by air turbulence (aircraft) and sea waves (ships). The glint spectrum is normally of the low-pass type, with bandwidth proportional to the spread in radial velocities within the target, divided by radar wavelength. The glint spectrum for one of Delano's target models, an infinite number of small reflectors in random phase distributed with constant density along a strip, has been studied for different types of motion by Muchmore.[8]

The influence of frequency agility on the glint spectrum can be studied without specifying any original (fixed frequency) glint spectrum in detail. Using the notation of Delano and Muchmore, glint is denoted by $U(t)$. The glint spectrum is denoted by $W_u(f)$, an example is shown in Fig. 2. $R_u(\tau)$ is the autocorrelation function of the glint $U(t)$ and also the cosine transform of glint spectrum $W_u(f)$. An example of this autocorrelation of glint is shown in Fig. 3.

If we consider the effect of sampling and holding in a pulse radar the autocorrelation function will be

$$R_{u1}(\tau) = \sum_{n=-\infty}^{\infty} R_u(nT_0)P(\tau - nT_0), \qquad (1)$$

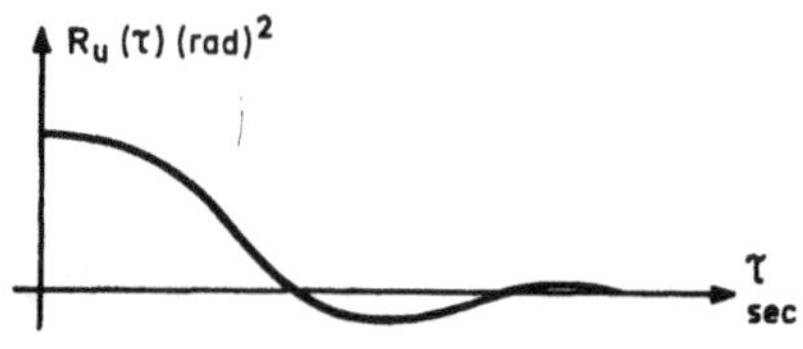

Fig. 3. Example of autocorrelation function of glint.

where $P(\tau)$ is a triangular function of unit height and base $2T_0$ and T_0 is the pulse repetition period. Sampling is assumed to be with short pulses, and holding is done with a boxcar function that ends immediately before the next pulse. This model is used throughout the paper.

Birkemeier and Wallace have shown that a regular frequency modulation, repeating itself every N pulses, with a frequency change between pulses wide enough to give independent glint values, results in the following autocorrelation function:

$$R_{u2}(\tau) = \sum_{n=-\infty}^{\infty} R_u(nNT_0)P(\tau - nNT_0). \qquad (2)$$

In (2) correlation has ceased to exist between pulses except where the interval is a multiple of NT_0, the period for the frequency modulation.

With frequency agility, conditions are somewhat different. As shown by Birkemeier and Wallace, complete decorrelation in the glint between two pulses only occurs if the frequency difference for the two pulses is greater than a certain frequency difference. This frequency difference, the critical frequency difference Δf_c, has been shown to be

$$\Delta f_c = c/2D, \qquad (3)$$

where c is the velocity of light and D is target depth. Here the Delano strip target model is used. See also Ray[3] and Simon,[5] where similar results are presented for slightly different conditions and target models.

With random frequency modulation from pulse to pulse within a frequency region Δf_A, where $\Delta f_A \gg \Delta f_c$, there is a certain probability p_c for a frequency interval between any two pulses being smaller than Δf_c. This means that in a sequence of pulses, pairs of pulses with any number of pulses between them will show completely correlated glint in p_c of the cases and no correlation in $1-p_c$ of the cases. This argument involves a certain approximation, which will be discussed in Section IV.

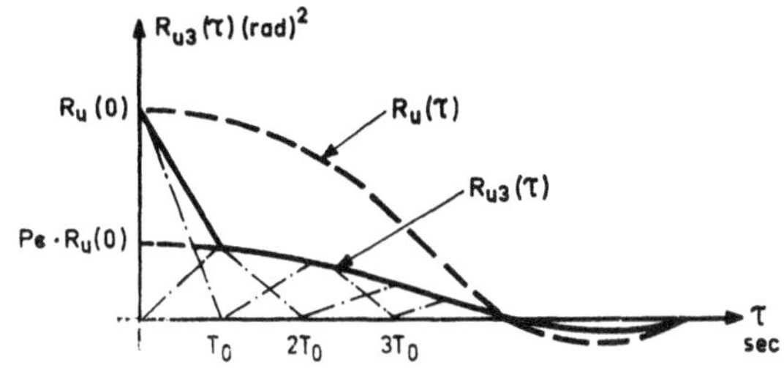

Fig. 4. Example of autocorrelation function for glint with frequency agility.

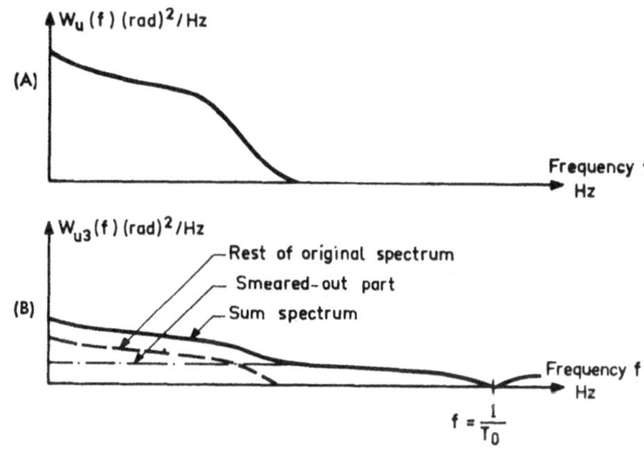

Fig. 5. Glint spectrum. (A) With fixed frequency. (B) With frequency agility introduced.

It is now clear, however, that the expression for frequency agility corresponding to (1) will be

$$R_{u3}(\tau) = (1 - p_c)R_u(0)P(\tau) + p_c \sum_{n=-\infty}^{\infty} R_u(nT_0)P(\tau - nT_0). \tag{4}$$

That is, with n equal to zero and correlation between one sample (pulse) and itself, the autocorrelation function with frequency agility equals that of fixed frequency. With n equal to nonzero and correlation between different samples (pulses), the autocorrelation function is p_c times the original (with fixed frequency). p_c, the probability for an undercritical frequency difference, is introduced because the autocorrelation function is defined as a time average. $R_{u3}(\tau)$ is illustrated in Fig. 4.

To retrieve the spectrum corresponding to $R_{u3}(\tau)$, the additive property of the Fourier transform is used. The first part of (4)

$$(1 - p_c)R_u(0)P(\tau),$$

gives a spectral term containing $(1-p_c)$ times the total glint power $R_u(0)$, sampled in independent samples with the pulse repetition frequency $1/T_0$. This part is "smeared out" in frequency over approximately half the PRF. The second part of (4) gives a spectral term containing the rest of the glint power $p_cR_u(0)$ with the original spectral form modified by sampling and holding. In practical radar systems the pulse repetition frequency is generally much higher than the bandwidth of the glint spectrum. The effect of sampling and holding on the glint autocorrelation function and spectrum is then very small and can therefore be neglected.

The spectrum for glint with frequency agility can now be written

$$W_{u3}(f) = 2T_0(1 - p_c)R_u(0)\frac{\sin^2(\pi f T_0)}{(\pi f T_0)^2} + p_cW_u(f). \tag{5}$$

Fig. 5(A) is an example of glint spectrum with fixed frequency, and 5(B) is an example of glint spectrum with frequency agility. p_c is assumed to be 0.3 in the example.

III. Fading Spectrum

Fading is the term for fluctuations in the received echo signal strength, caused by the same phenomena as glint, phase interference within the target. It is the cause of tracking errors in two ways. The first, which occurs both in monopulse and conical-scan systems, is often referred to as angular error noise. If there is an appreciable tracking error in angle, i.e., caused by servo lag, and the radar has a slow AGC system, the fading of the echo strength will cause the radar to report a varying tracking error. The tracking error is thus modulated by fading. The remedy for this is either to reduce the tracking lag or to use a fast AGC.

The second way in which tracking errors can be caused by fading occurs only in conical-scan or lobe-switching systems, but not in monopulse systems. Since the conical-scan or lobe-switching radar both use an artificial modulation on echo signal strength to measure angle deviation, it is only natural that a similar modulation caused by fading cannot be distinguished from the actual angle deviation. Only components in the fading spectrum near the conical-scan frequency cause slow tracking errors.

The fading spectrum for Delano's strip target model has been studied by Muchmore.[8] The fading spectrum is in general of a low-pass nature, having the same upper frequency limit as glint, but not necessarily the same spectral form. The total fading depth depends on the distribution in size of the small partial reflectors constituting a complex target. If they are numerous and reasonably equal in size, a Rayleigh distribution can be expected.

As with regular frequency modulation, frequency agility has the same influence on the fading spectrum as on the glint spectrum. The fading spectrum with frequency agility will consist of one part with the same form as for fixed frequency, multiplied by the probability for an undercritical frequency change p_c. The rest of the original fading power spectrum will be smeared out in $\sin^2 x/x^2$ pattern, with the first null at the PRF (see Fig. 5). The result of this is, however, not always a decrease in tracking error. For a conical-scan system the smeared-out part of the spectrum might even increase the fading spectral density around the conical-scan frequency. This happends especially in the case of very slow fading with little or no fading at conical-scan frequency. With a monopulse tracking system, the introduction of frequency agility always increases the tracking accuracy.

IV. The Probability of Undercritical Frequency Change

Given the distribution of transmitted frequencies in a frequency agile radar and assuming that the frequencies are independent from pulse to pulse, we first study the distribution of frequency differences for any pair of pulses. Let x denote the frequency of the first pulse, with probability density function $f(x)$. For the second pulse, the corresponding notations are y and $f(y)$. The functions $f(x)$ and $f(y)$ are identical except for the argument.

Let $z = |x-y|$ denote the frequency difference (always positive). The distribution of z will then be

$$F(z) = P(Z \leq z) = \int_{-\infty}^{\infty} \int_{x-z}^{x+z} f(x)f(y)dydx. \quad (6)$$

$$F(z) = \begin{cases} 0 & z < 0 \\ \frac{1}{2} + \arcsin(z-1) + \int_{\arcsin(z-1)}^{\pi/2} \arcsin(\sin y - z + 1)\, dy & 0 < z < 2 \\ 1 & 2 < z. \end{cases} \quad (11)$$

For our first example, with $f(x)$ a rectangular distribution between -1 and $+1$,

$$f(x) = \begin{cases} 0 & x < -1 \\ 0.5 & -1 < x < 1 \\ 0 & 1 < x, \end{cases} \quad (7)$$

and with the same function for $f(y)$, the result is

$$F(z) = \begin{cases} 0 & z < 0 \\ z - \frac{z^2}{4} & 0 < z < 2 \\ 1 & 2 < z. \end{cases} \quad (8)$$

The corresponding PDF for z is then

$$f(z) = \begin{cases} 0 & z < 0 \\ 1 - \frac{z}{2} & 0 < z < 2 \\ 0 & 2 < z. \end{cases} \quad (9)$$

In our second case, we chose (for convenience only) the same limits of distribution -1 and $+1$. The antisine distribution is defined:

TABLE I

z	$F(z)$	z	$F(z)$
0.1	0.1094	1.1	0.6738
0.2	0.1905	1.2	0.7149
0.3	0.2608	1.3	0.7546
0.4	0.3248	1.4	0.7929
0.5	0.3837	1.5	0.8300
0.6	0.4386	1.6	0.8660
0.7	0.4903	1.7	0.9009
0.8	0.5398	1.8	0.9348
0.9	0.5865	1.9	0.9679
1.0	0.6310		

$$f(x) = \begin{cases} 0 & x < -1 \\ \frac{1}{\pi\sqrt{1-x^2}} & -1 < x < 1 \\ 0 & 1 < x, \end{cases} \quad (10)$$

and the same function for $f(y)$. No closed expression has been found for $F(z)$ in this case, but the following form was found suitable for numerical evaluation on a computer:

The function above has been computed, and the result appears in Table I. For the PDF in this case, a closed expression exists[9]

$$g(z) = \frac{8}{(2+z)\pi^2} K\left(\frac{2+z}{2-z}\right) \quad 0 < z < 2 \quad (12)$$

and zero otherwise. K is the complete elliptic function of the second kind.

The functions $g(z)$ and $F(z)$ for the rectangular and anti-sine distributions of frequencies are shown in Fig. 6.

The model of glint and fading decorrelation from frequency changes that was introduced in Section II, with no decorrelation for frequency shifts less than Δf_c, and complete decorrelation when the shift is more than Δf_c, is crude. A more refined model would be the functions describing decorrelation as a function of frequency, which have been derived by Birkemeier and Wallace. They denote this function $\rho E(\Delta f)$ for glint and $\rho U(\Delta f)$ for fading. [2, eqs. (29) and (11)]

The average decorrelation for glint would then be

$$p_c' = \int_0^{\infty} \rho E(\Delta f) g(\Delta f) \Delta f, \quad (13)$$

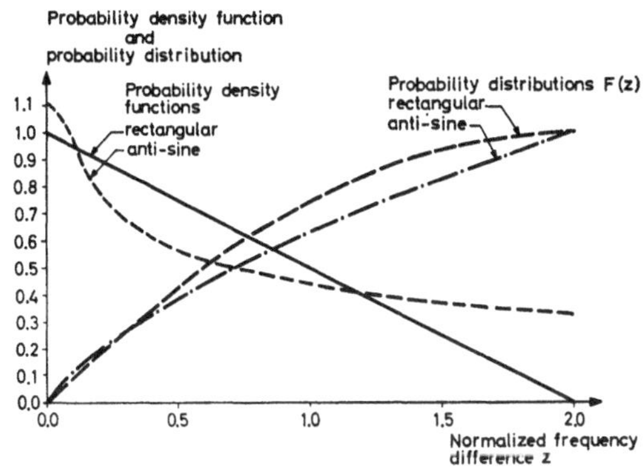

Fig. 6. Probability density function $g(z)$ and probability distribution $F(z)$ for frequency differences.

where $g(\Delta f)$ is the PDF for frequency differences. The same integral with $_\rho U(\Delta f)$ instead of $_\rho E(\Delta f)$ would apply to fading.

For frequency agility to be interesting at all in the situations discussed here, the band over which the frequency is varied would have to be much wider than Δf_c according to (3). As is shown in Birkemeier and Wallace[2] the functions $_\rho E(\Delta f)$ and $_\rho U(\Delta f)$ show very little correlation for $\Delta f > \Delta f_c$, and the PDF's vary only a little for small frequency changes (small z in (9) and (12)), so the model from Section II is a good approximation in all the interesting cases. For that model p_c is simply calculated:

$$p_c = F(\Delta f_c), \tag{14}$$

where F is the distribution function $F(z)$ for frequency differences according to (8) or (11), and Δf_c is the critical frequency change according to (3).

V. The Influence of Frequency Agility Bandwidth on Tracking Error Improvement

For a monopulse radar, with glint as the only source of tracking error, frequency agility gives an improvement in tracking accuracy. It is intuitively clear that this improvement must depend on the frequency agility bandwidth. No improvement exists, by definition, in a fixed frequency radar with agility bandwidth zero. With frequency agility, errors are reduced. In most cases the spectral density of glint at zero frequency is an adequate measure of tracking error, since the bandwidth of antenna servos is generally narrow.

The improvement in tracking accuracy I taken as the ratio of glint spectral density at zero frequency with fixed frequency to glint spectral density at zero frequency with frequency agility has been calculated and is shown in Figs. 7 and 8. Fig. 7 is valid for anti-sine frequency distribution, and Fig. 8 shows the improvement with rectangular frequency distribution.

The improvement I as defined above is given as a function of the normalized variable k, which is defined:

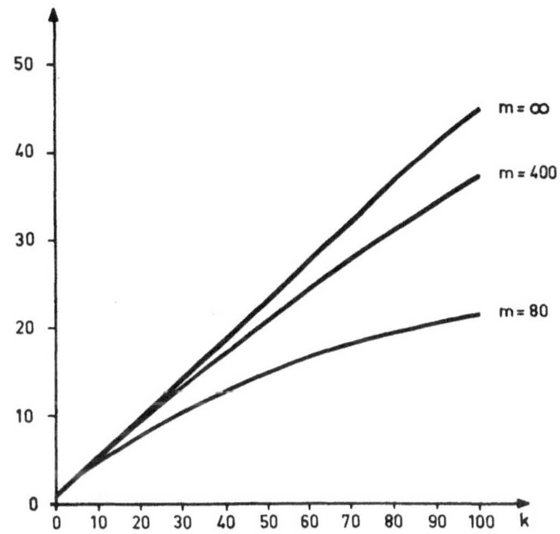

Fig. 7. Tracking improvement as a function of frequency agility bandwidth.

Fig. 8. Tracking improvement as a function of frequency agility bandwidth.

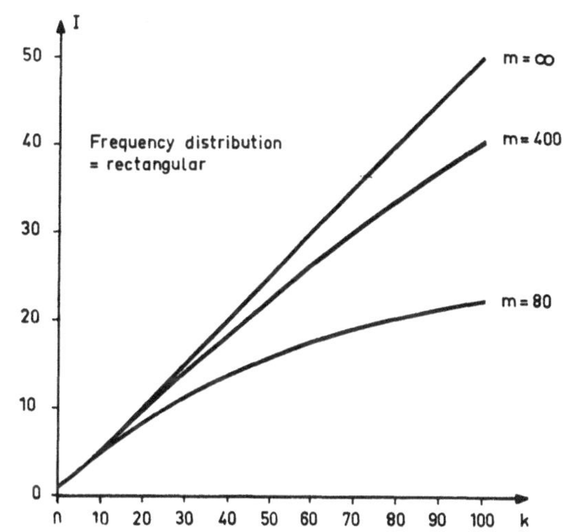

$$k = \frac{\text{frequency agility bandwidth}}{\text{critical frequency difference}}.$$

The critical frequency difference can be calculated from (3).

The efficiency of smearing out the glint spectrum with frequency agility is determined by comparing the PRF to the original glint bandwidth. In the figures different curves appear for different values of the normalized parameter m, which is defined:

$$m = \frac{\text{pulse repetition frequency}}{\text{original glint bandwidth}}.$$

The following values of m are represented: 80, 400, ∞.

The values 80 and 400 are chosen to be representative, e.g., glint bandwidths of 25 and 5 Hz, respectively, and PRF 2000 Hz. $m=\infty$ is given only to show the theoretical maximum of tracking improvement.

As can be seen from the figures, the tracking improvement is almost the same with the two frequency distributions. For infinite m, that is, infinite PRF, and for large k, that is, agility bandwidth much larger than the critical frequency difference, I approximates $k/2$. This is due to the simple fact that an undercritical frequency difference can be obtained in two directions.

For infinite k, that is, infinite agility bandwidth, the improvement is $m/2$. This is the same as the maximum improvement with regular frequency modulation according to Birkemeier.[2] In that case, the maximum is obtained when the frequency is changed in steps, each wide enough to ensure decorrelation. The number of frequencies must also be so large that no correlation exists over the period when the sequence of frequencies repeats itself.

VI. Examples

Two examples have been prepared showing the use of the theory in practical calculations. The spectral density of glint at zero frequency is calculated for the azimuth channel. The target in the first example is an aircraft on a straight course and in the second a ship moving in a seaway.

Example 1

Radar data:

Wavelength:	3 cm
PRF:	2000 Hz
AGC:	slow.

Target data:

Type:	Aircraft
Width:	12 meters
Depth:	5 meters (for significant radar reflectors)
Range:	3000 meters
Crossover distance:	1000 meters
Speed:	300 m/s.

The total glint power is calculated according to eq. (16b) in Delano[7] as $(12)^2/6\pi \approx 7.6$ m^2. From the geometry, the target aspect angle velocity is 0.033 rad/s. This gives, according to Muchmore,[8] the glint bandwidth $2\times 12\times 0.033/0.03 \approx 27$ Hz. The zero frequency spectral density with a fixed frequency radar is then approximately $7.6/27 = 0.28$ m^2/Hz.

Now assume frequency agility over a band of 450 MHz, anti-sine distributed. The critical frequency difference is, according to (3),

$$\Delta f_c = \frac{3 \times 10^8}{2 \times 5} = 30 \text{ MHz}.$$

This gives, when (14) is used

$$p_c \approx 0.14.$$

The spectral density at zero frequency with frequency agility can now be calculated. The smeared-out portion of $\sin^2 x/x^2$ type can be assumed white for low frequencies over a band from zero to half the PRF, in this case 1000 Hz. The result is, according to (5),

$$W_{u3}(0) = 0.86 \frac{7.6}{1000} + 0.14 \times 0.28 \approx 0.006 + 0.039$$

$$= 0.045 \text{ m}^2/\text{Hz}.$$

The tracking accuracy ratio can be estimated to be

$$\sqrt{\frac{0.28}{0.045}} \approx 2.5 \text{ times}.$$

Example 2

Radar data:

Wavelength:	3 cm
PRF:	1500
AGC:	slow.

Target data:

Type:	ship
Width:	75 meters
Depth:	15 meters (for significant radar reflectors)
Azimuth angle velocity due to course instability in sea waves:	5 mrad/s.

The total glint power is calculated as $(75)^2/6\pi \approx 300$ m^2. The target angle velocity is 0.005 rad/s, which gives the glint bandwidth:

$$\frac{2 \times 75 \times 5 \times 10^{-3}}{3 \times 10^{-2}} \approx 25 \text{ Hz}.$$

The zero frequency spectral density with a fixed frequency radar is then $300/25 \approx 12$ m^2/Hz.

The frequency agility is here assumed to cover 250 MHz with rectangular distribution. The critical frequency difference is

$$\Delta f_c = \frac{3 \times 10^8}{2 \times 15} = 10 \text{ MHz}.$$

Equation (14) then gives

$$p_c \approx 0.08.$$

With the same assumptions as in Example 1, the zero frequency glint spectral density is

$$W_{u3}(0) = 0.92 \frac{300}{750} + 0.08 \times 12$$

$$= 0.37 + 0.96 = 1.33 \text{ m}^2/\text{Hz}.$$

Here the tracking accuracy ratio is estimated as

$$\sqrt{\frac{12}{1.33}} \approx 3.0 \text{ times}.$$

VII. Conclusions

It has been shown that frequency agility gives a definite decrease of tracking error with a monopulse tracking radar. With a conical-scan tracking system glint errors are reduced, but an increase in error due to fading is possible. For the net result no general prediction can be given and a special study must be made in each case.

Glint and fading spectra are both influenced in the same way by frequency agility. It has been shown that the original glint or fading spectra with a fixed frequency radar is split into two parts under frequency agile conditions. One part has the same form as the original spectrum, but is decreased by multiplication by p_c. The other part, containing the rest of the original power, is spread out in a $\sin^2 x/x^2$ fashion with the first null at the pulse repetition frequency. p_c is the probability of undercritical frequency change. Methods for calculating this probability for two different types of frequency agility have been presented in Section IV.

With the preceding components a complete theory for calculating target-induced radar tracking errors with frequency agility can be built. Two examples of practical glint error calculation, with a tracking radar following an aircraft in Example 1 and a ship in Example 2, are presented in Section VI. With this model the influence of frequency agility bandwidth on tracking errors is treated in Section V. Charts are presented showing how tracking accuracy is improved as the frequency agility bandwidth is increased. Experimental tests have confirmed the validity of the theory presented here. The results, however, are not available for publication at this time.

REFERENCES

[1] J. H. Dunn, D. D. Howard, and A. M. King, "Phenomena of scintillation noise in radar-tracking systems," *Proc. IRE*, pp. 855–863, May 1959.

[2] W. P. Birkemeier and N. D. Wallace, "Radar tracking accuracy improvement by means of pulse-to-pulse frequency modulation," *IEEE Trans. Communications and Electronics*, January 1963; published in *AIEE Trans. (Communications and Electronics)*, vol. 81, pp. 571–575, 1962.

[3] H. Ray, "Improving radar range and angle detection with frequency agility," *Microwave J.*, pp. 63–68, May 1966.

[4] Kosowsky *et al.*, "The reduction of angle-of-arrival scintillation by a frequency shifting technique," presented at the 1963 Nat'l Electronics Conf.

[5] J. C. Simon, "Quelques problèmes de fluctuations en radioéléctricité," *Ann. Radioéléctricité*, January 1955.

[6] L. Thourel, *Les Radars de Veille Modernes*. Sofradel, 1965.

[7] R. Delano, "A theory of target glint or angular scintillation in radar tracking," *Proc. IRE*, pp. 1778–1784, December 1953.

[8] R. B. Muchmore, "Aircraft scintillation spectra," *IRE Trans. Antennas and Propagation*, vol. AP-8, pp. 201–212, March 1960.

[9] K. Beskow, private communication.

[10] B. G. Gustafson and B. O. Ås, "System properties of jumping frequency radar," *Philips Telecommun. Rev.*, vol. 25, no. 1, pp. 70–76.

Göran Lind was born in Luleå, Sweden, in 1934. He graduated in 1959 from the Royal Institute of Technology, Stockholm, where he obtained the M.E. degree in electrical engineering.

He joined the Research Institute of the Swedish National Defense in 1959, to work in the communications area. From 1960 till 1966 he was engaged in the design and projecting of gun fire control systems at AB Bofors, Bofors, Sweden. Since 1966 he has been doing project work on radar systems at the Philips Teleindustri AB, a Swedish subsidiary to the Philips concern.

Mr. Lind is a member of the Swedish Association of Electrical Engineers.

Tracking Errors in a Frequency Agility System

P.F. Guarguaglini

AGARD Conference Proceedings, No. 66 on Advanced Radar Systems (May 25-29, 1970)
Pages 9-1 to 9-11

SUMMARY

The present study evaluates the influence of frequency agility on tracking errors in a radar system. The use of frequency agility modifies the received spectrum. Therefore the magnitude of the glint and scintillating errors changes. In order to find the signal spectrum, the correlation, between succeeding pulses, has been evaluated. The correlation is dependent on frequency agility bandwidth, on target dimensions and on the law, by which the transmitted frequency is chosen.

Transforming the signal autocorrelation function, the received power spectrum has been obtained; a broadening of the spectrum due to the reduction of the correlation among pulses, occurs. Expressions of glint and scintillation errors have been obtained. For some specific cases, errors have been evaluated and plotted as a function of the correlation coefficient.

From the analysis of the results, it is possible to draw this conclusion: the use of frequency agility gives a reduction of tracking errors for a monopulse radar. For a conical-scan system, while the glint error is always reduced, the scintillation error may increase.

1. INTRODUCTION.

The angular accuracy has been completely evaluated for radars, using a fixed transmission frequency [1]. The main causes of error, for both a monopulse and a conical-scan system, are [2]: a) amplitude fluctuation, b) angle scintillation or glint, c) receiver noise, d) servo noise. While the latter two errors are independent of the transmitted signal spectrum, the former two are dependent on how a target looks when a pulse hits it.

Therefore in a frequency agility system, where the transmitted frequency is changed at every pulse, the tracking errors, due to angle and amplitude fluctuations, assume different values that in the conventional radar.

The purpose of this paper is to carry out a complete evaluation of the tracking errors for monopulse and conical scan radars, both employing frequency agility.

2. ANGULAR ERRORS IN A MONOPULSE RADAR

The system, here considered, is a traditional monopulse radar. A simplified block diagram of such a radar may be found in [1] or [2].

The errors, introduced by target scintillation, depend mainly on the AGC and servo characteristics and on the noise power spectra at the input of these two circuits. This requires to compute the received signal spectrum.

2.1. Received signal spectrum.

The signals, both in the sum and in difference channels, have spectra of equal shape. When frequency agility in not used, the echo spectrum in skechted in fig. 1.

The broadening in the lines is caused by target scintillation and its shape is dependent on target characteristics. A good fit for this shape is assumed, [1],:

$$W(f) = \frac{1}{\pi f_o} \frac{f_o^2}{f^2 + f_o^2} \qquad (1)$$

where f_o, the half-power frequency, depends on the maximum target dimension and on radar wavelength.

Without taking into consideration the carrier frequency, the spectrum is given by:

$$S(f) = \left[\frac{\sin(\pi\tau f)}{\pi\tau f}\right]^2 \sum_{n=-\infty}^{+\infty} W(f - nF_r) \tag{2}$$

where τ is the pulse length and F_r is the pulse repetition frequency.

In a frequency agility radar, the target echo fluctuates not only because of motions of the target but also because the trasmitted frequency is different from pulse to pulse. Therefore the r.f. power spectrum of the received signal is different from the one in the previous case.

The power spectrum is related to the correlation function by:

$$S(f) = 2\int_0^{\infty} R(\tau)\cos(2\pi\tau f)\, d\tau \tag{3}$$

Therefore, by evaluating the correlation function it is possible to obtain the signal spectrum.

Let $\Phi_{ji}(f_j, f_i)$ be the correlation between the i-th and the j-th pulses when f_i and f_j are the transmitted frequencies.

Considering every possible case, the correlation factor is given by:

$$\Phi_{ij} = \iint \Phi_{ij}(f_i, f_j)\, p(f_i, f_j)\, df_i\, df_j \tag{4}$$

where $p(f_i f_j)$ is the joint probability density function of f_i and f_j .

The criterion by which at every pulse the transmitted frequency is chosen depends on operative considerations. The criterion does not change during a single operative condition. Therefore the relation is valid:

$$p(f_q, f_{q+n}) = p(f_r, f_{r+n}) \qquad r \neq q \tag{5}$$

Eq. 4 may be written:

$$\Phi_n = \iint \Phi_n(f_i, f_{i+n})\, p(f_i, f_{i+n})\, df_i\, df_{i+n} \tag{6}$$

Let

$$\rho_s(t) = 2\int_0^{\infty} W(f)\cos\omega t\, dt \tag{7}$$

be the autocorrelation function of the internal fluctuations of the target. Taking into account eq. 6, except for a scaling factor, the envelope of the autocorrelation function of an infinite train of pulses is given by:

$$R_s(t) = \rho_s(t)\left[\sum_{n=-\infty}^{+\infty} \Phi_n \left(1 - \frac{|t - nT_r|}{\tau}\right) \cdot \mathrm{rect}\left(\frac{t - nT_r}{2\tau}\right)\right] \tag{8}$$

where rect(t) is given by [3]:

$$\mathrm{rect}(t) = \begin{cases} 1 & \text{for } |t| \leq 1/2 \\ 0 & \text{for } |t| > 1/2 \end{cases} \tag{9}$$

The presence of Φ_n in eq. 8 makes $R_s(t)$ decrease more rapidly than in the case of a single transmitted frequency. This means that the spectrum around the lines presents a broadening.

As eq. 6 shows, a complete evaluation of Φ_n is possible only if $\Phi_n(f_i, f_{i+n})$ and $p(f_i, f_{i+n})$ are known. There is no published experimental work on $\Phi_n(f_i, f_{i+n})$. A theoretical expression has been often used [4], [5]. In order to evaluate $p(f_i, f_{i+n})$ it is necessary define the criterion, employed for choosing the transmitted frequency. In a frequency agile radar, usually the frequency is randomly chosen within a band. Therefore the frequency of any two pulses are independent and the relation

$$p(f_i, f_{i+n}) = p(f_i)\, p(f_{i+n}) \tag{10}$$

holds.

Taking into consideration eq. 10, eq. 6 gives:

$$\Phi_n = \begin{cases} 1 & \text{for } n = 0 \\ \Phi & \text{for } n \neq 0 \end{cases} \tag{11}$$

The autocorrelation is now given by:

$$\begin{aligned} R_s(t) = {} & \rho_s(t) \left\{ \left(1 - \frac{|t|}{\tau}\right) \operatorname{rect}\left(\frac{t}{2\tau}\right) + \right. \\ & \left. + \Phi \cdot \sum_{\substack{n=-\infty \\ n \neq 0}}^{n=+\infty} \left(1 - \frac{|t - nT_r|}{\tau}\right) \operatorname{rect}\left(\frac{t - nT_r}{2\tau}\right) \right\} = \\ = {} & \rho_s(t) \left\{ (1 - \Phi)\left(1 - \frac{t}{\tau}\right) \operatorname{rect}\left(\frac{t}{2\tau}\right) + \right. \\ & \left. + \Phi \cdot \sum_{n=-\infty}^{n=+\infty} \left(1 - \frac{|t - nT_r|}{\tau}\right) \operatorname{rect}\left(\frac{t - nT_r}{2\tau}\right) \right\} \end{aligned} \tag{12}$$

Usually the pulse length is much shorter than the target correlation time and it may be assumed, without loss of generality $\rho_s(\tau) = 1$. Therefore

$$R_s(t) = (1 - \Phi)\left[1 - \frac{|t|}{\tau}\right] \operatorname{rect}\left(\frac{t}{2\tau}\right) + \Phi \cdot \rho_s(t) \sum_{n=-\infty}^{n=+\infty} \left(1 - \frac{|t - nT_r|}{\tau}\right) \operatorname{rect}\left(\frac{t - nT_r}{2\tau}\right) \tag{13}$$

Using eq. 3 it is possible to obtain the spectrum of the received signal:

$$S(f) = (1 - \Phi)\left[\frac{\sin(\pi f \tau)}{\pi \tau f}\right]^2 + \Phi \left[\frac{\sin(\pi \tau f)}{\pi \tau f}\right]^2 \sum_{n=-\infty}^{n=+\infty} W(f - nF_r) \tag{14}$$

As eq. 14 shows, the use frequency agility divides the spectrum in two terms. One of them has a shape equal to the spectrum of a convential radar, but is reduced by a factor Φ , which depends on target characteristics and on the width of the transmitted frequency band. The other term gives rise to a broad-band spectrum. If complete uncorrelation between any two pulses is obtained, only the broad-band term is present.

Fig. 2 shows a typical spectrum of the received signal for a frequency agile monopulse radar.

2.2. Error evaluation.

In order to evaluate the filtering effect of AGC and servo loops on scintillating noise, it is necessary to consider the presence of detectors both in the sum and difference channels. As mentioned before, the signal spectrum shape is equal in both channels. Therefore the signal spectrum at the output of a phase-sensitive detector is given by the convolution of the i.f. spectrum with itself, as for a square-law detector.

Regarding the envelope detector, the shape of detected spectra are slightly different for square-law and linear detection [6]. Therefore only the square-law case will be considered.

In Appendix the signal spectrum after detection has been computed. This spectrum is given by:

$$S_e(f) = k_1 \cdot \left[\Phi^2 \frac{F_r}{\pi f_g} \sum_{n=-\infty}^{n=+\infty} \frac{f_g^2}{f_g^2 + (f - nF_r)^2} \cdot \frac{2\pi n F_r \tau - \sin(2\pi n F_r \tau)}{(2\pi n F_r \tau)} + (1-\Phi^2)\,\frac{2\pi f \tau - \sin(2\pi f \tau)}{(2\pi \tau f)^3} \right] \tag{15}$$

where $f_g = 2f_o$

If k_1 is given in meter per sec or mils per sec, $S_e(f)$ gives directly the error spectrum.

The servo and AGC filters are low-pass; therefore, only the error spectrum components around $f=0$ are of interest.

This part, $S_e'(f)$, is given by:

$$S_e'(f) = k_2 \left\{ \Phi^2 \frac{\alpha}{\pi} \frac{f_g^2}{f_g^2 + f^2} + (1-\Phi^2) \right\} \tag{16}$$

where $\alpha = F_r/f_g$.

It is now possible to evaluate errors.

The error introduced by amplitude scintillation, is proportional to the bias error, θ_o. It is null when the target is on the boresight.

The spectrum of this error is [1]:

$$W_n(f) = S_e'(f) \left| Y_s(f) \right|^2 \theta_o^2 \tag{17}$$

where $Y_s(f)$, the scintillation error factor, is defined by:

$$Y_s(f) = \frac{Y_c(f)}{\left| 1 + Y_a(f) \right|} \tag{18}$$

$Y_a(f)$ is the open-loop voltage gain tranfer function of the AGC loop and $Y_c(f)$ is the closed-loop transfer function of the servo.

Integrating $W_n(f)$ from $-\infty$ to $+\infty$ and taking the square root of the integral the error, σ_s ,is obtained:

$$\sigma_{s1} = \theta_0 \left\{ k_2 \Phi^2 \frac{\alpha}{\pi} \int_{-\infty}^{+\infty} \frac{f_g^2}{f_g^2 + f^2} \left| Y_s(f) \right|^2 df + k_2 (1 - \Phi^2) \int_{-\infty}^{+\infty} \left| Y_s(f) \right|^2 df \right\}^{1/2} \tag{19}$$

The error, when frequency agility is not used, is:

$$\sigma_s = \theta_0 \left\{ k_2 \frac{\alpha}{\pi} \int_{-\infty}^{+\infty} \frac{f_g^2}{f_g^2 + f^2} \left| Y_s(f) \right|^2 df \right\}^{1/2} \tag{20}$$

Therefore eq. 17 may be written:

$$\sigma_{s1} = \sigma_s \left\{ \Phi^2 + \frac{\pi (1 - \Phi^2)}{\alpha \sigma_s} B_n \left| Y_s(0) \right|^2 \right\}^{1/2} \tag{21}$$

where B_n is the noise bandwidth.

Eq. 21 shows that, for Φ constant, the error reduces as B_n and $Y(0)$ increase. It reduces by a lesser amount as α and σ_s increase. The angular noise spectrum is like the scintillation spectrum. Therefore the glint error, σ_α , when AGC is absent, is expressed by eq. 21, where $|Y_s(0)|^2$ is substituted by $|Y_c(0)|^2$.

To compute the glint error, σ_{a1} , when a fast AGC is present, is very complex. However it is possible to use an approximation. It has been found, [6], [7], that, because of a correlation between the echo amplitude and the angular error, the presence of a fast AGC doubles σ_{a1} . Therefore, considering this approximation valid, it is possible to use eq. 21.

3. ANGULAR ERRORS IN A CONICAL-SCAN RADAR.

In a conical-scan radar another error must be added to the two errors, evaluated for monopulse systems. This error is produced by high frequency components within the passband which is centered at the scan frequency.

This error is [1] :

$$\sigma_c^2 = 0.45\, \theta_B^2\, S_e'(f)\, \beta_n \tag{22}$$

where θ_B is the 3 dB beamwidth and β_n is the servo bandwidth.

Taking into consideration eq. 16 we have:

$$S_e'(f) = k_2 \frac{\alpha}{\pi} \frac{f_g^2}{f_g^2 + f^2} \qquad \text{for } \Phi = 1 \tag{23a}$$

$$S_e'(f_s) = K_2 \left[\frac{\alpha}{\pi} \Phi^2 \frac{f_g^2}{f_g^2 + f_s^2} + (1 - \Phi^2) \right] \qquad \text{for } \Phi \neq 1 \tag{23b}$$

Therefore

$$\sigma_{c1} = \sigma_c \left\{ \Phi^2 + \frac{\pi}{\alpha} (1 - \Phi^2) \frac{f_g^2 + f_s^2}{f_g^2} \right\}^{1/2} \tag{24}$$

for small values of Φ and $f_g/f_s \ll 1$:

$$\sigma_{c1} \simeq \sigma_c \frac{f_s}{f_g} \sqrt{\frac{\pi}{\alpha}(1 - \Phi^2)} = \sigma_c \frac{f_s}{\sqrt{f_g F_r}} \sqrt{\pi (1 - \Phi^2)} \tag{25}$$

Therefore the use of frequency agility may increase this error, and the total error could be worse than that of a conventional radar, without frequency agility.

4. EXAMPLES.

In order to give an idea of the effect of frequency agility, errors have been evaluated, assuming for the servo and AGC loops, the transfer function, chosen in [1].The servo band-width and the target spectrum bandwidth are 5 cps, while three different AGC characteristics are taken into consideration (no AGC, slow AGC and fast AGC). A slow AGC usually in employed in a conical-scan radar, while a fast AGC is used in a monopulse radar.

By graphical integration, it is shown that:

$$\sigma_{s1}^2 = \sigma_s^2 \left[\Phi^2 + \frac{5{,}4}{\alpha} (1 - \Phi^2) \right] \qquad \text{for no AGC} \tag{25a}$$

$$\sigma_{s1}^2 = \sigma_s^2 \left[\Phi^2 + \frac{6}{\alpha} (1 - \Phi^2) \right] \qquad \text{for slow AGC} \tag{25b}$$

$$\sigma_{s1}^2 = \sigma_s^2 \left[\Phi^2 + \frac{8.8}{\alpha} (1 - \Phi^2) \right] \qquad \text{for fast AGC} \tag{25c}$$

In figs 3,4 and 5 $\gamma_o^2 = \sigma_{s1}^2/\sigma_s^2$ has been plotted versus Φ for three different values of α .

Comparing the results, the use of frequency agility reduces the conical-scan radar more than the monopulse error.

Eq. 26a allows us to obtain the glint error. It is possible to evaluate how a typical situation is modified by frequency agility.

The situation considered is shown in [8] and in [1] . Assuming $\Phi = 0{,}1$, the glint error is reduced by a factor 10. In fig. 6 the noise components vs.relative range are plotted, when the bias error θ_o is zero, for a monopulse radar. The amplitude noise is not present because $\theta_o = o$. Curve A represent the total error for a conventional radar, while curve B is the total error for a frequency agile radar.

The reduction of glint error is great and the total error is reduced for low values of R. The advantage of frequency agility disappears as R becomes large.

In the case of a conical-scan radar, assuming $f_s/f_g = 6$, we have:

$$\sigma_{c1} = 1.5\,\sigma_c \quad \text{for} \quad \alpha = 50$$
$$\sigma_{c1} = 1.06\,\sigma_c \quad \text{for} \quad \alpha = 100 \qquad (27)$$
$$\sigma_{c1} = 0.75\,\sigma_c \quad \text{for} \quad \alpha = 200$$

Then for $\alpha = 50$ the amplitude noise error increases.

In fig. 7 curve A shows the total error for a conventional conical-scan radar, curve B represents the total error for a frequency agile conical-scan radar when the amplitude fluctuation does not vary Curve C represents the case of $\alpha = 50$. Because the amplitude fluctuation increases, curve C is not always lower than curve A. Therefore there are some value of R, where the use of frequency agility worsens the tracking performance.

5. CONCLUSION.

The use of frequency agility reduces the glint error. This reduction may be great. The amplitude scintillation error, which is present only in a conical-scan radar, may increase or decrease. For small values of Φ , amplitude scintillation error increases as $f_s^2/(f_g \cdot F_r)$ increases.

Therefore the use of frequency agility increases monopulse system performance. For a conical-scan radar it is not possible to reach an univocal conclusion.

APPENDIX

Signal spectrum at the detector output

Apart a continuous term the autocorrelation function at the output of a square-law detector is:

$$R_u(t) = k\,R_s^2(t) \qquad (A1)$$

If $R_s(t)$ is given by eq. 12, we have

$$R_u(t) = k\left\{\varrho_s^2(t)\,\Phi^2 \sum_{n=-\infty}^{n=+\infty}\left[1-\frac{|t-nT_r|}{\tau}\right]^2 rect\left(\frac{t-nT_r}{2\tau}\right) + (1-\Phi^2)\left(1-\frac{|t|}{\tau}\right)^2 rect\left(\frac{t}{2\tau}\right)\right\} \qquad (A2)$$

The spectrum corresponding to the second term is:

$$S_{e2}(f) = 2k(1-\Phi^2)\int_0^{\infty} rect\left(\frac{t}{2\tau}\right)\cdot\left(1-\frac{t}{\tau}\right)^2\cos(\omega t)\,dt = 2k\,(1-\Phi^2)\int_0^{\tau}\left(1-\frac{t}{\tau}\right)^2\cos(\omega t)\,dt \qquad (A3)$$

Putting $t' = 1 - t/\tau$

$$S_{e2}(f) = 2k(1-\Phi^2)\int_0^1 t'^2 \cos(2\pi f\tau(1-t'))dt' = 4k(1-\Phi^2)\tau\,\frac{2\pi f\tau - \sin(2\pi f\tau)}{(2\pi f\tau)^3} \tag{A4}$$

Considering that :

$$\rho_s^2(t) = \exp[-4\pi f_0|t|] = \exp[-2\pi f_g|t|] \tag{A5}$$

the spectrum corresponding to $\rho_s^2(t)$ is :

$$W'(f) = \frac{1}{\pi f_g}\,\frac{f_g^2}{f_g^2 + f^2} \tag{A6}$$

Taking into account eqs. A4 and A6 and that :

$$\sum_{-\infty}^{+\infty} u(t - nT_r) \Longleftrightarrow F_r \sum_{-\infty}^{+\infty} U(nF_r)\,\delta(f - nF_r) \tag{A7}$$

we have:

$$S_{e1}(f) = 4k\,\frac{\Phi^2 F_r}{\pi f_g}\sum_{-\infty}^{+\infty}\frac{f_g^2}{f_g^2 + (f - nF_r)^2}\,\frac{2\pi nF_r\tau - \sin(2\pi nF_r\tau)}{(2\pi nF_r\tau)^3} \tag{A8}$$

The total error spectrum is

$$S_e(f) = k_1\left[\Phi^2\,\frac{F_r}{\pi f_g}\sum_{-\infty}^{+\infty}\frac{f_g^2}{f_g^2 + (f - nF_r)^2}\,\frac{2\pi nF_r\tau - \sin(2\pi nF_r\tau)}{(2\pi nF_r\tau)^3} + (1-\Phi^2)\,\frac{2\pi f\tau - \sin(2\pi f\tau)}{(2\pi f\tau)^3}\right] \tag{A9}$$

REFERENCES

[1] Barton D.K., "Radar System Analysis", Prentice-Hall, 1964.

[2] Skolnik M.I., "Introduction to Radar Systems", Mc Graw-Hill Book Co., 1962.

[3] Woodward P.M., "Probability and Information Theory, with Applications to Radar", Pergamon Press, 1960.

[4] Ray H., "Improving Radar Range and Angle Detection With Frequency Agility", 11th Annual East Coast Conf. on ANE, Oct. 1964.

[5] Grasso G.C. and P.F. Guarguaglini, "Valutazione delle prestazioni di un Sistema Radar a Diversità di Frequenza", Alta Frequenza vol. 36, n.9, Sept.1967 pp. 799-805.

[6] Davenport W.B. and W.L. Root, "Random Signals and Noise", Mc Graw-Hill Book Co.,1958.

[7] Dunn J.H. and D.D. Howard, "The Effects of Automatic Gain Control Performance on the Tracking Accuracy of Monopulse Radar Systems", Proc. IRE Vol. 47, pp. 430-435, March 1959.

[8] Dunn J.H. and al. "Phenomenon of Scintillation Noise in Radar Tracking Systems" Proc. IRE Vol. 47, pp. 855-863, May 1959.

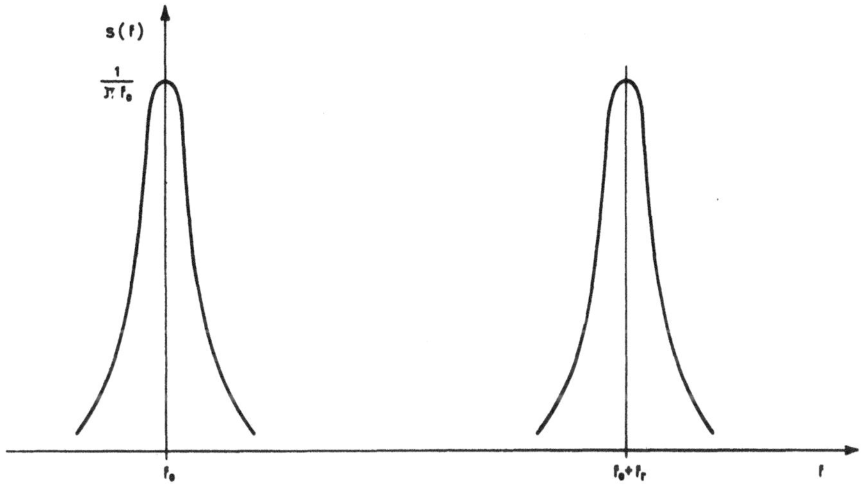

Figure 1

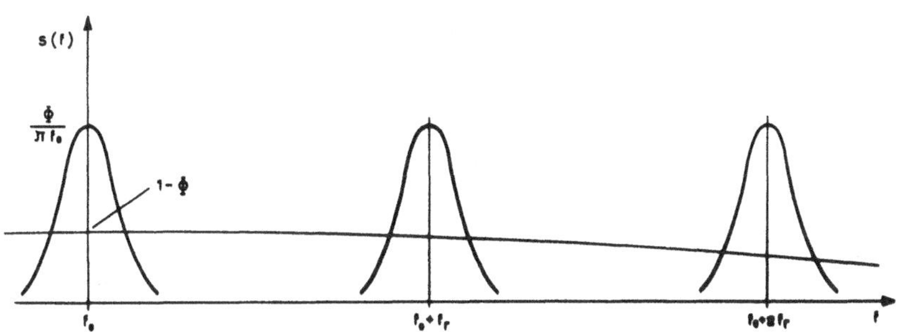

Figure 2

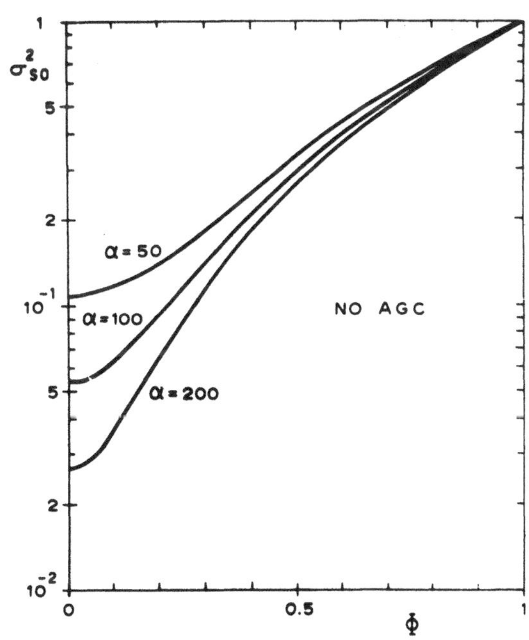

Figure 3

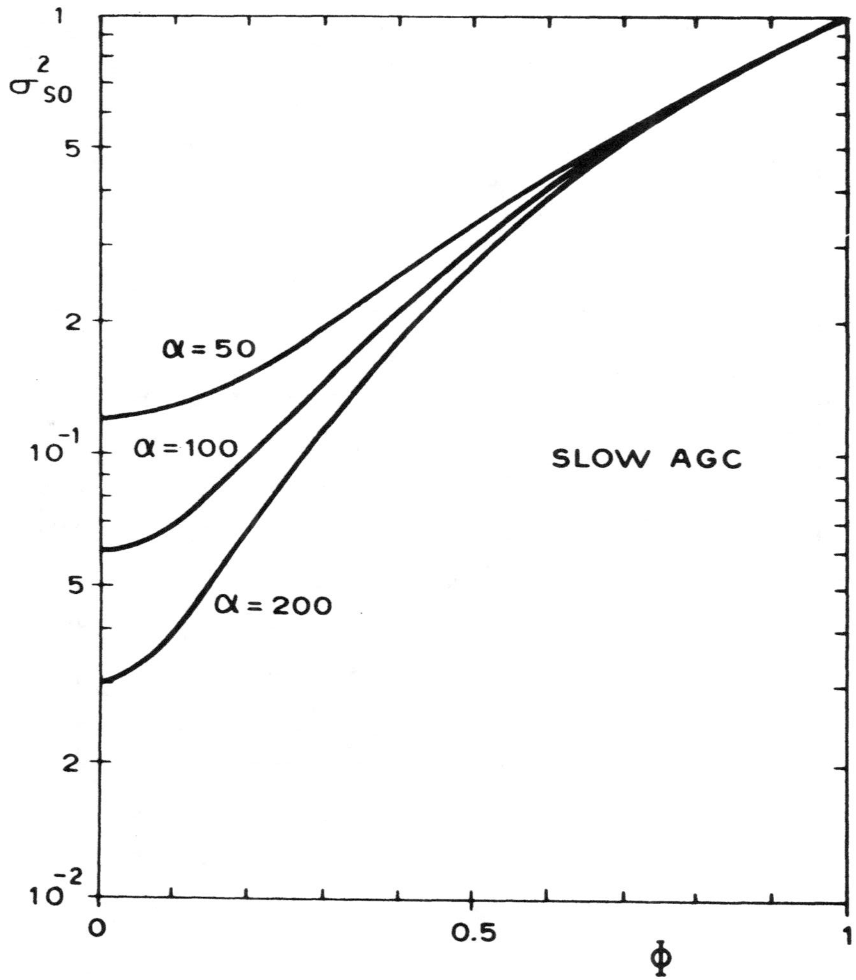

Figure 5

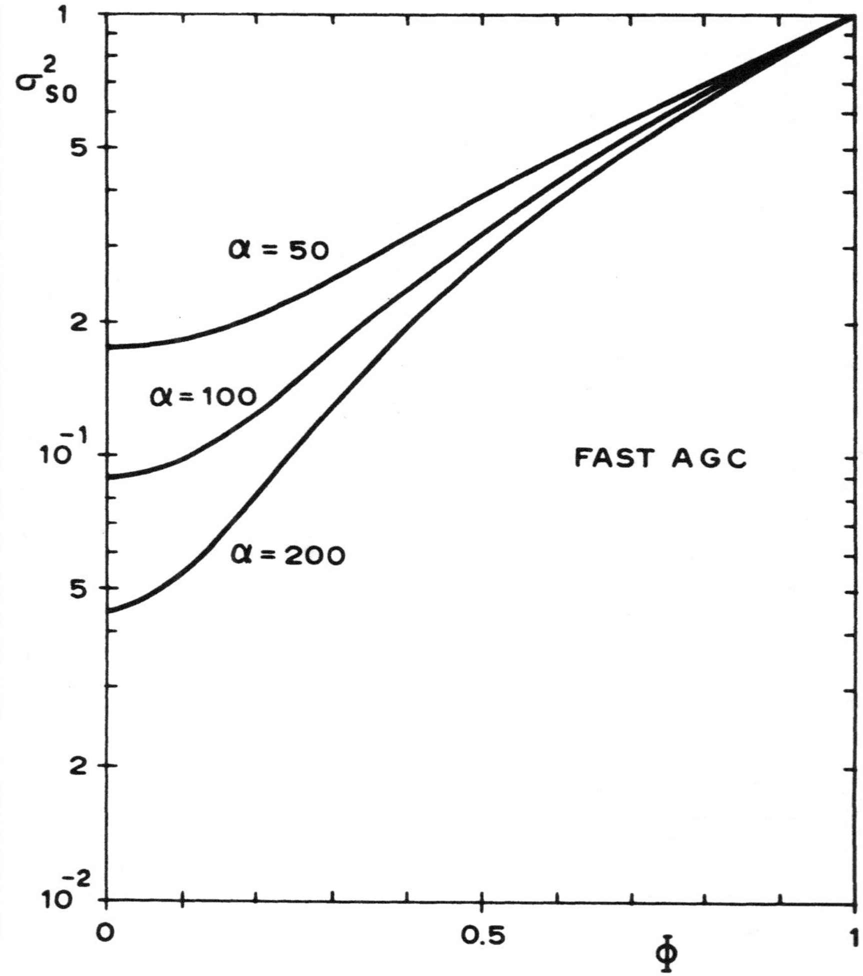

Figure 4

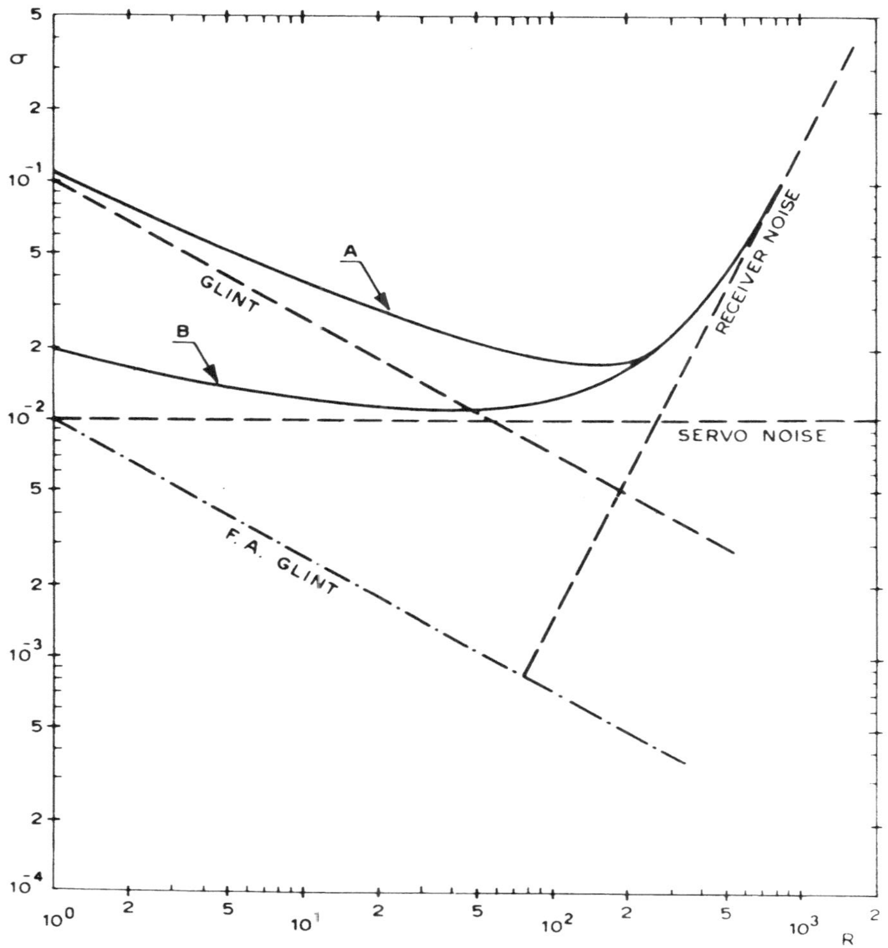

Figure 6

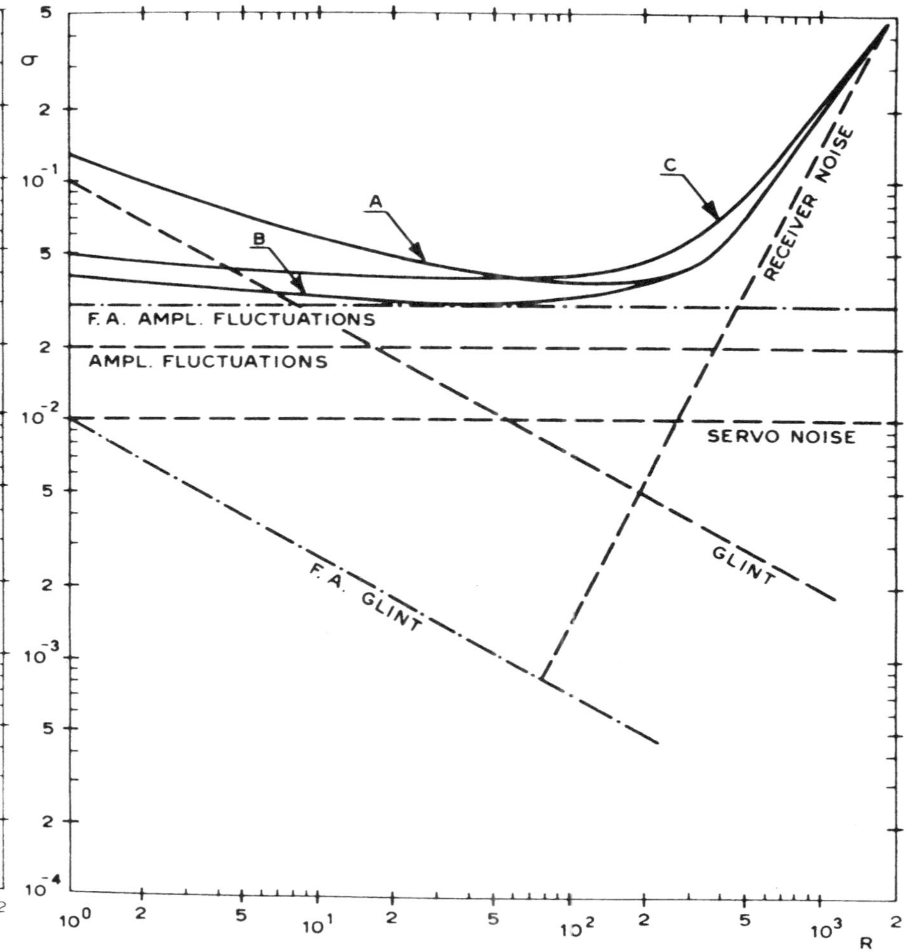

Figure 7

The Reduction of Radar Glint by Diversity Techniques

R.J. Sims and E.R. Graf

IEEE Transactions on Aerospace and Electronic Systems, Vol. AP-19, No. 4 (July 1971)
Pages 462-468

***Abstract*—Variation of the apparent angular location of a radar target, or glint, is a form of backscattered wave-interference phenomenon. The resulting error can be of such magnitude as to indicate the reflecting object to be many times the target span beyond the actual target extent.**

A survey of the literature dealing with glint is presented. This review reveals no analytical model which is adequate in describing glint for a complex target, and yet simple enough to permit reasonable implementation for evaluating diversity improvements.

A model is developed which uses the concept of computing the total backscattered field by coherent summation of the field contributions from nonisotropic scattering complexes. The glint error is determined by use of a simple geometric construction, together with values of total field phase in the region of the tracking aperture.

A technique of synthesizing nonisotropic amplitude and phase patterns for scattering complexes is described. The results, having used a synthesized target and applying frequency and space diversity in a computer analysis, are presented. These indicate considerable tracking improvement by using the simple scheme of choosing the tracking-error signal which corresponds to the diversity return having the largest signal amplitude.

I. Introduction

TARGET scintillation effects were first observed in the form of target fading during early search radar experiments. This scintillation, which was basically amplitude noise, effectively reduced the maximum detection range of the search radar equipment.

Scintillation was soon observed in the testing of experimental tracking radars. Throughout most of World War II, the effects of target noise on tracking performance were thought to be easily explained as a direct result of amplitude scintillation introducing an additional modulation component at the scanning or lobing frequency. However, experiments with high-speed lobing and monopulse radars at the Naval Research Laboratory showed that, at least at the short ranges tested, amplitude scintillation made a relatively small contribution to the total noise [1], [2]. This suggested that another mechanism might be responsible for part of the observed tracking noise. This effect, glint, was determined to originate in the changing relative phases of the received voltages from various parts of a complex target. These time-varying phase angles, in the phasor summation performed by the antenna aperture, resulted in a time variation in the apparent angle of arrival of the radar return.

Manuscript received June 22, 1970; revised December 16, 1970. This work was supported in part by the U. S. Army Missile Command under Contract DA-01-02-AMC 14525 (Z).

R. J. Sims is with the U. S. Army Satellite Communication Agency, Fort Monmouth, N. J.

E. R. Graf is with the Department of Electrical Engineering, Auburn University, Auburn, Ala. 36830.

While there was basic agreement as to the origin of glint, there evolved two schools of thought as to the appropriate modeling technique and interpretation of results: the statistical approach and the deterministic method. Statistical techniques were first developed by Delano [3] and later published in a paper which has served as the basis for many subsequent glint analyses. Using as a model a set of isotropic secondary radiation sources located near the tracking axis and having statistically independent amplitudes and phases, he obtained an expression for the probability density of the fluctuation of the output of an open-loop tracker. He assumed that gross changes in aspect were sufficiently slow to guarantee stationarity of the assumed statistics of the model, and that the number of random scatterers was large enough to produce a Rayleigh distribution of the echo amplitudes. Muchmore [4], [5] subsequently extended Delano's work to include the spectra of the reflected signals and presented a review of the available experimental data on both amplitude and angular scintillation.

Concurrently with the work using the statistical approach to target modeling, other investigators were developing deterministic techniques. At Ohio State University, Peters and Weimer [6], [7] used the coherent summation of contributions from a multipoint model and derived general expressions for the balance point of various types of radars. The Ohio State University group also made extensive measurements of balance point and cross section for various geometrical shapes. They were the first to demonstrate that the balance point could be outside the confines of the target. Lewis [8] further developed the technique of Peters and Weimer to show that conical scan and phase monopulse radars (ideally) track the balance point or phase center of the target. The balance point was seen to vary directly as the derivative of the phase with respect to change of aspect angle.

While Lewis did not state it as such, the derivative of phase with aspect angle is actually the slope of the phase front, or line of constant phase. The concept of visualizing angular scintillation as distortion of the echo phase front was reported by Howard [9] in late 1959. This concept has probably contributed more to the understanding of glint than any other work. In his report, Howard showed that the slope of the echo signal phase front is identical to the angular error for both dual- and multipoint targets. He further reasoned that since the distortion is a property of the backscattered wave, glint affects all angular locating devices, even search radar.

The first literature of the 1960's appeared in the form of a debate between the major advocates of the two differing approaches to glint modeling [10]–[15]. Because Delano reasoned that the Rayleigh statistics of observed data indicated that five or six scatters with equal amplitudes and random phases was an appropriate model, the absence of Rayleigh statistics (in an actual tracking situation) might invalidate such a model. The deterministic school contended that over observation intervals appropriate to tracking servo response times, Rayleigh statistics may not hold. The measurements of Povejsil *et al.* [16] and Moll [17] support this theory. There was also discussion of the observability of the predicted glint at amplitude minima.

There have been several glint-oriented papers published in the open Russian literature. Ostrovityanov [18] outlined a mathematical development which seems to indicate that Meade's result for the boresight angle [19] should be instead for the tangent of that angle. Gubonin [20] extended Delano's results to include the probability density function for glint when there is one sinusoidal deterministic reflector included among the random independent reflectors. For this case, the density function is more narrow than that for random reflectors alone.

The most recent contribution to the glint literature was from Dunn and Howard [21]. Their analysis of a target model consisting of a collinear nonuniform array of electrically short dipoles yielded a component of the time-averaged Poynting vector orthogonal to the radial direction. This demonstrated that the power flow is not necessarily radial, and at a given observation point, may be in such a direction as to indicate the target position to be outside the actual target span.

As early as 1956, Weimer and Peters [6] suggested two diversity techniques which might be used to reduce pointing errors. Both of these are based on the concept that maximum error occurs at amplitude minima. The first suggestion was to servo-control the radar frequency to maintain the amplitude of the return within certain limits. The other idea was to accept and clamp the error signal to the servo only when the amplitude is greater than a certain minimum value.

Birkemeier and Wallace [22] investigated the case of pulse-to-pulse frequency modulation for possible angle tracking accuracy improvement. They used a statistical target model consisting of elements uniformly distributed along a strip and having independent random phase angles to show that considerable tracking improvement could be obtained for a monopulse radar using frequency diversity. In 1966, Gaheen [23] reported on a study performed to verify experimentally the effects of frequency and polarization diversity on the statistics of the radar cross section of satellite-type targets. The analytical model used and the comments made paralleled those of Birkemeier and Wallace, and the conclusions reached were the same. It was also suggested that, due to the negative correlation between severe phase-front tilt and return amplitude, a technique of rejecting the error signals obtained at frequencies when the return amplitude was small might increase tracking accuracy.

Ray [24] analyzed the improvement in radar range and angle detection with frequency agility. While his discussion of angle detection was restricted to the angular locating accuracy of a search-type beam, much of his analysis is applicable to automatic tracking radars. The most pertinent discussion of diversity (as applied to improved radar tracking) to appear in the open literature was contained in the recent paper of Dunn and Howard [21]. The authors commented on space diversity, as achieved by locating several radar receivers at positions widely separated with respect to the phase-front distortion pattern. Each receiver then observes independent samples of angular scintillation. Frequency diversity was indicated as another technique for averaging target scintillation, either instantaneously or within a short period of time. The object is to provide sufficient separation between frequencies such that the scintillation at each frequency is uncorrelated with that at the other frequencies.

The review of the open literature just presented yields no model which is sufficiently general and realistic to illustrate the effectiveness of diversity techniques in reducing glint errors. This paper is devoted to the development of a model based upon the coherent summation of contributions from nonisotropic scattering complexes, and the use of the model to illustrate diversity effectiveness.

II. Development of Glint Model

The determination of total backscattered amplitude and phase by coherent summation of the measured contributions from nonisotropic scattering complexes was introduced in an unclassified part of a 1967 report [25]. However, the technique of determining glint from the data involves the gradient of the phase term. It thus requires not only the measured amplitude and phase patterns of the complexes, but the slope of these patterns in orthogonal directions at each point in space as well. The resulting expression for the direction of the normal to the phase front is rather complex. Moreover, the actual amplitude and phase of the total field is not a direct result of the process of evaluating this expression. It is reasonable to presume that a deterministic model which would characterize the amplitude and phase of the reflected wave, in addition to the normal to the phase front, might prove very useful in formulating a signal-processing scheme to reduce glint error. This can be achieved by forming the coherent sum of the contributions from the scattering complexes (with the phases modified for motion, if required) and then using simple geometrical relationships to determine the glint error. Information concerning dynamic effects, such as the spectra of amplitude and angle scintillations during a particular type of target or tracker maneuver, can be determined by an appropriate point-by-point analysis. Such a procedure permits the analyt-

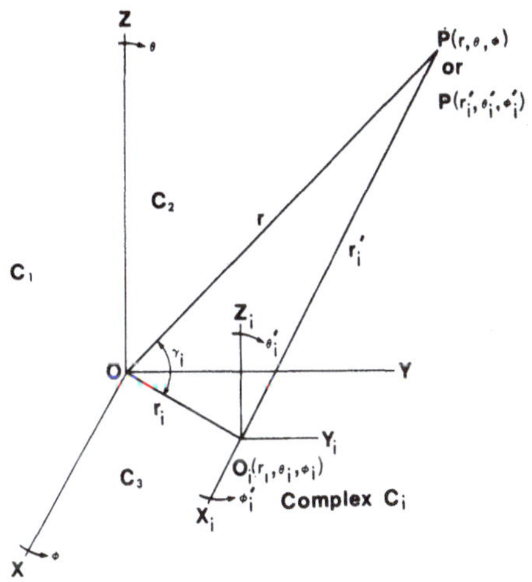

Fig. 1. Geometry for specifying relative locations of field point P and scattering complex C_i located at O_i.

ical evaluation of various compensation and smoothing schemes in a set of chosen environments.

It is assumed that the nonisotropic amplitude and phase patterns for the scattering complexes are available. These may be measured data obtained from actual target sections on a static coherent pattern range, or they may be synthesized in some analytical fashion.

The geometry for the coherent summation is indicated in Fig. 1. The total field is to be determined at the field point P; the complex C_i, one of the N scattering complexes which make up the radar target, is located at O_i. Letting (r,θ,ϕ) and (r_i,θ_i,ϕ_i) be the coordinates of P and O_i, respectively, the path length r_i' can be found from the law of cosines:

$$(r_i')^2 = r^2 + r_i^2 - 2rr_i \cos \gamma_i \tag{1}$$

where

$$\cos \gamma_i = \cos \theta \cos \theta_i + \sin \theta \sin \theta_i \cos (\phi - \phi_i). \tag{2}$$

The angular coordinates of P relative to the origin O_i, (θ_i',ϕ_i'), can be found from the transformations

$$\theta_i' = \cos^{-1} \frac{r \cos \theta - r_i \cos \theta_i}{r_i'} \tag{3}$$

$$\phi_i' = \cos^{-1} \frac{r \sin \theta \cos \phi - r_i \sin \theta_i \cos \phi_i}{r_i' \sin \theta_i}. \tag{4}$$

Let the amplitude and phase contributions from complex C_i, in the direction given by the angular coordinates of P relative to O_i, be A_i and α_i, respectively. These are the values, determined from the recorded or synthesized data, which represent the pattern of the complex C_i in this direction. Assume that the physical dimensions of C_i are sufficiently small for the contribution to have a far-field variation with range. Then the amplitude and phase contributions of complex C_i at the field point P, with the amplitude dependence on distance absorbed into A_i, are

$$A_i' = A_i \tag{5}$$

and

$$\alpha_i' = \alpha_i - 2kr_i' \tag{6}$$

respectively, where $k = 2\pi/\lambda$ is the wavenumber. The factor of two in the propagation path term arises because of the assumed monostatic configuration. Now let $E_i = A_i' \angle \alpha_i'$ be the phasor representation of a singular polarization component of the field at P contributed by the complex C_i. Then the total received signal at P is

$$E = \sum_{i=1}^{N} A_i' \angle \alpha_i' \tag{7a}$$

or

$$E = \sum_{i=1}^{N} A_i \angle (\alpha_i - 2kr_i'). \tag{7b}$$

The phase front is defined as the minimum range change increment F in linear units, required to reach a position having some reference phase. An approximation has been made in F as shown in Fig. 2 in that F is thus chosen parallel to OP rather than PC'. This causes a slight error since the wave phase front is perpendicular to PC' rather than OP. The phase α is related to F by

$$\alpha = -kF. \tag{8}$$

Thus the phase front F as measured from the range coordinate where α is determined, is simply

$$F = -\alpha/k. \tag{9}$$

In Fig. 2, F is measured at the points A and B to yield F_1 and F_2. It is assumed that the phase front is a well-behaved function of aspect angle in the vicinity of P. This will generally be true if a region the size of a typical airborne radar aperture (less than 1 m) spans no more than one full cycle of the distorted wave front. For the case of the two-reflector target, Dunn and Howard [21] showed that this requires a range greater than

$$R_{\min} = d\Delta/\lambda \tag{10}$$

where d is the reflector spacing and Δ is the aperture diameter. It is assumed that this relationship is also reasonably valid for more general target configurations of maximum dimension, and moreover, that the phase front is approximately a linear function of aspect angle over the region around P.

For this analysis, the glint error will be defined as the linear deviation of the instantaneous apparent radar center from the origin O. The glint error will be determined separately in the elevation and transverse tracking planes and then the results combined to find the total error. To determine the glint error, consider the geometry depicted

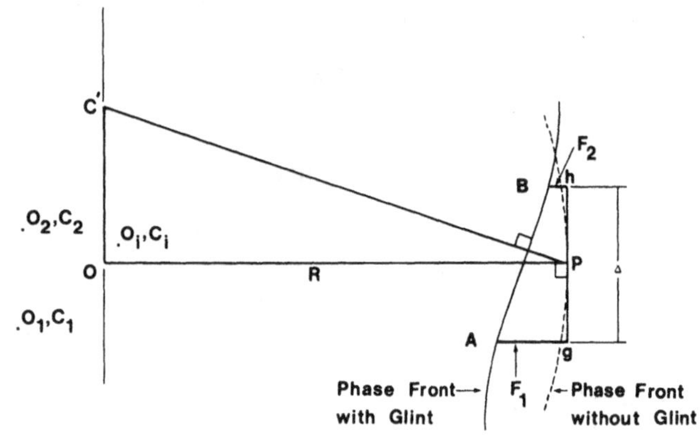

Fig. 2. Geometry for determination of glint.

in Fig. 2, where $OC'P$ lies in either the elevation or the transverse plane, and where C' is the projection of the apparent radar center onto a line passing through O and parallel to the tangent gh. The range R is assumed to be much larger than Δ, and thus the pattern contributions A_i and α_i may be considered constant over the solid angle subtended by Δ. F_1 and F_2 are the values of the phase front at the two extremes of Δ as defined by (9). The linear glint error OC' is the parameter to be determined.

By definition of the apparent target center, the line PC' is normal to the phase front. The slope of the phase front is

$$m_1 = (F_1 - F_2)/\Delta \tag{11}$$

while the slope of PC' with respect to OP is

$$m_2 = OC'/R. \tag{12}$$

By similar triangles, these slopes are identical, therefore

$$OC' = R(F_1 - F_2)/\Delta. \tag{13}$$

Let α_A and α_B be the total field phases at the extremes of the sampling region. Then substituting from (9) for the phase front in terms of these phases, (13) becomes

$$OC' = -R(\alpha_A - \alpha_B)/k\Delta. \tag{14}$$

However, in determining these total field phases, the phase contribution α_i' from each complex must be obtained in a slightly different manner than that indicated by (6). The illumination path from P to the complex C_i is still r_i', but the path for the radar return from the complex is now the distance from O_i to the extreme of the sampling region Δ where the field is to be determined. Denoting this distance by r_i'', the propagation path over which phase retardation takes place is $(r_i' + r_i'')$. Substitution of this value in place of $2r_i'$ in (6) yields

$$\alpha_i' = \alpha_i - k(r_i' + r_i''). \tag{15}$$

The contributions can then be summed according to (7a) to find the total field amplitude and phase at each region extremity, and the results used to find the linear glint error from (14). The total linear glint error in the plane normal to OP is found simply by forming the square root of the sum of the squares of the components in the elevation and transverse planes.

III. Illustrations of Diversity Effectiveness

A. Program Philosophy

The glint model was implemented in Fortran IV for an IBM 360/50 computer. An effort was made to keep the programs sufficiently general to permit their use in describing the effect of such parameters as tracking aperture size, target motion, and diversity techniques. The entry of the amplitude and phase data for the scattering complexes was implemented through a separate subroutine to facilitate use of the programs with measured static range patterns, should they become available. The space trajectory over which glint or other field parameters are determined was also thus incorporated by preparing a subroutine describing the desired series of observation points, in spherical coordinates, as a function of time. The field parameters were then computed at each of these sampling points.

The tracker trajectory subroutine used for these illustrations implements the coordinate equations

$$r = 4650 - 3.75t \tag{16}$$

$$\theta = 90° + \tan^{-1}\left\{\frac{50 \sin [2\pi(r - 900)/3750]}{r}\right\} \tag{17}$$

$$\phi = 45° + \tan^{-1}\left\{\frac{150 \sin [2\pi(r - 900)/3750]}{r}\right\}. \tag{18}$$

With t incremented in steps of 0.025 from 0 to 25 seconds, this trajectory provides a series of sample points having a maximum range of 4165 km, a minimum range of 900 m, and aspect angles which vary about nominal values of $\theta = 90°$, $\phi = 45°$. The projected synthesized target extent from this aspect is approximately 20 m.

B. Synthesis of Scattering Complex Patterns

It is desirable to generate some form of nonisotropic amplitude and phase patterns to represent those which might be obtained from actual scattering complexes. Because measured data for the complexes will probably display considerable amplitude and phase sensitivity to changes in aspect angle, any synthesizing technique which might be used should produce patterns with a corresponding fine-grain lobe structure. One of the pattern-generation techniques in this study was to describe the desired function of aspect angle by an algebraic equation. The product of two sinusoidal functions can be used to generate a large variety of polar patterns. Expressed as a logarithmic (power) amplitude function, such a product is

$$A = 10 \log_{10} [K \cos (n_1\phi) \cos (n_2\phi)]^2, \quad \text{dB.} \tag{19}$$

TABLE I
AMPLITUDE AND PHASE GENERATING FUNCTIONS FOR SYNTHESIZED SCATTERING COMPLEXES

Complex	Generating Functions
1	$A_1 = 10 \log_{10} [\sin\theta \sin\phi \cos 360\phi]^2$
	$\alpha_1 = 180 \cos 720\phi$
2	$A_2 = 10 \log_{10} [0.9 \sin\theta \sin\phi \cos 720\phi]^2$
	$\alpha_2 = 180 \cos 360\phi$
3, 4	$A_{3,4} = 10 \log_{10} [\sin\theta \cos\phi \cos 360\phi]^2$, $(-\pi/2 < \phi < \pi/2$
	$A_{3,4} = 0, \phi > \pi/2$
	$A_{3,4} = 180 \cos 720\phi$

TABLE II
SPHERICAL COORDINATES FOR SCATTERING COMPLEXES IN SAMPLE TARGET CONFIGURATION

Complex	r_i (m)	ϕ_i (deg)	θ_i (deg)
1	20	180	90
2	5	180	90
3	10	90	90
4	10	−90	90

TABLE III
LINEAR CORRELATION COEFFICIENTS BETWEEN SETS OF GLINT ERROR DATA FOR VARIOUS RADAR FREQUENCIES

Frequency (GHz)	Frequency (GHz) 10.0	9.98	9.96	9.94	9.92	9.90
10.0	1.0	0.138	0.404	0.186	0.326	0.163
9.98		1.0	−0.023	0.611	−0.076	0.163
9.96			1.0	0.025	0.457	0.156
9.94				1.0	0.079	0.008
9.92					1.0	0.102
9.90						1.0

TABLE IV
RMS ERROR IN METERS RESULTING FROM AVERAGING DATA AT TWO RADAR FREQUENCIES

Frequency (GHz)	Frequency (GHz) 10.0	9.98	9.96	9.94	9.92	9.90
10.0	11.9	8.3	8.9	8.9	9.8	12.3
9.98		10.1	6.9	9.6	7.8	11.5
9.96			9.6	7.6	9.5	11.3
9.94				11.4	8.8	11.1
9.92					12.5	11.9
9.90						19.2

TABLE V
LINEAR CORRELATION COEFFICIENTS BETWEEN MAGNITUDE OF GLINT ERROR AND RELATIVE FIELD AMPLITUDE

Frequency (GHz)	Correlation Coefficient
10.0	−0.605
9.98	−0.599
9.96	−0.621
9.94	−0.630
9.92	0.622
9.90	0.438

Here K is some maximum amplitude, n_1 is some small integer, and n_2 is a large integer. Thus the latter cosine function provides the desired degree of fine-grain lobe structure, while the former results in a slowly varying amplitude envelope. A phase pattern can be generated in a similar manner. This function is not realistic for a complex scatterer, but is convenient for the analyses performed. Table I summarizes the amplitude and phase generating functions for four synthesized scattering complexes which were used in the studies. The additional sine functions involving θ were included in the amplitude equations to provide a degree of variation in the elevation plane. Table II lists the spherical coordinates of each of these four complexes, relative to a primary origin, which were used for the computer data runs presented here.

C. Frequency Diversity

Table III presents linear correlation coefficients between sets of error data with frequency as a parameter. A correlation of 0.5 or less was chosen as a reasonable condition for independence of the data. Ray [24] has used a similar value. Judging from the values in the table, it would seem advantageous to investigate the use of the glint data obtained at two or more of these frequencies to obtain more accurate tracking information. The magnitude of rms error before and after diversity processing will be used as a measure of accuracy improvement (or deterioration).

The first technique which might be implemented is to simply average the two values obtained with the two frequencies at each sample point along the trajectory. The results of this operation are presented in Table IV, where the rms values listed are in meters, and the values along the diagonal are the rms values at the corresponding single frequencies. Note that in most cases, the result of averaging at two frequencies is a value less than the value at either single frequency.

It may be recalled from the discussion earlier that several investigators indicated a negative correlation between regions of large glint error and total field amplitude. This hypothesis was tested at several different radar frequencies. For each case, the linear correlation between the total field amplitude and the magnitude of the linear glint error was computed. The results presented in Table V show the correlation coefficients to be negative and quite large, significant indeed for correlation between two different types of field parameters. This strong negative correlation is illustrated graphically in Fig. 3, in which the glint error waveform for $f = 10$ GHz is superimposed with a relative plot of the corresponding total field amplitude. Note that the regions of large positive and negative error coincide exactly with the regions of small amplitude.

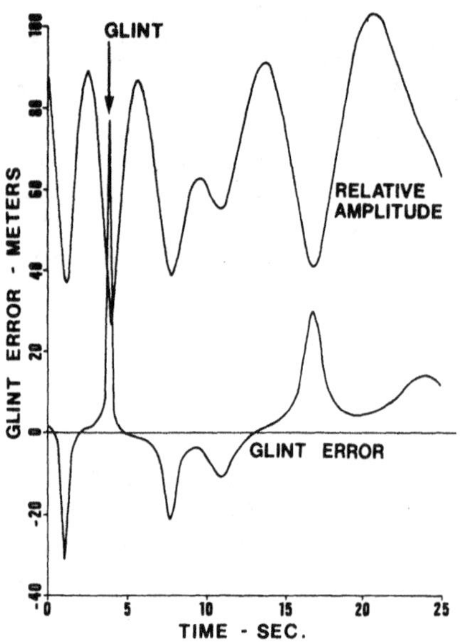

Fig. 3. Typical relationship between relative field amplitude and glint.

TABLE VI
RMS Error in Meters Resulting from Selecting Value at Each Sample Point Which Corresponds to Largest Total Field Amplitude

Frequency (GHz)	Frequency (GHz)					
	10.0	9.98	9.96	9.94	9.92	9.90
10.0	11.9	7.3	8.0	5.6	7.2	4.8
9.98		10.1	7.7	7.4	6.0	5.8
9.96			9.6	5.5	7.1	5.0
9.94				11.4	5.7	7.5
9.92					12.5	5.9
9.90						19.2

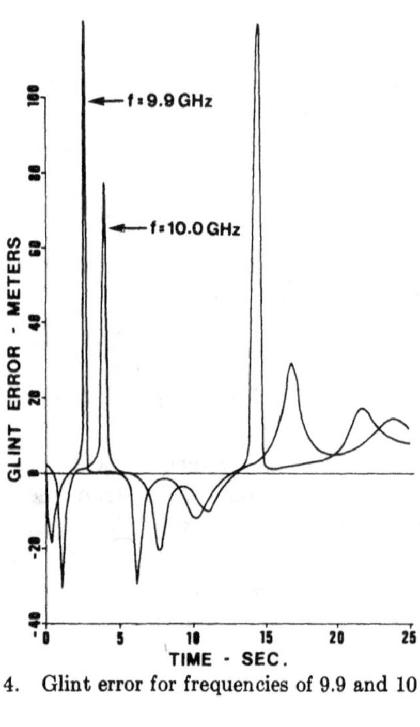

Fig. 4. Glint error for frequencies of 9.9 and 10 GHz.

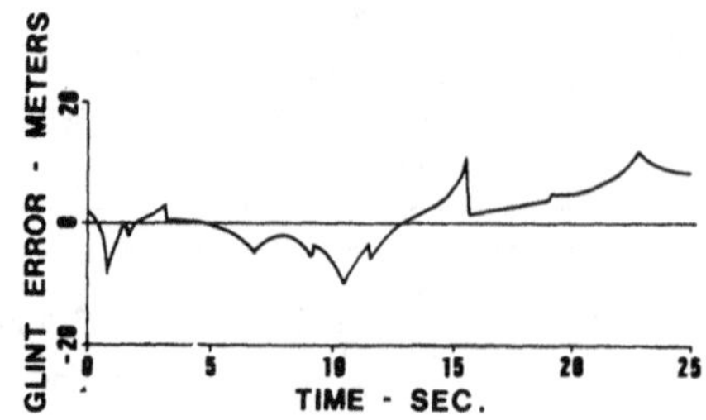

Fig. 5. Glint error using error-selection frequency diversity for frequency combination of 10 and 9.9 GHz.

Another technique of data processing which might be implemented is based upon this negative correlation between glint error and total field amplitude, and might be termed error selection diversity. Here the computer program selects from the pair of glint values at each sample point that value which corresponds to the frequency having the largest field amplitude. The resulting rms error values are presented in Table VI, where the values on the diagonal are again the single-frequency rms errors. This processing technique appears to provide improved tracking over simple averaging.

To illustrate the nature of the improvement provided by error selection diversity, it is instructive to compare a resulting glint waveform with the corresponding waveforms at the single frequencies. The frequency combination in Table VI resulting in the lowest rms value is 10–9.9 GHz. The individual waveforms for these frequencies appear in Fig. 4, while Fig. 5 shows that the large values of error are eliminated by the error-selection technique.

As a test of slightly more sophisticated diversity processing schemes, the error-selection technique was implemented on the basis of the amplitudes at all six of the radar frequencies previously used. This resulted in an rms value of only 4.6 m.

D. *Space Diversity*

Several runs were made at a frequency of 10 GHz and using a dual tracker to gain insight concerning the degree of decorrelation achieved by spaced antennas. Each dual tracker run was actually implemented by making two data runs of the main glint program. At each observation point, the value of the azimuth coordinate was increased or decreased by an amount corresponding to a fixed linear distance at the value of range for the observation point. A linear correlation was then performed on the pairs of glint error values obtained for the two data runs, and rms computations made for each data set and also on the result of using the error-selection technique based upon the received amplitude at each tracker. The results of these computations are presented in Table VII, where rms_1 and rms_2 are the rms error values for the individual trackers, and rms_{div} is the corresponding value using error-selection diversity. It appears from the data that even rather small spacings can be quite effective in reducing glint error.

TABLE VII
EFFECT OF SPACE DIVERSITY FOR VARIOUS DUAL-TRACKER SPACINGS

Spacing (m)	Correlation Coefficient	rms_1	rms_2	rms_{div}
1	0.383	12.0	11.8	6.1
3	0.001	12.9	11.7	5.5
6	0.045	15.5	11.7	6.4

IV. CONCLUSIONS

In summary, a review of prior glint investigations reveals no analytical model which is adequate in describing glint for a complex target, and yet simple enough to permit implementation for preliminary evaluation of diversity techniques. The technique developed here uses the concept of coherent summation of the field contributions from various nonisotropic scattering complexes. The glint error is determined by use of a simple geometric construction together with values of total field phase in the region of the tracking aperture.

The model was implemented on the digital computer using synthesized nonisotropic amplitude and phase patterns for the scattering complexes. The effects of using frequency and space diversity with a synthesized target were studied. The results indicate that a considerable tracking improvement can be achieved by using the easily implemented data processing scheme of choosing the tracking-error signal which corresponds to the diversity return having the largest signal amplitude. However, care must be exercised in the interpretation of the results with respect to real targets since the model used was relatively quite simple.

ACKNOWLEDGMENT

The authors wish to thank C. Cash for his aid in performing this work.

REFERENCES

[1] J. H. Dunn, D. D. Howard, and A. M. King, "Phenomena of scintillation noise in radar-tracking systems," *Proc. IRE*, vol. 47, May 1959, pp. 855–863.
[2] J. E. Meade, A. E. Hastings, and H. L. Gerwin, "Noise in tracking radars," Naval Research Lab., Rep. 3759, AD 90 876, 1950.
[3] R. H. Delano, "A theory of target glint or angular scintillation in radar tracking," *Proc. IRE*, vol. 41, Dec. 1953, pp. 1778–1784.
[4] R. B. Muchmore, "Aircraft scintillation spectra," *IRE Trans. Antennas Propagat.*, vol. AP-8, Mar. 1960, pp. 201–212.
[5] ——, "Review of scintillation measurements," Hughes Aircraft Co., Tech. Memo. TM 272, AD 4849, 1952.
[6] F. C. Weimer and L. Peters, Jr., "Study of pointing errors in conically scanning and monopulse tracking radars for multipoint targets," Antenna Lab., Ohio State Univ., Columbus, Rep. 601-12, AD 123 476, 1956.
[7] L. Peters, Jr., "Accuracy of tracking radar systems," Antenna Lab., Ohio State Univ., Columbus, Rep. 601-29, AD 200, 1957.
[8] D. E. Lewis, "On tracking radar systems and the target phase center," Antenna Lab., Ohio State Univ., Columbus, Rep. 777-4, AD 211 187, 1958.
[9] D. D. Howard, "Radar target angular scintillation in tracking and guidance systems based on echo signal phase front distortion," Naval Research Lab., Rep. 5430, 1959.
[10] L. Peters, Jr., and F. C. Weimer, "Concerning the assumption of random distribution of scatterers as a model of an aircraft for tracking radars," *IRE Trans. Antennas Propagat.* (Commun.), vol. AP-9, Jan. 1961, pp. 110–111.
[11] R. B. Muchmore, "Reply to comments by Leon Peters, Jr., and F. C. Weimer," *IRE Trans. Antennas Propagat.* (Commun.), vol. AP-9, Jan. 1961, pp. 112–113.
[12] R. H. Delano, "Further reply to comments by Leon Peters, Jr., and F. C. Weimer," *IRE Trans. Antennas Propagat.* (Commun.), vol. AP-9, Mar. 1961, pp. 227–228.
[13] L. Peters, Jr., and F. C. Weimer, "Reply to comments by R. H. Delano," *IRE Trans. Antennas Propagat.* (Commun.), vol. AP-9, Mar. 1961, p. 228.
[14] ——, "Reply to comments by R. B. Muchmore," *IRE Trans. Antennas Propagat.* (Commun.), vol. AP-9, Mar. 1961, p. 229.
[15] K. M. Siegel, "Comparison between theoretical and experimental radar cross sections of aircraft," *IRE Trans. Antennas Propagat.* (Commun.), vol. AP-9, Mar. 1961, p. 228.
[16] D. J. Povejsil, R. S. Raven, and P. Waterman, *Airborne Radar*. New York: Van Nostrand Reinhold, 1961.
[17] J. W. Moll, "On the radar echo from aircraft," *IEEE Trans. Aerosp. Electron. Syst.* (Corresp.), vol. AES-3, May 1967, pp. 574–577.
[18] R. B. Ostrovityanov, "Angular noise," *Radiotekh. Elektron.*, vol. 4, 1966, p. 507.
[19] A. S. Locke, *Guidance*. New York: Van Nostrand Reinhold, 1955.
[20] N. S. Gubonin, "Fluctuation of the phase front of the wave reflected from a complex target," *Radiotekh. Electron.*, vol. 10, 1965, p. 717.
[21] J. H. Dunn and D. D. Howard, "Radar target amplitude, angle, and Doppler scintillation from analysis of the echo signal propagating in space," *IEEE Trans. Microwave Theory Tech.*, vol. MTT-16, Sept. 1968, pp. 715–728.
[22] W. P. Birkemeier and N. D. Wallace, "Radar tracking accuracy improvement by means of pulse-to-pulse frequency modulation," *AIEE Trans. (Commun. Electron.)*, vol. 81, 1962, pp. 571–575 (Jan. 1963 sec.).
[23] A. F. Gaheen, J. T. McDonough, and D. P. Tice, "Frequency diversity radar study," Westinghouse Defense and Space Center, Rep. AFAL-TR-66-128, vol. 1, AD 480 358, 1966.
[24] H. K. Ray, "Improving radar range and angle detection with frequency agility," *Microwave J.*, vol. 9, 1966, p. 63.
[25] J. T. Coleman and F. W. Someroski, "A survey and analysis of radar glint as applied to air defense missile system design," Battelle Memorial Inst., Columbus, Ohio, Rep. RSIC-699, AD 385 494, 1967.

Frequency-Agile Processing to Reduce Radar Glint Pointing Error

J.M. Loomis and E.R. Graf

IEEE Transactions on Aerospace and Electronic Systems, Vol. AES-10, No. 6 (November 1974)
Pages 811-820

Abstract

A technique to reduce radar pointing errors due to glint using frequency agility and amplitude weighting is presented. The reduction in rms tracking error is developed into an equation dependent upon the original glint tracking error, σ_g, and the number of returns weighted, N. The rms tracking error is thereby reduced approximately by a factor of N. Finally, the equation formulated allows one to evaluate the reduction in glint error versus the number of frequencies chosen for frequency agility.

I. Introduction

The advent of complex radar systems with stringent mission requirements has generated an intense interest in the solution to the problem of glint or angular scintillation. Weimer and Peters [1], [2], Dunn *et al.* [3], Delano [4], and others [5]-[8] have studied this phenomenon in order to improve the performance of tracking radars. Glint, which denotes a phase perturbation of the echo signal wavefront, causes an apparent wandering of the target center and can be of such a magnitude as to indicate the reflecting object to be many times the target span beyond the physical target extent [3], [4].

This paper addresses monopulse tracking error and the means to reduce the error due to glint in the case where glint is the most predominant source of error. Glint becomes the limiting factor when tracking complex targets at short ranges. In these situations, adequate signal-to-noise ratio is present to essentially negate the effects of noise from other sources [3]. The conditions of large S/N ratios and short ranges are normally present in missile seeker applications. Furthermore, in the case of a missile, velocities are very high; consequently, the rapid determination of target position is imperative. Typical seeker radar tracking antennas have small apertures, thereby reducing any averaging inherent to antennas. For these reasons, missiles are especially prone to glint tracking errors. These errors can be of such a magnitude as to cause the missile to miss the target completely [5].

In an effort to reduce glint errors, frequency agility was chosen as an initial step to implement the tracking improvement. Various methods have been proposed to implement frequency agility into radar systems. Lind [9] has proposed using random frequency variation from pulse to pulse. This method yields good results; however, an important advantage of frequency diversity, that is, decorrelation of the echo signal, is not always present. Sims and Graf [10] proposed transmitting 5 separate frequencies, the return of each being decorrelated from one another, and choosing the position corresponding to the return with the largest amplitude. This method significantly reduces the rms glint error.

In this work we address ourselves to several crucial questions concerning the implementation of frequency agility and its improvement in rms tracking error as compared to fixed-frequency radar systems. Prominent among the questions to be answered are: Is there a preferred sequence of frequencies to be transmitted? What number of decorrelated frequencies are necessary? What type of signal processing would be most effective? Based on the results of these questions, what is the resulting reduction in rms tracking error?

In an effort to resolve the above questions, a computer-simulated radar system, which incorporates glint into the target model, was employed. The data available from the simulation, along with the experimental results of previous researchers, are used to formulate an equation which describes the reduction in rms tracking error.

Manuscript received October 25, 1973. Copyright 1974 by *IEEE Trans. Aerospace and Electronic Systems*, vol. AES-10, no. 6, November 1974.

This work was supported in part by the U.S. Army Missile Command under Contract DA-AH01-71-C1303.

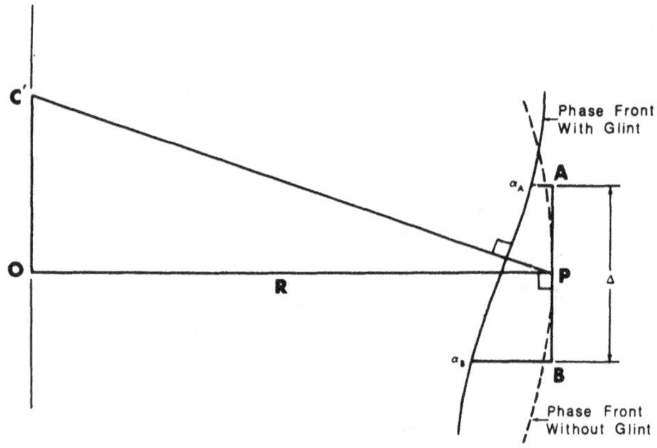

Fig. 1. Basic geometry associated with glint.

Fig. 2. Radar cross section of a typical model at two different frequencies. Solid line = 5.35 GHz; dashed line = 5.50 GHz.

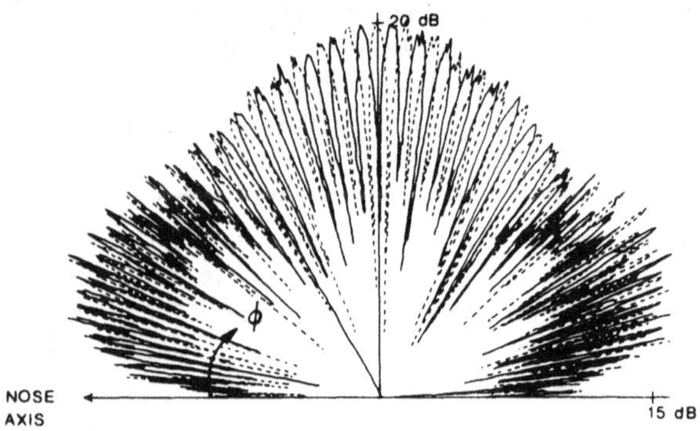

II. Radar System Model

A computer program, similar to that devised by Sims [11], was used to model the radar system, the target, and the target's flight path. This program also calculated the pointing error due to glint, which is examined next.

The determination of glint error is carried out in one plane at a time. The amplitude and phase of the echo signal is first calculated by coherently summing the returns from each scattering point. This sum is computed at the center and at each edge of the aperture in the plane of consideration. The phases at each extreme of the aperture, represented by α_A and α_B in Fig. 1, are used to find the average phase difference across the aperture. The normal to the reflected wave, PC', is then calculated from the average phase difference. Finally, glint error OC', in linear units of displacement at the target, is the slope of the reflected wave M_{PC}, multiplied by the target range R; thus, $OC' = R \cdot M_{PC}$.

The target model is composed of four nonisotropic scattering complexes representing the fuselage, each wing, and the stabilizer. Experimental radar cross-section data, as presented by Skolnik [12], were employed to assign typical cross-section magnitudes to the various complexes. The fine lobe structure, characteristic of aircraft radar cross sections, was furnished by multiplying the amplitude function by a rapidly varying sinusoidal function. In addition, the target-scattering model is dependent upon the transmitted frequency. This was accomplished by making the argument of the sinusoids a function of the phase constant ($k = 2\pi/\lambda$). The cross section of a typical scattering complex is assumed to be

$$\sigma(\Phi) = \left\{[50/(\pi/2)^2]\,\Phi + 8\right\} \cos(k\Phi/2) \qquad m^2 \tag{1}$$

for the purpose of the simulation performed, where Φ is the azimuth angle measured from the nose axis of the aircraft.

The radar cross section of a typical target model at two different frequencies is shown in Fig. 2. Note that the scattering pattern for 5.50 GHz, plotted as the dashed line, tends to fill in the nulls that occur in the pattern at 5.35 GHz. This filling of the nulls usually provides at least one echo signal with large amplitude if both frequencies are transmitted. As will be shown later, a large amplitude will insure a reasonably accurate estimate of target position.

Several flight patterns of aircraft were implemented in the simulation; these included crossing, straight-in, and fly-by trajectories. In addition, Doppler phase changes due to yaw were also included.

The entire radar model computer program was implemented in FORTRAN IV for an IBM 360/50 digital computer. Glint error and echo-signal amplitude data were obtained for a number of different target models. These data were determined for a particular model on a simulated flight path at time increments corresponding to typical radar pulse repetition rates, 0.2 to 2 ms. In addition, data were calculated at each position along the flight path at 25 different frequencies, which were chosen to be in a typical tracking radar frequency band, 5.35 to 5.85 GHz.

The frequency spacing was selected to be 20 MHz, thus fulfilling the requirements for echo-signal decorrelation, namely, that

$$\Delta f_s = 150/D \qquad \text{MHz}, \tag{2}$$

with D being the target depth in meters [9], [13]. The target models considered had a depth of at least 7.5 meters. The Δf_s requirement is the most stringent quoted in the literature for low-return signal decorrelation. For example, Ray [14] proposed using $45/D$ MHz; however, the correlation coefficient is only 0.5 for this value of Δf_s. In all simulations considered in this study the return signal correlation coefficient between frequencies was less than 0.1. The variation in signal amplitude and tracking error as a function of frequency can now be applied to determine the effect of frequency agility on tracking error.

III. Signal Processing

The concept that maximum glint error occurs at amplitude minima has been the foundation for processing radar returns to diminish glint errors by Weimer and Peters [1], Dunn and Howard [13], and others [10], [11]. The large negative correlation between error and echo-signal amplitude suggests that by excluding those returns of low amplitude, the radar performance should be improved. This concept can be extended to include several signal processing schemes that weight the indicated position of the target according to the amplitude of the return.

An analysis of the data obtained from the computer simulation illustrates the fact that maximum glint error occurs at amplitude minima. Fig. 3 depicts a two-dimensional probability density function of amplitude and error analogous to the density functions obtained from the computer-simulated data. The curves, $g_1, g_2 \cdots g_n$, are Gaussian density functions, all having the same mean value

$$\mu_1 = \mu_2 = \cdots = \mu_n. \tag{3}$$

The variance of these curves decreases as the return amplitude increases,

$$\sigma_1^2 > \sigma_2^2 > \cdots > \sigma_n^2. \tag{4}$$

The curves $r_1, r_2, \cdots, r_n$ represent the single-dimensional probability density function of amplitude.

In as much as the variance of error and amplitude of the echo signal are inversely proportional, the probability of obtaining an excessive glint error accompanying a large amplitude is remote, but nonzero. In addition, many returns of median amplitude correspond to proper target locations, so that they cannot be disregarded completely. Consequently, linear weighting processes that utilize all echo signals and weight the position according to the magnitude of return might seem most feasible. However, as will be shown in the next section, choosing the position corresponding to the largest amplitude in any particular sequence will provide the greatest reduction in rms tracking error.

The method of linear weighting of target position is based on the signal-to-noise ratio in the tracking antenna being of such a magnitude that the echo-signal amplitude is large and glint is the only contributor to tracking inaccuracies. Receiver noise, amplitude fading, and other noise terms are neglected. This is a valid assumption only when the target is close to the radar antenna. In each case a single return pulse is adequate to make a determination of target position. Fig. 4. is a representation of target position as determined from a sequence of pulses. The points x_1, y_1), $(x_2, y_2, \cdots,$ (x_N, y_N) are the positions corresponding to each of the pulses received by the radar. Each pulse has a definite return amplitude A_i associated with it, either a peak amplitude or an integrated pulse over a constant duration. The point (x_m, y_m) is the weighted center of the N returns; that is,

$$\sum_{i=1}^{N} W(A_i)[(x_i - x_m) + (y_i - y_m)] = 0 \tag{5}$$

at the point (x_m, y_m). The term $W(A_i)$ is the weighted amplitude at the general point (x_i, y_i).

Equation (5) can be decomposed into two separate equations in x and y. The solutions for x_m and y_m are then

$$x_m = \frac{[w(A_1)x_1 + w(A_2)x_2 + \cdots + w(A_N)x_N]}{[w(A_1) + w(A_2) + w(A_3) + \cdots + w(A_N)]} \tag{6a}$$

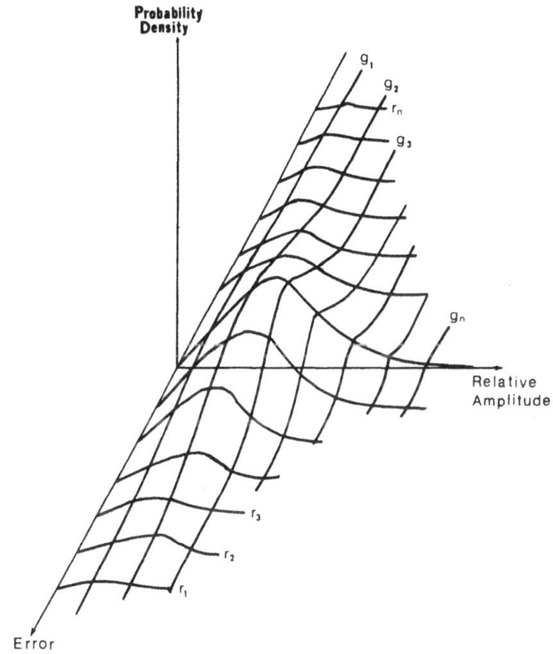

Fig. 3. Typical two-dimensional probability density function of amplitude and glint error.

Fig. 4. Return from a series of radar pulses, indicating a position corresponding to each return pulse.

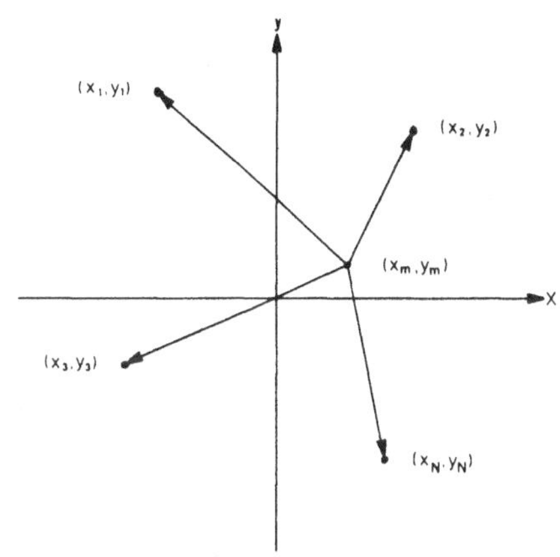

$$y_m = \frac{[w(A_1)y_1 + w(A_2)y_2 + \cdots + w(A_N)y_N]}{[w(A_1) + w(A_2) + w(A_3) + \cdots + w(A_N)]} \tag{6b}$$

Any constant divided into the weighted amplitudes, $w(A_1)$, $w(A_2)$, $w(A_3)$, $\cdots$, $w(A_N)$, can be factored from both the numerator and denominator without altering the results of x_m or y_m. The weighting functions $w(A)$ are thus normalized to the echo signal with the largest amplitude A_{max}; hence, $w(A_{max}) = 1$.

The number of possible weighting functions $w(A)$ is infinite. In as much as the larger echo-signal amplitudes have been shown to provide a better estimate of target position than small amplitude returns, weighting functions that weight small amplitude returns more than large amplitude

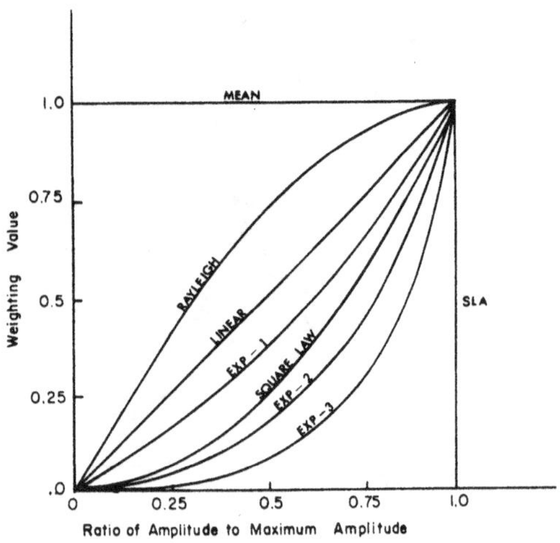

Fig. 5. Comparison of the weighting methods examined.

returns will not be considered. The weighting methods examined encompass a variety of processes that range between the arithmetic mean and selection of the largest amplitude (SLA), as given in Table I. These weighting functions are graphed in Fig. 5 as a function of normalized amplitude. The arithmetic-mean method weights all return signals equally, regardless of amplitude. Each of the other weighting functions assign a small weight to the echo signals with lower amplitude until the limit is reached, and only the return with the largest amplitude is used as an estimate of target position.

The next consideration is the implementation of frequency agility into the simulated radar system. Several different frequency sequencing methods were incorporated into the analysis. A fixed-frequency sequence was provided to obtain a basis of comparison between fixed-frequency and frequency-agile radar systems. A sequencing method which uses a random-number generating routine to randomly choose all or any group of the 25 available frequencies was also considered. In addition, three sequencing methods that assign frequencies, spaced 20, 40, or 180 MHz apart, to consecutive transmit pulses were furnished. Finally, a frequency sequence was devised which randomly chooses the first frequency of the sequence; the rest of the frequencies are assigned to be one step (20 MHz) from the previous frequency used. Once the total number of pulses is given sequence was transmitted, a new starting frequency is chosen randomly and the sycle commenced again. For example, in a particular application, sixteen pulses were to be transmitted in a given scan in groups of four transmit pulses each. The transmit pulses of the scan were transmitted with frequencies $f_1, f_2, \cdots, f_{16}$. Now, f_1 was chosen randomly to 5.45 GHz; then f_2 was 5.47 GHz, f_3 was 5.49 GHz, and f_4 was 5.51 GHz. Upon transmission of the f_4, the first cycle was completed; f_5 is then also chosen randomly; f_6 will be 20 MHz greater than f_5; f_7 will be 20 MHz greater than f_6; and so on until f_{16} is transmitted and the scan is completed.

The frequency selection methods were combined with the the weighting functions to simulate the operation of both fixed-frequency and frequency-agile radar systems. The computer program implemented the simulation as follows. All the N pulses of a particular scan were transmitted with transmission frequencies selected according to the frequency selection method employed. The first j return pulses ($j < N$) are weighted according to the weighting method being utilized. Upon reception of the next echo signal, the second pulse (the second pulse of the first group of j pulses) through the $(j+1)$st pulses are now weighted, allowing a continuous, overlapping procedure that will process j pulses at a time. This scheme is carried out until the last j pulses, $(N-j+1)$ to N, are weighted. Utilizing this procedure, continuous target tracking information can be provided once the first j pulses of any particular scan have been received. The linear rms glint error is defined as

$$\sigma_{rms} = \sqrt{\sum_{i=j}^{N} E_w^2/k} \tag{7}$$

where E_w is the weighted linear error and K is the total number of target position determinations made, $K = N - j + 1$. The value of σ_{rms} is computed as a function of both pulses processed and the number of frequencies employed.

IV. Tracking Error Reduction

The computation of linear rms tracking error σ_{rms} was performed for various weighting and frequency sequence combinations. In order to formulate a process to reduce σ_{rms}, the results of these calculations were examined to

TABLE I

Weighting Functions Applied to the Radar Echo Signal

Designation	Weighting Function
Mean weighting	$W(A) = 1$
Linear weighting	$W(A) = A/A_{max}$
Square-Law weighting	$W(A) = (A/A_{max})^2$
Rayleigh weighting	$W(A) = e^{-(1/2)}(A/A_{max}) \exp - [(A/A_{max})^2/2]$
Exponential weighting #1	$W(A) = [1/(1-e)](1 - e^{A/A_{max}})$
Exponential weighting #2	$W(A) = [1/(1-e^3)](1 - e^{3A/A_{max}})$
Exponential weighting #3	$W(A) = [1/(1-e^5)](1 - e^{5A/A_{max}})$
Select largest amplitude (SLA)	$W(A) = \delta(A - A_{max})$

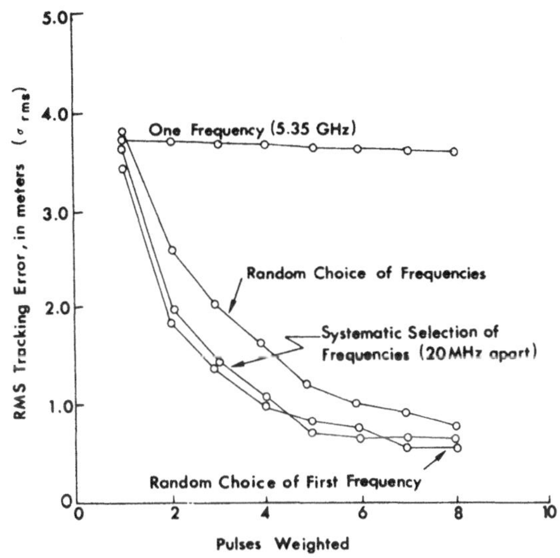

Fig. 6. Rms tracking error reduction as a function of pulses processed for various frequency selection methods. Frequencies used were in a band between 5.35 and 5.85 GHz. The weighting method employed was to choose the pulse with the largest amplitude (SLA method).

determine the optimum combination of amplitude weighting and frequency selection.

A tracking radar predominantly encounters targets that have characteristics similar to those of a Swerling type I or III target [15], [16]. The echo signals from this class of targets are nearly constant on a pulse to pulse basis, but are assumed to be independent from scan to scan. In this case, frequency agility can be used to obtain independent samples of target position which are necessary to fully utilize the capability of the radar system. The most effective implementation of frequency agility must now be determined.

The primary requirement for a frequency-agile radar system is that the echo signals provide independent target information; electronic countermeasures afforded by frequency agility are considered to be of secondary importance. Fig. 6 depicts a set of typical SLA curves of σ_{rms} for various frequency selection methods. The curve corresponding to the systematic selection of frequencies 20 MHz apart is representative of those curves obtained when a systematic frequency selection method is employed. Systematic sequences assigned the transmit frequencies in a manner such that the probability of transmitting two or more consecutive pulses with the same frequency is zero. The spacing between consecutive frequencies, Δf_s, is large enough to insure virtual decorrelation of the echo signal. Consequently, the probability of receiving two or more nonindependent samples of target position is essentially zero. This selection method yields a large amount of tracking error improvement. In contrast, the curve corresponding to the use of monochromatic transmit pulses shows no significant reduction in rms tracking error due to the use of nonindependent returns; it illustrates the typical performance of a conventional fixed-frequency radar system employing the SLA weighting function.

Electronic countermeasure (ECM) protection can be furnished by choosing each frequency randomly. However, the rms tracking error may be increased compared to that which could be obtained by a systematic selection of frequencies, as shown in Fig. 6. When random selection is used there is a finite probability that the echo signal will not be independent. For example, if N_f is the number of properly spaced frequencies available for transmission, then the probability that an n sequence of pulses will all be the same frequency is $(1/N_f)^n$.

The selection method that provides both ECM protection and a small probability of nonindependent returns selects only the first frequency of a given group randomly. The curve in Fig. 6 that corresponds to such a sequence shows that the rms tracking error reduction using this method is essentially equal to that of the systematic frequency sequences.

If N_p is the number of pulses in a given group before resequencing, then only once is N_p is there nonzero probability of transmitting identical frequencies. This occurs only if the last frequency of the group is the same as the randomly selected frequency. Since there are N_f possible transmitting frequencies, the probability of transmitting two successive pulses with the same frequency is $(1/N_p)(1/N_f)$. Transmission of three successive coincident frequencies is impossible.

The essential factors to be considered when implementing frequency agility are:

1) The available transmission frequencies must be separated by at least

$$\Delta f_s = 150/D \qquad \text{MHz} \tag{8}$$

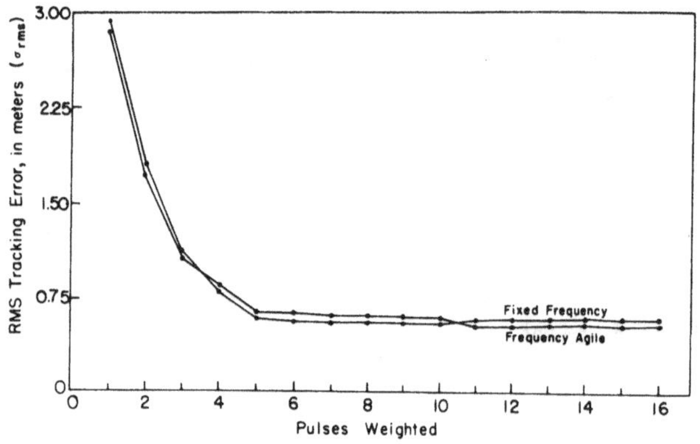

Fig. 7. Comparison of reduction in rms tracking error for a fixed-frequency radar and a frequency-agile radar when tracking a Swerling type II target (SLA weighting method).

where D is the target depth in meters.

2) There should be a minimal probability of transmitting two or more pulses in a single scan with the same frequency.

Any frequency sequence method that meets the above criteria will be an effective method to obtain independent samples of target position. Independent echo signals are necessary to fully utilize the weighting methods presented herein.

In the previous discussion, the targets were assumed to be of Swerling type I or III. In the event the echo-signal amplitude is independent from pulse to pulse (a Swerling type II or IV target), frequency agility is not necessary to insure echo-signal decorrelation, but what be valuable as an electronic countermeasure. The rms tracking error reduction of a Swerling type II or IV target with and without frequency agility is compared in Fig. 7. In each case the tracking error is the same. Observe that some type of amplitude signal processing is still necessary, as demonstrated by the reduction in σ_{rms} when several pulses are weighted. Since frequency agility insures independent samples of target position, it is the first step necessary to use the amplitude weighting methods previously presented.

The reduction in rms tracking error achieved by the various amplitude weighting methods is depicted in Fig. 8. Of paramount importance in this discussion is the rapidity of convergence on the true target position afforded by these schemes. The curves, typical of those generated by the computer simulation, demonstrate that taking the arithmetic mean of the instantaneous apparent target positions yields the smallest amount of error reduction for a small number of pulses. The greatest reduction in rms tracking error is obtained by the weighting method which fixes target position solely from returns with the largest amplitude.

At this time it would be well to discuss the mean weighting method. This is the type of processing that is normally employed in radar systems. Lind [9] showed that a substantial improvement in tracking error accuracy would be manifest using such a tracking scheme coupled with the use of frequency agility. In fact, our similations showed that if N samples corresponding to N target returns, are made on a Gaussian-distributed error function, such as that shown in Fig. 3, the rms value of target tracking error is reduced to $1/\sqrt{N}$ of its original value. In the limit, $N = \infty$, the mean weighting method would point to the actual target's radar centroid. The radar centroid, or center of gravity, is the average target position indicated by the radar over either a long time span or for all aspect angles. Thus, the curve corresponding to mean weighting in Fig. 8 is monotonically decreasing and will continue to do so. In many circumstances the processing of a large number of returns is not feasible; therefore, a method to more rapidly determine target position is necessary.

If Fig. 8 is compared with Fig. 5, it should be noted that as the echo signals with small amplitude are assigned correspondingly less weight, the tracking error improvement increases. For example, the square-law weighting method assigns a lower weight to returns with a small amplitude than either the Rayleigh or the linear weighting methods. In fact, these curves point out that the method which achieves the greatest reduction in rms tracking error is one that selects the largest amplitude (SLA). Since SLA weighting achieves the greatest amount of error reduction, it will be examined more closely.

In an effort to gain some further insight into the anticipated pointing error obtained by using both frequency agility and the SLA weighting method, some actual aircraft glint-error measurements obtained by Dunn and Howard [3], [13] and condensed in [17] should be noted. The results of their experiments show that the instantaneous glint tracking error expressed in linear units from the actual target's radar centroid has four important characteristics:

1) Glint error is essentially independent of range, except for extremely short ranges, where

$$R < L\Delta/\lambda \tag{9}$$

where L is the maximum target extent and Δ is the radar antenna aperture diameter.

2) The error is essentially a Gaussian distribution, with a mean value of zero and an rms value of σ_g, as measured from the radar center of gravity.

3) The glint error is essentially constant, regardless of RF frequency, provided a target span of several wavelengths is assumed.

4) The rms value of linear tracking error lies somewhere in the range between 0.1 and 0.3 of the maximum target extent.

The rms tracking error σ_g is dependent upon the distribution of the major scattering centers, such as the engines, wing tanks, etc., and the angle of observation. For example, a small jet aircraft viewed from nose-on would have a σ_g approximately $0.1L$, whereas if there were outboard engines on the aircraft, the value of σ_g would approach $0.3L$. In addition, with an aircraft viewed from the side, where the

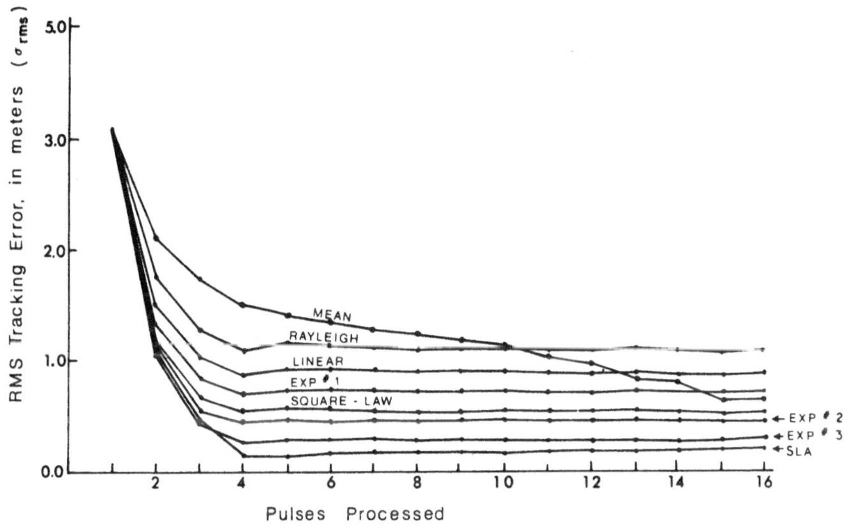

Fig. 8. Rms tracking error using the weighting methods presented in Fig. 5. (Frequencies chosen systematically 40 MHz apart between 5.35 and 5.85 GHz.)

distribution of reflecting areas is nearly uniform, the value of σ_g would approach $0.3L$. These results provide guidelines by which an initial estimate of σ_g can be made.

The rms tracking error reduction using the SLA weighting scheme will now be formulated. If the assumption is made that the echo signal with the largest amplitude is always the most accurate estimate of target position, then the probability of receiving at least one return with magnitude of error between 0 and e_1 is

$$P[-e_1 < e < e_1] = 1 - [1 - \int_{-e_1}^{e_1} p(e)\, de]^N \tag{10}$$

where $p(e)$ is the probability density function of the instantaneous tracking error and N is the number of returns processed. Defining

$$Q(e_1) = \int_0^{e_1} p(e)\ de = \int_0^{e_1} (1/\sqrt{2\pi}\sigma_0) \cdot \exp\,[-e^2/2\sigma_0{}^2]\,de,$$

where σ_0 is the standard deviation of the original tracking error, (10) can be rewritten as

$$P[e'] = P[-e_1 < e < e_1] = 1 - [1 - 2Q(e_1)]^N. \tag{11}$$

The probability density function corresponding to the distribution function $P[e']$ is plotted in Fig. 9 for $N = 1, 4$, and 8. Note that as N increases, the probability of obtaining at least one estimate of target position that is very close to the actual target position increases rapidly. The probability density function associated with $P[e']$ was found to be approximately Gaussian, with a standard deviation given by

$$\sigma = \sigma_0 N^{-0.9} \tag{12}$$

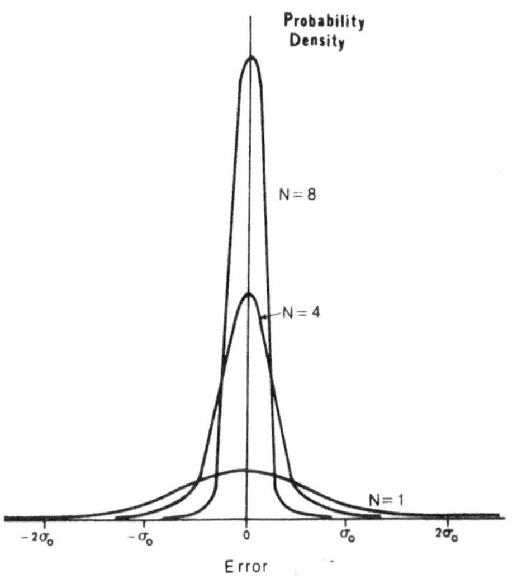

Fig. 9. Probability density functions of error corresponding to the use of the SLA weighting method; 1, 4, and 8 pulses processed.

for N a positive integer. Since the tracking error is measured from the radar center of gravity, the mean value of tracking error is zero. In this case, the rms tracking error is equal to the standard deviation of the tracking error. Therefore, σ_0 in (12) becomes σ_g, which is the rms tracking error for the target without using either frequency agility or a weighting scheme. Equation (12) may then be rewritten as

$$\sigma_{rms} = \sigma_g N^{-0.9} \tag{13}$$

where N, as before, is the number of pulses processed. Also, we note that σ_g, the rms error without agility and weighting, can be estimated as previously described.

391

[6] J.E. Meade, A.E. Hastings, and H.L. Gerwin, "Noise in tracking radars," Naval Research Lab., Washington, D.C., Rept. 3759, 1950; ASTIA Doc. AD90876.
[7] R.B. Muchmore, "Aircraft scintillation spectra," *IRE Trans. Antennas and Propagation,* vol. AP-8, p. 201, 1960.
[8] "Review of scintillation measurements," Hughes Aircraft Co., Tech. Memo. TM 272, 1952; ASTIA Doc. AD4849.
[9] G. Lind, "Reduction of radar tracking errors with frequency agility," *IEEE Trans. Aerospace and Electronic Systems,* vol. AES-4, pp. 410-16, May 1968.
[10] R.J. Sims and E.R. Graf, "The reduction of glint by diversity techniques," *IEEE Trans. Antennas and Propagation,* vol. AP-19, pp. 462-68, July 1971.
[11] R.J. Sims, "The determination of glint for a complex radar target," Doctoral Dissertation, Auburn University, Auburn, Ala. 1969.
[12] M.I. Skolnik, *Introduction to Radar Systems.* New York: McGraw-Hill, 1962.
[13] J.H. Dunn, and D.D. Howard, "Radar target amplitude, angle, and Doppler scintillation from analysis of the echo signal propagating in space," *IEEE Trans. Microwave Theory,* vol. MTT-16, pp. 715-28, September 1968.
[14] H.K. Ray, "Improving radar range and angle detection with frequency agility," *Microwave J.,* vol. 9, pp. 63-68, May 1966.
[15] P. Swerling, "Probability of detection for fluctuating targets," *IRE Trans. Information Theory,* vol. IT-6, pp. 269-308, April 1960.
[16] M. Schwartz, "Effects of signal fluctuation on the detection of pulse signals in noise," *IRE Trans. Information Theory,* vol. IT-2, pp. 66-71, June 1956.
[17] M.I. Skolnik, Ed., *Radar Handbook.* New York: McGraw-Hill, Hill, 1970, pp. 28-8-28-15.
[18] W.P. Birkemeir and N.D. Wallace, "Radar tracking accuracy improvement by means of pulse-to-pulse frequency modulation," *AIEE Trans. Commun. Electron.,* vol. 81, p. 571, 1963.
[19] K. L. Horn and N.D. Wallace, "Detection of slowly fading targets with frequency agility," *Proc. IEEE,* vol. 57, pp. 817-818, May 1969.

Jester M. Loomis, III (S'69–M'72) was born in Oklahoma City, Okla., on September 5, 1946. He received the B.E.E., M.S., and Ph.D. degrees from Auburn University, Auburn, Ala., in 1969, 1970, and 1972, respectively.

While doing graduate work he was employed as a Graduate Teaching/Research Assistant with the Auburn University Electrical Engineering Department. During this time he participated in antenna and radar research activities and taught undergraduate courses in electromagnetic field theory. He is currently with the U.S. Army, assigned to the U.S. Army Missile Command at Redstone Arsenal, Ala.

Dr. Loomis is a member of Eta Kappa Nu and Tau Beta Pi.

Edward R. Graf (M'58–SM'69) was born in Cullman, Ala., on September 26, 1931. He received the B.S. and M. S. degrees from Auburn University, Auburn, Ala., in 1957 and 1958, respectively, and the Ph.D. degree in electrical engineering from the University of Stuttgart, Stuttgart, Germany, in 1963. His doctoral research was related to crystal production for maser applications.

Since 1963 he has been with the Department of Electrical Engineering of Auburn University, where he is a Professor teaching and doing research in the area of electromagnetic field theory. He has served as Project Leader on a number of government and private foundation research grants and contracts.

Dr. Graf is a member of Sigma Xi and Phi Kappa Phi.

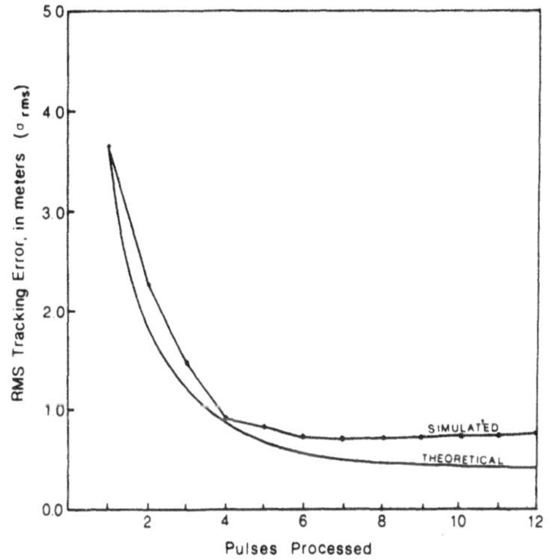

Fig. 11. Comparison of theoretical and simulated rms tracking error reduction.

The reduction in rms tracking error for this model, using the SLA weighting scheme, is shown in Fig. 11. The original rms tracking error, σ_g, of the target is 3.65 meters, which is 0.23*D*. This value of σ_g compares well with the results obtained by Dunn *et al.* [3], [13]. The simulated reduction in tracking error provides a σ_{rms} that closely approximates the theoretical σ_{rms} for all values of *N* between 1 and 4. After 4 pulses are processed, the simulated tracking error curve reaches a minimum; therefore, $\sigma_g/4$ or 0.9 meters is near the lower limit on tracking error for this case.

VI. Conclusions

With the inception of more accurate radar systems, glint error has been found to be a limiting factor on tracking accuracy. Previous researchers [14], [18], [19] have determined that frequency agility is an effective method to reduce glint error. This reduction in tracking error has been primarily expressed in a qualitative manner.

This paper shows that glint error can be reduced substantially after processing only a few pulses. The processing is based on incorporating frequency agility and the weighting scheme which establishes target position as a function of the echo-signal amplitude.

Frequency agility serves to obtain independent samples of target position. Any method of choosing frequencies, spaced apart by at least 150/*D* MHz, where *D* is the target depth in meters, will provide decorrelated estimates of target position. These frequencies should be selected so that there is a minimum probability of transmitting two or more pulses with the same frequency in a given pulse group.

The processing scheme which fixes the target position solely from the echo signal with the largest amplitude, SLA weighting, provides the greatest amount of error reduction. This method has an added advantage of being easily implemented. Moreover, this weighting scheme, when used to track targets in the Swerling type I or III class, requires as many different transmit frequencies as there are pulses being processed. When Swerling type II or IV targets are tracked, frequency agility is not necessary. The rms tracking error for a target using a fixed-frequency radar without frequency agility or amplitude weighting processing has been developed by Dunn *et al.* [3], [13]. If the σ_g is the original rms tracking error, then the reduced rms tracking error is

$$\sigma_{rms} \cong \sigma_g/N, \qquad 1 \leqslant N \leqslant 4 \tag{15}$$

where *N* is the number of pulses SLA weighted. The upper limit of *N* is dependent upon the degree of negative correlation between amplitude and error, the magnitude of σ_g, and the target velocity characteristics.

Acknowledgment

The authors wish to thank H. G. Woods and H. B. Buie for their aid in performing this work.

References

[1] F.C. Weimer and L. Peters, Jr., "Study of pointing errors in conically scanning and monopulse tracking radars for multipoint targets," Antenna Lab., Ohio State University, 1950; ASTIA Doc. 123476.

[2] L. Peters, Jr., "Accuracy of tracking radar systems," Antenna Lab., Ohio State University, pp. 601-629, 1957; ASTIA Doc. AD200.

[3] J.H. Dunn, D.D. Howard, and A.M. King, "Phenomena of scintillation noise in radar tracking systems," *Proc. IRE*, vol. 47, pp. 855-63, May 1959.

[4] R.H. Delano, "A theory of target glint or angular scintillation in radar tracking," *Proc. IRE*, vol. 41, pp. 1778-84, December 1953.

[5] J.E. Lindsay, "Angular glint and the moving, rotating, complex radar target," *IEEE Trans. Aerospace and Electronic Systems*, vol. AES-4, pp. 164-173, March 1968.

Section VI Multifrequency Radar Systems

The areas least adequately covered by the available literature are the details of implementation and examples of actual applications of diversity and agility to radar systems. In this section, three exceptions to this general condition are reprinted and a few others are covered by the annotated bibliography.

The principles of design of rapidly-tuned local oscillators, needed in frequency-agile radar, are discussed in all three papers reprinted here. Backmark et al. describe the ingenious approach by which the approximate frequency at which the magnetron will transmit is sensed prior to the transmitted pulse, permitting the local oscillator to be prepositioned within the bandwidth of the conventional IF AFC discriminator. The AFC loop then can provide a rapid vernier adjustment during the recovery time of the normal receiver system. This subject is explored further by Weissglas and Svensson. Fuller discusses systems which coarse-tune the local oscillator by electromechanical coupling to the magnetron tuning elements. Again, an AFC loop is used for fine tuning after the transmission.

Magnetron transmitters for frequency-agile radars represent another special subject, covered by Backmar et al. and by Fuller in the reprinted papers. The relative merits of spin-tuned and bellows-tuned magnetrons will be debated for some time, and other electromechanical tuning methods have been developed. Radars using parallel diversity channels can be implemented either with separate magnetrons or power amplifiers, or with a single broadband RF amplifier driven by a frequency-stepping exciter. The first option provides complete redundancy, and has been used for many years in radars requiring very high reliability, such as the air traffic control radars of the FAA. The second provides for any number of frequency channels without an RF multiplexer and its associated losses and tuning limitations. System application examples of both options are described in the paper by Ward, Fowler, and Lipson (see bibliography) in connection with the Air Force AN/TPN-19 landing control system, in which the ASR uses dual magnetrons and separate receivers while the PAR uses a broadband-amplifier transmitter and parametric-amplifier receiver. Considerations of multiplexer and antenna design for diversity radar are discussed in the paper by Perry (see bibliography).

ANNOTATED BIBLIOGRAPHY

Backmark, N.; Krim, J. E. V.; and Sellberg, F.; "Frequency-Agile Radar," *Philips Technical Review, Vol. 28, No. 11,* 1967, pp. 323-328. Presents a summary of advantages of frequency agile radar, along with discussion of the spin-tuned magnetron and the coarse-fine AFC loop used to tune the receiver accurately over a broad band. Reprint 6-1.

Fuller, J. B.: "Implementation of Radar Performance Improvements by Frequency Agility," *IEE Conference Publication No. 105,* October 1973, pp. 56-61. Summarizes the advantages of frequency agile radar and discusses the Accutune plunger-tuned magnetron, which provides fast tuning of the transmitter and coarse control output for the receiver AFC loop. Reprint 6-2.

Fuller, J. B.: "Frequency-Agile Local Oscillator," *Microwave Journal, Vol. 17, No. 5,* May 1974, pp. 31-33, 41. Discusses implementation of a fast-tuning local oscillator using a coarse open-loop setting derived from the agile source and a closed discriminator loop for fine tuning.

Ward, H. R.; Fowler, C. A.; and Lipson, H. I.: "GCA Radars: Their History and State of Development," *Proceedings of the IEEE, Vol. 62, No. 6,* June 1974, pp. 705-716. Contains descriptions of AN/TPN-19 PAR and ASR, both of which use dual diversity search waveforms to improve detection in noise and clutter.

Perry, K. H.: "Frequency Diversity in Radar Systems," *Microwave System News,* February/March 1975, pp. 38-41. Reviews advantages of frequency diversity radar and describes L-band diplexer and antenna implementation for search radar.

Weissglas, P.; and Svensson, S.: "Local Oscillators for Frequency Agile Systems," *Microwave Journal,* January 1977, pp. 35-38. Describes requirements for fast-tuned local oscillators capable of following agile magnetron frequency shifts and solid-state implementation capable of low noise and high stability. Reprint 6-3.

Frequency-Agile Radar

N. Backmark, J.E.V. Krim, and F. Sellberg

Philips Telecommunication Review, Vol. 28, No. 11 (1967)
Pages 323-328

Radar systems are used for detecting the presence of certain objects — usually ships or aircraft — and determining their position. These observations are beset by various inaccuracies because of the complexity of the surfaces of these targets, whose attitude relative to the radar aerial considerably affects the strength of the received echo signal. These inaccuracies can be eliminated by making the frequency of the radar transmitter jump in an irregular way from one pulse to another ("frequency agility"). By using the Philips spin-tuned magnetron a light and compact frequency-agile radar has been achieved.

Introduction

The purpose of a radar system is to detect the presence of certain objects of interest — usually aircraft or ships — and then to determine their co-ordinates. Until recently it was the general practice to assign to each radar system its own fixed frequency. As we shall show in this article, however, there is much to be gained if instead the frequency of the radar transmitter is made to jump irregularly from pulse to pulse.

This method of operation has been chosen — even though it requires a more complicated equipment — on account of certain variations which can occur in radar echoes. These variations, which may occur even with targets that change their position relatively little, can both reduce the probability of detection and affect adversely the accuracy with which the angular co-ordinate is measured [1]. Variations of this kind arise because structures such as aircraft and ships do not have simple shapes, but present to the radar beam a large number of reflecting surfaces of different size and at varying angles to the beam.

Owing to interference between the echoes from the different surfaces of the target the effective reflecting surface, and hence the amplitude of the echo reaching the radar receiver, depends critically on the attitude of the target. It was found in certain cases, for example, that a rotation of the target by only 1/3° caused a variation in signal amplitude of 15 dB. This effect, referred to as "scintillation", results therefore in a kind of fading, making a target which is otherwise within the range of the radar temporarily "invisible". Since with a radar system it is generally desirable to keep the target under continuous observation, the effect can obviously be troublesome.

Another phenomenon, "glint", is due to interference between the various component echoes from a target and gives an effect in which the apparent direction of the received signal depends on the attitude of the target. We know from practical experience with radars that the point from which the echo seems to be coming need not even lie within the "optical" confines of the target.

This point from which the echo seems to be coming is known as the scattering centre of the target. When a target is continuously illuminated at constant frequency by a radar system, the changing attitude of the target will cause the centre of the scattering cross-section to wander over its surface. As the attitude of the target only changes slowly, so too does the scattering centre. This slow displacement can easily be followed by automatic tracking radars, such as those used in fire con-

Civilingenjör N. Backmark and Teknologie Licentiat F. Sellberg are with Svenska Philips A.B., Stockholm; Ir. J. E. V. Krim is with N.V. Hollandse Signaalapparaten, Hengelo, Netherlands.

[1] See: J. I. Marcum, A statistical theory of target detection by pulsed radar, IRE Trans. on information theory **IT-6**, 59-267, 1960, and P. Swerling, Probability of detection for fluctuating targets, IRE Trans. on information theory **IT-6**, 269-308, 1960.

trol systems on warships, so that an apparent movement of the target is observed. Although it is not of first importance which part of the target is followed, the drift of the scattering centre over the target surface can cause serious errors in the measurement of target speed. The drift speed, which is added vectorially to the true speed of the target, causes inaccuracy in the prediction of target location, and in fire control, this prediction is of course the very thing that has to be correct.

As both of the phenomena described — "scintillation" and "glint" — result from interference between component echoes, they are very strongly affected by the radar frequency. One may therefore expect that a particular target will return a strong echo at one frequency and a weak one at another, even if its attitude does not change. Moreover, the apparent direction of the echo can deviate one way at one frequency, and the opposite way at another.

Since the radar beam has a certain width, a number of pulses — say 20 — will arrive at the target every time the beam sweeps over it. If the frequency of the transmitter is now made different at every pulse, then in the first place we may expect that interference between the echoes will not result in cancellation of the signal at all the frequencies. In the second place we may expect that the apparent directions of the echoes from the different pulses will be statistically distributed about the true bearing. In fact, with the bearing, the result of the frequency changes is that the scattering centre shifts to and fro at great speed over the target surface. An automatic tracking system is *not* able to follow such *fast* displacements, but takes the mean observation over a period depending on the time constant of the tracking system. The use of frequency agility thus increases the probability of detection and also improves the accuracy of direction finding and the prediction of target location.

Another advantage of using frequency agility is that the chance of jamming or interference from other transmitters is reduced. In an area where there are several radar stations with fixed frequencies it is necessary to allocate these frequencies carefully so as to avoid inadvertent interference between them. If, however, the transmitter frequency jumps irregularly from pulse to pulse, there is very little chance that the receiver frequency, which is tied to that of the transmitter, will in consecutive receiving intervals be equal to, or close to, the frequency of a pulse received from some other transmitter. This holds for both accidental and deliberate interference.

Finally, the use of frequency agility does away with the difficulty of "second-trace" echoes. These are echoes reflected from targets so far distant that the echoes reach the receiver only after the next pulse has been transmitted; they combine with the echoes of the second pulse reflected from nearby targets, producing an incorrect and confusing indication on the radar screen. Clearly, if transmitter and receiver jump to an entirely different frequency after a pulse has been sent out, the receiver will be insensitive to these "second-trace" echoes.

These considerations, which have been confirmed by experiment [2], prompted Philips to design a frequency-agile radar system, some particulars of which are given in this article. The main problems were in the generation of the transmitter frequency and the design of the receiver.

Generation of the transmitter frequency

In microwave radar systems magnetrons are very often used in the transmitter as the source of high-frequency energy, particularly when the equipment has to be transportable and thus as simple and light as possible. This is because a magnetron possesses a unique combination of features: though small and light it has high output power and high efficiency, it is very sturdy in construction and it generates the output power at a sufficiently constant frequency to make external frequency control superfluous.

The frequency at which a magnetron oscillates is chiefly determined by the resonant frequency of its cavities. A simple magnetron has no elements which can be used to tune it and therefore oscillates at one unalterable frequency. Nowadays, however, owing to the need to be able to avoid jamming or unintentional interference from other radar transmitters by changing the operating frequency, nearly all radar transmitters have tunable magnetrons. The means of changing the magnetron frequency generally consists in the insertion of adjustable rods into the resonant cavities. To maintain a high efficiency there has to be a rod in each cavity and all rods must be moved simultaneously and by the same amount. This is done by means of a screw mechanism and a gear transmission system. Although this method of tuning is very effective and enables a wide range of frequencies to be covered, it is far too slow for a frequency-agile radar. With this method a complete tuning cycle takes a few seconds, whereas in order to be able to change the operating frequency to the required extent between two successive pulses a tuning speed about a thousand times greater is needed.

This explains why up till now magnetrons have not been used in frequency-agile radars. One solution of the problem that has been found is to employ a frequency control system comprising a number of crystal oscillators, frequency dividers and fre-

[2] See: B. G. Gustafson and B-O. Ås, System properties of jumping-frequency radars, Philips Telecomm. Rev. **25**, 70-76, 1964.

quency multipliers, followed by amplifiers and terminated by a high-power aperiodic output stage. The design of a light and compact frequency-agile radar unit was however made possible by the use of the *spin-tuned magnetron* developed for this work in the Valve Laboratory of Svenska Philips A.B., Stockholm.

The spin-tuned magnetron

This magnetron, whose construction is shown in simplified form in *fig. 1*, contains a rotating element. This consists of a disc which is situated above the anode block and whose axis coincides with that of the magnetron.

The disc, which spins above the resonant cavities, has a toothed edge, and there are as many teeth as there are resonant cavities. The shaft of the disc runs on ball-bearings, and is driven by means of a rotating magnetized yoke which is situated outside the vacuum and connected to the shaft of an electric motor.

When a tooth of the disc moves over a cavity, it causes a change in the resonant frequency. The two effects are additive, resulting in a wide frequency swing. Since the teeth alter the tuning of all the resonant cavities identically, optimum efficiency is obtained. In every revolution of the disc as many tuning cycles are completed as there are cavities in the anode block, and therefore the tuning speed is proportional to the number of cavities as well as to the speed of revolution of the disc.

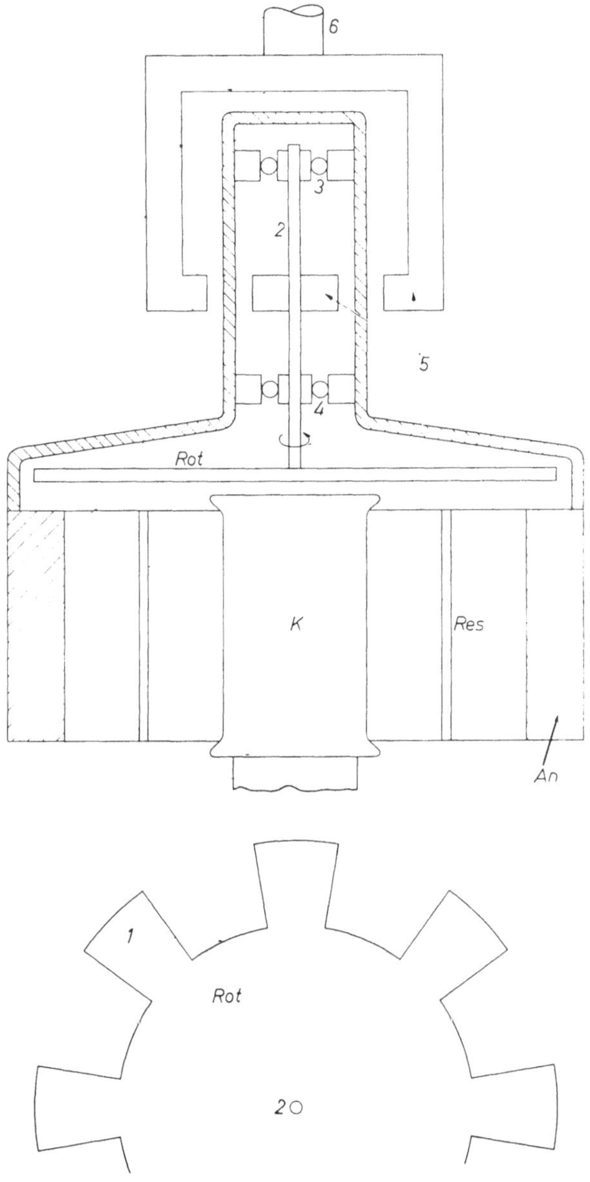

Fig. 1. Simplified cross-section and horizontal projection of the spin-tuned magnetron. *K* cathode, *An* anode block and *Res* one of the resonant cavities of the magnetron. *Rot* rotating disc with teeth *1*, driven by shaft *2*, mounted in ball-bearings *3*, *4* and driven by a magnetic coupling *5*, whose external yoke is connected to the shaft *6* of an electric motor.

Fig. 2. Spin-tuned magnetron type YJ 1181. The motor which drives the tuning disc is located on the upper side.

In one type of spin-tuned magnetron, type YJ 1181 (see *fig. 2*), which operates in the frequency band 8.5 to 9.6 GHz and has 16 resonant cavities, the frequency varies almost sinusoidally with time and, at a speed of 4500 revolutions per minute, one tuning cycle is completed in 900 microseconds. The frequency swing is 450 MHz and the maximum tuning speed is 1.4 MHz per microsecond. No simple relation will exist between the speed of rotation of the disc and the pulse repetition frequency — especially if the latter is made subject to small random fluctuations — and therefore the transmitter frequency jumps in an irregular manner from one pulse to another (see *fig. 3*).

As there are no reciprocating movements, no vacuum-tight bellows are required. This means that the life of the tuning mechanism is good and does not limit the life of the magnetron as a whole. Moreover, the simplicity of the construction guarantees a high degree of reliability.

The power required to rotate the tuning disc is very small: only 18 W for the YJ 1181 magnetron. If it is required to transmit a fixed frequency, the disc can be held at any desired position by means of an electromechanical device.

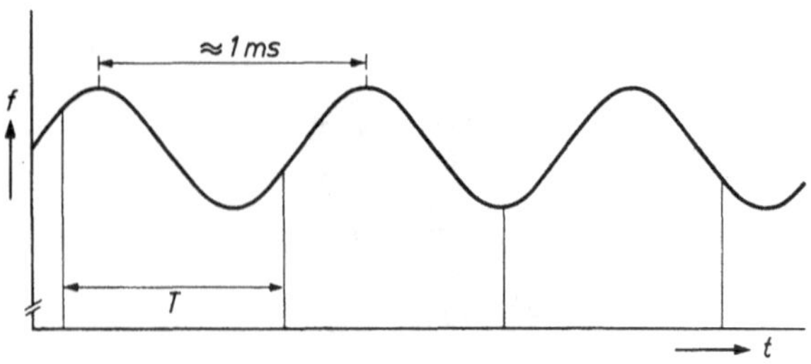

Fig. 3. Rotation of the tuning disc causes the magnetron frequency *f* to vary with time in a regular pattern which is approximately sinusoidal. Each magnetron pulse is so short that it appears on the time scale as a point. There is no simple relation between the number of revolutions per second of the tuning disc and the pulse repetition frequency (the pulse interval *T* is not necessarily constant from pulse to pulse) and therefore the transmitter frequency jumps in an irregular way from pulse to pulse.

The receiver

In the receiver the incoming echo signals, whose frequency is determined by the transmitter, are mixed with a local-oscillator signal and the resultant difference-frequency signal is amplified in an i.f. amplifier before being detected. For frequency-agile operation it is therefore necessary to solve the problem of how to keep the difference in frequency between the transmitter and the local oscillator exactly at the value to which the i.f. amplifier is tuned.

This requires that the local oscillator should be tunable over the same range of frequencies and that it should be capable of being tuned at the same speed as the magnetron. In fact, as we shall show later, it requires to be tuned at a much higher speed than the magnetron. Furthermore the output power of the oscillator should be high enough over the whole frequency band to guarantee a satisfactory i.f. signal. These requirements can be satisfied by a backward-wave oscillator (BWO), which has two advantages for such an application. Firstly, owing to its non-resonant character, it can deliver sufficient power over a very wide frequency range; and secondly, the frequency control is entirely electronic, so that there are no speed restrictions of mechanical origin.

To permit the reception of echoes from targets very close to the transmitter, the local oscillator must already be set to the correct frequency at the moment the pulse transmission has ended. So as to be able to detect long-range targets as well, this frequency setting should remain fixed, preferably until just before the moment at which the next pulse is generated. This means, therefore, that only a fraction of the time between two successive transmitted pulses is available for setting the local-oscillator frequency.

Since such fast and accurate control of the local oscillator frequency could not be achieved with conventional automatic frequency-control circuits, an entirely new control system was designed. The principle, which we shall now describe, is illustrated by the block diagram in *fig. 4*.

The output signal of the local oscillator *LO* is fed to the mixer stage of the receiver *R* and also to a circulator *Circ*; the signal emerging from the next port of the circulator is applied to the output of the magnetron and thence injected into the magnetron resonant structure.

At the beginning of each control period (this is equal to the time interval between two successive pulses) the resonant frequency of the magnetron will differ from the oscillator frequency, so that the local-oscillator signal is reflected back to the circulator. It then continues through the circulator, following the same path as the high-power pulse which appears at a certain instant from the magnetron, and arrives in the transmitter aerial waveguide. Incorporated in this waveguide is an automatic, instantly acting switch that conducts high-power signals, i.e. the outgoing pulse, to the aerial, and

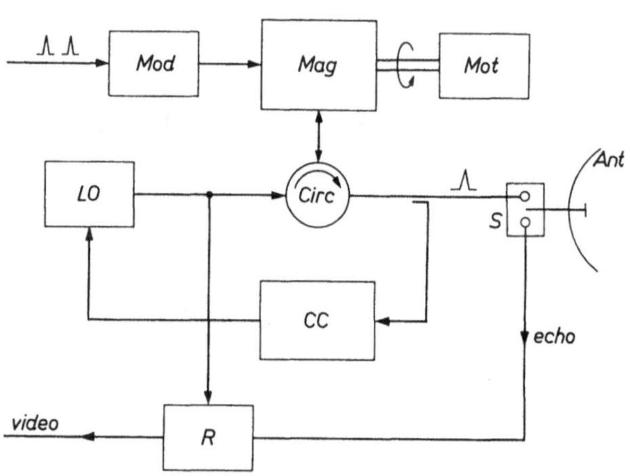

Fig. 4. Simplified block diagram of the frequency control circuit. *Mag* spin-tuned magnetron, tuned by electric motor *Mot*. The magnetron is driven by modulator *Mod*. *LO* local oscillator (backward-wave oscillator). *Circ* circulator. *S* transmit-receive switch (duplexer). *Ant* transmitting-receiving aerial. *CC* control circuit. *R* receiver. The control circuit controls the local oscillator in such a way that the oscillator frequency during each receiving interval differs from the frequency at which the last radar pulse was transmitted by an amount equal to the (fixed) intermediate frequency *IF*.

low-power signals, i.e. the reflected BWO signal, to the control circuit *CC*, where they are detected. During the presence of the d.c. voltage (obtained by detection) the control circuit delivers a signal that causes the local-oscillator frequency to drop rapidly along line *b* (see *fig. 5*) from the value *a* it had at the beginning of the control period. The value *a* is greater than the maximum tuning frequency of the magnetron: the magnetron frequency follows the curve *e*.

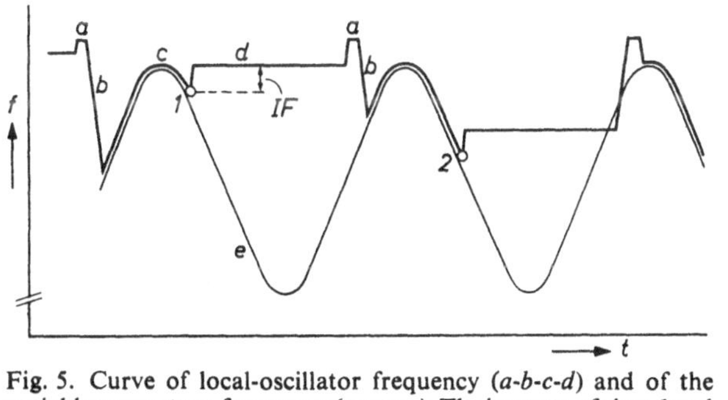

Fig. 5. Curve of local-oscillator frequency (*a-b-c-d*) and of the variable magnetron frequency (curve *e*). The instants of time *1* and *2* mark the magnetron pulses. During the receiving interval (*d*) which follows *1*, *2*, etc., the local-oscillator frequency remains constant and higher by a fixed amount *IF* than the frequency of the transmitted pulse. For clarity the length of the control period *a-b-c* is somewhat exaggerated here in relation to the receiving interval, and *c* and *e* are separated.

At the instant that line *b* meets curve *e*, in other words as soon as the frequency of the local oscillator becomes equal to the tuned frequency of the magnetron structure, the latter absorbs part of the oscillator energy. There is then very little reflection and the detected signal drops abruptly to almost zero. At this abrupt drop in the detector signal the control circuit stops reducing the oscillator frequency.

After switching off the "search" along line *b*, the next step is to switch to "following" the magnetron frequency, as after termination of the search the magnetron frequency continues to follow curve *e*.

At the same time as the search is switched off an auxiliary oscillator is therefore switched on, which is used to modulate the frequency of the local oscillator. Since this frequency is very close to that of the magnetron, the effect of modulating the oscillator frequency is a perceptible amplitude modulation of the residual reflected and detected signal. Moreover, every time the oscillator frequency passes the magnetron frequency there is a phase shift in the modulation of the detected signal. By comparing the phase of the modulation component of the detected signal with that of a signal derived straight from the auxiliary oscillator it can be established whether the oscillator frequency is higher or lower than the tuned frequency of the magnetron. A control signal can thus be derived which adjusts the oscillator frequency in the right direction. With this method the oscillator frequency is kept close to the magnetron frequency and therefore follows it along curve *c* which practically coincides with curve *e*.

At a given moment, determined by the modulator (*Mod* in fig. 4), the magnetron begins to oscillate, e.g. at the moment its tuned frequency passes the value at point *1* in fig. 5. The control circuit, being linked to the transmitter waveguide, reacts to this oscillation by altering the value which the voltage controlling the local oscillator has at that moment. The magnitude of this step in voltage is arranged to make the local-oscillator frequency differ from the magnetron oscillating frequency by an amount equal to the intermediate frequency (*IF*) of the receiver. The control circuit then freezes the control voltage, and hence the local-oscillator frequency remains at this changed value (*d*) for a period (the "receiving interval") which is somewhat shorter than the pulse interval of the modulator. At the end of the receiving interval the control circuit returns the oscillator frequency to the value *a* and a moment later switches on the search *b* again, thus starting the next control cycle.

As with this method of adjusting the local-oscillator frequency there is still a chance of slight errors at the moment the magnetron starts oscillating, the i.f. section of the receiver contains a modified conventional AFC circuit for fine adjustment of the local-oscillator frequency if it differs from the correct value. This adjustment takes place while the pulse is being transmitted.

Summarizing, the following phases may be distinguished in the tuning cycle:

1) Searching for the instantaneous tuned frequency of the magnetron along line *b*.
2) Following the changing magnetron frequency along curve *c*.
3) Stopping after reception of the transmitter pulse, immediately followed by a controlled frequency step *IF*.
4) Stopping at the frequency attained for the duration of the receiving interval, line *d*.
5) Return to the starting frequency *a*.

The control circuit has to ensure that the BWO completes the phases of this cycle correctly and in the right order. To enable it to do this, the control circuit receives information from various sources. Recapitulating, these are successively: the drop in the detected reflected high-frequency power, the output voltage of the phase-sensitive detector, the transmitter pulse and the output voltage of the frequency discriminator in the i.f. part of the receiver as a correction for the controlled step in BWO frequency.

The electronic functions for the return to the frequency *a*, the search, the step *IF* and for marking off the receiving interval are performed by appropriate circuits in the control system.

Future developments

The special features of the spin-tuned magnetron and of its associated control circuits make it suitable for other applications that lie outside the scope of conventional systems.

First, we may mention a system in which the transmitter frequency, instead of being changed in an arbitrary manner from one pulse to another, is varied to follow a predetermined programme.

Next, a frequency-agile transmitter can also be combined with a FRESCAN aerial [3], i.e. an aerial where the direction of the principal axis of the beam is a function of the frequency. An aerial of this type can be used to scan a sector, without mechanically moving the aerial, by electronically controlled variation of the frequency. Obviously, much higher scanning speeds are possible with this system, producing a much less cluttered picture on the screen and eliminating troublesome "tails".

Finally, the variable-frequency magnetron can be used in military systems for target analysis and gives good results in the determination of the optimum frequency for a reliable maximum return at any given moment.

---

[3] See, for example: J. Croney, Doubly dispersive frequency scanning antenna (for two plane scanning), Microwave J. **6**, No. 7, 76-80, 1963.

Summary. The article discusses certain inaccuracies that can occur in radar detection and tracking and explains how these can be eliminated by making the frequency of the radar transmitter jump in an irregular way from pulse to pulse. This procedure also offers protection against jamming or unintentional interference from other radar transmitters. The design of a frequency-agile radar system was greatly facilitated by the development of the Philips spin-tuned magnetron, which is briefly described. The article then discusses the principles of a control circuit which, during each receiving interval, keeps the frequency difference between the transmitter and the local oscillator in the receiver equal to the intermediate frequency of the receiver.

Implementation of Radar Performance Improvements by Frequency Agility

J.B. Fuller

IEE Conference Publication, No. 105 (October 1973)

Introduction:

The use of frequency agility in the generation of radar signals will produce improvements in the performance of the system with respect to probability of detection, reduction of clutter returns, reduction of tracking errors and resistance to ECM. This paper indicates some of the improvements available through the use of frequency agility and discusses subsystem approaches to the implementation of such capability.

Frequency Agility Benefits

The operational performance of a pulsed radar system can be improved by the use of frequency agility where the transmission frequency is changed on a pulse to pulse basis. The effect is to reduce the correlation between various undesired radar returns such as clutter. In this manner statistical processing can be used to improve system performance.

Target Detection

The typical radar target for a fixed frequency system consists of a number of discrete scatterers producing a radar return amplitude which is a random variable with an amplitude function which can be described by a Rayleigh distribution. For a frequency agile system, the amplitude of the radar return can also be described by a Rayleigh density function. Where N return pulses are considered, the variance of the frequency agile density function is reduced by $\sqrt{N}$ with respect to the fixed frequency case as illustrated in Figure 1. The probability of detection for a given amplitude threshold level is greater for the frequency agile compressed density function than for the non-agile case.

Clutter Reduction

The use of frequency agility will improve the effective signal to clutter ratio by decorrelation of clutter returns. The use of frequency agility will "speed up" the time base for clutter state changes and clutter decorrelation can be achieved over a single radar scan in N = 15 to 20 pulses with an effective 10 to 12 dB reduction in clutter intensity.

Tracking Error Reduction

The accuracy of a tracking system to determine the deviation of the target axis from the antenna axis is effected by target glint and fading, refraction errors in radomes and variations in the propogation media. These sources of angular resolution error can be averaged down by rapid variation of the transmission frequency.

James B. Fuller is with Varian Associates, Solid State Division, Beverly, Massachusetts, USA

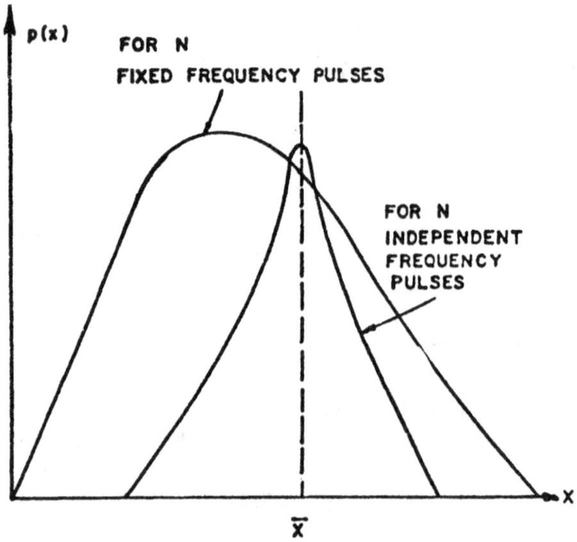

Figure 1. - Echo Amplitude Density Functions

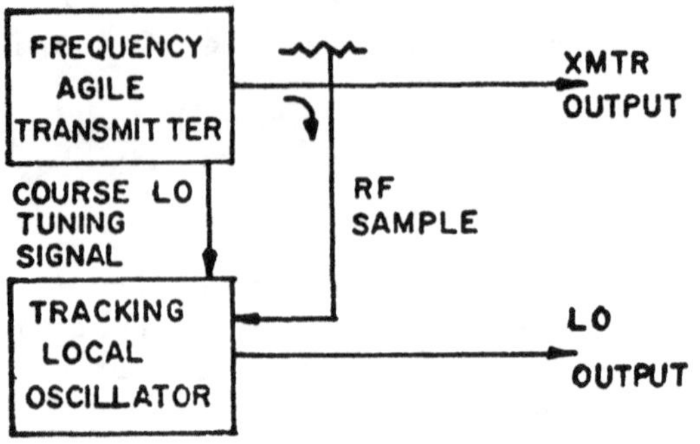

Figure 2. - Frequency Agile Subsystem

Resistance to Jamming

The effect of varying the operating frequency of a radar on a pulse to pulse basis causes the system to be much less susceptible to narrow band jamming or deceptive ECM. Little improvement is possible in a broadband jamming environment but this requires large amounts of total power to obtain effective power spectral density.

Frequency Agility Subsystem

The practical implementation of frequency agility into a radar system is to utilize a subsystem consisting of a frequency agile transmitter source and a matching tracking local oscillator as illustrated in Figure 2. The transmitter produces a pulse to pulse frequency variation, $\Delta f_T(p-p)$, with a frequency agility excursion $\Delta f_T(max)$ at a frequency agility rate fm. The tracking local oscillator provides a pulse to pulse stabilized frequency separated from the transmitter frequency by the system IF.

Frequency Agile Source

Frequency Agile Coaxial Magnetron

A convenient source of frequency agile power for radar transmitter application is the Frequency Agile (dither) coaxial Magentron or FAM. The FAM, which functions so as to produce pulsed RF power with a carrier frequency which changes on a pulse to pulse basis, is illustrated in Figure 3.

Programmable Frequency Agile Magnetron

The frequency agile operation of the coaxial magnetron can be expanded by use of the Accutune® servo control loop indicated in Figure 4. The loop uses a servo motor to drive the agile magnetron and the applicable readout circuit output is fed to a servo amplifier along with the servo control input. The Accutune magnetron functions as a voltage controlled high power pulse oscillator capable of slewing over 400 MHz of frequency range in less than 60 milliseconds at X-Band. A tracking local oscillator (TLO) is required as part of the frequency agility subsystem to provide an LO signal whose frequency is adjusted on a pulse by pulse basis so as to enable radar returns received over each interpulse period to be within a designated IF passband. There are both open loop and closed loop tracking methods which will enable the LO to track the transmitter frequency. The TLO is achieved in two steps on a pulse by pulse basis: (1) the course tuning or preset of the local oscillator prior to the transmitter pulse and (2) the fine tuning of the local oscillator by the IF preset error signal which occurs when the transmitter is pulsed.

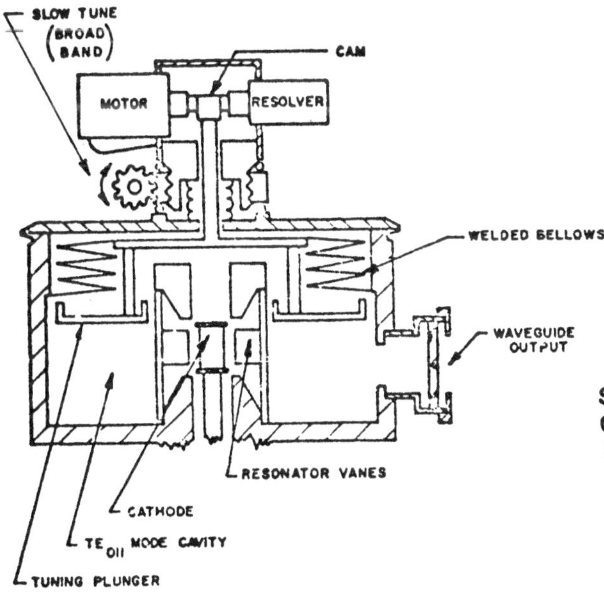

Figure 3. - Cross Section of FAM

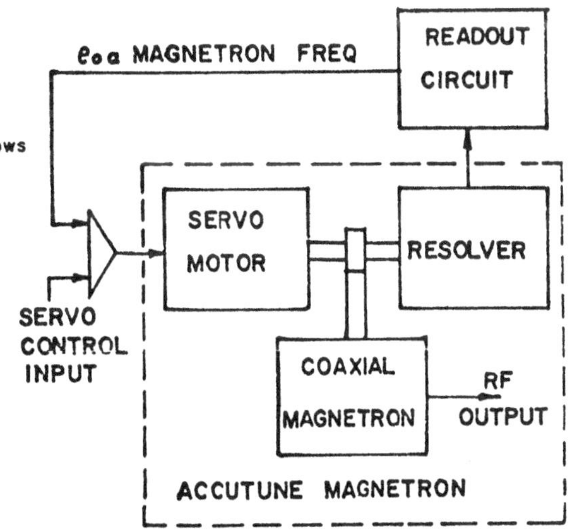

Figure 4. - Accutune® Servo Control Loop

Readout Circuit LO Preset

The frequency agile magnetron is provided with a readout circuit producing an analog voltage proportional to the instantaneous transmitter frequency deviation from its agile mean, $\bar{f}_T$. If this analog voltage is sampled just prior to each transmitter pulse, applied to a VCO and held constant for the ensuing pulse period, the LO frequency can be preset so that the IF produced when the transmitter is pulsed will be

$$f_{IF} = f_T - f_{LO} = \bar{f}_{IF} \pm \epsilon_{IF1} \quad (1)$$

The maximum IF preset error can be expressed as

$$\begin{aligned} |\epsilon_{IF1}|_{max} = {} & (K_T - K_o)\,\Delta T && \text{(temperature coefficient drift)} \quad (2) \\ & + \Delta f_T(\max) \sum_{i=1}^{3} \delta K_i && \text{(preset gain error)} \\ & + \delta f_T && \text{(magnetron/readout absolute error)} \\ & + \delta f_{LO} && \text{(LO) tuning accuracy)} \end{aligned}$$

where K1 K2 K3 = 1 for linear frequency preset of the LO and second order sources of error are excluded.

Figure 5 indicates the worst case maximum IF preset error, for agility peak excursions of 50, 100 and 300 MHz, as a function of the cumulative gain error in the preset channel. In this figure the absolute frequency accuracy of the magnetron/readout circuit and VCO are assumed at 3 MHz and 5 MHz respectively. The IF preset error must be corrected by the fine tuning circuitry to a level consistent with the receiver bandwidth requirements.

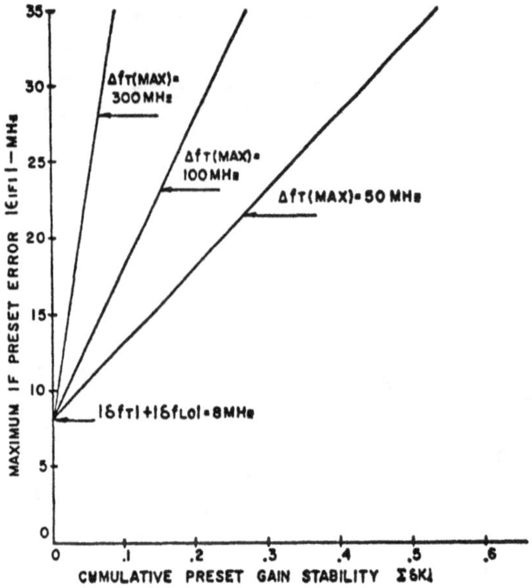

Figure 5. - IF Preset Frequency Error

Fine Tuning - Open Loop

The fine tuning, or correction of the IF preset error, can be accomplished by using a linear IF discriminator to generate a pulsed error signal proportional to ϵ_{IF1} as indicated in Figure 6. The amplitude of this error signal is sampled, stored and applied to the VCO after the transmitter pulse. The resultant IF correction, ignoring sample and hold droop over the pulse period, can be expressed by

$$f_{IF} = f_{IF} \pm \epsilon_{IF2}. \qquad (3)$$

The maximum value of the fine tuning correction error is, neglecting second order error terms and discriminator crossover accuracy

$$|\epsilon_{IF2}| \max = |\epsilon_{IF1}| \max \times \sum_{i=3}^{5} \delta K_i + \delta f_{LO} \qquad (4)$$

where K3 K4 K5 = 1.

Consequently, the residual open loop corrected error is effected by the cumulative transfer gain stabilities or linearities and the VCO frequency accuracy.

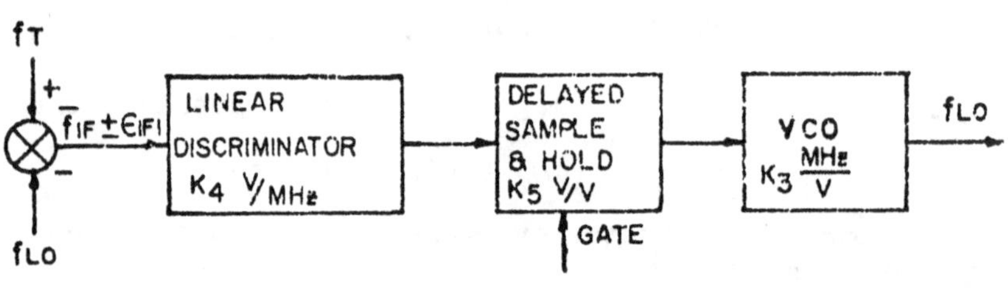

Figure 6. - TLO Open Loop Fine Tuning

Fine Tuning - Closed Loop (Discriminator)

The correction of the IF preset error can be achieved by using the closed loop discriminator loop illustrated in Figure 7. The sample/hold circuit, acting as a switch, is closed during the time of the transmitter pulse. The loop is required to correct the IF preset error during the time of the transmitter pulse and sample/hold switch is then used to maintain the closed loop error correction signal at the VCO input after the transmitter pulse, with an error of

$$|\epsilon_{IF2}|_{max} = \frac{|\epsilon_{IF1}|_{max}}{1 + \prod_{i=3}^{5} K_i} + \delta_{IF}. \qquad (5)$$

This correction error is essentially limited by the error compression ratio of the closed loop and the crossover accuracy of the discriminator.

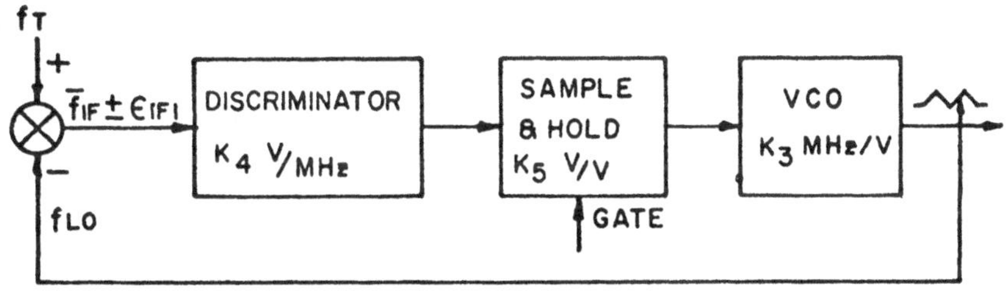

Figure 7. - TLO Discriminator Closed Loop Fine Tuning

Fine Tuning - Closed Loop (Phase Lock)

The correction of the IF preset error can be achieved by using the offset phase lock loop illustrated in Figure 8. The sample/hold circuit is switched to sample during the transmitter pulse time. During this time, the loop will lock if ϵ_{IF1} is within the instantaneous acquisition range of the loop. The sample/hold, is than used to maintain the closed loop error correction signal at the VCO input after the transmitter pulse.

The maximum value of fine tuning correction of the phase lock loop is essentially $|\epsilon_{IF2}|_{max} = \delta f_R$.

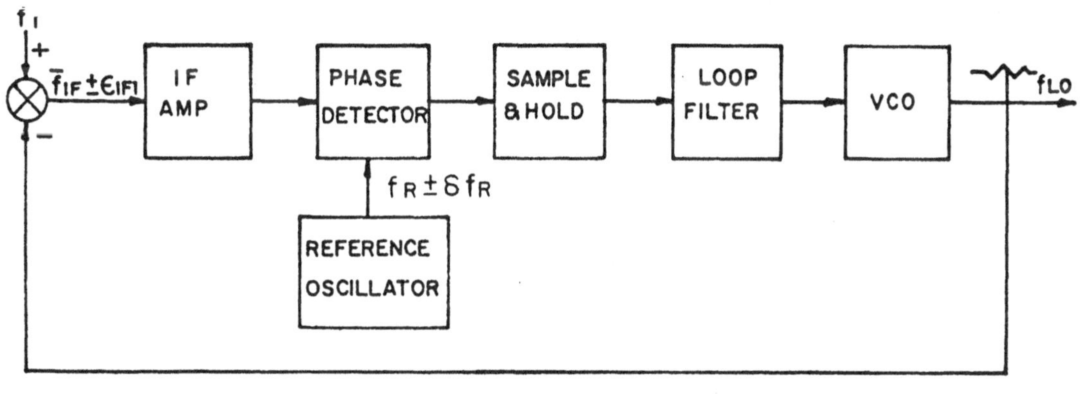

Figure 8. - TLO Phase Lock Loop Fine Tuning

TLO Comparison

The following table summarizes the relative methods of producing a tracking local oscillator for frequency agility use. The methods are listed in order of increasing difficulty of implementation.

Table 1

TLO METHOD COMPARISON

Method	Typical IF Accuracy	Implementation Problems
Open Loop	Readout Preset ±10 to ±30 MHz	Requires linear VCO and readout. Sensitive to gain changes.
	Discriminator ± 5 to ±10 MHz	Requires linear VCO and discriminator. Sensitive to gain changes.
Closed Loop	Discriminator ±.5 to ± 2 MHz	Requires very wide band loop. Requires very high slew rates.
	Phase Lock Loop ±100 KHz	Requires very wide bandwidth. Requires very high slew rates. Sensitive to reference oscillator leakage into loop due to wide bandwidth.

References

1. DiFranco, J.V. and W. L. Rubin, "Radar Detection," Electro Technology, April 1964, pp 64-90.

2. Barton, D.K.,"Simple Procedure for Radar Detection Calculations," IEEE Trans. on Aerospace and Electronic Systems, Vol. AES-5, No. 5, Sept. 1969, pp 837-846.

3. Ivey, H.D., W.V. Neisius, I.E. Perlin, "Clutter Cross-Correlation in a Dual Frequency Radar," Georgia Institute of Technology Report, 1952.

415 + 416 - BLANK

Local Oscillators for Frequency Agile Systems

P. Weissglas and S. Svensson

Microwave Journal, Vol. 20, No. 1 (January 1977)
Pages 35-38

SYSTEM CONSIDERATIONS

The accutune magnetron is electrically tunable and may offer a tuning range of up to about 1 GHz. The spin tuned magnetron has a tuning range of about 0.5 GHz at X-Band but it is not electrically tunable *i.e.*, it cannot be made to follow the frequency of a source with arbitrarily varying frequency. Nevertheless, the spin tuned magnetron is of great interest because of its high tuning speed. As an example, the maximum frequency with which an accutune magnetron can sweep a 1 GHz band is about 20 Hz *i.e.*, df/dt ~ 40 kHz/μs whereas the df/dt of a spin tuned magnetron is about 1 MHz/μs.

If electric tuning of the magnetron is available, it is possible to design systems where the magnetron follows the LO frequency. If a limited number of discrete frequencies are used, crystal controlled LOs can offer full MTI capability.[1] In this paper, we shall concentrate on systems where the LO follows the electronically or mechanically tuned magnetron. We will show that satisfactory performance can be achieved even with the widest available tuning range and that an MTI mode may also be built into the system.

The block diagram of **Figure 1**[2] shows a practical system for a frequency agile radar. We shall first consider the requirements on the tracking LO in order that the received signal remain within the IF amplifier bandwidth throughout the listening time after a transmitted pulse. Referring to **Figure 2**, the system is designed such that the LO frequency returns to point 1 at the end of a listening period. The LO frequency is then made to increase linearly until the instantaneous cold frequency of the magnetron is located. At this time, the local oscillator locks to the cold magnetron.

At point 3 (**Figure 2**) a voltage is applied to the tuning input of the LO causing a frequency deviation equal to the net sum of the IF frequency and the frequency difference between the hot and the cold magnetron. During the transmit phase a fine AFC further corrects the LO with respect to the

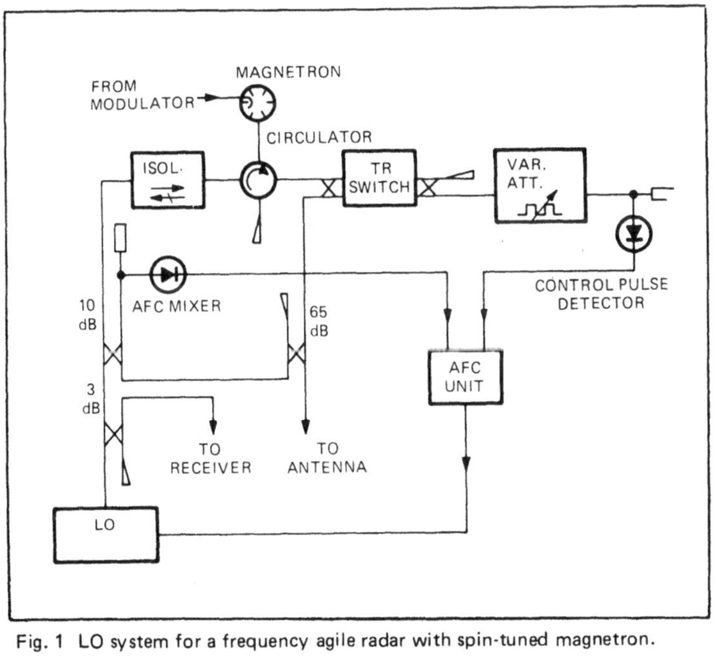

Fig. 1 LO system for a frequency agile radar with spin-tuned magnetron.

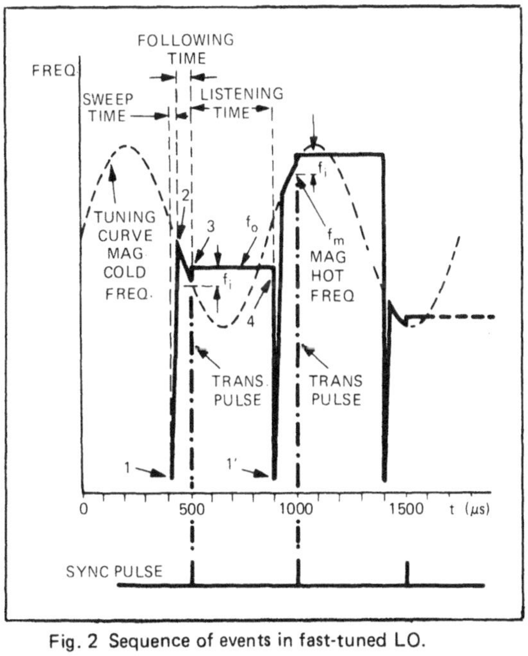

Fig. 2 Sequence of events in fast-tuned LO.

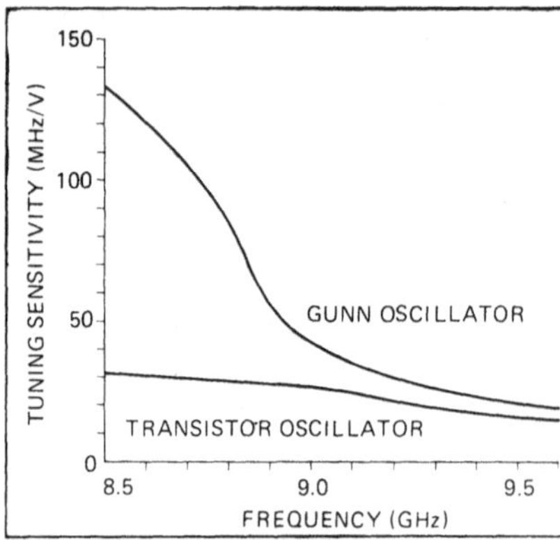

Fig. 3 Tuning sensitivity vs frequency.

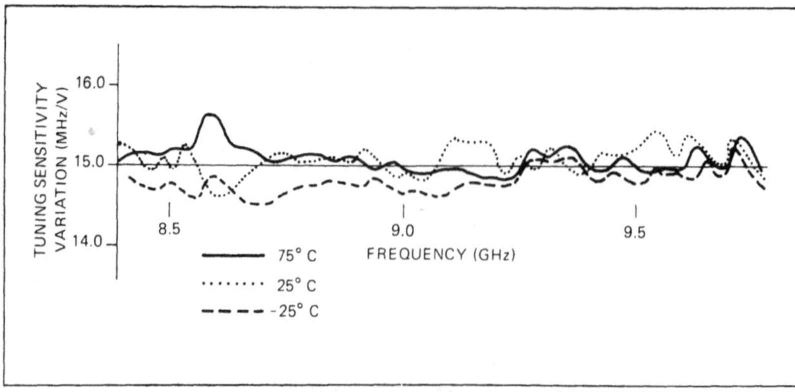

Fig. 4 Tuning sensitivity for linearized transistor source.

magnetron frequency. Throughout the listening phase the LO must then remain at this frequency with an accuracy better than the IF half bandwidth of approximately 1 MHz. The following requirements are therefore imposed on the local oscillator.

- The tuning curve must be monotonic and nearly linear. The detailed specification is determined primarily by tracking servo stability considerations. The tuning must also be sufficiently linear that a constant voltage step gives a correct IF +hot − cold frequency deviation independent of magnetron frequency.
- The transient response of the frequency to a fast tuning pulse must be rapid and free of overshot or lag.
- No frequency drift must occur during the listening phase (~1 ms).
- The noise performance must be acceptable in spite of the wide band nature of the LO.

LINEARITY

We shall consider two different types of solid state LOs. The first one is a fundamental frequency, varactor tuned Gunn oscillator, the second a varactor tuned transistor oscillator followed by a multiplier. Starting with the linearity requirements we give the tuning sensitivity of two practical sources in **Figure 3**. It is clear that none of the sources comes close to a constant tuning sensitivity but the transistor LO has a much smaller sensitivity variation. It will be necessary to include a linearizing network but the design of this network is much simpler when using the transistor source.

A linearized transistor oscillator will behave as shown in **Figure 4** where the temperature dependence is also illustrated. Several points are worth noting. First, in order to facilitate the design of the linearizer, the varactor voltage should be high. Such a high voltage varactor is usually excluded in Gunn oscillators because it will have a lower Q value. Second, the nonlinearity in the tuning curve must be reproducible otherwise it will require excessive work to trim the linearizer. Third, an additional fixed dc voltage equal to the highest tuning voltage of the varactor will be required for the linearizer. Finally, the nominally higher tuning sensitivity of a Gunn source measured in MHz/V will impose more stringent requirements on the filtering of the tuning voltage supply.

All these considerations point towards the transistor oscillator as the preferred alternative. This statement remains true in spite of the fact that the tuning sensitivity of the Gunn oscillator in **Figure 3** could be very much improved by designing the oscillator to have a much larger tuning range than actually required.

It is possible to design an oscillator tunable from 5 to 10 GHz and then to

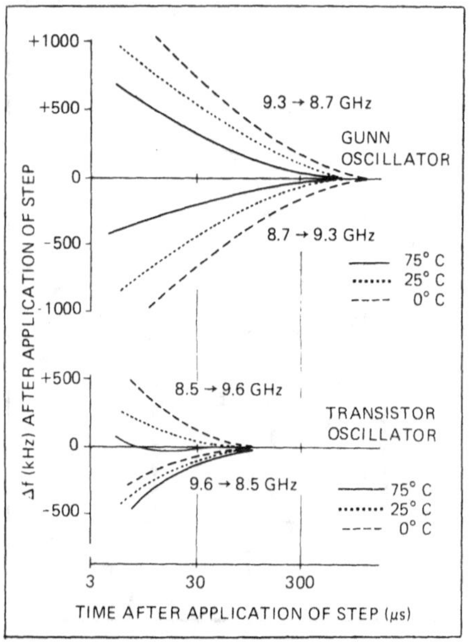

Fig. 5 Post tuning drift for Gunn and transistor oscillators.

TABLE I

SPECIFICATION FOR AN X-BAND FREQUENCY AGILE LO

ELECTRONIC TUNING RANGE	8.4-9.7	FM NOISE Δf_{rms} IN 1 kHz AT 100 kHz	50 Hz
POWER OUTPUT OVER FREQ. & TEMP. RANGE	20-60 mW	AM NOISE IN 1 MHz AT 30 MHz	< 90 dBc
FREQ. VARIATION OVER TEMP. RANGE	< 0.5 MHz/°C	FREQ. PULLING SWR 1.5	< 0.5 MHz
TUNING SENS. VARIATION OVER FREQ. & TEMP. RANGE	< ± 10%	FREQ. PUSHING	< 15 MHz/V
TUNING SPEED	> 150 MHz/μs	POST TUNING DRIFT	SEE FIG. 5
SPURIOUS SIGNALS	< 60 dBc	SUPPLY VOLTAGE	24V, 280 ± 60 mA 95V, 25 ± 1 mA
HARMONICS BELOW CARRIER ABOVE CARRIER	 < 15 dB < 20 dB	TEMP. RANGE	25 ÷ 75 C

use only the interval from 8.5 to 9.6 GHz which would result in a tremendous improvement in linearity. In practice however, this requires a very low Q oscillator which would be entirely unacceptable with respect to FM noise, pushing, pulling and temperature stability. We shall return to the FM noise in discussions below.

POST TUNING DRIFT

The frequency stability requirements imposed on the LO are even more difficult to achieve. In a transistor oscillator this stability is almost exclusively dictated by the varactor. In a Gunn oscillator the Gunn diode itself will also make a contribution to the frequency drift. Most important are, however, surface charges in the varactor which will cause a capacitance variation after a step in voltage.

Qualitatively it has been found that two types of such charges may be present. One type has a short time constant and will give rise to an overshot in the capacitance change at room temperature. The other type has a longer time constant and results in a lagging capacitance response. Sometimes it may be possible to make these effects partially cancel but the apparently positive result is deceptive. The reason is that the two effects have a substantially different temperature dependence and therefore compensation is possible at one temperature only.

The varactor design must therefore aim at eliminating charging effects by keeping the surface as clean as possible. A specially developed silicon varactor is used to produce the results shown in **Figure 5**. GaAs varactors would be preferable from a linearity point of view but appear to be grossly inferior with respect to post tuning drift.

It is instructive to make a comparison between the requirements on surface "cleanliness" imposed here with similar requirements in high quality MOS devices. We therefore assume that all surface charges in the varactor have time constants such that they will contribute to the post tuning drift within the considered time interval 3 to 300 μs. If the allowed post tuning drift Δf caused by a capacitance drift ΔC is 500 kHz at 10 GHz, we find

$$\frac{\Delta f}{f} = \frac{1}{2}\frac{\Delta C}{C} \approx \frac{500 \times 10^{-3}}{10\,000} \approx 0.05 \times 10^{-3} \ .$$

Typically the nominal capacitance C is of the order of 10 pF and the voltage step is 70 V. The amount of surface charge required to produce ΔC then is

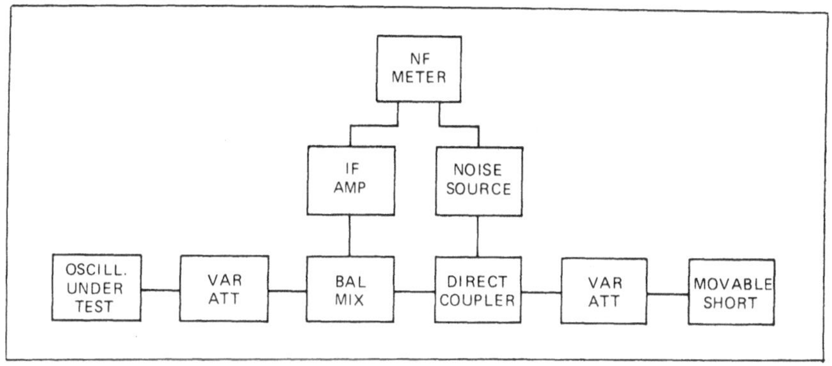

Fig. 6 Test set-up for noise measurement.

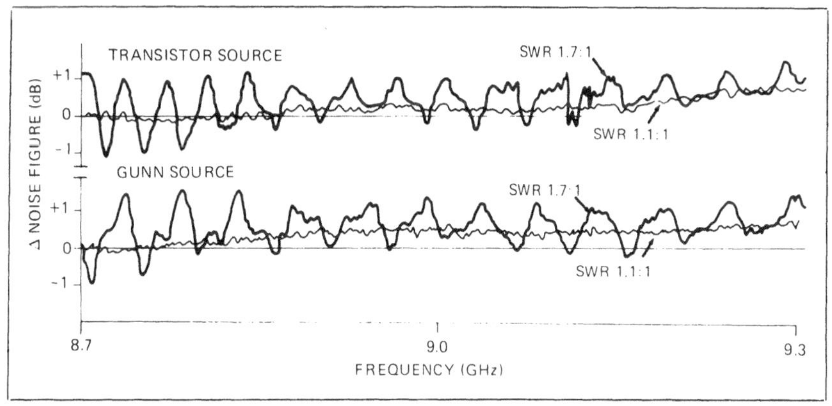

Fig. 7 Noise figure for balanced mixer (SWR on signal input port).

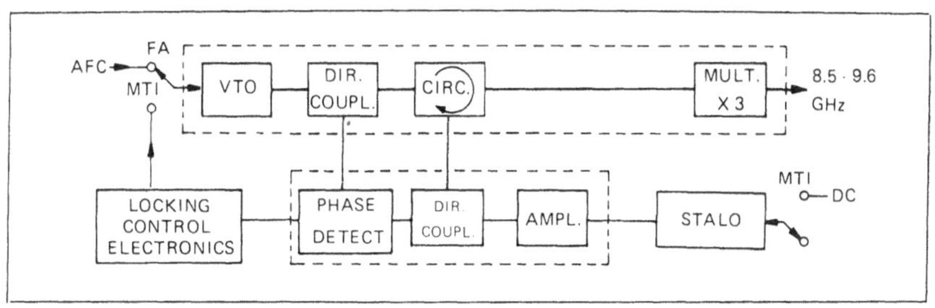

Fig. 8 Block diagram of LO with frequency agile mode and MTI mode.

$$\Delta Q = V\Delta C = (70V)\,(10pF)\,(2)\,(0.05 \times 10^{-3}) = 7 \times 10^{-14}\ C.$$

The corresponding total number of surface charges ΔN_{tot} is given by

$$\Delta N_{tot} = \Delta Q/q = \frac{7 \times 10^{-14}}{1.6 \times 10^{-19}} \approx 4.4 \times 10^{5}$$

These charges are assumed to be uniformly distributed over the increase in varactor surface area. If we take the mesa diameter to be 300 μm and the change in depletion layer width corresponding to a voltage step of 70 V to be 1 μm this surface increase ΔA is given by

$$\Delta A = 2\pi\,(300 \times 10^{-8})\ cm^2 \approx 1.9 \times 10^{-5}\ cm^2 \ .$$

The number of surface charges per unit area is then

$$N_{ss} = \frac{\Delta N_{tot}}{\Delta A} = 2.3 \times 10^{10}\ cm^{-2} \ .$$

This N_{ss} is of the same order of magnitude as the lowest possible N_{ss} on a perfect 100 silicon-oxide interface. It is true that the charges on such a perfect surface are not mobile with time constants in our considered time interval. Nevertheless, practical experience tells that even high quality oxides contain mobile surface charge in concentrations at least comparable to the ideal value.

In Gallium Arsenide, presently available technology does not nearly allow the same kind of perfection in MOS devices.[3] It is therefore only natural that Gallium Arsenide varactors should be inferior with respect to post tuning drift. Taking into account the problem of realizing a high quality oxide even on silicon without deteriorating other varactor properties or metallization it is not surprising that the varactor problem is very difficult indeed. Our own practical experience leads to the conclusion that only very carefully manufactured, unpassivated varactors are satisfactory. Any attempt to passivate the device to date results in increased post tuning drift.

A practical system specification is typically that 10 μs after the application of a step voltage the frequency must be within 500 kHz of its final value. It is seen that this may be achieved with a transistor oscillator but not quite with a Gunn source. Note also that the effect of the Gunn diode on post tuning drift is of opposite sign to that of the varactor (**Figure 5**).

LO NOISE

The nominal noise of a Gunn or transistor oscillator 30 or 60 MHz away from the carrier is low enough so that it does not contribute to the noise figure of a good balanced mixer. However, practical experience has shown that caution must be exercised to avoid any extra noise contribution. This statement can be explained with the aid of a noise measurement test set-up as shown in **Figure 6**.

This block diagram of the receiver and noise meter shows the antenna replaced by a variable attenuator and a movable short. LO power leaking through the mixer will be reflected by such a mismatch and the frequency dependence due to the long-line effect will act as a discriminator thus converting LO FM noise into receiver noise. The practical importance of this effect is shown in **Figure 7** which gives the measured effect on noise of an antenna mismatch. This effect is especially important in frequency agile systems as the wide band nature of the local oscillator enhances FM noise. An isolator close to the mixer in the antenna branch will diminish this problem.

Since even an isolator will have a certain SWR and electrical length, it will still be necessary to specify the FM noise of the LO. In a practical radar an FM noise of 1 $Hz/\sqrt{Hz}$ is tolerable at an IF frequency of 30 MHz. If we assume that a typical noise measure of both a Gunn diode and a transistor is about 30 dB this means that the smallest acceptable Q value of the oscillator is about 100 as calculated by the usual formula

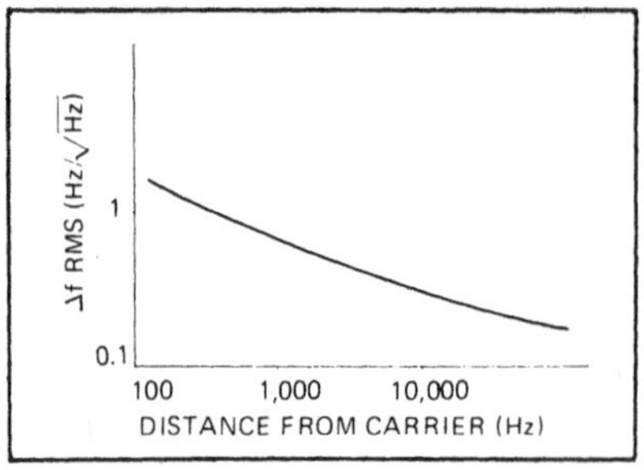

Fig. 9 FM-noise for a YIG-tuned stalo.

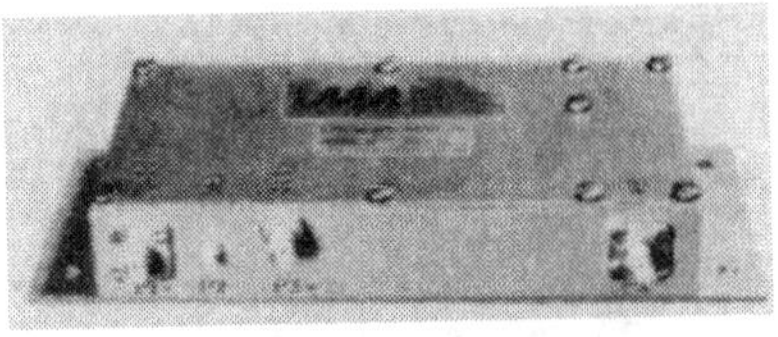

Fig. 10 X-Band transistor LO.

$$\Delta f_{rms} = \frac{f}{Q}\sqrt{\frac{MkTB}{P}} .$$

On the other hand, a Q value of 100 is about the maximum that can be tolerated for an oscillator tunable from 8.5 to 9.6 GHz.

MTI

A wide band tunable LO typically has too much FM noise to be useful in MTI operation. We have, therefore, developed a unit in which the LO may be switched into a mode when it is phase locked to a YIG-tuned, high-stability oscillator with sufficiently high Q and low noise to admit MTI operation. A block diagram of such a unit is shown in **Figure 8**. It should be noted that the YIG oscillator locks the LO at the output frequency of the transistor oscillator thus avoiding an additional multiplier chain. The noise performance of the YIG oscillator shown in **Figure 9** is sufficiently good unless extreme subclutter visibility is required.

Figure 10 shows a transistor LO without MTI. Similar units both with and without MTI have also been developed for X- and Ku-Band.

A summary of the most important parameters of a transistor LO is given in **Table I**.

REFERENCES

1. Scherer, E. F. "Compact Multi-Frequency State Sources in Frequency Agile Systems," *Microwave Journal*, March 1976, pp. 41-44.
2. *Spin Tuned Magnetrons for Frequency Agile Radar*, Philips Product Information 61, June 24, 1974.
3. Kohn, E., et al. "Characterisation of anodic GaAs-oxides for MOS applications," *6th European Solid State Device Research Conference Munich*, Sep 13-16, 1976.

www.ingramcontent.com/pod-product-compliance
Lightning Source LLC
LaVergne TN
LVHW080010210926
848481LV00011B/2985
9780890060674